AF579989

Lecture Notes in Physics

The Lecture Notes in Physics

The series Lecture Notes in Physics (LNP), founded in 1969, reports new developments in physics research and teaching – quickly and informally, but with a high quality and the explicit aim to summarize and communicate current knowledge in an accessible way. Books published in this series are conceived as bridging material between advanced graduate textbooks and the forefront of research and to serve three purposes:

- to be a compact and modern up-to-date source of reference on a well-defined topic
- to serve as an accessible introduction to the field to postgraduate students and nonspecialist researchers from related areas
- to be a source of advanced teaching material for specialized seminars, courses and schools

Both monographs and multi-author volumes will be considered for publication. Edited volumes should, however, consist of a very limited number of contributions only. Proceedings will not be considered for LNP.

Volumes published in LNP are disseminated both in print and in electronic formats, the electronic archive being available at springerlink.com. The series content is indexed, abstracted and referenced by many abstracting and information services, bibliographic networks, subscription agencies, library networks, and consortia.

Proposals should be sent to a member of the Editorial Board, or directly to the managing editor at Springer:

Christian Caron
Springer Heidelberg
Physics Editorial Department I
Tiergartenstrasse 17
69121 Heidelberg / Germany
christian.caron@springer.com

M. Plionis
O. López-Cruz
D. Hughes (Eds.)

A Pan-Chromatic View of Clusters of Galaxies and the Large-Scale Structure

Manolis Plionis
Institute of Astronomy &
Astrophysics
National Observatory of Athens
Palaia Pendeli 152 36, Athens
Greece
and
Inst. Nacional de Astrofísica
Óptica y Electrónica
C/Luis Enrique Erro 1, Tonantzintla
72840 Puebla, México

Omar López-Cruz
Inst. Nacional de Astrofísica
Óptica y Electrónica
C/ Luis Enrique Erro 1, Tonantzintla
72840 Puebla, México

David Hughes
Inst. Nacional de Astrofísica
Óptica y Electrónica
C/Luis Enrique Erro 1, Tonantzintla
72840 Puebla, México

M. Plionis et al. (Eds.), *A Pan-Chromatic View of Clusters of Galaxies and the Large-Scale Structure*, Lect. Notes Phys. 740 (Springer, Dordrecht 2008), DOI 10.1007/ 978-1-4020-6941-3

ISBN: 978-1-4020-6940-6 e-ISBN: 978-1-4020-6941-3

Lecture Notes in Physics ISSN: 0075-8450

Library of Congress Control Number: 2007939886

Cover design: eStudio Calamar S.L., F. Steinen-Broo, Pau/Girona, Spain

Printed on acid-free paper

9 8 7 6 5 4 3 2 1

springer.com

Preface

The study of clusters of galaxies has advanced tremendously in recent years due to the advent of large or dedicated ground-based telescopes, the increasingly sensitive space observatories and the significant advances in numerical astrophysics and cosmology. The current generations of large spectroscopic and wide-field imaging surveys and ongoing multi-wavelength studies are making major breakthroughs in our understanding of galaxy aggregation and transformation processes in different environments, on the properties of the tenuous ICM gas, on the starburst activity and on revealing environmental effects on galaxy formation and evolution.

We therefore felt that it was timely to update previous reviews on the physical nature of clusters of galaxies, their evolution, their galaxy, dark-matter and gas content and the cosmological constraints that they can provide.

This book is the selection of invited reviews, presented during the 2005 Guillermo Haro Advanced School (GH2005) on *"A Panchromatic view of Clusters of Galaxies and the LSS"*, organized by the *Instituto Nacional de Astrofísica, Óptica y Electrónica* in Tonantzintla, México. As the title of the school indicates, a variety of cluster physics themes were discussed: the physics of the ICM gas, the internal cluster dynamics, the detection of clusters using different observational techniques, the great advances in analytical or numerical modeling of clusters, weak and strong lensing effects, the large-scale structure as traced by clusters, the cosmological significance of clusters as well as the formation and evolution of clusters and the cosmic-web within the new cosmological paradigm.

The GH2005 advanced summer-school provided the opportunity to disseminate the new results and methodologies of cluster research to approximately a hundred senior graduate students and post-docs from all over the world.

The organizers of the school are deeply indebted to the distinguished lecturers for their excellent presentations and contributions to this book, the students of the school for their interest and inquisitive attitude which helped deepen the discussions, the INAOE which hosted this summer school and the Mexican government which through the *Consejo Nacional de Ciencia y*

Technología dealt with the financial and logistic aspects of our endeavour. We would also like to thank the *The American Astronomical Society* and *The Harlow Shapley Visiting Lectureship Program* for supporting one of our lecturers (Christine Jones) and last but not least, the *Talavera de La Luz* for allowing us to use their artwork for our poster.

Manolis Plionis
Omar López-Cruz
David Hughes

Gas Dynamics in Clusters of Galaxies

C. L. Sarazin

Department of Astronomy, University of Virginia, P. O. Box 3818, Charlottesville, VA 22903-0818, USA
sarazin@virginia.edu

1 Introduction

One of the more surprising results from X-ray astronomy is that the great volumes of space between galaxies in clusters of galaxies are not empty, as they appear in optical images. Instead, they are filled with a diffuse, hot plasma, with typical temperatures of $T \sim 10^7$–10^8 K. At this temperature, the sound speed in the gas is comparable to the orbit velocities of the galaxies in the cluster, which is consistent with the gas being in hydrostatic equilibrium with the same gravitational potential as binds the galaxies. This intracluster medium (ICM) is highly rarefied, with electron number densities of $n_e \sim 10^{-4}-10^{-2}$ cm^{-3}. At least on large scales, the gas is stably stratified, with the density decreasing with increasing radius r. The gas extends out to distances of $r \gtrsim$ Mpc from the cluster center. The total mass of hot gas is typically $M_{\rm gas} \sim 10^{14}\,M_\odot$; this mass exceeds the total mass of all the galaxies in a typical rich clusters, although even more of the mass is in the form of unseen "dark matter."

At temperatures of 10^6–10^8 K, the dominant radiation mechanism of a plasma is X-ray emission. As a result, clusters of galaxies are generally very luminous X-ray emitters, with luminosities of $L_X \sim 10^{43} - 10^{45}$ ergs s^{-1}. Clusters are second only to quasars as the most luminous X-ray sources in the Universe, and are the most luminous extended sources. While X-ray emission is the primary observational diagnostic for the intracluster medium, the ICM has a number of other important physical effects. It confines and distorts radio galaxies within the cluster. The cosmic ray and magnetic field components of the intracluster medium can also produce diffuse radio emission (see Feretti & Giovannini this volume). The ICM can strip interstellar gas from galaxies as they move through the cluster. Intracluster gas cools at the centers of many clusters, producing lower temperature gas. If the ICM contains dust, the dust will be strongly heated by the plasma, and may emit strongly in the infrared. The ICM also has a number of opacity effects; for example, it scatters and

C. L. Sarazin: *Gas Dynamics in Clusters of Galaxies*, Lect. Notes Phys. **740**, 1–30 (2008)
DOI 10.1007/978-1-4020-6941-3_1

heats the cosmic background radiation which passes through it. The magnetic field in the ICM leads to Faraday rotation and depolarization (Sect. 4.2).

In this chapter, I will review the physical state and X-ray emission processes of the ICM (Sect. 2 and 3). The ICM is shown to act as a fluid in Sect. 4. In Sect. 5, the transport properties of the gas, particularly thermal conduction, are discussed. The hydrodynamical equations for the ICM are given in Sect. 6. Models for the distribution of the gas, and the use of the gas to determine the total mass distributions of clusters are described in Sect. 7 and Sect. 8. The heating and cooling processes in the ICM are discussed in Sect. 9. Much of the heating is due to the hierarchical formation of clusters, and cluster mergers are introduced in Sect. 10. The thermal effects of merger shocks are discussed in Sect. 11. In Sect. 12, the effects of mergers on cluster cooling cores and the phenomena of cold fronts are described.

As much as possible, comparisons to observations in this chapter assume the standard WMAP cosmology [2], with a Hubble constant of $H_0 = 71$ km s^{-1} Mpc^{-1}, a ratio of the mass density to the critical density of $\Omega_m = 0.27$, and the ratio of the dark energy density to the critical density of $\Omega_\Lambda = 0.73$.

2 Physical State of the Intracluster Gas

2.1 Local Thermal State

At the very high temperatures of the intracluster gas, the gas is very highly ionized, but not completely so for the heavy elements. Thus, to describe the local thermal state of the gas, we need to specify three things. First, there are the motions of free particles (electrons and ions), or the kinetic state of the gas. Then, we need to give the ratios of electrons which are free to those which are bound to ions, or the ionization state of the gas. Finally, for the bound electrons, we need to which energy levels they occupy; this is the excitation state of the gas.

Kinetic Equilibrium

If Coulomb collisions are sufficiently rapid, the free particles in the gas (free electron, free proton, and ions) will be brought into kinetic equilibrium and develop a Maxwellian distribution. The time scale for a particle of mass m_1 and charge $Z_1 e$ to collide with field particles of mass m_2 and charge $Z_2 e$ with a number density of n_2 in a Maxwellian distribution at a temperature T is [43]:

$$t_{\rm eq}(1,2) = \frac{3 m_1 \sqrt{2\pi}\,(kT)^{3/2}}{8\pi \sqrt{m_2} n_2 Z_1^2 Z_2^2 e^4 \, \ln \Lambda} \,. \tag{1}$$

Here, $\ln \Lambda \equiv \ln(b_{\rm max}/b_{\rm min}) \approx 40$ is the Coulomb logarithm, and $b_{\rm min}$ and $b_{\rm max}$ are the minimum and maximum impact parameters for Coulomb collisions in the gas. The Coulomb logarithm has a weak (logarithmic) dependence on density and temperature, but is nearly constant under ICM conditions. Coulomb collisions between electrons will bring the electrons into equilibration (an isotropic Maxwellian velocity distribution) on a time scale of roughly

$$t_{\rm eq}(e,e) \approx 3 \times 10^5 \, {\rm yr} \left(\frac{T}{10^8 \, {\rm K}} \right)^{3/2} \left(\frac{n_e}{10^{-3} \, {\rm cm}^{-3}} \right)^{-1} {\rm yr} \, . \tag{2}$$

The time scale for protons to equilibrate among themselves is $t_{\rm eq}(p,p) \approx (m_p/m_e)^{1/2} \, t_{\rm eq}(e,e)$, or roughly 43 times longer than the value in (2). Following this time, the protons and ions would each have Maxwellian distributions, but generally at different temperatures. The time scale for protons to collide with electrons and exchange energy is $t_{\rm eq}(p,e) \approx (m_p/m_e) t_{eq}(e,e)$, or roughly 1870 times the value in (2). The time scale for the electrons and protons to come into equipartition (equal temperatures) is roughly one half of $t_{\rm eq}(p,e)$ [43].

Under typical conditions in the intracluster gas, these time scales are $t_{\rm eq}(e,e) \sim 10^5$ yr, $t_{\rm eq}(p,p) \sim 4 \times 10^6$ yr, and $t_{\rm eq}(p,e) \sim 2 \times 10^8$ yr. Most clusters have existed for $\gtrsim 10^9$ yr, so one would expect the gas to generally be in kinetic equilibrium, with the distributions of free particles being isotropic Maxwellians. Moreover, the electrons and ions should generally be in equipartition, with a common kinetic temperature $T = T_e = T_p$. Possible exceptions might be the outermost regions of clusters (where the gas density is low), or regions where the gas properties have changed rapidly, such as shocks [26].

Collisional Ionization Equilibrium

The main ionization process in the intracluster gas is collisional ionization,

$$e^- + X^{+i} \rightarrow e^- + e^- + X^{+i+1} \, . \tag{3}$$

The main recombination processes are radiative and dielectronic recombination,

$$e^- + X^{+i+1} \rightarrow X^{+i} + {\rm photon(s)} \, . \tag{4}$$

Here, X^{+i} represents some element X which has been ionized i times. Note that neither radiative nor dielectronic recombination (4) are the inverse of collisional ionization (3), which implies that the ionization state in the intracluster gas is not that in thermodynamic equilibrium (the Saha equation).

Let $C(X^i,T)$ be the rate coefficient for collisional ionization out of the ion X^i (3), while $\alpha(X^i,T)$ is the rate coefficient for recombination to the ion X^i (4). If the gas starts in a lower ionization state than in equilibrium, it will be ionized up towards equilibrium on a time scale of roughly:

$$t_{\rm ion} \approx \left[C(X^i,T) n_e \right]^{-1} \approx 3 \times 10^8 \left[\frac{C(X^i,T)}{10^{-13} \, {\rm cm}^3 \, {\rm s}^{-1}} \right]^{-1} \left(\frac{n_e}{10^{-3} \, {\rm cm}^{-3}} \right)^{-1} {\rm yr} \, . \tag{5}$$

In general, the collisional ionization rates are high enough that one would expect he ICM to generally be in collisional ionization equilibrium. Again, possible exceptions might be the outermost regions of clusters (where the gas density is low), or regions where the gas properties have changed rapidly, such as shocks.

In collisional ionization equilibrium, the rates of collisional ionization and radiative and dielectronic recombination balance, which implies that

$$\left[C(X^i, T_g) + \alpha(X^{i-1}, T_g)\right] n(X^i) = C(X^{i-1}, T_g) n(X^{i-1}) + \alpha(X^i, T_g) n(X^{i+1}) . \quad (6)$$

Here, $n(X^i)$ is the number density of the X^i ion. Note that, unlike thermodynamic equilibrium (Saha equilibrium), the state of ionization in collisional ionization equilibrium is independent of density, and only depends on the electron kinetic temperature T. Generally, each ionization fraction reaches a maximum at a temperature that is some fraction of its ionization potential. At the temperatures which predominate in clusters, iron is mainly in the fully stripped, hydrogenic, or heliumlike stages.

Excitation Equilibrium

For ions with bound electrons, the population of excited states are determined mainly by a balance between collision excitation by free electrons and radiative de-excitation. In general, the spontaneous radiative de-excitation rates are much higher than the excitation rates, and the electrons are almost always found in the ground level. The population of excited states are much lower than would be expected in thermodynamic equilibrium (Boltzmann distribution). Collisional de-excitation rates are much lower than radiative de-excitation rates; this means that there are no X-ray spectral diagnostics which determine the local density in the gas.

3 X-ray Emission

The X-ray emission of the intracluster gas is mainly due to thermal bremsstrahlung and line emission. There are smaller contributions of continuum from bound-free (recombination) emission and from two-photon decays of 2s levels in hydrogenic and helium-like ions.

The emissivity due to thermal bremsstrahlung (free–free emission) is given by

$$\epsilon_\nu^{\rm ff} = \frac{2^5 \pi e^6}{3 m_e c^3} \left(\frac{2\pi}{3 m_e k} \right)^{1/2} n_e T^{-1/2} \exp(-h\nu/kT) \sum_i Z_i^2 n_i g_{\rm ff}(Z_i, T, \nu) , \quad (7)$$

where the emissivity ϵ_ν is defined as the emitted energy per unit time, frequency, and volume. The sum is over the various ions in the plasma, but is dominated by hydrogen and helium for Solar abundances. The Gaunt factor $g_{\rm ff}(Z_i, T, \nu)$ corrects for quantum mechanical effects and for the effect of distant collisions, and is a slowly varying function of frequency and temperature. As a result, the dominant dependence of the free–free emissivity on frequency is the Boltzmann exponential factor, and the main dependences on temperature are this factor and the square–root factor $T^{-1/2}$. Thermal bremsstrahlung produces a roughly exponential continuum component in the X-ray spectrum. At high temperatures $T \gtrsim 3 \times 10^7$ K, thermal bremsstrahlung is the dominant emission mechanism.

At lower temperatures, the main X-ray radiation is from lines. The strongest line feature observed from most clusters of galaxies is the complex of iron Fe Kα lines at about 6.7 keV. This line feature is actually a blend of lines from iron ions (mainly Fe^{+24} and Fe^{+25}) and weaker lines from nickel ions. The notation "Kα" gives the principal quantum number n of the lower level of the transition and the change in the principal quantum number $\Delta n \equiv n' - n$, where n' is the principal quantum number of the upper level of the transition. K indicates that the lower level is in the K-shell ($n = 1$), L indicates the lower level is in the L-shell ($n = 2$), and so on, while α indicates that $\Delta n = 1$, β indicates that $\Delta n = 2$, etc. In addition to the Fe K line complex, the X-ray spectra of clusters of galaxies contain a large number of lower energy lines. These include the K lines of the common elements lighter than iron, such as C, N, O, Ne, Mg, Si, S, Ar, and Ca, as well as the L lines of Fe and Ni. These lines become very strong at lower temperatures ($T \lesssim 10^7$ K).

As an illustration, Fig. 1 shows the predicted X-ray spectrum of an X-ray cluster [53]. The model cluster is isothermal in its outer regions (with a temperature of 8×10^7 K), and has a cooling core at its center. The figure shows the overall exponential continuum from thermal bremsstrahlung, the Fe K lines at about 7 keV (which come mainly from the region of the cluster outside of the cooling flow), and the lower energy lines from the cooling core.

Most X-ray lines are excited by collisional excitation by electrons, although radiative and dielectronic recombination and inner shell collisional ionization also play a role. The emissivity due to a collisionally excited line is usually written [36]:

$$\int \epsilon_\nu^{\rm line} {\rm d}\nu = n(X^i) n_e \frac{h^3 \nu \Omega(T) B}{4 \omega_{gs}(X^i)} \left[\frac{2}{\pi^3 m_e^3 kT} \right]^{1/2} e^{-\Delta E/kT} \,, \tag{8}$$

where $h\nu$ is the energy of the transition, ΔE is the excitation energy above the ground state of the excited level, B is the branching ratio for the line (the probability that the upper state decays through this transition), and Ω is the 'collision strength', which is often a slowly varying function of temperature.

The intracluster gas is almost certainly in collisional ionization equilibrium (Sect. 2.1); under these circumstances, the ionization fractions depend only on the electron temperature T, and are independent of the density of the gas. Then, the density of any ion is just proportional to the proton density in the

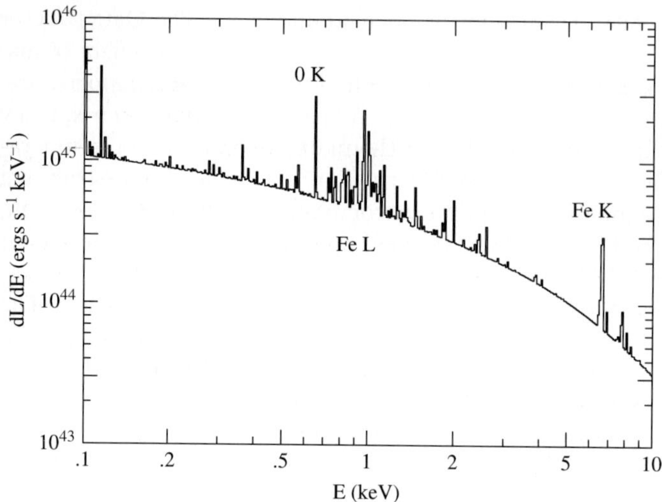

Fig. 1. Model X-ray spectrum of a cluster of galaxies. The cluster was assumed to be isothermal at $T = 8 \times 10^7$ K in its outer regions, and to have a large amount of cool gas (a cooling core) in its inner regions

gas times the abundance of the relevant element relative to hydrogen. Thus, all of the X-ray emission processes in the gas scale with the product $n_p n_e$ of the proton and electron densities, respectively. If L_ν is the X-ray luminosity per unit frequency emitted by a cluster, then this can be written as

$$L_\nu = \Lambda_\nu(T, \text{Abundances}) \int n_e n_p \, dV \,, \tag{9}$$

where the integral is over the volume V of the cluster. The total emissivity $\epsilon_\nu = \Lambda_\nu n_e n_p$, where Λ_ν depends only on the temperature and the abundances of the heavier elements relative to hydrogen. Similarly, the X-ray surface brightness is given by

$$I_\nu = \Lambda_\nu(T, \text{Abundances}) \int n_e n_p \, dl \,. \tag{10}$$

Here, the integral is along the line of sight distance l through the cluster.

The emissivity of a line is then proportional to the square of the density and to the abundance of the relevant element, and depends significantly on the electron temperature. Because the thermal bremsstrahlung emissivity also is proportional to the square of the density (7), the ratio of line emission to thermal bremsstrahlung continuum emission is independent of density. Line ratios or the shape of the X-ray continuum spectrum can be used to derive a temperature for the gas in a cluster. Then, the ratio of line emission to thermal bremsstrahlung continuum emission can be used to determine the abundance of the heavy element responsible for the line.

4 The Intracluster Medium as a Fluid

4.1 Mean Free Paths

The mean free paths of electrons and ions in a plasma without a magnetic field are determined by Coulomb collisions [43]. The electrons in a Maxwellian plasma undergo Coulomb collisions in a time which is a factor of $\sqrt{m_e/m_p}$ shorter than the protons (Sect. 2.1). On the other hand, the electrons move faster by the inverse of this factor. Thus, the mean free paths of electrons and protons are essentially equal, with

$$\lambda_p = \lambda_e = \frac{3^{3/2}(kT)^2}{4\pi^{1/2} n_e e^4 \ln \Lambda} \approx 23 \left(\frac{T}{10^8\,\mathrm{K}} \right)^2 \left(\frac{n_e}{10^{-3}\,\mathrm{cm}^{-3}} \right)^{-1} \mathrm{kpc}\,. \tag{11}$$

These mean free paths are smaller than most scales of interest in clusters; they are only about 1% of the radius of a cluster (~2 Mpc). Thus, it is reasonable to treat the ICM as a fluid under most circumstances. The fluid approximation might breakdown in the outer parts of a cluster (where the lower density increases λ_e), in interactions with galaxies (whose sizes are comparable to λ_e), if the ICM is very inhomogeneous, or in sharp transitions in the ICM properties at shocks or cold fronts (Sects. 11 and 12.2).

4.2 Magnetic Fields and Gyroradii

In any case, the ICM apparently contains a significant magnetic field, with typical values of $B \sim 1\ \mu$G. (See the chapter by Feretti & Giovannini for more details concerning the magnetic field in clusters.) Stronger fields occur in some smaller volumes of clusters. These fields are probably too weak to be very important dynamically, as the magnetic pressure, $P_B = B^2/(8\pi)$, is much smaller than the typical gas pressures. However, the magnetic field does very strongly effect the microscopic motions of electrons and ions. In the presence of a magnetic field, electron and ions follow helical orbits, gyrating about magnetic field lines. The gyroradii of electrons and ions in cluster magnetic fields are very small. For example, the gyroradius of a typical electron is

$$r_g \approx 3 \times 10^8 \left(\frac{T}{10^8\,\mathrm{K}} \right)^{1/2} \left(\frac{B}{1\,\mu\mathrm{G}} \right)^{-1} \mathrm{cm}\,. \tag{12}$$

These very small gyroradii probably insure that the ICM acts as a fluid even when the Coulomb mean free paths are long.

5 Transport Processes

The fact that the mean free paths are small but finite implies that the local properties of the gas will be influenced by the properties of the surrounding gas through diffusive processes, also called transport processes. These include the thermal conduction of heat energy in non-isothermal gases, the viscous transport of momentum, and the diffusion and settling of heavy elements within the intracluster gas. I will concentrate on thermal conduction here; viscosity and diffusion are discussed in [40].

5.1 Thermal Conduction

In a plasma with a gradient in the electron temperature, heat is conducted down the temperature gradient. If the scale length of the temperature gradient $l_T \equiv T/|\boldsymbol{\nabla} T|$ is much longer than the mean free path of electrons, $l_T \gg \lambda_e$, then the heat flux is given by

$$\boldsymbol{Q} = -\kappa \boldsymbol{\nabla} T \,, \tag{13}$$

where the thermal conductivity for a hydrogen plasma is [43]:

$$\begin{aligned} \kappa &= 1.31 n_e \lambda_e k \left(\frac{k T_e}{m_e} \right)^{1/2} \\ &\approx 4.6 \times 10^{13} \left(\frac{T_e}{10^8\,\mathrm{K}} \right)^{5/2} \left(\frac{\ln \Lambda}{40} \right)^{-1} \mathrm{erg\,cm^{-1}\,s^{-1}\,K^{-1}} \,. \end{aligned} \tag{14}$$

Because of the inverse dependence on the particle mass, thermal conduction is primarily due to electrons. If the very weak dependence of $\ln \Lambda$ on density is ignored, then κ is independent of density but depends very strongly on temperature.

If heat conduction operates at this "Spitzer" rate, then the gas in the central regions of clusters is likely to be isothermal. In addition, heat conduction would be very important at and would tend to eliminate any large local temperature gradients, such as appear to occur in the cooling core of clusters or near cold fronts (Sect. 12). On the other hand, the rate of thermal conduction along a thermal gradient perpendicular to the magnetic field is very low, as a result of the small gyroradii of electrons (12). Thus, transverse or tangled magnetic fields may be able to suppress thermal conduction in clusters, at least in some regions. The existence of very steep temperature gradients in cold fronts has been used to argue that heat conduction is suppressed by a factor of $\gtrsim 10^2$ in these regions (Sect. 12.2).

6 Hydrodynamics

In the fluid limit, the ICM can be characterized by the local values of the gas density ρ, the gas pressure P, the gas temperature T or internal energy, and the gas velocity $\boldsymbol{v}$. The gas pressure is determined by the ideal gas law:

$$P = \frac{\rho k T}{\mu m_p} \,, \tag{15}$$

where μ is the mean mass per particle in terms of the mass of a proton m_p. The dynamical equation for a single component fluid is [20]:

$$\rho \frac{D\boldsymbol{v}}{Dt} + \boldsymbol{\nabla} P + \rho \boldsymbol{\nabla} \Phi = 0 \, , \tag{16}$$

where Φ is the gravitational potential, and D/Dt is the Lagrangian derivative with respect to time. Equation (16) ignores non-gravitational forces, such as magnetic stresses or viscosity. The continuity equation (mass conservation) is [20]:

$$\frac{\partial \rho}{\partial t} + \boldsymbol{\nabla} \cdot (\rho \boldsymbol{v}) = 0 \, . \tag{17}$$

There is also an equation giving the variation in the energy in the fluid. However, it is simpler to give this equation in terms of the entropy in the gas, S. A useful quantity to consider is the specific entropy per particle in the gas, $s \equiv S/N$, where N is the total number of particles. To within additive constants, the specific entropy of an ideal gas is

$$s = \frac{3}{2} k \ln \left(\frac{P}{\rho^{5/3}} \right) = \frac{3}{2} k \ln \left(\frac{T}{\rho^{2/3}} \right) \, . \tag{18}$$

To avoid the logarithmic character of the entropy, it is conventional to define an "entropy parameter" K as

$$K \equiv \frac{kT}{(n_e)^{2/3}} \tag{19}$$

with units of keV cm^2. Thus, $s \propto \ln K$. Then, the equation for the change in the gas entropy can be written [20]:

$$\frac{\rho}{\mu m_p} k \frac{Ds}{Dt} = \mathcal{H} - \mathcal{L} \, , \tag{20}$$

where $\mathcal{H}$ and $\mathcal{L}$ are the rate of heating and cooling per unit volume in the gas. In the absence of irreversible processes like heating or cooling or shocks, the specific entropy of a parcel of gas is constant.

7 Hydrostatic Equilibrium

Unless it is disturbed in some way, one would expect the gas in a cluster to relax into hydrostatic equilibrium on roughly the sound crossing time of the cluster,

$$t_s \equiv \frac{D}{c_s} \approx 6.6 \times 10^8 \, \mathrm{yr} \left(\frac{T}{10^8 \, \mathrm{K}} \right)^{-1/2} \left(\frac{D}{1 \, \mathrm{Mpc}} \right) \, . \tag{21}$$

Here, D is the diameter of the cluster, and c_s is the sound speed. Since this time scale is shorter than the age of a typical cluster, which is a fraction of the Hubble time, the gas in many clusters should be close to hydrostatic equilibrium. Exceptions would include clusters which are undergoing or have recently undergone a major merger, and regions of a cluster where an AGN has injected energy recently.

In hydrostatic equilibrium, the pressure forces balance gravity:

$$\nabla P = -\rho \nabla \Phi \ , \qquad \frac{1}{\rho} \frac{\mathrm{d}P}{\mathrm{d}r} = -\frac{GM(r)}{r^2} \ , \tag{22}$$

where $M(r)$ is the total cluster mass within r, and the second form assumes spherical symmetry. Because (22) gives a single relation for two gas properties (density and pressure), one must also specify the entropy distribution of the gas to determine its distribution.

7.1 Isothermal Models

A very simple model follows if the gas is assumed to be isothermal (T = constant); isothermality might result if thermal conduction were efficient in the cluster (Sect. 5.1). Then, the solution of the hydrostatic equation is

$$\ln \left[\frac{\rho(r)}{\rho_0} \right] = \frac{\mu m_p}{kT} \left[\Phi_0 - \Phi(r) \right] \ , \tag{23}$$

where ρ_0 and Φ_0 are the central values of the the gas density and gravitational potential, respectively. Note that the gas density will generally go to a finite value as $r \rightarrow \infty$.

Numerical simulations suggest that the dark matter distribution in clusters should have a power-law drop off at large radii, and a flatter power-law at small radii [35]. Thus, the dark matter distribution should have a cusp at the center of the cluster. The NFW dark matter profile [35] has:

$$\rho_{\mathrm{DM}}(r) = \rho_s \left[\left(\frac{r}{r_s} \right) \left(1 + \frac{r}{r_s} \right)^2 \right]^{-1} \ , \tag{24}$$

where r_s and ρ_s are the characteristic scaling radius and density, respectively. If this distribution applies to the sum of all the matter in a cluster, then the potential is

$$\Phi(r) = \Phi_0 \, \frac{\ln\left(1 + r/r_s\right)}{r/r_s} \ , \tag{25}$$

and the central potential is $\Phi_0 = -4\pi G \rho_s r_s^2$.

However, in the past the dark matter and/or galaxy distributions in clusters were modeled using a function with a constant density core,

$$\rho_{\mathrm{DM}}(r) = \rho_{\mathrm{DM},0} \left[1 + \left(\frac{r}{r_c} \right)^2 \right]^{-3/2} \ , \tag{26}$$

where $\rho_{\mathrm{DM},0}$ is the central density and r_c is the core radius. If this form is assumed for the total matter density in a cluster, or if it applies to the galaxy distribution, and the galaxies have an isotropic gaussian velocity distribution, then the resulting gas density distribution is the "beta model" [4]:

$$\rho(r) = \rho_o \left[1 + \left(\frac{r}{r_c} \right)^2 \right]^{-3\beta/2} . \tag{27}$$

If the gas is isothermal, then this density distribution gives an X-ray surface brightness distribution of the form

$$I_X(r) = I_X^o \left[1 + \left(\frac{r}{r_c} \right)^2 \right]^{-3\beta+1/2} . \tag{28}$$

This beta-model provides a reasonable fit to the X-ray surface brightness in the outer regions of many cluster, with a typical value of $\beta \approx 2/3$. However, it does not fit the inner parts of cooling core clusters.

7.2 Adiabatic or Polytropic Models

The temperature profiles in clusters of galaxies are generally more consistent with a gradual decline with radius at large radii, rather than isothermal gas [48]. A simple alternative would be if the gas in clusters was adiabatic (had a constant specific entropy); then the pressure and density would vary together as $P \propto \rho^\gamma$ with $\gamma = 5/3$. Often, one also considers distributions with the same pressure-density relationship, but for values of γ in the range $1 \leq \gamma \leq 5/3$. We will refer to these distributions as "polytropic." Then, the hydrostatic equation can be solved to give

$$\frac{T(r)}{T_0} = 1 + (\alpha - 1) \left[1 - \frac{\Phi(r)}{\Phi_0} \right] , \tag{29}$$

$$\frac{\rho(r)}{\rho_0} = \left[\frac{T(r)}{T_0} \right]^{1/(\gamma-1)} . \tag{30}$$

Here, T_0 is the central temperature, and $\alpha \equiv T(\infty)/T_0$. The temperature profiles in the outer parts of clusters can generally be fit with intermediate values of $\gamma \sim$1.2–1.3 [24].

7.3 Surface Brightness Deprojection

The gas distributions in clusters can be derived directly from observations of the X-ray surface brightness of the cluster, if the shape of the cluster is known and if the X-ray observations are sufficiently detailed and accurate. The X-ray surface brightness at a photon frequency ν and at a projected distance b from the center of a spherical cluster is

$$I_\nu(b) = \int_{b^2}^{\infty} \frac{\epsilon_\nu(r)\mathrm{d}r^2}{\sqrt{r^2 - b^2}} \,, \tag{31}$$

where ϵ_ν is the X-ray emissivity. This Abel integral can be inverted to give the emissivity as a function of radius,

$$\epsilon_\nu = -\frac{1}{2\pi r}\frac{\mathrm{d}}{\mathrm{d}r}\int_{r^2}^{\infty} \frac{I_\nu(b)\mathrm{d}b^2}{\sqrt{b^2 - r^2}} \,. \tag{32}$$

The emissivity ϵ_ν is proportional to the square of the density, and its spectral dependence is determined by the gas temperature and abundances (7), (8) and (9). Thus, the radial dependence of the spectrum and intensity of X-rays can be de-projected to the the gas density $\rho(r)$, gas temperature $T(r)$, and abundances as a function of physical radius. The gas pressure is then given by the ideal gas law (15).

8 Cluster Masses

Once the gas density has been determined by either model fitting or de-projection, the gas mass can be derived simply as

$$M_{\mathrm{gas}}(r) = 4\pi \int_0^r \rho(r')(r')^2 \, \mathrm{d}r' \,. \tag{33}$$

Here, $M_{\mathrm{gas}}(r)$ is the gas mass interior to the radius r.

The total gravitational mass can be derived from the condition of hydrostatic equilibrium (22), which can be written as

$$M(r) = -\frac{r^2}{G\rho(r)}\frac{\mathrm{d}P}{\mathrm{d}r} \,, \tag{34}$$

where $M(r)$ is the total mass interior to r. This equation can also be written as

$$M(r) = -\frac{kT(r)r}{\mu m_p G}\left[\frac{\mathrm{d}\ln\rho(r)}{\mathrm{d}\ln r} + \frac{\mathrm{d}\ln T(r)}{\mathrm{d}\ln r}\right] \,. \tag{35}$$

Optical observations of the galaxies can be used to estimate the total mass of galaxies interior to r, $M_{\mathrm{gal}}(r)$. Any diffuse stellar light can be included in this; although these values can be difficult to determine, the stars and galaxies constitute only a small fraction of the mass, so this correction is not so important. Then, the mass of dark matter in the cluster (interior to r) is given by

$$M_{\rm DM}(r) = M(r) - M_{\rm gas}(r) - M_{\rm gal}(r) \, . \tag{36}$$

In typical clusters, the masses of stars and galaxies are much smaller than those of the hot gas, with $M_{\rm gal} \approx 0.15 M_{\rm b}$ at large radii [49]. Thus, hot plasma is the dominant form of baryonic matter in clusters of galaxies, with $M_{\rm gas} \approx 6 M_{\rm gal}$ at large radii. It appears that the same may be true on large scales throughout the present day Universe; it seems that most of the baryons in the Universe today are in hot, diffuse intergalactic gas (often called WHIM, or Warm Hot Intergalactic Medium), rather than stars and galaxies (e.g., [8]). In this sense, cluster represent the tip of the iceberg. With their very high densities, they are the one place it has been easy to detect the bulk of the baryons, which are in intergalactic gas.

The gas mass fraction $f_{\rm gas}(r)$ and baryon fraction $f_{\rm b}(r)$ are then

$$f_{\rm gas}(r) = \frac{M_{\rm gas}(r)}{M(r)} \, , \qquad f_{\rm bary}(r) = \frac{M_{\rm gas}(r) + M_{\rm gal}(r)}{M(r)} \, . \tag{37}$$

The observations of most clusters show an increase in $f_{\rm gas}(r)$ with radius r in the inner parts of clusters [1]. Thus, the gas is more broadly distributed than the dark matter in clusters. The gas fractions level out at radii which are $r \gtrsim 0.2 r_{\rm vir}$. Rich clusters have gas fractions which average $\langle f_{\rm gas}(r_{2500}) \rangle = 12\%$ at a radius where the mean interior density is 2500 times the critical density [1]. For typical clusters, $r_{2500} \approx 0.25 r_{\rm vir}$. The total gas fraction within $r_{\rm vir}$ will be a bit larger than this. Thus, clusters appear to consist of about 2–3% stars and galaxies, ~14% hot gas, and ~84% dark matter. Although these are recent values, clusters of galaxies provided some of the earliest evidence that the mass in the Universe was predominantly dark matter.

Clusters of galaxies are very useful cosmological probes. Arguably, they are the largest objects in the Universe which are dynamically relaxed. On the other hand, they are probably the smallest objects which formed from a sufficiently large volume that they contain a fair sample of the material in the Universe. Thus, the ratio of baryons to dark matter in clusters should be close to the universal value. Numerical simulations do indicate that the baryon fraction in clusters is nearly the general value in the Universe; even at r_{2500}, $f_{\rm bary}$ is about 82% of the universal value [10].

When combined with the density of baryons inferred from Big Bang nucleosynthesis, the observed baryon fraction in clusters indicates that the total mass density in the Universe is $\Omega_m \approx 0.3$ [52]. Thus, cluster have provided some of the earliest and strongest evidence that we live in a low density Universe, with too little matter to close the Universe and reverse the expansion of the Big Bang.

The measured values of $f_{\rm gas}$ and $f_{\rm bary}$ depend on the distance d to a cluster as $d^{3/2}$. On the other hand, if clusters are fair samples of the materials in the Universe, then $f_{\rm bary}$ should be independent of redshift or distance. Thus, a comparison of $f_{\rm bary}$ in low redshift and high redshift clusters provides a measure of the distance to the clusters which is independent of the redshift.

Such measurements provide evidence that we live in an accelerating Universe, with an effective cosmological constant of $\Omega_\Lambda \approx 0.7$ [1]. This is in concordance with the results from WMAP [2] and supernova Type Ia observations at high redshifts.

9 Heating and Cooling of Intracluster Gas

9.1 Why Is the ICM So Hot?

When it was first observed in X-rays, one of the most surprising features about the intracluster gas was its very high temperature. Why is this gas so hot? In fact, this is one of the easiest aspects of the ICM to understand. At least in rich clusters, most of the heating is gravitational in origin. The basic idea is that clusters have huge masses, and very deep gravitational potential wells. Essentially, any means of introducing the gas into a cluster will cause it to move very rapidly, and collide with other gas, and be shocked. For example, if the gas fell into the cluster (either at the same time as the dark matter, or subsequently), cluster gravitational potentials imply that the gas would fall in at a speed $\gtrsim 1000$ km s^{-1}. Unless the gas motions were very carefully controlled, the gas would encounter other gas moving at similar velocities, and the intersecting gas streams would collide and shock. Since the ICM has heavy elements, a portion of it came out of galaxies. If it did so after the clusters formed, then the galaxies would be moving at orbital speeds of $\gtrsim 1000$ km s^{-1} in the cluster, and gas ejected from different galaxies would collide and shock at these sorts of speeds. (If the gas came out of galaxies before clusters formed, then it had to fall into a cluster, and was shocked as described previously.) Thus, it is likely that essentially all of the gas in the ICM medium shocked at speeds of $\gtrsim 1000$ km s^{-1}, and was heated in this way.

In actually, we believe that clusters form hierarchically from the merger of smaller groups and clusters. Such mergers are discussed extensively below (Sect. 10). Thus, the specific mechanism for much of the heating of the ICM is likely to be merger shocks.

9.2 Simple Scaling Laws for Gravitational Heating

If one assumes that gravitational heating dominates in clusters, and makes a few other simple assumptions, it is possible to derive a number of simple scaling laws for the X-ray properties of clusters [18]. If the gas in clusters is in hydrostatic equilibrium and is distributed similarly to the dark matter, then the typical gas temperature should be $kT \sim \mu m_p GM/R$, where M is the total cluster mass, and R is the cluster radius. If one can treat the formation of a cluster as equivalent to the collapse of a isolated, spherical region of overdensity in the Universe, then the post-collapse average density in the cluster should be $\langle \rho_{\rm tot} \rangle \sim 180 \rho_{\rm crit}(z_{\rm form})$, where $\rho_{\rm tot}$ is the total mass density (dark matter and baryons), and $\rho_{\rm crit}(z_{\rm form})$ is the critical density for the Universe to collapse at the epoch of formation of the cluster. If most clusters have formed recently, then one could approximate $z_{\rm form} \sim 0$. Finally, clusters are large enough to contain a fair sample of the material in the Universe, and thus it is reasonable to assume that the baryon fraction in clusters (which is predominantly in the hot gas) is the universal value (Sect. 8). Then, the radii of clusters should scale with mass as

$$R \propto M^{1/3} \, . \tag{38}$$

The gas temperature would scale as

$$T \propto M^{2/3} \, , \tag{39}$$

and the X-ray luminosity vs. temperature relationship would be

$$L_X \propto T^2 \, . \tag{40}$$

The latter scaling assumes that the X-ray emission is mainly due to thermal bremsstrahlung, which is true for hot clusters.

9.3 Non-Gravitational Heating

There are a number of indications that non-gravitational heating or cooling processes may affect the ICM, particularly in smaller clusters and groups. First, the observed cluster X-ray properties do not agree very well with the scaling relations for purely gravitational heating (38), (39) and (40). Probably, the most significant deviation is that the measured X-ray luminosity–temperature relation is much steeper than (40) [23]. The departures for the scaling relations are particularly strong for cooler clusters and groups. Second, the observed gas distributions in clusters are more extended than would be expected from purely gravitational heating. The gas distributions often have central cores. This suggests that some non-gravitational heating processes have occurred and have puffed up the gas distributions, particularly in the poorer clusters. This would lower the average density in the gas, and thus reduce the X-ray luminosity. An alternative possibility is that inhomogeneous cooling has removed the cooler ICM, increasing the average temperature of the gas which remains. Presumably, this cooling would also lead to star and galaxy formation. These topics have been reviewed extensively in [50].

If the non-gravitational heating occurred just prior to the collapse of a cluster, then the amount of heat needed is ~ 2 keV per particle [21]. However, a more useful quantity to describe the preheating is probably the extra entropy per particle Δs (18). As noted in Sect. 6, the specific entropy is a Lagrangian quantity which moves with the gas, and which remains constant for reversible changes. As discussed in Sect. 6, it is conventional to use the entropy parameter K (19) rather than s. For purely gravitational heating, the scaling laws described above (Sect. 9.2) imply that the entropy parameter is expected to scale as

$$K \propto T \propto M^{2/3} \, . \tag{41}$$

Observations of clusters and groups initially suggested that preheating produced an extra entropy of $\Delta K \approx 135$ keV cm^{-2} [22, 37]. It now appears that such an "entropy floor" may be to simplified to explain the detailed variations in entropy between clusters and the radial variations within clusters. Also, the existence of the Lyman alpha forest and other quasar absorption lines indicates that not all of the intergalactic gas underwent the same level of preheating. Nonetheless, this value provides a useful value in assessing models for the thermal history of the ICM.

The radial variation of the entropy in the ICM also appears to be inconsistent with purely gravitational heating. Gravitational heating models predict that the entropy vary roughly as $K \propto r^{1.1}$. The observed entropy profiles in clusters are much flatter in the center [38].

Supernovae could provide a significant source of heating of the ICM. These would include core collapse supernova associated with the deaths of massive stars. Since the galaxies in clusters today contain very few such stars, this would have occurred during the epoch of star formation and galaxy formation. The supernovae might have driven galactic winds. The second type of supernovae are Type Ia's, which are produced by older binary star systems. They would provide a more continuous source of heating.

Supernovae also eject heavy elements. Thus, the abundances in clusters can be used to limit the total number of supernova which have occurred. The observed abundances suggest that the extra energy added is probably $\sim$0.3 keV per particle [21]. This is a bit low to explain the required preheating, but might be possible. However, this mechanism would also require that a large fraction of the supernova explosion energy be converted into heat in the ICM, which may also be a difficulty.

Active galactic nuclei (AGNs) within clusters might also provide a significant amount of heating. As with the supernovae, it is difficult to determine what fraction of the energy produced by AGNs goes into heating the surrounding medium. In this regard, it is only the AGN output in kinetic energy in jets or in relativistic particles which is likely to be useful. It may be important that the early-type galaxies found in clusters generally host radio galaxies and radio quasars, which are more likely to deposit energy into the ICM.

One way to limit the total energy input by AGNs is to look at the total masses of supermassive black holes contained in clusters today. In general, all large bulges appear to contain supermassive black holes, and there is a strong correlation of black hole mass with bulge mass or velocity dispersion. If the growth of black holes occurred largely by accretion (rather than merging of existing massive black holes), then the total accretion energy from black holes can be derived from their total mass. This could provide a significant level of heating for the ICM if the fraction of accreted energy which goes into heating is $\gtrsim$10% [5].

9.4 Cooling in the Intracluster Medium

The primary cooling process for the ICM is the emission of X-ray radiation. The emission is proportional to the square of the density and varies with temperature (7), (8), and (9). Thus, the total cooling rate per unit volume $\mathcal{L}$ in the gas can

$$\mathcal{L} = \Lambda(T, \text{Abundances}) n_e n_p \, , \tag{42}$$

where Λ depends only on the temperature and the abundances of the heavier elements relative to hydrogen. At high temperature ($kT \gtrsim 2$ keV), the dominant radiation is thermal bremsstrahlung, and $\Lambda \propto T^{1/2}$. At lower temperature, line emission becomes dominant, and Λ decreases with increasing temperature.

At high temperatures where thermal bremsstrahlung dominates, the time required for gas to cool to low temperatures at constant pressure is

$$t_{\rm cool} = 69 \left(\frac{n_e}{10^{-3}\,{\rm cm}^{-3}} \right)^{-1} \left(\frac{T}{10^8\,{\rm K}} \right)^{1/2} \ {\rm Gyr} \, . \tag{43}$$

Note that cooling accelerates as the gas cools; this tendency is even stronger below $kT \lesssim 2$ keV due to line emission. The cooling time is much longer than the Hubble time in the outer parts of clusters. However, it can be quite short ($t_{\rm cool} \sim 300$ Myr) in the inner regions of cooling core clusters.

It is interesting to write the cooling time as a function of the entropy and temperature rather than the density and temperature:

$$t_{\rm cool} = 17 \left(\frac{K}{130\,{\rm keV\,cm}^{-2}} \right)^{3/2} \left(\frac{kT}{2\,{\rm keV}} \right)^{-1} \ {\rm Gyr} \, . \tag{44}$$

Note that the cooling time is less than the Hubble time for $K \lesssim 130$ keV cm^{-2} for $kT \sim 2.5$ keV. If clusters start with gas with a wide range of entropies, the lower entropy gas will cool rapidly and be removed from the ICM. Thus, cooling can increase the average entropy of the gas and provide an effective "floor" to the ICM entropy [51]. The cooled gas presumably goes into forming galaxies and stars. Feedback heating from supernovae, galactic winds, and AGNs (Sect. 9.3) might result in some of the cooled gas actually becoming hotter ICM. However, the result is to remove the cooler gas and raise the average entropy of the ICM.

Thus, while the bulk of the heating of the ICM in large clusters is due to gravitational heating, mainly by merger shocks, smaller clusters and the centers of clusters show evidence for the effects of non-gravitational heating and cooling. The cooling leads to star and galaxy formation, and which leads to possible heating by supernovae, galactic winds, and AGNs. Thus, the ICM (and intergalactic medium more generally) preserve a unique record of the thermal history of the Universe.

10 Cluster Mergers

Major cluster mergers are the most energetic events in the Universe since the Big Bang. Cluster mergers are the mechanism by which clusters are assembled. In these mergers, the subclusters collide at velocities of $\sim$2000 km/s, releasing gravitational binding energies of as much as $\gtrsim 10^{64}$ ergs. During mergers, shocks are driven into the intracluster medium. In major mergers, these hydrodynamical shocks dissipate energies of $\sim 3\times 10^{63}$ ergs; such shocks are the major heating source for the X-ray emitting intracluster medium. The shock velocities in merger shocks are similar to those in supernova remnants in our Galaxy, and we expect them to produce similar effects. Mergers shocks should heat and compress the X-ray emitting intracluster gas, and increase its entropy. We also expect that particle acceleration by these shocks will produce relativistic electrons and ions, and these can produce synchrotron radio, inverse Compton (IC) EUV and hard X-ray, and gamma-ray emission. (See the chapter by Feretti & Giovannini for more details relativistic particles and non-thermal emission in clusters.)

11 Thermal Physics of Merger Shocks

The intracluster medium (ICM) is generally close to hydrostatic equilibrium in clusters which are not undergoing strong mergers. The virial theorem then implies that the square of the thermal velocity (sound speed) of the ICM is comparable to the gravitational potential. During a merger, the infall velocities of the subclusters are comparable to the escape velocity, which implies that the square of the infall velocity is larger (by roughly a factor of two) than the gravitational potential. Thus, the motions in cluster mergers are expected to be supersonic, but only moderately so. As a result, one expects that cluster mergers will drive shock waves into the intracluster gas of the two subclusters. Let v_s be the velocity of such a shock wave relative to the preshock intracluster gas. The sound speed in the preshock gas is $c_s = \sqrt{(5/3)P/\rho}$, where P is the gas pressure and ρ is the density. Then, the Mach number of the shock is $\mathcal{M} \equiv v_s/c_s$. Based on the simple argument given above, one expects shocks with Mach numbers of $\mathcal{M} \lesssim 2$. Stronger shocks may occur under some circumstances, such as in the outer parts of clusters, or in low mass subclusters merging with more massive clusters.

Shocks are irreversible changes to the gas in clusters, and thus increase the entropy S in the gas. A useful quantity to consider is the specific entropy per particle in the gas, s (18). Observations of X-ray spectra can be used to determine T, while the X-ray surface brightness depends on ρ^2. Thus, one can use X-ray observations to determine the specific entropy in the gas just before and just after apparent merger shocks seen in the X-ray images. Since merger shocks should produce compression, heating, pressure increases, and entropy increases, the corresponding increase in all of these quantities (particularly the entropy) can be used to check that discontinuities are really shocks (e.g., not "cold fronts" or other contact discontinuities, Sect. 12.2).

In [26], this test was applied to ASCA temperature maps and ROSAT images of Cygnus-A and Abell 3667, two clusters which appeared to show strong merger shocks. Recent Chandra images have shown that the feature in Abell 3667 is a cold front [47]. In Cygnus-A, the increase in specific entropy in the shocked regions is roughly $\Delta s \approx (3/2)k$. The specific heat per particle q which must be dissipated to produce this change in entropy is $q \approx T\Delta s \approx (3/2)kT$, or about the present specific heat content in the shocked gas. Thus, these observations provide a direct confirmation that merger shocks contribute significantly to the heating of the intracluster gas.

The most dramatic merger shock which has been seen with Chandra is in the "Bullet Cluster" 1E0657-56 [7, 29, 30]. This is a very high velocity ($\sim$4500 km s^{-1}) merger occurring nearly in the plane of the sky, with a merger bow shock located ahead of a "cold front" (Sect. 12.2). Another prominent merger shocks with a Mach number of $\mathcal{M} \approx 2.1$ is seen in Abell 520 [25]. In both cases, the merger shocks appear to have associated diffuse radio emission (See the chapter by Feretti & Giovannini for more details.)

11.1 Shock Kinematics

The variation in the hydrodynamical variables in the intracluster medium across a merger shock are determined by the standard Rankine–Hugoniot jump conditions [20], if one assumes that all of the dissipated shock energy is thermalized. Consider a small element of the surface of a shock (much smaller than the radius of curvature of the shock, for example). The tangential component of the velocity is continuous at the shock, so it is useful to go to a frame which is moving with that element of the shock surface, and which has a tangential velocity which is equal to that of the gas on either side of the shock. In this frame, the element of the shock surface is stationary, and the gas has no tangential motion. Let the subscripts 1 and 2 denote the preshock and post-shock gas; thus, $v_1 = v_s$ is the longitudinal velocity of material into the shock (or alternative, the speed with which the shock is advancing into the preshock gas). Conservation of mass, momentum, and energy then implies the following jump conditions

$$\begin{aligned} \rho_1 v_1 &= \rho_2 v_2 \, , \\ P_1 + \rho_1 v_1^2 &= P_2 + \rho_2 v_2^2 \, , \\ w_1 + \frac{1}{2} v_1^2 &= w_2 + \frac{1}{2} v_2^2 \, . \end{aligned} \tag{45}$$

Here, $w = P/\rho + \epsilon$ is the enthalpy per unit mass in the gas, and ϵ is the internal energy per unit mass. If the gas behaves as a perfect fluid on each side of the shock, the internal energy per unit mass is given by: $\epsilon = /(\gamma - 1)\, P/\rho$, where γ is the ratio of specific heats (the adiabatic index) and is $\gamma = 5/3$ for fully ionized plasma. The jump conditions can be rewritten as:

$$\begin{aligned} \frac{P_2}{P_1} &= \frac{2\gamma}{\gamma+1}\mathcal{M}^2 - \frac{\gamma-1}{\gamma+1} \\ \frac{v_2}{v_1} = \frac{\rho_1}{\rho_2} \equiv \frac{1}{C} &= \frac{2}{\gamma+1}\frac{1}{\mathcal{M}^2} + \frac{\gamma-1}{\gamma+1}\,, \end{aligned} \tag{46}$$

where $C \equiv \rho_2/\rho_1$ is the shock compression.

If one knew the velocity structure of the gas in a merging cluster, one could use these jump condition to derive the temperature, pressure, and density jumps in the gas. At present, the best X-ray spectra for extended regions in clusters of galaxies have come from CCD detectors on ASCA, Chandra, and XMM/Newton. CCDs have a spectral resolution of $>$100 eV at the Fe K line at 7 keV, which translates into a velocity resolution of $>$4000 km/s. Thus, this resolution is (at best) marginally insufficient to measure merger gas velocities in clusters. In a few cases with very bright regions and simple geometries, the grating spectrometers on Chandra and especially XMM/Newton may be useful.

At present, X-ray observations can be used to directly measure the temperature and density jumps in merger shocks. Thus, one needs to invert the jump relations to give the merger shock velocities for a given shock temperature, pressure, and/or density increase. If the temperatures on either side of the merger shock can be measured from X-ray spectra, the shock velocity can be inferred from [26]:

$$\Delta v_s = \left[\frac{kT_1}{\mu m_p}(C-1)\left(\frac{T_2}{T_1} - \frac{1}{C}\right)\right]^{1/2}, \tag{47}$$

where $\Delta v_s = v_1 - v_2 = [(C-1)/C]v_s$ is the velocity change across the shock, and μ is the mean mass per particle in units of the proton mass m_p. The shock compression C can be derived from the temperatures as

$$\frac{1}{C} = \left[\frac{1}{4}\left(\frac{\gamma+1}{\gamma-1}\right)^2\left(\frac{T_2}{T_1} - 1\right)^2 + \frac{T_2}{T_1}\right]^{1/2} - \frac{1}{2}\frac{\gamma+1}{\gamma-1}\left(\frac{T_2}{T_1} - 1\right). \tag{48}$$

Alternatively, the shock compression can be measured directly from the X-ray image. However, it is difficult to use measurements of the shock compression alone to determine the shock velocity, for two reasons. First, a temperature is needed to set the overall scale of the velocities; as is obvious from (46), the shock compression allows one to determine the Mach number $\mathcal{M}$ but not the shock velocity. The second problem is that temperature or pressure information is needed to know that a discontinuity in the gas density is a shock, and not a contact interface (e.g., the "cold fronts" discussed in Sect. 12.2 below).

X-ray temperature maps of clusters have been used to derive the merger velocities using these relations. Reference [26] used ASCA observations to determine the kinematics of mergers in three clusters (Cygnus-A, Abell 2065, and Abell 3667). Because of the poor angular resolution of ASCA, these analyses were quite uncertain. More recently, possible shocks have been detected in Chandra images of a number of merging clusters (e.g., Abell 85 [19], Abell 665 [27], Abell 3667 [47]), and the shock jump conditions have been applied to determine the kinematics in these clusters.

The simplest case is a head-on symmetric merger ($b = 0$ and $M_1 = M_2$) at an early stage when the shocked region lies between the two cluster centers. Reference [26] suggests that the Cygnus-A cluster is an example. If the gas within the shocked region is nearly stationary, then the merger velocity of the two subclusters is just $v = 2\Delta v_s$. Applying these techniques to the ASCA temperature map for the Cygnus-A cluster, Reference [26] found a merger velocity of $v \approx 2200$ km/s. This simple argument is in reasonable agreement with the results of numerical simulations of this merger [39].

One can compare the merger velocities derived from the temperature jumps in the merger shocks with the values predicted by free-fall from the turn-around radius. In the case of Cygnus-A, [26] found good agreement with the the free-fall velocity of ~2200 km/s. This consistency suggests that the shock energy is effectively thermalized, and that a major fraction does not go into turbulence, magnetic fields, or cosmic rays. Thus, the temperature jumps in merger shocks can provide an important test of the relative roles of thermal and non-thermal processes in clusters of galaxies.

11.2 Nonequilibrium Effects

Cluster mergers are expected to produce collisionless shocks, as occurs in supernova remnants. As such, nonequilibrium effects are expected, including non-equipartition of electrons and ions and nonequilibrium ionization [26, 44, 45]. Collisionless shocks are generally not as effective in heating electrons as ions. Assuming that the post-shock electrons are somewhat cooler than the ions, the time scale for electron and protons to approach equipartition as a result of Coulomb collisions in a hot ionized gas is (2) [43]:

$$
\begin{aligned}
t_{\rm eq} &= \frac{3 m_p m_e}{8\sqrt{2\pi} n_e e^4 \ln \Lambda} \left(\frac{kT_e}{m_e} \right)^{3/2} \\
&\approx 2.1 \times 10^8 \left(\frac{T_e}{10^8\,{\rm K}} \right)^{3/2} \left(\frac{n_e}{0.001\,{\rm cm}^{-3}} \right)^{-1} {\rm yr} \,,
\end{aligned} \tag{49}
$$

where n_e and T_e are the electron number density and temperature, respectively, and Λ is the Coulomb factor. The relative velocity between the post-shock gas and the shock front is $(1/4)v_s$; thus, one would expect the electron temperature to reach equipartition a distance of

$$d_{\rm eq} \approx 160 \left(\frac{v_s}{3000\,{\rm km/s}}\right) \left(\frac{T_e}{10^8\,{\rm K}}\right)^{3/2} \left(\frac{n_e}{0.001\,{\rm cm}^{-3}}\right)^{-1} {\rm kpc} \qquad (50)$$

behind the shock front. Of course, it is the electron temperature (rather than the ion or average temperature) which determines the shape of the X-ray spectrum. This distance is large enough to insure that the lag could be spatially resolved in X-ray observations of low redshift clusters. Similar effects might be expected through non-equilibrium ionization.

On the other hand, it is likely that the nonequilibrium effects in cluster merger shocks are much smaller than those in supernova blast wave shocks because of the low Mach numbers of merger shocks. That is, the preshock gas is already quite hot (both electrons and ions) and highly ionized. Moreover, a significant part of the heating in low Mach number shocks is due to adiabatic compression, and this would still act on the electrons in the post-shock gas in merger shocks, even if there were no collisionless heating of electrons. For example, in a $\mathcal{M} = 2$, $\gamma = 5/3$ shock, the total shock increase in temperature is a factor of 2.08 (46). The shock compression is $C = 2.29$, so adiabatic compression increases the electron temperature by a factor of $C^{2/3} = 1.74$, which is about 83% of the shock heating.

11.3 Mergers and Basic Gravitational Physics Effects

Merging clusters also provide several very direct tests of basic gravitational physics. These tests are possible because of the dynamical nature of mergers, and the difference in the behavior of collisional and non-collisional components of clusters. The gas is clusters is a collisional fluid (Sect. 4) with a mean-free-path which is small compared to the scale of clusters (11). Thus, when clusters collide, the motion of the gas will be retarded by ram pressure and shocks. On the other hand, the galaxies in clusters are essentially collisionless. When clusters collide, the galaxies will fly by one another. Thus, the galaxies in a merging subcluster will often be found ahead of the gas from the same subcluster. This is particularly obvious in late stage mergers with "cold fronts" (Sect. 12.2), where the gas which was initially at the center of a subcluster will be found lagging behind the central dominant and other galaxies from the subcluster. Perhaps the most prominent example of this is the in the "Bullet Cluster" 1E0657-56 [7, 29, 30], where the cold front and dense gas from the subcluster are clearly separated for the galaxies from the same subcluster.

In the most widely accepted model for "dark matter" (Sect. 8), the dominant component of the mass of the Universe is made up of collisionless elementary particles. For example, this would be true of Cold Dark Matter (e.g., [2]). If this is the case, one would expect that the dark matter would be located in the same regions of merging clusters as the galaxies. An alternative idea is that the law of gravity or laws of motions differ from the Newtonian form at large distances or small accelerations [33]. In these MOdified Newtonian Dynamics (MOND) theories, there is no dark matter component of the Universe, and gravity is just due to ordinary baryonic matter. In clusters of galaxies, the vast majority of the ordinary baryonic matter is in the hot X-ray gas (Sect. 8). Thus, MOND theories predict that in a merger, the gravity (or apparent dark matter) should mainly be located where the gas is located. In an advanced merger, the gas is located behind the galaxies.

The location of the gravity (or apparent dark matter) can be determined from weak gravitational lensing observations of the cluster [7, 30]. This test has been performed on the "Bullet Cluster" 1E0657-56 [7, 30], where the weak lensing measurement show that the gravity of the merging subcluster is centered on the galaxies, and is clearly displaced from the located of the subcluster gas. Thus, these measurements provide what is arguably the strongest proof of the existence of dark matter, rather than a change in the laws of gravity.

Another alternative to the conventional Cold Dark Matter hypothesis is that the dark matter consists of weakly interacting elementary particles, but the particles do have a small but significant cross-section for self-interaction [42]. If this were the case, the dark matter would act as a collisional fluid, and would be displaced from the position of the galaxies in a merging cluster towards the center of the subcluster gas. In the "Bullet Cluster" 1E0657-56 no such displacement is evident [30], and the lack of such a displacement can be used to set an upper limit on the self-interaction cross-section per unit mass of dark matter of < 1 cm^2 g^{-1}. This is a very serious constraint on models of self-interacting dark matter.

12 Mergers and Cool Cluster Cores

12.1 Cooling Flows vs. Mergers

The centers of a significant fraction of clusters of galaxies have luminous cusps in their X-ray surface brightness known as "cooling flows" (see [12] for an extensive review). In every case, there is a bright (cD) galaxy at the center of the cooling flow region. The intracluster gas densities in these regions are much higher than the average values in the outer portions of clusters. X-ray spectra indicate that there are large amounts of gas at low temperatures (down to $\sim 10^7$ K), which are much cooler than those in the outer parts of clusters. The high densities imply rather short cooling times $t_{\rm cool}$ (the time scale for the gas to cool to low temperature due to its own radiation). The hypothesis is that the gas in these regions is cooling from higher intracluster temperature ($\sim 10^8$ K) down to these lower temperatures as a result of the energy loss due to the X-ray emission we observe. Typical cooling rates are $\sim 100\ M_\odot$ yr^{-1}. The cooling times, although much shorter than the Hubble time, are generally much longer than the dynamical (i.e., sound crossing time) of the gas in these regions. As a result, the gas is believed to remain nearly in hydrostatic equilibrium. Thus, the gas must compress as it cools to maintain a pressure which can support the weight of the overlying intracluster medium.

The primary observational characteristics of cooling flows are very bright X-ray surface brightnesses which increase rapidly toward the center of the cluster. The high surface brightnesses imply high gas densities which also increase rapidly towards the cluster center. These regions contain cooler cluster gas.

Empirically, there is significant indirect evidence that mergers disrupt cooling flows. There is a strong statistical anticorrelation between cooling flows and/or cooling rates, and irregular structures in clusters as derived by statistical analysis of their X-ray images [3]. Looked at individually, very strong cooling cores are almost never associated with very irregular or bimodal clusters, which are likely merger candidates [9, 17]. There are some cases of moderate cooling flows in merging clusters; in most cases, these appear to be early-stage mergers where the merger shocks haven't yet reached the cooling core of the cluster. Examples may include Cygnus-A [26] and Abell 85 [19]. There also are a large number of merging clusters at a more advanced stage with relatively small cooling cores; Abell 2065 [26] may be an example. Recently, Chandra Observatory X-ray images have shown a number of merging clusters with rapidly moving cores of cool gas (the "cold fronts" discussed below in Sect. 12.2). In these systems, the cooling flows appear to have survived, at least to the present epoch in the merger.

It is unclear exactly how and under what circumstances mergers disrupt cooling flows. The cooling flows might be disrupted by tidal effects, by shock heating the cooler gas, by removing it dynamically from the center of the cluster due to ram pressure, by mixing it with hotter intracluster gas, or by some other mechanism. Numerical hydrodynamical simulations are needed to study the mechanisms by which cooling flows are disrupted. This is a relatively unexplored area, largely because the small spatial scales and rapid cooling time scales in the inner regions of cooling flows are still a significant challenge to the numerical resolution of hydrodynamical codes. McGlynn and Fabian [32] argued that mergers disrupted cooling flows, but this was based on purely N-body simulations. Hydrodynamical simulations have been made of the effects of head-on mergers with relatively small subclusters (1/4 or 1/16 of the mass of the main cluster) on a cooling flow in the main cluster [16]. They find that the mergers disrupt the cooling flow in some cases, but not in others. Their simulations suggest that the disruption is not due to tidal or other gravitational effects.

Another possibility is that the merger shocks heat up the cooling flow gas and stop the cooling flow. In the simulations, this does not appear to be the main mechanism of cooling flow disruption. There are a number of simple arguments which suggest that merger shocks should be relatively inefficient at disrupting cooling flows. First, it is difficult for these shocks to penetrate the high densities and steep density gradients associated with cooling flows, and the merger shocks would be expected to weaken as they climb these steep density gradients. Even without this weakening, merger shocks have low Mach numbers, and only produce rather modest increases in temperature ($\lesssim$ a factor of 2). These small temperature increases are accompanied by significant compressions. As a result, shock heating actually decreases the cooling time due to thermal bremsstrahlung emission for shocks with Mach numbers $\mathcal{M} \leq (21 + 12\sqrt{3})^{1/2} \approx 6.5$. It is likely that the shocked gas will eventually expand, and adiabatic expansion will lengthen the cooling time. However, even if the gas expands to its preshock pressure, the increase in the cooling time is not very large. For a $\mathcal{M} = 2$ shock, the final cooling time after adiabatic expansion to the original pressure is only about 18% longer than the initial cooling time.

The simulations by [16] suggest that the main mechanism for disrupting cooling flows is associated with the ram pressure of gas from the merging subcluster. The gas in the cooling flow is displaced, and may eventually mix with the hotter gas [39]. Earlier, [13] had argued that ram pressure, rather than shock heating, was the main mechanism for disrupting cooling flows. Assuming this is the case, one expects that the merger will remove the cooling flow gas at radii which satisfy

$$\rho_{\rm sc} v_{\rm rel}^2 \gtrsim P_{\rm CF}(r) \,, \tag{51}$$

where $P_{\rm CF}(r)$ is the pressure profile in the cooling flow, $\rho_{\rm sc}$ is the density of the merging subcluster gas at the location of the cooling flow, and $v_{\rm rel}$ is the relative velocity of the merging subcluster gas and the cooling flow. Reference [16] finds that this relation provides a reasonable approximation to the disruption in their simulations.

The pressure profile in the cooling flow gas prior to the merger is determined by the condition of hydrostatic equilibrium. If the cluster gravitational potential has a wide core within which the potential is nearly constant (e.g., as in a King model), then the cooling flow pressure will not increase rapidly into the center. In this case, once the merger reaches the central regions of the cluster, if the ram pressure is sufficient to remove the outer parts of the cooling flow, it should be sufficient to remove nearly all of the cooling flow. On the other hand, if the cluster potential is sharply peaked as is predicted by numerical simulations [35], the merger may remove the outer parts of the cooling flow but not the innermost regions. Thus, the survival and size of cool cores in merging clusters can provide evidence on whether clusters have sharply peaked potentials [26].

12.2 Cold Fronts

One of the more dramatic early discoveries with the Chandra X-ray Observatory was the presence of very sharp surface brightness discontinuities in merging clusters of galaxies. A pair of such discontinuities were first seen in the public science verification data on the Abell 2142 cluster [28]. Initially, it seemed likely that these were merger shocks. However, temperature measurements showed that this was not the case. The high X-ray surface brightness regions were both dense and *cool*; thus, the gas in these regions had a lower specific entropy than the gas in the less dense regions. The lack of a pressure jump and the incorrect sign of the temperature and entropy variations showed that these features could not be shocks [28]. Instead, they appear to be contact discontinuities between hot, diffuse gas and a cloud of colder, denser gas [28]. In [47], these contact discontinuities were named "cold fronts." Reference [28] argues that the source of the cold clouds are the cooling cores of one or both of merging subclusters. As noted above, cooling flows do appear to be able to partially survive in mergers, at least for some period. Subsequently, cold fronts have been observed in a number of other clusters; for an extensive review of the observations of these cold fronts, see [14].

Kinematics of Cold Fronts

As discussed extensively in [47], the variation in the density, pressure, and temperature of the gas in a cold front can be used to determine the relative velocity of cold core. This technique is analogous to that for merger shocks discussed above (47) and (48). The geometry is illustrated in Fig. 2, which is drawn in the rest frame of the cold core. We assume that the cold core has a smoothly curved, blunt front edge. The normal component of the flow of hot gas past the surface of the cold core will be zero. There will be at least one point where the flow is perpendicular to the surface of the cold core, and the flow velocity of the hot gas will be zero at this stagnation point ("st" in Fig. 2). Far upstream, the flow of the hot gas will be undisturbed at the velocity of the cold core relative to the hotter gas, v_1. Let c_{s1} be the sound speed in this upstream gas, and $\mathcal{M}_1 \equiv v_1/c_{s1}$ be the Mach number of the motion of the cold core into the upstream gas. If $\mathcal{M}_1 > 1$, a bow shock will be located ahead of the cold front.

The ratio of the pressure at the stagnation point to that far upstream is given by [20]

$$\frac{P_{\rm st}}{P_1} = \begin{cases} \left(1 + \frac{\gamma-1}{2}\mathcal{M}_1^2\right)^{\frac{\gamma}{\gamma-1}} \, , & \mathcal{M}_1 \le 1 \\ \mathcal{M}_1^2 \left(\frac{\gamma+1}{2}\right)^{\frac{\gamma+1}{\gamma-1}} \left(\gamma - \frac{\gamma-1}{2\mathcal{M}_1^2}\right)^{-\frac{1}{\gamma-1}} & , \mathcal{M}_1 > 1 \, . \end{cases} \tag{52}$$

The ratio $(P_{\rm st}/P_1)$ increases continuously and monotonically with $\mathcal{M}_1$. Thus, in principle, measurements of P_1 and $P_{\rm st}$ in the hot gas could be used to determine $\mathcal{M}_1$. The pressures would be determined from X-ray spectra and images. In practice, the emissivity of the hot gas near the stagnation point is likely to be small. However, the pressure is continuous across the cold front, so the stagnation pressure can be determined just inside of the cold core, where the X-ray emissivity is likely to be much higher. Once $\mathcal{M}_1$ has been determined, the velocity of the encounter is given by $v_1 = \mathcal{M}_1 c_{s1}$.

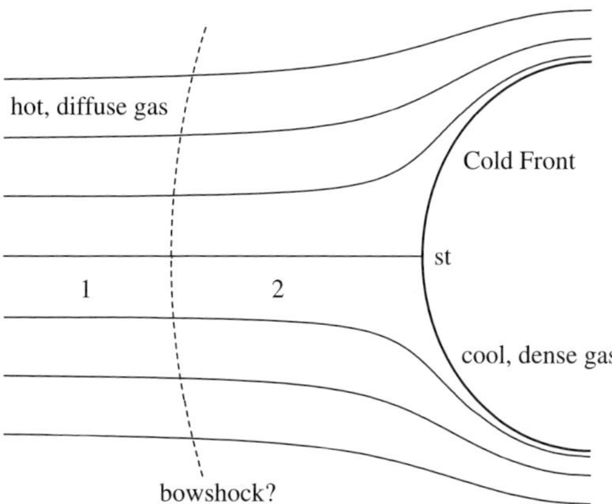

Fig. 2. A schematic diagram of flow around a "cold front" in a cluster merger. The heavy solid arc at the right represents the contact discontinuity between the cold, dense cold core gas, and the hotter, more diffuse gas from the outer regions of the other cluster. The cold core is moving toward the left relative to the hotter gas. The narrow solid lines are streamlines of the flow of the hotter gas around the cold core. The region labelled "1" represent the upstream, undisturbed hot gas. If the cold front is moving transonically ($\mathcal{M}_1 > 1$), then the cold front will be preceded by a bow shock, which is shown as a dashed arc. The stagnation point, where the relative velocity of the cooler dense gas and hotter diffuse gas is zero, is marked "st"

If the motion of the cold core is transonic ($\mathcal{M}_1 > 1$), one can also determine the velocity from the temperature and/or density jump at the bow shock (47) and (48). If the bow shock can be traced to a large transverse distance and forms a cone, the opening angle of this Mach cone corresponds to the Mach angle, $\theta_M \equiv \csc^{-1}(\mathcal{M}_1)$. However, variations in the cluster gas temperature may lead to distortions in this shape.

The distance between the stagnation point and the closest point on the bow shock (the shock "stand-off" distance d_s) can also be used to estimate the Mach number of the motion of the cold front [47]. The ratio of d_s to the radius of curvature of the cold front $R_{\rm cf}$ depends on the Mach number $\mathcal{M}_1$ and on the shape of the cold front. Figure 3 shows the values of $\mathrm{d}_s/R_{\rm cf}$ as a function of $(\mathcal{M}_1^2 - 1)^{-1}$ for a spherical cold front [41]. Although there is no simple analytic expression for the stand-off distance which applies to all shapes of objects, a fairly general approximate method to calculate d_s has been given by [34]. and some simple approximate expressions exist for a number of simple geometries. The stand-off distance increases as the Mach number approaches unity; thus, this method is, in some ways, a very sensitive diagnostic for the Mach number for the low values expected in cluster mergers. On the other hand, the stand-off distance also depends strongly on the shape of the cold front as the Mach number decreases. The application of this diagnostic to observed clusters is strongly affected by projection effects. Because the radius

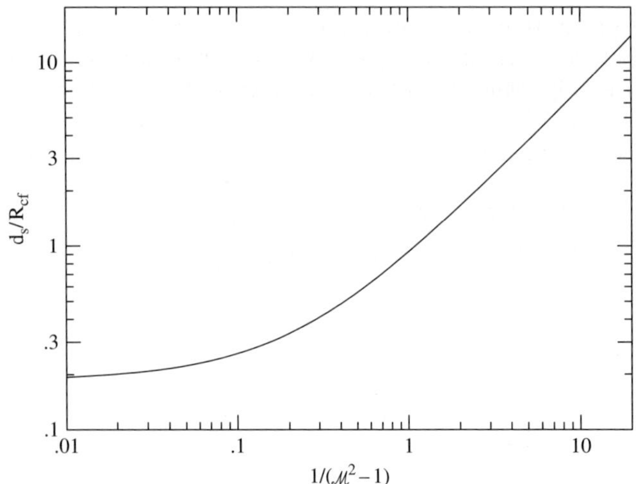

Fig. 3. The ratio of the stand-off distance of the bow shock d_s to the radius of curvature $R_{\rm cf}$ of the stagnation region of the cold front, as a function of $1/(\mathcal{M}_1^2 - 1)$, where $\mathcal{M}_1$ is the Mach number. This is for a spherical cold front and $\gamma = 5/3$

of curvature of the bow shock is usually greater than that of the cold front, projection effects will generally cause d_s to be overestimated and $\mathcal{M}_1$ to be underestimated. Projection effects also make the true shape of the cold front uncertain.

These techniques have been used to determine the merger velocities from cold fronts in Abell 3667 [47], RXJ1720.1+2638 [31], and Abell 85 [19]. The most spectacular application is the "Bullet Cluster" 1E0657-56 [7, 29, 30], which contains a very high Mach number merger.

Width of Cold Fronts

One remarkable aspect of the cold fronts observed with the Chandra Observatory in several clusters is their sharpness. In Abell 3667, the temperature changes by about a factor of two across the cold front [47], and the accompanying change in the X-ray surface brightness occurs in a region which is narrower than 2 kpc [47]. This is less than the mean-free-path of electrons in this region. The existence of this very steep temperature gradient and similar results in other merging clusters with cold fronts requires that thermal conduction be suppressed by a large factor [11, 46, 47] relative to the classical value in an unmagnetized plasma (14) [43]. It is likely that this suppression is due to the effects of the intracluster magnetic field. It is uncertain at this point whether this is due to a generally tangled magnetic field (in which case, heat conduction might be suppressed throughout clusters), or due to a tangential magnetic field specific to the tangential flow at the cold front [46].

Because of the tangential shear flow at the cold front (Fig. 2), the front should be disturbed and broadened by the Kelvin–Helmholtz (K–H) instability. Reference [46] argues that the instability is suppressed by a tangential magnetic field, which is itself generated by the tangential flow. This suppression requires that the magnetic pressure P_B be a non-trivial fraction of the gas pressure P in this regions, $P_B \gtrsim 0.1P$. The required magnetic field strength in Abell 3667 is $B \sim 10\ \mu$G. Alternatively, cold fronts might be stabilized by gravity [15] or acceleration along the front [6].

References

1. Allen, S.W., Schmidt, R.W., Ebeling, H., Fabian, A.C., van Speybroeck, L.: MNRAS **353**, 457 (2004)
2. Bennett, C.L., et al.: ApJS **148**, 1 (2003)
3. Buote, D.A., Tsai, J.C.: ApJ **458**, 27 (1996)
4. Cavaliere, A., Fusco-Femiano, R.: A&A **49**, 137 (1976)
5. Cavaliere, A., Lapi, A., Menci, N.: ApJ **581**, L1 (2002)
6. Churazov, E.: Physics of cluster cores. In: The X-ray Universe 2005, (in press)
7. Clowe, D., Gonzalez, A., Markevitch, M., et al.: ApJ **604**, 596 (2004)
8. Davé, R.: ApJ **552**, 473 (2001)
9. Edge, A.C., Stewart, G.C., Fabian, A.C.: MNRAS **258**, 177 (1992)
10. Eke, V.R., Navarro, J.F., Frenk, C.S.: ApJ **503**, 569 (1998)
11. Ettori, S., Fabian, A.C.: MNRAS **317**, L57 (2000)
12. Fabian, A.C.: ARA&A **32**, 277 (1994)
13. Fabian, A.C., Daines, S.J.: MNRAS **252**, 17p (1991)
14. Forman, W., Jones, C., Markevitch, M., Vikhlinin, A., Churazov, E.: High angular resolution cluster observations with Chandra: a new view. In: Merging Processes in Galaxy Clusters, pp. 109–132. Kluwer, Dordrecht (2002)
15. Fujita, Y., et al.: ApJ **575**, 764 (2002)
16. Gómez, P.L., Loken, C., Roettiger, K., Burns, J.O.: ApJ **569**, 122 (2002)
17. Henriksen, M.J.: ApJ **407**, L13 (1988)
18. Kaiser, N.: MNRAS **222**, 323 (1986)
19. Kempner, J., Sarazin, C.L., Ricker, P.R.: ApJ **579**, 236 (2002)
20. Landau, L.D., Lifshitz, E.M.: Fluid Mechanics, Pergamon, Oxford (1959)
21. Loewenstein, M.: ApJ **532**, 17 (2000)
22. Lloyd-Davies, E.J., Ponman, T.J., Cannon, D.B.: MNRAS **315**, 689 (2000)
23. Markevitch, M.: ApJ **504**, 27 (1998)
24. Markevitch, M., Forman, W.R., Sarazin, C.L., Vikhlinin, A.: ApJ **503**, 77 (1998)
25. Markevitch, M., Govoni, F., Brunetti, G., Jerius, D.: ApJ **627**, 733 (2005)
26. Markevitch, M., Sarazin, C.L., Vikhlinin, A.: ApJ **521**, 526 (1996)
27. Markevitch, M., Vikhlinin, A.: ApJ **563**, 95 (2001)
28. Markevitch, M., et al.: ApJ **541**, 542 (2000)
29. Markevitch, M., et al.: ApJ **567**, L27 (2002)
30. Markevitch, M., et al.: ApJ **606**, 542 (2004)
31. Mazzotta, P., Markevitch, M., Vikhlinin, A., Forman, W.R., David, L.P., Van Speybroeck, L.: ApJ **555**, 205 (2001)

32. McGlynn, T.A., Fabian, A.C.: MNRAS **208**, 709 (1984)
33. Milgrom, M.: ApJ **270**, 365 (1983)
34. Moekel, W.E.: Approximate Method for Predicting Forms and Location of Detached Shock Waves Ahead of Plane or Axially Symmetric Bodies, NACATechnical Note 1921 (1949)
35. Navarro, J.F., Frenk, C.S., White, S.D.M.: ApJ **490**, 493 (1997)
36. Osterbrock, D.E.: Astrophysics of Gaseous Nebulae and Active Galactic Nuclei, pp. 53–65. University Science, Mill Valley (1989)
37. Ponman, T.J., Cannon, D.B., Navarro, J.F.: Nature **397**, 135 (1999)
38. Ponman, T.J., Sanderson, A.J., Finoguenov, A.: MNRAS **343**, 331 (2003)
39. Ricker, P.M., Sarazin, C.L.: ApJ **561**, 621 (2001)
40. Sarazin, C.L.: X-ray Emission from Clusters of Galaxies, Cambridge University Press, Cambridge (1988)
41. Schreier, S.: Compressible Flow, pp. 182–189. Wiley, New York (1982)
42. Spergel, D.N., Steinhardt, P.J.: Phys. Rev. Lett. **84**, 3760 (2000)
43. Spitzer Jr., L.: Physics of Fully Ionized Gases, Interscience, New York (1956)
44. Takizawa, M.: ApJ **520**, 514 (1999)
45. Takizawa, M.: ApJ **532**, 182 (2000)
46. Vikhlinin, A., Markevitch, M., Murray, S.M.: ApJ **549**, L47 (2001a)
47. Vikhlinin, A., Markevitch, M., Murray, S.M.: ApJ **551**, 160 (2001b)
48. Vikhlinin, A., Markevitch, M., Murray, S.S., Jones, C., Forman, W., Van Speybroeck, L.: ApJ **628**, 655 (2005)
49. Voevodkin, A., Vikhlinin, A.: ApJ **601**, 610 (2004)
50. Voit, G.M.: RMP **77**, 207 (2005)
51. Voit, G.M., Bryan, G.L.: Nature **414**, 425 (2001)
52. White, S.D.M., Frenk, C.S.: ApJ **379**, 52 (1991)
53. Wise, M.W., Sarazin, C.L.: ApJ, **415**, 58 (1993)

Dynamics of the Hot Intracluster Medium

C. Jones[1], W. Forman[1], A. Vikhlinin[1], M. Markevitch[1], M. Machacek[1] and E. Churazov[2]

[1] Center for Astrophysics, Cambridge, MA 02138, USA
cjones@cfa.harvard.edu
[2] MPA, Garching, Germany & IKI, Moscow, Russia
churazov@mpa-garching.mpg.de

1 General Properties of Clusters

Clusters of galaxies might well have been called something else, if they had been seen first in other than visible light, since the optically luminous galaxies make up only a small fraction (<5%) of the cluster mass. In clusters, most of the baryons are in the hot intracluster gas and most of the matter is in the form of dark matter.

The distribution of galaxies on the sky shows that galaxies are often concentrated in groups or clusters. In the 1930's Zwicky [145] measured the velocities of galaxies in the Coma cluster and used the virial theorem to estimate the total amount of mass in the cluster. He determined a total cluster mass of about 400 times the mass he estimated by adding up the mass in all the Coma galaxies. Since the galaxies themselves did not provide enough matter, Zwicky postulated that "missing mass" or dark matter must be present in the cluster to gravitationally bind the galaxies. Around 1970, a new cluster component, a diffuse hot gas, was found both from the observations of tailed radio sources in Perseus [122] and from Uhuru observations that showed the X-ray emission from the Coma, Perseus, and Virgo clusters was spatially extended [49, 59, 76].

Today we know that clusters of galaxies are complex, multi-component systems with hundreds of galaxies, a hot intracluster medium, radio plasmas and dark matter evolving in tightly coupled ways. Clusters are the largest gravitationally bound systems in the Universe with total masses of about 10^{14}–10^{15} $M_{\odot}$. Since clusters form from large volumes (~20 Mpc radius), their mass components are representative of the Universe as a whole. Most of the matter in clusters is dark matter, with only about 15% of the matter being baryonic. Most of the baryons are in the form of hot X-ray gas, with the stars in galaxies contributing only about 20% of the baryonic mass in rich clusters, and up to half the baryonic mass in groups.

Clusters of galaxies are very bright in X-rays, with luminosities as high as several 10^{45} ergs s^{-1}, which allows them to be observed to relatively high

C. Jones et al.: *Dynamics of the Hot Intracluster Medium*, Lect. Notes Phys. **740**, 31–69 (2008)
DOI 10.1007/978-1-4020-6941-3_2

redshifts. Their X-ray emission is due primarily to thermal bremsstrahlung emission from the hot gas, along with line emission. The intracluster medium (ICM) is heated primarily through shocks produced as matter falls into the deep gravitational potential of the cluster. If clusters form through simple gravitational collapse, then their bolometric luminosity should scale as the square of the gas temperature (e.g. [73, 74]). However, observations have long shown that the X-ray luminosity increases as the cube of the temperature, and even more steeply for poor clusters or groups, due to the increased entropy of the gas caused by non-gravitational processes [62, 112]. The additional gas heating, beyond that expected from gravitational collapse alone, is at the level of 1–3 keV per particle and is too large to be supplied by supernovae, but may be supplied by active galactic nuclei (AGN).

As the Chandra X-ray image of the rich cluster A1413 shows (Fig. 1), the hot gas fills the gravitational well of the cluster and often is peaked on a central bright galaxy. A massive galaxy lies at the centers of many clusters and groups and at the center of this galaxy lies a supermassive black hole. The high gas density in the cluster cores results in gas cooling and accreting onto the supermassive black hole. This accretion results in AGN outbursts that reheat the cooling gas and substantially reduce the amount of cool gas available for star formation and for accretion. This cyclical process of gas cooling and feedback is described in Sect. 2.

In addition to using the X-ray imaging to study the morphology and structure of the cluster, for symmetric clusters, we can measure the density distribution of the hot gas directly from the X-ray surface brightness (Fig. 2). The hot gas is generally in hydrostatic equilibrium and thus traces the gravitational potential of the cluster. The cluster gas temperature and heavy ele-

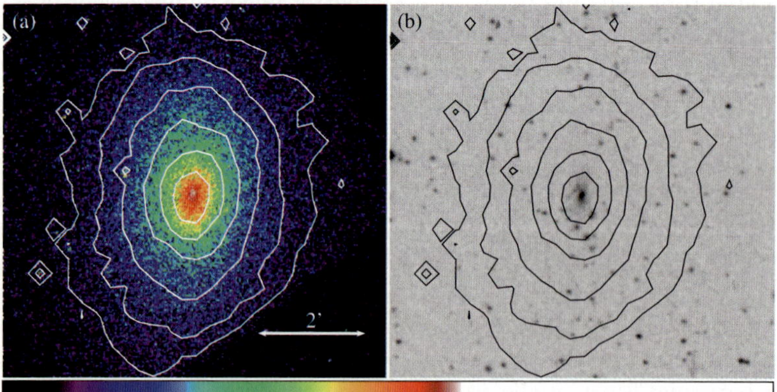

Fig. 1. (**a**) The Chandra image of the cluster A1413 ($z = 0.14$) shows that the X-ray emission is strongly peaked in the central region. (**b**) Isointensity contours from the X-ray image are superposed on the optical image. A cD galaxy lies at the cluster center

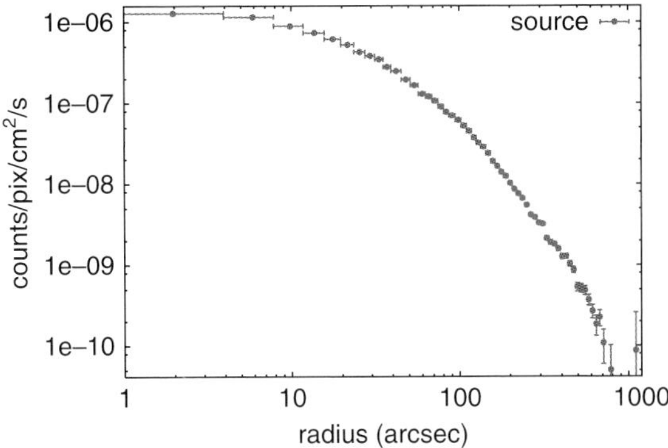

Fig. 2. The X-ray surface brightness profile for A1413 is centrally peaked

ment abundances can be measured through X-ray spectroscopy (see Fig. 3). By measuring the temperature and density distributions in the gas, one can map the total cluster mass which is dominated by dark matter (see Fabricant et al. [45] for an early application of this technique to map the dark matter around M87 in the Virgo cluster).

At the high gas temperatures of the ICM (10^7–10^8 K), the gas in nearly fully ionized. Hydrogen and helium are fully stripped of their electrons, while heavier elements retain only a few electrons. The hot ICM is not pristine primordial material, but has been enriched in heavy elements. Outside the cluster cores, the heavy element abundance in the ICM is typically about 0.25 solar, while in the dense cluster cores, particularly around a central bright galaxy,

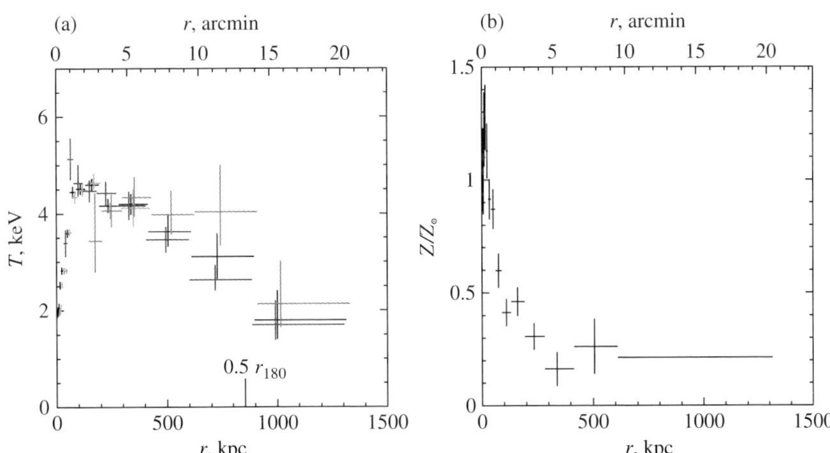

Fig. 3. (**a**) The gas temperature often decreases in the cluster core and in the cluster outskirts as shown for A133. (**b**) The heavy element abundance generally increases toward the cluster center for clusters with a central cD galaxy [137]

abundances can increase to roughly solar (see Fig. 3). It was the detection of Fe K_α emission at 6.5 keV in the X-ray spectrum of the Perseus cluster that firmly established that cluster X-ray emission arises from hot gas [96].

1.1 The Epoch of Cluster Formation

Until the 1970's, clusters were generally thought to be dynamically relaxed systems that were evolving slowly after an initial, short-lived episode of violent relaxation. However, in 1972, Gunn and Gott argued that, while the dynamical timescale ($\tau \sim G\rho^{-1/2}$) for the Coma cluster was comfortably less than the Hubble time, the dynamical timescale in other less dense clusters would be comparable to or longer than the age of the Universe. Thus they concluded that "The present is the epoch of cluster formation" [58].

Launched in 1978, the Einstein Observatory with its X-ray imaging capability showed that nearly half of all rich clusters had significant substructures reflecting complex, unrelaxed gravitational potentials [52, 68, 69]. The Einstein images, as shown in Fig. 4, changed our view of clusters from one in which they were virially relaxed systems to one in which even many present epoch clusters are undergoing subcluster mergers.

Today the generally accepted view is that structure in the Universe grows through the gravitational amplification of small scale instabilities in the early Universe. Large scale filaments containing gas and galaxies form around voids. Rich clusters form at the intersections of these filaments and grow through the accretion of galaxies and groups that fall along the filaments into the deeper cluster potential. A beautiful example of this is the 4 Mpc long filament of X-ray bright gas and groups apparently falling into the A85 cluster [32–34]. Thus clusters form through the mergers of small systems, a process called hierarchical clustering. Much of the accretion is expected to occur at very large radii in clusters and has not been directly observed. However both Chandra and XMM-Newton have provided significant insight into the physical processes associated with major mergers, the mergers of nearly equal mass components involving kinetic energies as large as $\sim 10^{64}$ ergs. Both supersonic, as well as sound speed, cluster mergers are discussed in Sect. 3.

If cluster dynamics are governed primarily by the gravity of the dark matter, then relaxed clusters should have similar gas density and temperature profiles. Using ASCA observations of nearby clusters, Markevitch et. al. found that the gas temperature profiles for symmetric, relaxed clusters were remarkably similar when the temperature is plotted against the cluster radius in units of the virial radius [85] (see [26] for similar BEPPOSAX results). Recently, Vikhlinin et al. and Piffaretti et al. measured the temperature profiles to large radii for samples of nearby relaxed clusters observed with Chandra [137] and XMM-Newton [110]. As Fig. 5 shows, outside the central cluster core where cooling and AGN feedback affect the gas temperature, clusters exhibit a "universal" temperature profile. Comparing cluster density profiles, Vikhlinin et al. also found that, outside the cluster core, the gas and the total density

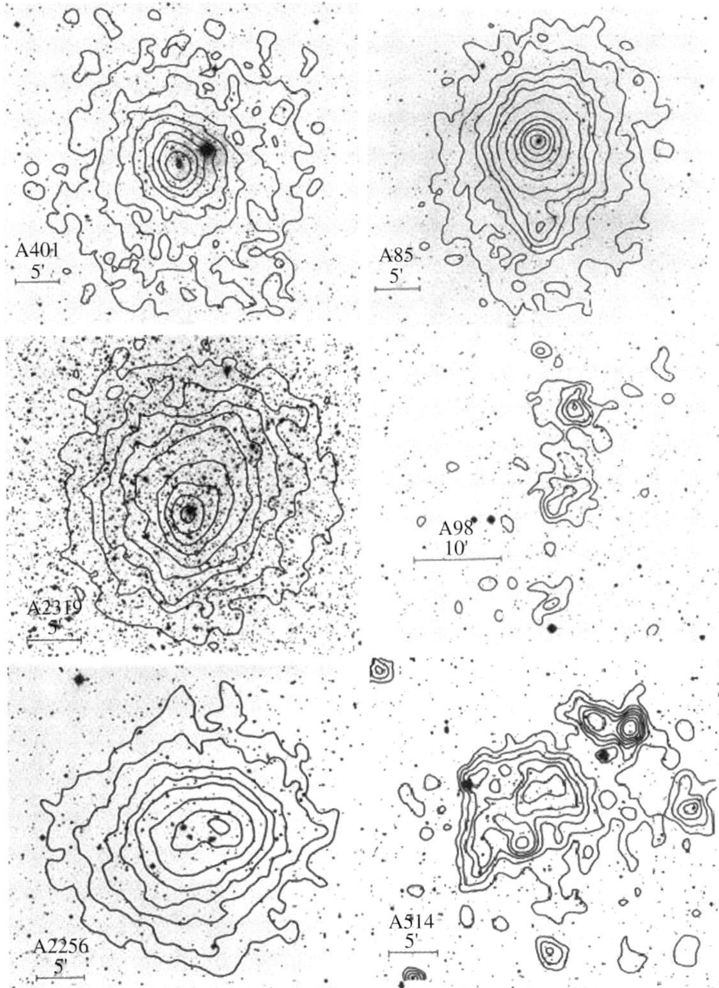

Fig. 4. Each panel shows isointensity contours of the cluster's X-ray emission superposed on an optical image from Einstein observations (see Jones & Forman [70] for details). The clusters illustrate the variety of structures and morphologies seen through their X-ray emission

profiles of hot ($kT > 4$ keV) clusters are similar, when the densities are plotted against the radius measured in units of r/r_{180} (see Fig. 5; r_{180} is the radius within which the mean density is 180 times the critical density). The total density profile agrees with the NFW profile [97, 98]. Thus at the present epoch, outside the cores, hot relaxed clusters have "universal" temperature and density profiles and appear self-similar.

Since structure grows hierarchically, clusters of galaxies, as the most massive, quasi-relaxed systems, are dynamically young and "remember" the con-

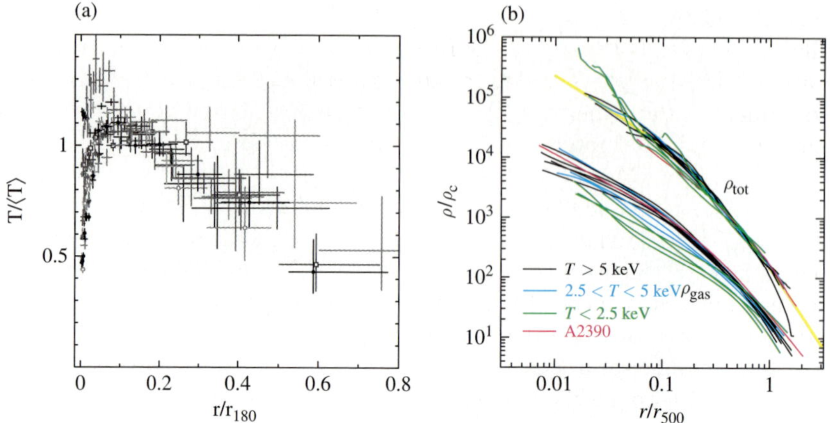

Fig. 5. (**a**) When plotted in units of r_{180}, the ICM in the outer regions of clusters has a similar temperature profile. (**b**) When plotted in units of r_{500}, the gas and total mass profiles for hot clusters are similar in the outer regions [137]

ditions from which they formed. Thus they are sensitive to the underlying cosmological parameters. Beginning with the Einstein observations, there was strong indication that the most luminous clusters were undergoing significant evolution at modest redshifts ($z \sim 0.6$), such that there was a substantial deficit of X-ray luminous clusters at high redshifts [64]. Large ROSAT surveys confirmed this deficit [95, 131]. Chandra and XMM-Newton observations also show that clusters at high redshifts had higher gas densities and were hotter and more X-ray luminous for a given mass than present epoch clusters [77, 135]. Cluster observations, particularly Chandra observations of distant clusters, place strong constraints on the cosmological parameters Ω_m, Ω_Λ and σ_8 as well as the equation of state for dark energy. The use of clusters to study the cosmology of the Universe is discussed in Sect. 4.

Cluster environments also influence the cluster galaxies. It has long been known that the cores of rich clusters are predominantly populated by early type galaxies–ellipticals and lenticulars–that show little evidence of recent star formation, while late type galaxies–spirals and irregulars–are found in less dense environments [29, 103]. In this morphology-density relation, low density regions (filaments and the field) have mostly blue, star-forming spirals, as well as galaxies with more dust, and higher numbers of AGN. By contrast in high density regions (rich clusters and dense groups), there are mostly galaxies with little ongoing star formation, red ellipticals and lenticulars and some "anemic" spirals. Work by Kauffmann et al. [75] shows high star formation rates for most disks and some bulges in low density regions and low star formation rates for most bulges and roughly half the disks in high density regions.

A number of physical mechanisms and processes probably act in concert to deplete the gas in galaxies and regulate their star formation. When galaxies form in low density regimes, the cooling, star forming gas is not shock heated, while in high density systems, the intracluster is shock heated to virial temperatures, after which very little of this gas will cool to form stars. In dense environments galaxy mergers, galaxy harassment due to high velocity galaxy encounters and tidal interactions with the cluster potential can truncate or destroy disks (see Machacek et al. [81] for a discussion of the interaction of the Virgo spiral NGC4438 with its companion galaxy NGC4435). Interactions of the galaxy interstellar medium with the hot gas in the potential well of the group or cluster can strip the galaxy of its gas. The hot ICM also can evaporate and compress the cooler gas in the cores of elliptical galaxies. In Sect. 3, we describe X-ray observations of the stripping of the interstellar medium for both elliptical and spiral galaxies in groups and in clusters. We also describe the small cool coronae that have been found in the cores of several massive ellipticals in rich clusters.

2 Cooling and Feedback in Cluster Cores

Early results from Chandra showed that the central regions of cooling flow clusters, as well as early type galaxies, are morphologically very complex. Observations of X-ray cavities and weak shocks demonstrate the impact of the central AGN, with its associated radio emission, on the hot cluster gas. The X-ray, as well as the radio, observations show not only the current state of the AGN, but the reflections of those outbursts in the surrounding gas show the history of AGN outbursts.

In many clusters, the X-ray emission is strongly peaked at the cluster center on a bright cD galaxy (see 1). From the earliest X-ray observations, Fabian and Nulsen, as well as Cowie and Binney, realized that "Cooling gas in the cores of clusters can accrete at significant rates onto slow-moving central galaxies" [22, 41]. Einstein and ROSAT imaging observations showed the dramatic contrast in cluster cores between the very peaked surface brightness distribution in cooling flow clusters and the flatter distribution in non-cooling flow clusters. These imaging observations showed that $\sim$70% of clusters were centrally peaked, with central gas densities as large as 0.1 cm^{-3}. Thus the radiative cooling times of the gas in their cores were as short as a few 10^8 years, much less than the age of the cluster. If the dense X-ray gas in the core is not reheated, it will cool and in order to maintain pressure balance (hydrostatic equilibrium) with the hotter cluster gas outside the core, this cool gas will be compressed by the overlying cluster gas and flow toward the central galaxy at rates up to 100–1000 solar masses per year (e.g. [42]).

Although spectroscopic measurements from Einstein, ROSAT and ASCA found that these "cooling flow" clusters had cooler X-ray gas in their cores, the repository of the expected large amounts of cold gas in the form of HI or recent star formation was not found. Instead star formation and gas at temperatures of about $10^{5.5}$ K seen in FUSE observations of clusters [102] implied cooling flow rates at only 10–20% of the standard cooling flow model. The question of the fate of the cooling gas was finally resolved through the XMM-Newton high resolution spectroscopy that did not show the strong iron and oxygen X-ray emission lines expected as gas cools through about 10^7 K [107, 109], and through the Chandra spatially resolved spectra that showed that the cluster gas in the core cools only to a temperature of about 1/3 of the outer gas temperature (e.g. [25]). Since the cooling time depends on the gas temperature and density, it is amazing that just when the coolest, densest gas in cluster cores should be cooling the most rapidly, instead most of the gas appears not to be cooling, so must be reheated.

Since recent star formation in cD galaxies and cold gas were seen only at low levels, compared to those expected from the standard cooling flow model, many mechanisms to reheat the cooling gas have been proposed. These include cluster-subcluster mergers which produce vast amounts of energy ($\sim 10^{64}$ ergs). However although cooling cores are not found in merging systems, the irregular time intervals between merger events makes cluster mergers an unreliable source of reheating the cooling gas in cluster cores. The large amounts of hot gas in the outskirts of clusters has long made thermal conduction a popular method of reheating cluster cores (e.g. [23, 144]). However, while conduction may be able to provide some energy to the outer regions of the cores, in the most centrally peaked cooling flow clusters, conduction is not effective in transporting energy to the cluster center. Finally, reheating the gas by AGN, particularly for clusters with radio emission, had long been suggested (e.g. [8, 10, 119, 130]). New Chandra observations show that jets, bubbles and shocks produced by periodic outbursts from the supermassive black hole in the central cluster galaxy, along with possible gas "sloshing" also driven by the central AGN, can reheat the surrounding gas in the cluster center. X-ray observations of the effects of AGN outbursts on the cluster gas have radically changed our views on cooling flows. The cooling and feedback process that is observed in present epoch clusters may be the same process that leads to a the observed cutoff at the bright end of the galaxy luminosity function.

The sections below review first the X-ray observations of the Perseus and M87 clusters, then Hydra A, Hercules A and MS0735.6+7421, three clusters which show very energetic AGN outbursts, and finally the impact of outbursts on the hot interstellar medium in early type galaxies.

2.1 Cavities and Shocks in the Perseus Intracluster Gas

As the brightest cluster in the X-ray sky, Perseus has been observed by every X-ray mission from Uhuru to Chandra and XMM-Newton. Figure 6 shows the deep (900 ksec) Chandra observation [44]. Two X-ray cavities located north and south of the central giant galaxy NGC1275 and inflated by radio jets were first recognized in the ROSAT images [11]. As is generally true for the bright X-ray rims surrounding radio lobes, the gas around the cavities is cooler than the surrounding ICM. North of the northern inner lobe is a sharp

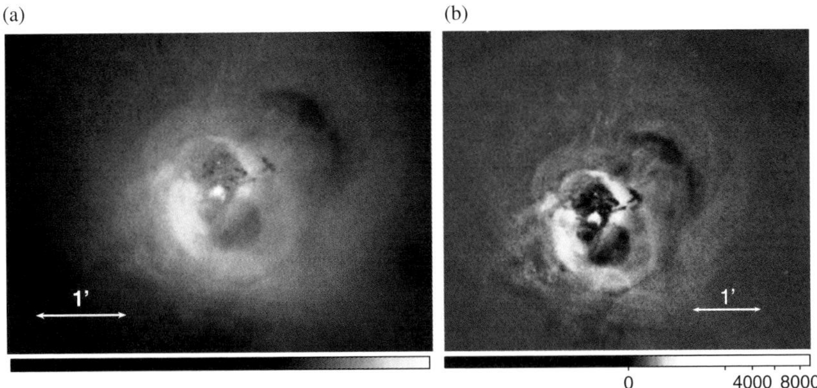

Fig. 6. (**a**) The 900 ksec Chandra image of Perseus shows the bright nucleus centered on NGC1275, X-ray cavities to the north and south, dark regions of absorbing gas stripped from an infalling galaxy, outer ghost cavities and sharp surface brightness edges marking weak shocks [44]. (**b**) Following the analysis procedure of Fabian et al., the result of subtracting a $10''$ smoothed image from the unsmoothed data shows multiple ripples in the gas. These are especially prominent northeast of the northern cavity [44]

surface brightness edge and corresponding pressure jump in the gas, which Fabian et al. recognized as a shock [43]. Additional ripples at larger radii also can be seen in the Chandra image. Figure 6, in which the Chandra image smoothed with a $10''$ gaussian is subtracted from an unsmoothed image, shows nearly regularly spaced ripples in the surface brightness distribution that also correspond to small variations in the density and pressure profiles [44]. If the ripples are the result of earlier nuclear outbursts and move at the sound speed (~ 1200 km s^{-1} in 5 keV gas), their typical separation of 11 kpc corresponds to an outburst frequency of 10^7 years. As Fabian et al. noted, if these weak shocks deposit half of their energy in the surrounding gas within the central 50 kpc cooling region, this energy input will balance that lost through radiative cooling [44]. In addition to the central cavities, at larger radii the Chandra image (Fig. 6) shows additional cavities or "ghost bubbles" which have not been detected in the radio [43]. These cavities could have been produced by earlier nuclear outbursts.

The hot intracluster gas in Perseus shows an unusual spiral structure as seen in the ROSAT [16] and XMM-Newton images [18], as well as in the deep Chandra observations [44]. This structure may result from "sloshing" of the gas in the core due to the past merger history of the cluster or may have been caused by the passage of a subcluster through the core [4].

2.2 AGN Outbursts in M87

At a distance of only 16 Mpc, M87 (NGC4486), the giant elliptical at the center of the Virgo cluster, provides a unique laboratory to study the interactions of the hot intracluster gas with the energy generated by the 3×10^9 $M_\odot$ black hole at its core. Like many central cluster galaxies, M87 has long been considered a classic example of a "cooling flow" system (e.g. [13, 128]). In fact, using radio studies, Owen et al. pioneered the idea that the mechanical energy produced by the SMBH was more than sufficient to compensate for the energy radiated in X-rays [104]. Chandra, XMM-Newton and ROSAT HRI observations of M87, as well as radio VLA maps, show rich structures on many angular scales, including a bright nucleus, knots in the jet, jet cavities, radio and X-ray "arms" and weak shocks in the hot gas. As described in this chapter (see also Forman et al. [53]), X-ray and radio observations of M87 are illuminating the energy input mechanisms from the AGN into the cooling gas.

The Chandra image (Fig. 7) shows a bright central region, X-ray arms that correspond to those seen in the radio image [104], and outer rings or shells that Forman et al. interpret as shocks [53]. On the largest scales, the VLA map (Fig. 11; also see [104]) shows outer lobes or "pancakes" which formed 10^8 years ago and, although these are the oldest visible radio structures, they require the continual injection of energy [104]. M87 also has bright radio arms to the East and Southwest, and in the core lies the famous jet (see Fig. 8).

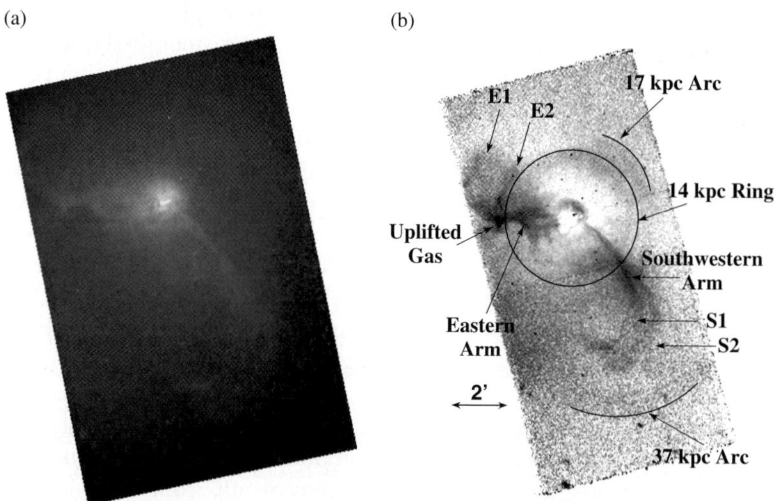

Fig. 7. (**a**) The Chandra image of M87 shows a bright central region with X-ray arms to the east and southwest and two sharp surface brightness edges. (**b**) The M87 image with a model for the gas distribution subtracted from the X-ray data. The (data-model) is then divided by the model [53]

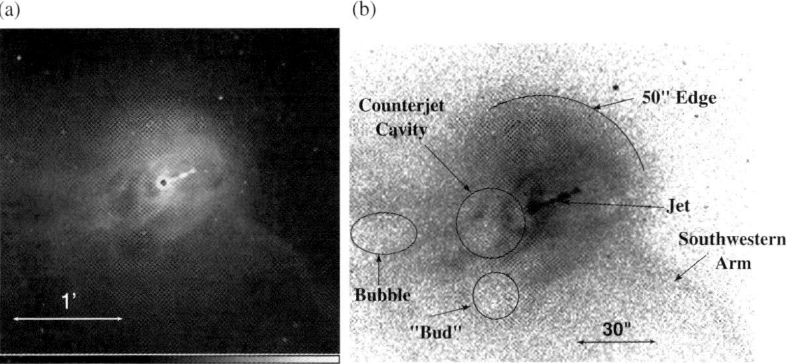

Fig. 8. (**a**) The Chandra image of the core of M87 shows the nucleus and jet as well as bright rims around the jet and counterjet cavities. Small cavities are seen at the base of the eastern arm. (**b**) Labeled X-ray image of the core of M87. Small cavities are labeled "bubble" and "bud" [53]

The central region of M87, containing the jet and inner radio lobes (the cocoon region), originated in an episode of recent AGN activity. As shown in Fig. 8, the central core is very complex. At the center is the jet and around it an X-ray cavity. An X-ray cavity also surrounds the region of the counter jet. There is radio emission and an X-ray cavity associated with another bubble southeast of the nucleus that is not aligned with the jet. Several additional bubbles can be seen in the Eastern arm. Each of these bubbles is about 1 kpc in radius. The composite image of the X-ray and radio emission (Fig. 11) shows that the X-ray cavities surrounding the jet and counterjet are filled with radio emitting plasma. Assuming subsonic expansion for the radio plasma, the age of these lobes is about 1.7×10^6 years. The energy required to remove the gas from these cavities is about 10^{56} ergs.

Outside the core, the prominent features are the X-ray arms and two shells or rings of emission (see Fig. 7). The inner shell at 14 kpc ($3'$) can be seen nearly all the way around M87, while the 17 kpc shell is most prominent to the northwest. To better examine the structure in the gas, Forman et al. modeled M87's overall halo of X-ray emission with a smooth model and subtracted that from the Chandra image. Figure 7(right) shows the deviations from the smooth model–the X-ray arms and 14 and 17 kpc arcs are prominent. The sharpness of the shells, as well as the completeness of the 14 kpc ring, suggests they are due to shocks likely driven by the initial rapid expansion of the core radio lobes. Both X-ray arms brighten at about the radii of the shocks, suggesting that they lie nearly in the plane of the sky. Beyond the radii of the shocks, the X-ray arms curl and each one splits into two filaments.

The radial surface brightness and temperature profiles across the X-ray arcs can be modeled as shocks [53]. An instantaneous outburst with an energy of nearly 10^{58} ergs occurring 10^7 years ago provides a good match to the observations of the 14 kpc ring. The shock is mildly supersonic. The outer partial ring would have been caused by an explosion of similar energy occurring about 4 million years earlier.

Shocks appear to be a significant channel for energy from the AGN to reheat the cooling gas. But shocks are not the whole story. Bubbles rising through the arms also play a role in heating and transporting cool gas. The amount of energy associated with the observed outbursts will balance the energy lost through radiative cooling, if outbursts occur every 10^7 years.

Churazov et al. modeled the structure in the arms as buoyant bubbles. Figure 9 shows a schematic of the three episodes of bubbles corresponding to AGN outbursts that occurred 10^6–10^8 years ago [17]. Radio bubbles, generated in the core, rise rapidly and lose about half their energy through adiabatic expansion. In Fig. 10, the panels show the evolution of the gas temperature in and around the bubble, since the AGN outburst. Initially a hot bubble is created by the AGN. As it rises buoyantly through the atmosphere, it entrains cool gas behind it. This rising, expanding gas produces the arms seen in the radio and X-ray observations. Several solar masses of material can be uplifted this way. In this model, the arms should be cool. In fact the X-ray gas temperatures of the arms has been found to be cooler than the surrounding gas from XMM-Newton observations [5].

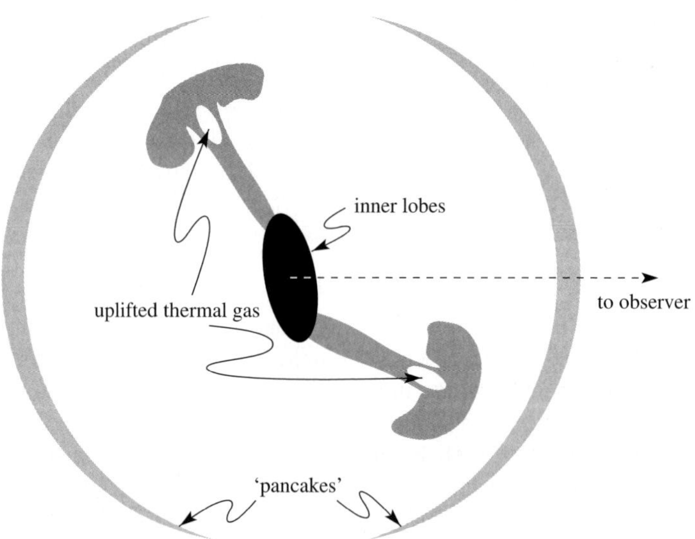

Fig. 9. A schematic of the Churazov et al. model for buoyant bubbles in the core, arms and outer lobes of M87, which were produced by three episodes of AGN outbursts with timescales of 10^6–10^8 years [17]

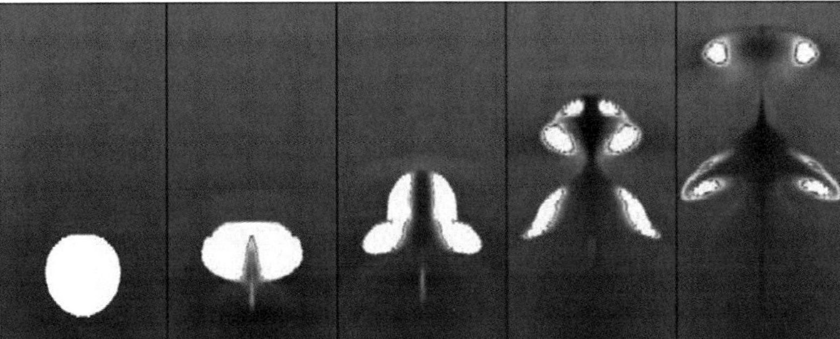

Fig. 10. A simulation of a buoyant bubble rising in the atmosphere of M87 shows the evolution in the temperature with time [17]. Dark blue is the coolest gas. The first panel shows the hot bubble created by the AGN. As the bubble rises, it expands and entrains cool gas behind it. The size of each panel is 20 × 40 kpc

As can be seen in Fig. 11, the X-ray emission from M87's arms appears quite different. The eastern arm begins with bubbles and ends in the radio "ear" that lies between the 14 and 17 kpc shocks. Simulations by Ensslin and Bruggen have shown that a shock passing through a bubble of relativistic plasma can produce this strong vorticity [37]. By contrast the southwestern arm begins as a narrow filament ($10''$ or 0.8 kpc at its narrowest) and shows little correlation with the radio emission. Magnetic tension, as well as thermal pressure from the surrounding gas, could confine the cooler gas in the arm. As shown by the composite X-ray and radio image (Fig. 11), the southwest radio arm could envelop the X-ray filament. This image also emphasizes the complex nature of the gas emission and of the radio and X-ray interaction.

2.3 Powerful Outbursts in the Hydra A, Hercules A and MS0735.6+7421 Clusters

Chandra images show X-ray cavities with radii of a few to several tens of kiloparsecs in the hot gas of many clusters and groups of galaxies, as illustrated in the earlier sections by Perseus and M87. Perseus and M87 show the effects of weak shocks in their central 20–50 kpc cores. By contrast, the X-ray emission from the three clusters Hydra A, Hercules A and MS0735.6+7421 shows giant cavities, along with a single large shock with a radius of 160–300 kpc.

At a redshift $z = 0.22$, the optically poor Zwicky cluster MS0735.6+7421 has an X-ray luminosity of 8×10^{44} ergs s^{-1} [12], a central cD galaxy with an optical emission nebula extending 20 kpc in its core [28] and a radio source (4C 74.13) extending 550 kpc in diameter [94]. As McNamara et al. found from Chandra observations (see Fig. 12), the X-ray cavities corresponding to the radio lobes are huge, 200 kpc in diameter, and surrounding these cavities, is a cocoon with a sharp X-ray surface brightness edge marking the location of the shock. McNamara et al. estimate an age and required energy for the

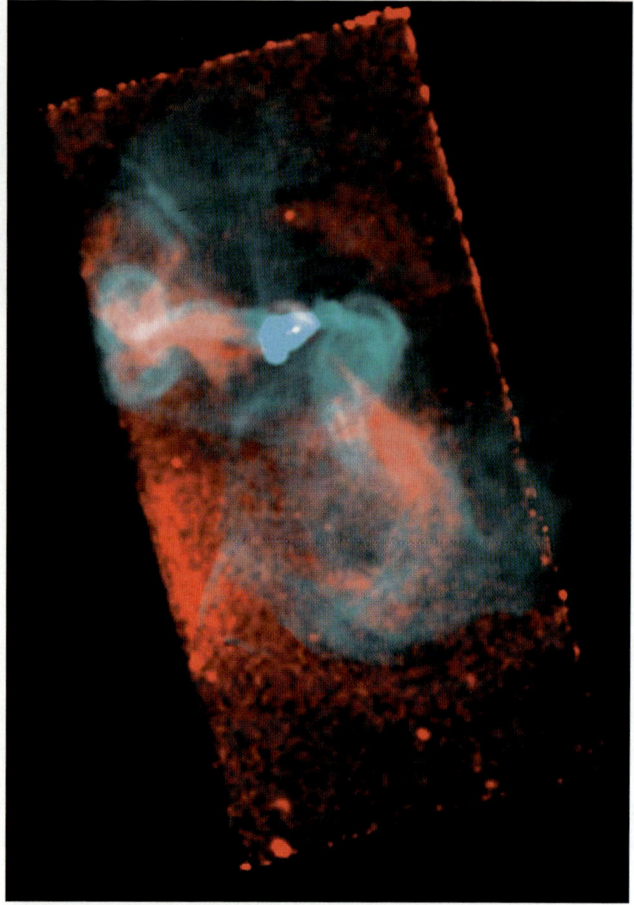

Fig. 11. The X-ray (red) and radio (cyan) image of M87 shows the complex interaction of the X-ray gas and radio plasma and the striking differences between the eastern and southwestern arms. The outlined red region of the X-ray emission is 8′ by 16′ [53]

shock of 10^8 years and 5.7×10^{61} ergs. The pV work required to inflate each giant cavity is $\sim 10^{61}$ ergs. The resulting enthalpy ($\sim 8 \times 10^{61}$ ergs) is close to the energy estimated for the shock. Unlike most of the cavities seen in other systems, the faint X-ray emission from these cavities is hotter than the surrounding cluster gas, suggesting that the gas around the cavities was heated by the shock. Inverse Compton emission from the radio also may contribute to the hard X-ray emission from the cavities.

As shown in Fig. 13, Chandra observations of the Hydra A cluster show a pair of X-ray cavities associated with 1.4 GHz radio lobes extending about 40 kpc from the nucleus, a second pair of cavities lying about 100 kpc from the

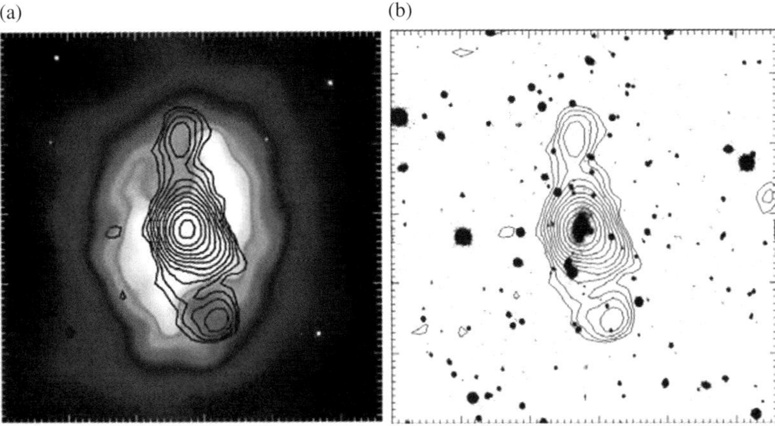

Fig. 12. (**a**) The Chandra image of MS0735.6+7421 with the radio contours superposed. (**b**) The optical image of MS0736.6+7421 showing a cD galaxy in an optically poor cluster, with contours of radio emission superposed [94]

cluster center and an edge in the surface brightness distribution 200–300 kpc from the center [25, 93, 100]. The surface brightness edge is most prominent northeast of the nucleus, where the shock is farther from the cluster center and the pressure jump is greater, resulting in a stronger shock. As Nulsen et al. noted, the difference in the eastern extension of the shock compared to the western extension is more likely due to differences in the large scale gas density in the cluster and not to asymmetries in the outburst. Based on modeling the western surface brightness distribution, Nulsen et al. find an age for the shock of about 10^8 years and an energy of nearly 10^{61} ergs. The mean

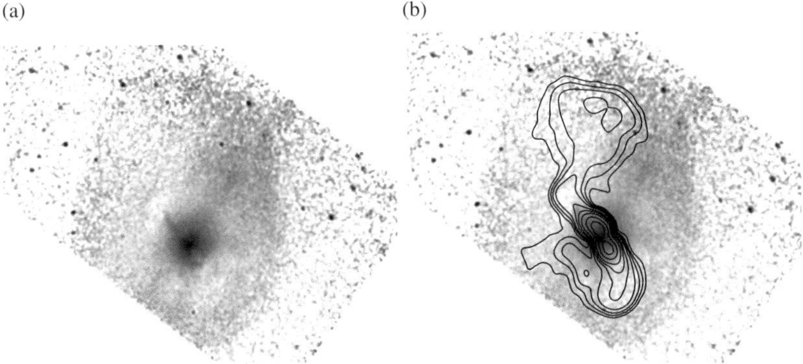

Fig. 13. (**a**) The Chandra image of Hydra A shows inner cavities, larger outer cavities and, at large radii from the cluster center, an edge in the surface brightness distribution. (**b**) Radio contours superposed on the X-ray image [100]

mechanical power of the outburst is $\sim 2 \times 10^{45}$ ergs s^{-1}, typical of quasar luminosities.

A third powerful nuclear outburst has been found in the Hercules A cluster [101]. The Chandra image shows a bright X-ray region extending ∼160 kpc, similar in size to the radio lobes. A sharp surface brightness edge surrounding this region marks the shock front location. Through modeling the surface brightness distribution, Nulsen et al. measure a Mach number of 1.65 for the shock and determine that it was produced 6×10^7 years ago by an outburst with an energy of 3×10^{61} ergs. Of the three clusters known to have powerful outbursts, Hercules A has the strongest shock.

Although all nuclear outbursts are likely powered by accretion onto a supermassive black hole lying in the core of the central giant galaxy, as McNamara et al. emphasize, the minimum accreted mass required to produce the shock energy seen in MS0735.6+7421, assuming a mass to energy conversion of 0.1 Mc^2, is $\sim 3 \times 10^8 M_\odot$ [94]. The three large outbursts that have been found all imply significant growth in the mass of their supermassive black holes in recent times. If the Magorrian relation [86] between the galaxy bulge size and the mass of the black hole is tightly maintained, the stellar mass of the bulge must grow in concert with the black hole, presumably through star formation from the reduced cooling flow.

2.4 Nuclear Outbursts in Elliptical Galaxies

While the outbursts in clusters can affect the dense atmospheres in the cores of clusters, nuclear outbursts can have even more dramatic effects on the hot ISM in individual elliptical galaxies. Centaurus A and M84, an elliptical galaxy in the Virgo cluster, are two of the best studied examples of nearby galaxies with radio emission. Their Chandra and XMM-Newton images are shown in Figs. 14 and 15.

At a distance of only 3.4 Mpc, Cen A is our nearest active galactic nucleus, hosting a low-power FRI radio source,[1] as well as the brightest "steady" extragalactic Gamma-ray source. Cen A appears as an elliptical galaxy crossed by a dust lane, believed to be the result of a merger with a small spiral galaxy about 10^9 years ago. Radio observations show a nucleus, a subarcsecond jet and counter jet, a predominantly one-sided jet on kiloparsec scales, inner lobes, a middle lobe and diffuse emission on scales up to 250 kpc, which together provide strong evidence for repeated nuclear outbursts (see Israel for a review [67]). The Chandra X-ray image of Cen A (Fig. 14) shows the nucleus and jet, a warm (0.29 keV) interstellar medium, hundreds of point X-ray sources, and an X-ray shell surrounding the southern inner radio lobe [78, 79].

[1] In the standard model for FRI radio sources, a supersonic jet from the nucleus forms a radio lobe, while in the more powerful FRII sources, the lobes are expanding supersonically. If the radio lobe has a leading bow shock, the surrounding X-ray gas will be shocked and heated by the passage of the bow shock.

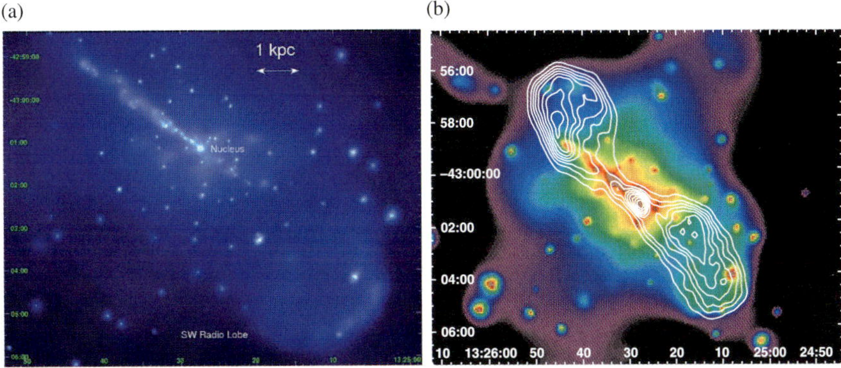

Fig. 14. (**a**) The Chandra image of Centaurus A shows emission from the nucleus and jet, the southern radio lobe, the interstellar medium and hundreds of low mass X-ray binaries. (**b**) Radio contours superposed on the smoothed XMM-Newton image

The Cen A shell has a mass of 3×10^6 $M_\odot$ and is likely interstellar material swept up by the southern lobe as it expands. This is the best known example of a shell of gas compressed and shock heated by the supersonic expansion of a radio lobe. As Kraft et al. measured, the hot gas in the X-ray shell is ten times hotter and twelve times denser than the surrounding interstellar gas. Thus the thermal pressure in the shell is about 100 times the pressure in the surrounding medium. If the pressure in the shell that exceeds the ISM pressure is balanced by the ram pressure of the expanding lobe, this highly over-pressured shell can

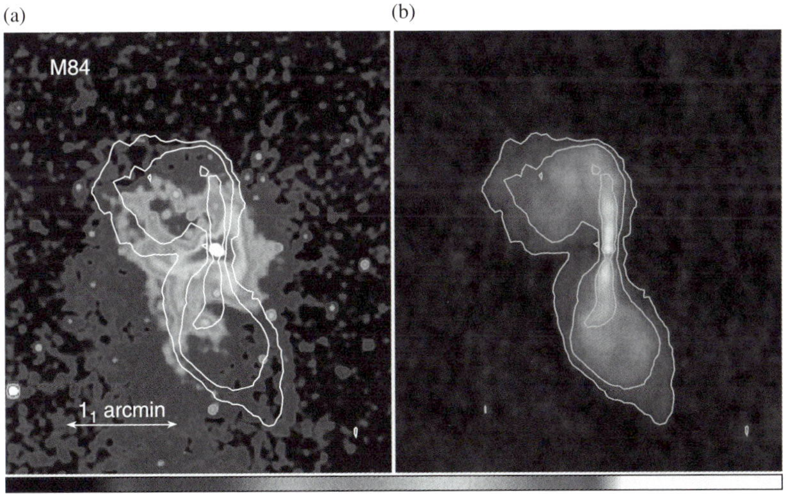

Fig. 15. (**a**) The Chandra image of M84 with the FIRST radio contours superposed. Cool X-ray rims outline the radio lobes. (**b**) The FIRST radio image of M84

be maintained. The required ram pressure corresponds to the shell expanding at 2400 km s^{-1}, equivalent to Mach 8.5. The kinetic energy in the shell is about six times its thermal energy ($\sim 4.2 \times 10^{55}$ ergs) and also greater than the thermal energy of the hot ISM within the central 15 kpc. Most of the shell's kinetic energy eventually should be converted into heating the ISM. The energy in the shell would be transferred to the ISM through conduction.

M84 (NGC 4374), a luminous elliptical galaxy in the core of the Virgo cluster hosts the radio source 3C272.1. As Finoguenov and Jones described, the radio lobes define the structure of M84's X-ray gas [48], as shown in Fig. 15. Unlike Cen A, the brightest X-ray emission is seen along the sides of the radio lobes and not at the ends where the lobes are expanding, implying that the expansion of the radio lobes is not supersonic. As seen in Fig. 20, M84 has a relatively small X-ray corona, compared to the elliptical galaxy M86, as well as in comparison to other optically luminous early type galaxies. Its small size may be due to gas being stripped from M84 as it moves through the denser gas in the Virgo core, since it does show an X-ray tail toward the southeast and compression of its northern radio lobe. Alternatively, the smaller gas mass in M84's corona may be due to the effects of past nuclear outbursts which heated and expelled much of its interstellar medium. As stars in the galaxy continue to shed mass, they will replenish the gas in the galaxy's corona.

While the presense of bright lobed radio sources in Cen A and M84 should have suggested that the gas in these galaxies would be morphologically disturbed, the general expectation for other "normal" elliptical galaxies was that their X-ray emission would be composed of "relaxed", symmetric, hot gaseous coronae, punctuated with the brighter X-ray binaries distinguished as point sources. However observations of early-type galaxies often showed X-ray cavities in their interstellar gas. Examples include NGC4636 [71], NGC4472 [7], NGC4552 [84], and NGC507 [80].

Figure 16(**a**) shows the Chandra image of NGC4636. The bright central region is surrounded by arms that extend ~ 8 kpc in a pinwheel shape. The sharp edges along the inner edges of the arms, as well as modest increases in gas temperature in the southwest arm are characteristics associated with a shock. These features can be produced by a nuclear outburst that occurred 3 million years ago, with a total energy driving the shocks of 6×10^{56} ergs. The central region of bright X-ray emission appears relatively undisturbed and suggests that the energy input was not primarily deposited in the center, but was transferred to larger radii, probably through jets. However, the cavities in NGC4636, as well as those in NGC4552, are unusual in that they have not been detected as radio lobes, although both of these ellipticals do show some radio emission from their nuclei. In comparison to other galaxies, NGC4636 and NGC4552 had two of the most recent outbursts, suggesting that the bright radio lobe phase may begin later than the initial expansion.

In early type galaxies, although the gas in the core has a high density and the radiative cooling times are short, the cooling flow rates, even in the stan-

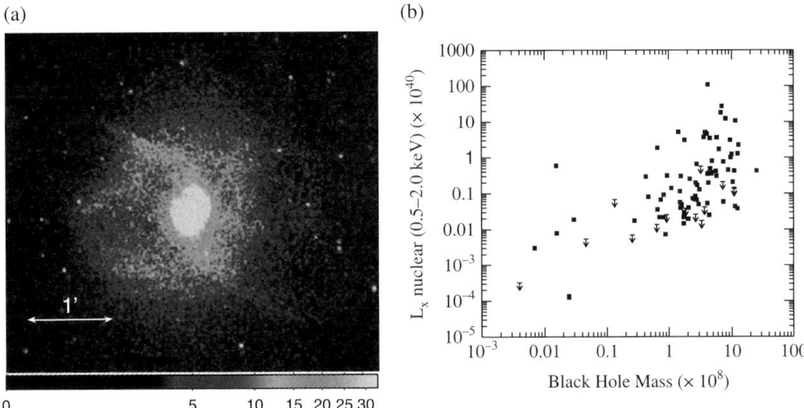

Fig. 16. (**a**) The Chandra image of the elliptical galaxy NGC4636 shows 8 kpc long X-ray arms [71]. (**b**) The X-ray luminosity of nuclear sources in early type galaxies plotted against the SMBH mass estimated from the velocity dispersion [72]

dard cooling flow model, are modest, typically one solar mass per year [129]. Outbursts like those seen in NGC4636 and other early type galaxies would need to occur only every 50 million years or so to prevent the accumulation of large amounts of cold gas in their cores. Such galaxies would be in a shock phase ∼10% of the time or less. In our sample of 160 early type galaxies, we identify 26 galaxies (15% of the sample) as having X-ray cavities or jets and thus are undergoing a current outburst that is likely to influence the surrounding gas [72].

Figure 16(**b**) shows the nuclear X-ray luminosity plotted against the mass of the SMBH, as estimated from the galaxy velocity dispersion. X-ray emission from the nucleus is detected in ∼80% of the galaxies [72]. While there is several orders of magnitude dispersion in the nuclear luminosity, the level of X-ray emission from SMBHs in "normal" early type galaxies is quite low, especially compared to that seen in quasars. If one compares the central luminosity to the Eddington luminosity (generally the maximum accretion allowed without beaming for a black hole of a given mass), one finds ratios of about 0.1 or more for quasars, while for the "normal" galaxies, the ratios are very small (10^{-5}–10^{-9}). The accretion that is ongoing in these galaxies, most likely from cooling gas, results in very modest emission. The SMBH in "normal" galaxies are practically starving and their growth through accretion is very modest.

In summary the energy input to reheat the gas in clusters and in galaxies comes from two sources. The first is the mechanical energy of the lobes/cavities that are inflated by jets from the AGN, and generally are observed as symmetric pairs located on opposite sides of the nucleus. The low density of the cavity causes it to rise buoyantly in the cluster gas, while it continues to expand adiabatically. When the bubble expansion becomes less than its rate of ascension in the ICM, it will detach from the jet as it continues to rise. If the jet is still active, a new bubble will form. This pattern of "effervescent" bubbles [120] may explain the filamentary structure seen in the eastern arm of M87 [54]. Birzan et al. found that, in nine of the 18 systems they studied, which included 16 clusters, one group and one galaxy, the mechanical energy input was sufficient to balance cooling [9]. Recently Dunn et al. reported X-ray cavities in twelve of seventeen clusters with cooling cores [31].

The second mechanism to reheat the gas is through the generation of shocks by the initial relativistic expansion of the radio lobes as described above for Perseus, M87, Hydra A, MS0735.6+7421 and Hercules A. The expanding radio jets drive a roughly spherical shock into the surrounding gas [60]. Observing the effects of the weak shocks in the X-ray gas requires deep X-ray observations, so that the number of clusters in which these large scale shocks have been found is far fewer than the number of systems with X-ray cavities.

While the X-ray rims around the cavities are cooler than the surrounding gas in most clusters and galaxies, in a few systems (e.g. Cen A, NGC4636, NGC4552 and Cygnus A), the X-ray emission around the cavities has been shock heated by the supersonic expansion of the lobes and is hotter than the surrounding medium and therefore can heat the surrounding gas through conduction and mixing.

It is illustrative to compare the energy and timescales of the shocks that have been seen in systems ranging from the elliptical galaxy NGC4636 [71] to the most energetic shock in the cluster MS0735.6+7421 [94]. Table 1 lists the radius of the shock, the time since the AGN outburst, the initial outburst energy and the mean power of the outburst. As long as the energy from the outburst can be transferred to the gas, the amount of energy in the outbursts is enough to replenish the heat lost by the gas through radiative cooling. While Mathews et al. [92] and Fujita and Suzuki [55] have argued that shocks will only reheat the inner cores of clusters, the simulations by Ruszkowski et al. and by Heinz and Churazov show that the energy from the AGN outbursts can

Table 1. SMBH outburst parameters: from clusters to galaxies

Source	Shock radius (kpc)	Age (My)	Energy (10^{61} erg)	Mean power (10^{46} erg s^{-1})	Mass swallowed (10^8 $M_\odot$)
NGC4636	~ 5	3	0.00006	0.0007	0.00003
M87	14	10.6	0.0008	0.0024	0.0005
Hydra A	210	136	0.9	0.2	0.5
Hercules A	160	59	3.0	1.6	1.7
MS0735.6+7421	240	104	5.7	1.7	3

be distributed throughout the cooling core [61, 120, 121]. The last column in Table 1 gives the mass that would need to be accreted by the SMBH in order to produce the outburst, if one assumes that the mass to energy conversion efficiency is 0.1. As noted by McNamara et al. and Nulsen et al., the large accreted masses required to produce the giant outbursts in clusters like MS0735.6+7421 imply a significant growth in black hole masses during the current epoch.

AGN outbursts cannot be so large that they drive the gas from the cluster core (or entirely remove the X-ray gas from the poor groups or early-type galaxies). On the other hand, the outbursts must be large enough to reheat the gas and significantly reduce the cooling rates to the low residual cooling seen in most clusters at 10%–20% of the standard cooling flow rate. It now seems likely that nuclear outbursts and cooling flows are strongly connected as supported by the correlation between cavity mechanical power and cooling luminosity found by Birzan et al. [9]. Through a feedback cycle, cooling gas can fuel the outbursts, while the outbursts can reheat and reduce the amount of cooling gas. However the occasional galaxy merger, as in Cen A and more frequently at earlier epochs, also can provide the necessary fuel to produce outbursts.

The SMBHs seen in present epoch galaxies are the relics of those in quasars. To explain the dramatic changes in luminosity between the bright quasar phase and the present, quiet phase, Churazov et al. have suggested that the black hole energy release is a function of the infalling mass accretion rate, such that the luminosity is low at low accretion rates and reaches a fixed fraction of $\dot{M}c^2$ at accretion rates above 0.01–0.1 of the Eddington value [19] . Initially, when the black hole is small, feedback from accretion onto the black hole is not sufficient to reheat the cooling gas. In this quasar phase, there is near-Eddington accretion resulting in rapid black hole growth and high luminosity, but weak feedback. As the black hole grows, eventually mechanical feedback reheats the cooling gas, so that at the present time, the radiative efficiency of accretion is very low and the black hole growth is very slow.

Through X-ray and radio observations, we are beginning to understand the outbursts from SMBH at the centers of galaxies and the energy transfer mechanisms between the SMBHs and the gas. The X-ray, as well as the radio observations, show not only the current state of the AGN, but by looking at the reflections of these outbursts in the surrounding gas, we can chronicle the history of AGN outbursts. The cooling and feedback process may lead to the observed cutoff at the bright end of the galaxy luminosity function, as well as to significant growth of the black holes.

3 Cluster Formation and Evolution

The growth of structure in the Universe proceeds through gravitational amplification of small scale instabilities in an hierarchical manner in which clusters form and grow through the gravitational infall and mergers of smaller subclusters. The intracluster gas is heated to 10^7–10^8 K by shocks generated during these mergers. The early Einstein observations showed significant structures in 40% of clusters [69], demonstrating that clusters were not yet relaxed systems, but were still forming. Matter continues to accrete onto clusters, preferentially along large scale filaments. While much of the growth is through the accretion of small groups, at the extreme is the merger of nearly equal mass components. These major mergers can be spectacular events involving kinetic energies as large as 10^{64} ergs. Cluster mergers convert the kinetic energy of the gas in the colliding clusters into thermal energy by driving shocks and turbulence in the cluster gas. A small fraction of this energy may be diverted into nonthermal phenomena, including magnetic field amplification and the acceleration of relativistic particles that are seen in synchrotron radio halos (e.g. [46, 57]) and in inverse Compton X-ray emission. Merger driven turbulence is likely the main process responsible for generating the ultrarelativistic electrons that produce the diffuse radio emission. However shock acceleration or the compression of the magnetic fields along with an increase in the density of pre-existing relativistic electrons due to gas compression at the shock is likely important in the cap or edge of the radio halo, near the shock front [90].

For the first time, with Chandra's sensitivity and spatial resolution, it is possible to observe the classical bow shocks generated by cluster mergers. While bow shocks are rare, "cold fronts"–sharp contact discontinuities between gas regions with different temperatures and densities–are often seen in clusters. Cold fronts are associated with the bulk motion of a cool dense subcluster moving through the hotter cluster gas. They delineate the boundaries of the subcluster and while the subcluster and cluster have different entropies, the pressure in the dense cool subcluster is in balance with the thermal plus ram pressure of the surrounding cluster gas. Cold fronts also are found in many cool cluster cores where "sloshing" of the cool gas in the core can be caused by any minor merger, even one with an infalling gasless subcluster, one stripped of its gas earlier in the merger [4]. While the origins of cold fronts associated with subcluster mergers and sloshing cores are different, the physical interpretation of X-ray edges as contact discontinuities between gases with different entropies that are moving with respect to each other holds for both.

3.1 Shock Fronts in Supersonic Mergers

The only two clusters known to exhibit a shock front with both a sharp density edge and temperature jump are 1E0657-56, a Mach 3 merger (see Fig. 17) [88] and A520, a Mach 2 merger [90]. Observations of shock fronts are rare, since one must catch the merger before the shock has moved to the outer, lower

Fig. 17. The 500 ksec Chandra image of 1E0657-56 shows the "bullet" subcluster after it has passed through the main cluster. In front (west) of the "bullet" is the shock front [91]

surface brightness regions, and the merger must be nearly in the plane of the sky for the shock region to be visible.

The cluster 1E0657-56 has the highest X-ray luminosity (3×10^{45} ergs s^{-1}), the highest gas temperature (14 keV), and the most luminous radio halo of all known clusters. The 500 ksec Chandra image (Fig. 17) shows a spectacular merger occurring almost exactly in the plane of the sky, with a prominent bow shock preceding the small, cool "bullet" subcluster [88]. From the abrupt, factor of three density jump at the shock, Markevitch et al. determine that the "bullet" subcluster is moving at Mach 3.0 ± 0.4, which corresponds to a velocity of 4700 km s^{-1}. Gas from the cooler subcluster that had a lower pressure than that of the combined ram pressure and cluster thermal pressure has been stripped and swept back, leaving only the core of the "bullet" to continue to travel supersonically through the cluster.

The bow shock offers the opportunity to determine whether the electrons in the ICM are heated directly by shocks or compressed adiabatically and then heated by collisions with protons. From the gas density jump across the shock front and the pre-shock temperature, one can predict the post-shock adiabatic and shock-heated electron temperature and compare that to the observed temperature. Markevitch et al. measure electron temperatures exceeding 20 keV in the shock, which are consistent with instant heating by the shock and rule out the collisional electron-ion equilibration, which occurs on a Spitzer timescale.

While dark matter dominates the mass in clusters and in the Universe, its nature remains a mystery. From laboratory work, interaction cross-sections between dark and baryonic matter are very small [6]. Models with self-interacting dark matter have been suggested to better predict the mass profiles in galaxies. 1E0656-56 also allows constraints to be placed on the dark matter self-interaction cross-section [89]. As Fig. 18 shows, the gas in the "bullet" subcluster has been ram pressure stripped and lags behind both the subcluster galaxies, which are effectively collisionless, and the subcluster's dark matter peak determined from the weak lensing analysis [20, 21]. As the "bullet" subcluster passes through the main cluster, its dark matter is subjected to a flow of dark matter particles associated with the main cluster. From their analysis of 1E0657-56, Markevitch et al. place a limit of the dark matter collisional cross-section, $\sigma/m < 1$ cm^2 g^{-1} [89].

3.2 Cold Fronts in Cluster Mergers

A sharp discontinuity in the gas density had been observed in ROSAT images of A3667 [87] and was expected to be a shock front produced by a merger.

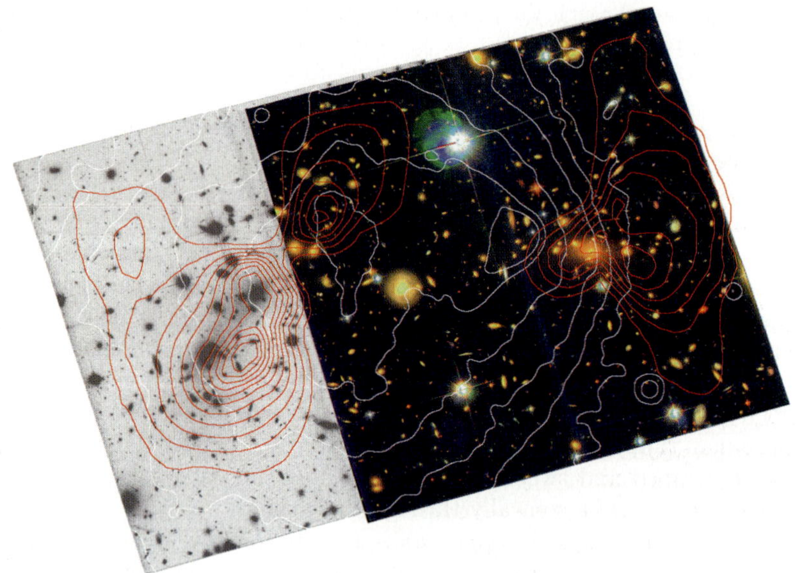

Fig. 18. The HST ACS image of 1E0657-56 with superposed "red" contours of the total mass distribution derived from lensing [21] as well as "white" contours of the X-ray intensity [14]

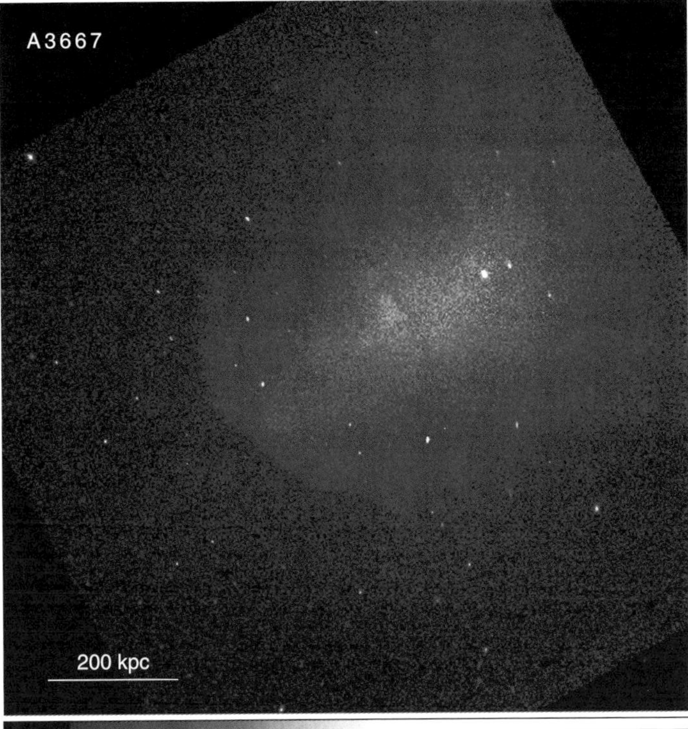

Fig. 19. The 500 ksec Chandra image of A3667 shows a sharp edge southeast of the cluster core which is the leading edge of the subcluster [138]

However the Chandra observation (Fig. 19) showed that this was not a shock front, but was instead a cold front [132]. From measurements of the gas density and the gas temperature, the gas pressure on both sides of the cold front can be accurately calculated. The higher pressure on one side of the cold front must be compensated by the ram pressure of the hot ICM on the moving cold front. Thus by measuring the difference between the thermal pressures of the gas inside the front and downstream from it in the free-flow region, Vikhlinin et al. determined the required ram pressure of the ambient flow and found the subcluster velocity in A3667 to be 1430 ± 290 km s^{-1}, Mach 1.0 ± 0.2 [132].

Cold fronts are remarkably sharp, in terms of both their gas density and temperature jumps. Thus thermal conduction across cold fronts must be suppressed from the classical collisional, Spitzer value by two orders of magnitude [39, 132]. In particular in A3667, Vikhlinin et al. found a limit on the width of the density jump of $3.5''$ (3.5 kpc), which is several times less than the 11 kpc electron mean free path for Coulomb collisions and thus required that transport processes across the edge be suppressed [132].

To suppress transport processes in plasmas, the presence of either a strongly tangled magnetic field or a large-scale field perpendicular to the likely direction of heat conduction or diffusion is generally invoked. If the magnetic field is tangled, the observed 3.5 kpc limit on the width of the front is also the upper limit on the size of magnetic loops. This magnetic field would effectively isolate the cold gas cloud and prevent the sharp density gradient across the cold front from dissipating [133].

4 The Effects of Clusters on Galaxies

X-ray images have shown that clusters are often complex systems with extensive structure and with mergers occurring on timescales of a few billion years. In these dynamically rich cluster environments, galaxies are subject to both tidal and hydrodynamic interactions that can significantly affect their evolution. The ram pressure of the intracluster medium (ICM) acting on the interstellar gas as a galaxy moves in a cluster can produce long X-ray tails, trailing wakes or debris tails of gas swept from the galaxy. Ram pressure stripping also has been invoked to explain the HI deficiency in spiral galaxies in clusters as well as the lower star formation activity seen in cluster spirals (e.g. [30, 111] As pointed out by Nulsen [99], stripping by transport processes (e.g. Kelvin-Helmholtz instabilities) can be much more effective than ram pressure stripping. Stripping of the galaxy gas, along with early epoch galactic winds, removes the gas enriched in heavy elements from the galaxies and disburses it throughout the cluster. Study of the tails can determine the thermal history of the gas as it is stripped from the galaxies and incorporated into the hot ICM.

In this section we first discuss M86 and NGC4552 in the Virgo cluster and NGC1404 in Fornax, all examples of elliptical galaxies moving at sonic or supersonic velocities through the ICM. Next we describe the three known examples of spiral galaxies currently undergoing significant ram pressure stripping in rich clusters (A1367, A2122 and A3627). We also discuss the interactions of the large spiral NGC 6872 with the hot diffuse gas in the Pavo group. Finally we describe the existence of very small, cool X-ray coronae in the cores of elliptical galaxies in the Coma, A1367 and A1060 clusters.

4.1 Ram Pressure Effects on Elliptical Galaxy Coronae

Although its optical appearance is that of a normal giant elliptical, of all the Messier galaxies, M86, a giant elliptical in Virgo, has the highest blue-shifted velocity, $-260\,kms^{-1}$. A comparison to the line of sight velocity of M87 (1300 km s^{-1}) which is assumed to be nearly at rest in the center of the cluster, suggests that M86 is moving supersonically through the Virgo cluster. The X-ray emission from M86, studied with Einstein, ROSAT, Chandra and XMM-Newton, shows a long (125 kpc on the sky) stripped tail or plume extending

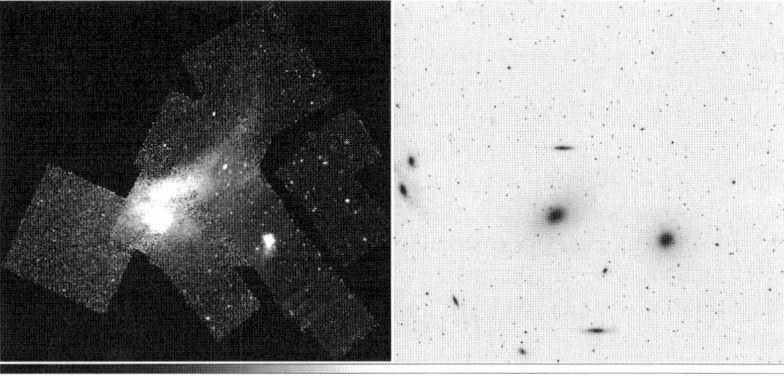

Fig. 20. (**a**) This Chandra mosaic of the M86 region of the Virgo cluster shows the bright core of the elliptical M86 and its very long tail (125 kpc on the sky) as well as the small corona and short X-ray tail associated with the elliptical M84. (**b**) The optical image on the same scale as the X-ray image

northwest from the galaxy, as well as a hot corona of gas around the galaxy itself [47, 50, 114, 141]. Figure 20 shows the X-ray emission, as seen with Chandra, and the optical field for M86.

A second Virgo elliptical that appears to be moving supersonically through the Virgo ICM is NGC4552 (M89). Compared with M87, its line of sight velocity of 340 km s^{-1} implies a supersonic velocity of nearly 1000 km s^{-1} towards us. Figure 21 shows the Chandra image of the diffuse gas surrounding NGC4552 [84]. The X-ray emission shows a sharp edge $40''$ (3 kpc) north of the galaxy, horns of emission extending southeast and southwest of the northern

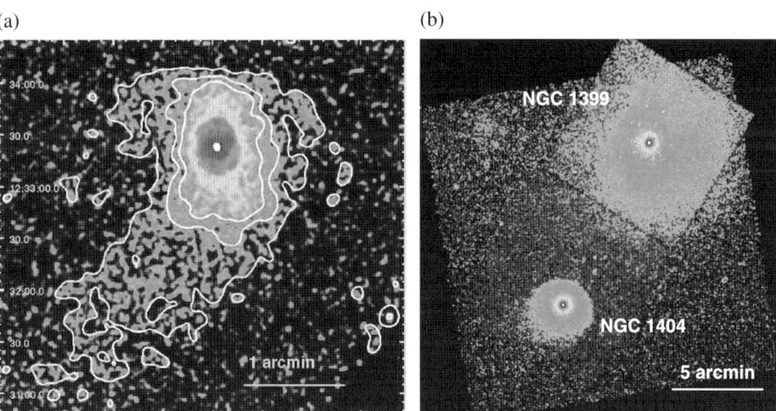

Fig. 21. (**a**) The X-ray tail and bright central region of the elliptical NGC4552 moving supersonically in the Virgo ICM [84]. (**b**) The cold front and tail of the elliptical NGC1404 as it falls toward NGC1399 in Fornax [81]

edge and a 2′ (10 kpc) tail of X-ray emission to the southeast. These are all features of the supersonic ram-pressure stripping of galaxy gas by the cluster ICM that have been found in simulations. As discussed in Sect. 3 and by Vikhlinin et al. [132], by measuring the gas temperature and the density on the galaxy side and on the cluster side of the surface brightness edge, one can measure the pressure jump across the edge, which determines the velocity of the galaxy through the cluster gas. For NGC4552, Machacek et al. found a velocity of $\sim$1600 km s^{-1} corresponding to Mach 2. Combining this velocity with NGC4552's line of sight velocity with respect to M87, they determined that NGC4552 was moving toward us at an angle of $\sim$35 degrees with respect to the plane of the sky [84].

In the Fornax cluster (Fig. 21), the elliptical galaxy NGC1404 also shows a sharp surface brightness edge characteristic of ram pressure stripping by the cluster ICM. The galaxy gas has a temperature of 0.55 keV, nearly three times cooler than the surrounding Fornax cluster gas (1.5 keV), making NGC1404 a galaxy-sized analog of the "cold front" seen in Abell 3667 [132]. From measuring the gas density and temperature across the edge and determining the pressure jump, Machacek et al. found that NGC1404 was moving through the cluster gas with a velocity of 600 km s^{-1}, approximately Mach 1 [82].

4.2 Stripped X-ray Tails from Late-type Galaxies in Merging Clusters

Since luminous ellipticals have hot gaseous coronae [51], it should not be surprising that one sees an X-ray tail as gas from an early-type galaxy is stripped by the motion of the galaxy through the ICM. However since much of the gas in late-type galaxies is cold, finding X-ray emitting tails behind spiral galaxies should be a rare event, since this requires that a gas rich spiral must penetrate deeply into the cluster core. Three spectacular examples of long X-ray tails associated with spiral galaxies have been found. All three occur in clusters undergoing major mergers. These three galaxies exhibit extreme examples of processes that probably influenced the evolution of many cluster galaxies, particularly at earlier epochs of cluster formation. Star formation rates can be increased when the galactic ISM is compressed, but not yet fully stripped by the intracluster gas. Compression of the galaxy ISM by the cluster ICM, as spiral galaxies first fall into the cluster potential, has been suggested as the mechanism for star formation that results in the high fraction of blue galaxies in some z$\sim$0.4 clusters, the "Butcher–Oemler effect" [15]. The three galaxies described here may be low redshift examples of the blue galaxies found in Butcher–Oemler clusters.

The first late-type galaxy with an extended tail that was recognized from Chandra observations was the disturbed spiral C153 in the richness class 4, $z = 0.247$ cluster A2125 [105, 140]. The X-ray tail stretches 80 kpc and has a luminosity of 5×10^{41} ergs s^{-1}. Owen et al. found the galaxy C153 to be bright in the U-band ([OII]) with evidence of star formation in the last 10^8 years.

While star-formation at the leading edge of the galaxy disk appears to be cut off, perhaps because the gas has been stripped from this region, Owen et al. report evidence for young stars in the stripped tail. C153's radio emission suggest both an AGN and a starburst have been triggered as the galaxy passed through the ICM at high velocity.

The starburst galaxy UGC6697 in the northwest merging subcluster of A1367 exhibits numerous giant HII regions, shocked gas and a radio "trail" [56], as well as a 60 kpc long X-ray tail detected in Chandra images by Sun and Vikhlinin [125]. These authors find that ram pressure alone is probably not sufficient to remove the galactic gas and that Kelvin-Helmholtz instabilities [99] and other mechanisms are likely important. However as Gavazzi et al. suggested [56], it is likely the ram pressure that compresses the ISM producing the starburst. Sun and Vikhlinin argue that the correspondence of the X-ray, radio and Halpha edges point to the region outside the front where star formation is truncated due to stripping, while the starburst activity inside the front is triggered by the galaxy's interaction with the ICM.

Recently Sun et al. found a 70 kpc long X-ray tail associated with a small late-type galaxy ESO 137-001 in the merging cluster A3627 [126]. The tail as seen by XMM-Newton and Chandra is shown in Fig. 22. Unlike the X-ray tail associated with UGC6697 in A1367, which does not extend beyond the optical galaxy, the ESO 137-001 tail stretches far behind the galaxy. As Sun et al. argue, the source of the material in the X-ray tail ($\sim 10^9$ $M_\odot$) is likely cold ISM from the galaxy, mixed with the hot (6.5 keV) cluster gas. The cool (0.7 keV) gas temperature in the tail implies that the tail is primarily composed of cold gas stripped from the galaxy. As Sun et al. conclude "ESO 137-001, in its first

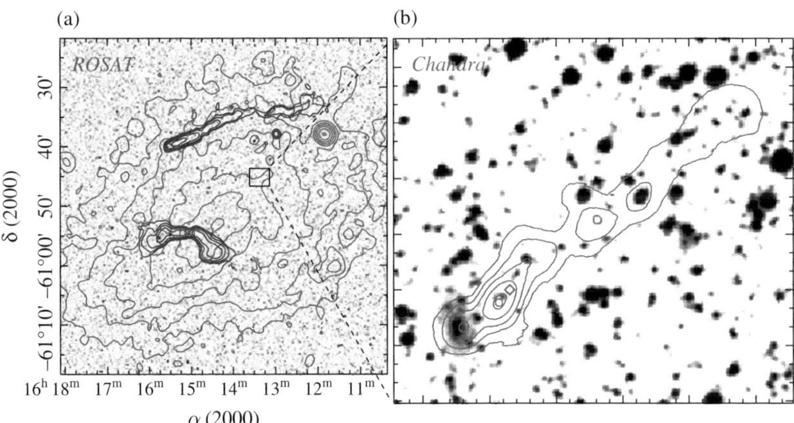

Fig. 22. (**a**) X-ray contours from ROSAT show the distribution of hot gas in the cluster A3627. The blue contours outline radio sources (**b**) The Chandra image from the region in the left panel marked by a small box shows a 70 kpc long ram pressure stripped tail from the small late-type galaxy ESO 137-001 in the cluster A3627 [126]

passage through the cluster core, is being converted into a gas-poor galaxy (likely an E+A galaxy) after a possible initial starburst, all by the interaction with the dense ICM."

4.3 X-ray Trails and Wakes in Poor Clusters

Interactions of galaxy gas with the surrounding medium have been seen in groups as well as in rich clusters. For example, in the Pavo group, Machacek et al. detected a bright trail of X-ray emission extending 90 kpc between the disturbed, large spiral galaxy NGC 6872 and the dominant elliptical NGC 6876 at the center of the group [83]. Figure 23 shows the XMM-Newton image as well as the DSS optical image of the same region. Unlike the three long tails behind late type galaxies that have been found in rich, merging clusters, all of which have temperatures much cooler than the surrounding hot ICM, the 1 keV gas temperature of the tail in the Pavo group is significantly hotter than the 0.5 keV Pavo gas. Machacek et al. suggest that the X-ray trail may be due to mixing of Pavo IGM gas with ISM gas that was stripped from the spiral NGC 6872 by turbulent viscosity, as the spiral moves supersonically (velocity $\sim$1300 km s^{-1}) through Pavo, although gravitational focusing of the group gas into a Bondi-Hoyle wake, due to the supersonic motion of the spiral NGC 6872, also can be significant.

In summary, the observations of X-ray tails, trails and wakes behind both elliptical and spiral galaxies in clusters and groups provides the opportunity to determine the dynamical motions of galaxies in these systems – the direction and total velocity that the galaxies are moving with respect to the surrounding cluster gas – and the opportunity to investigate the processes through which gas is stripped from the galaxy and incorporated into the surrounding medium as well as the effects of stripping on the galaxy.

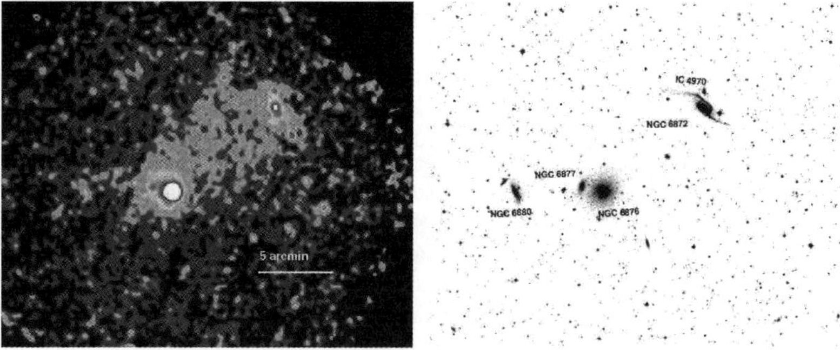

Fig. 23. (**a**) The XMM-Newton image of the Pavo group shows a trail of wake of emission between the spiral NGC 6872 and the elliptical NGC 6876. (**b**) The optical field of the Pavo group on the same scale as panel **a**

4.4 Pressure Confined Coronae

In addition to removing the interstellar medium from galaxies, as they move through the hot intracluster medium, the hot intracluster gas can confine the interstellar gas in the the centers of large ellipticals found in the cores of clusters [123, 134, 143]. In the hot dense cores of Coma, the northwest subcluster in A1367 and in A1060 (the Hydra cluster), gas clouds with temperatures of 1–2 keV, masses of $\sim 10^8$ $M_\odot$, and extents of only a few kpc have been found at the centers of large ellipticals. The top panels in Fig. 24 show the relative scale of the small X-ray coronae to the Coma galaxies NGC4889 and NGC4874. The very existence of cool gas surrounded by hot intracluster gas was a surprise, given that evaporation and stripping of the ISM make the survival of coronae difficult in dense cluster environments. The radiative cooling timescales for these coronae are $\sim 10^8$ years, while the time to evaporate the small cool coronae by the hot ICM is even shorter, only a few times 10^6 years. Either of these mechanisms could destroy the coronae, unless the heat influx driving the evaporation and the radiative cooling balance each other (stellar mass loss also can compensate for some of the mass lost from the coronae). To achieve this balance, the thermal conduction from the hot ICM must be suppressed at the boundary of the corona by a factor of 30 or more compared to the classical Spitzer value [134]. In two galaxies (NGC4874 in Coma and

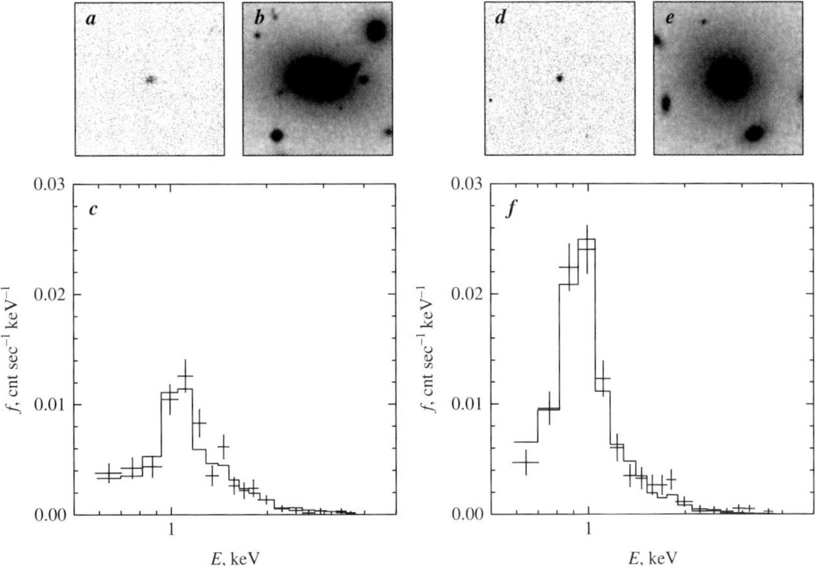

Fig. 24. The top panels show the small X-ray coronae embedded in the Coma galaxies on the same scales. The lower panels show the X-ray spectra of the coronae and that the emission is from a 1 keV gas [134]

NGC3842 in A1367), Sun et al. noted that their double-lobed radio sources must deposit most of their mechanical energy outside the coronae or the small coronae would be disrupted [123].

5 Clusters and Cosmology

During the twentieth century, scientific views changed from believing that all the matter in the Universe is in stars to understanding that the stars and cold gas make up only about 1% of the mass, with hot baryons contributing 3%, while dark matter makes up about 23% and dark energy about 73% [127]. Thus the two dominant components of the Universe are dark matter, a non-baryonic form of matter whose gravity is responsible for the formation of structure, and dark energy, whose pressure is apparently causing the expansion of the Universe to accelerate. However it is only through observations of the 4% of matter which is luminous that we can learn about the distribution of dark matter and the nature of dark energy.

In addition to being the most massive, gravitationally bound objects in the Universe, rich clusters are very luminous (X-ray luminosities as high as several 10^{45} ergs s^{-1}) and thus can be detected and studied at high redshifts. For a particular cosmology, theory can accurately predict what the cluster mass function[2] should be at any redshift. By measuring changes in the cluster mass function with redshift, one can constrain Ω_m, Ω_Λ and the equation of state w for dark energy, as well as determine an accurate measurement of the primordial power spectrum on the scale of cluster masses.

Thus clusters provide an independent and complimentary method of determining cosmological parameters compared to methods using type 1a supernovae (e.g. [108, 117]) or fluctuations in the Cosmic Microwave Background (CMB) [127]. The problem then is how to determine the cluster mass or a reasonable proxy for it. While in theory, it is straight forward to determine the cluster mass from the gas temperature and density profiles, it is very difficult to measure the cluster gas temperature at large cluster radii (i.e. at the virial radius) where the X-ray emission is extremely faint. Similarly weak lensing methods also yield poor determinations of the cluster mass at large radii. Instead a number of proxies for the cluster mass function have been used.

For clusters in hydrostatic equilibrium, the gas temperature should be closely related to the depth of the cluster potential well, therefore allowing the cluster X-ray temperature function to be used as a proxy for the cluster mass function. Henry and Arnaud measured the amplitude of density fluctuations and the slope of the perturbation spectrum on cluster scales from an analysis of the X-ray temperature function for a complete sample of nearby

[2] the cluster mass function is the number density of clusters with masses greater than M, in a comoving volume element.

clusters [63]. As samples of distant cluster temperatures were obtained, constraints on Ω_m were made using the evolution of the X-ray temperature function [27, 36, 65, 66]. Assuming the cluster baryon fraction approximates the cosmic mean, many researchers have estimated Ω_m [1, 24, 38, 40, 124, 142], generally finding $\Omega_m < 1$. Early attempts also were made to use the gas fraction as a distance indicator [38, 118]. However all of these early studies were limited by uncertainties in the cluster mass, since accurate temperature profiles could not be obtained in distant clusters or at large radii in nearby clusters. Chandra and XMM-Newton now provide accurate temperature profiles and thus mass measurements at large radii ($\sim 0.5\ r_{200}$) for low-redshift clusters, which provide good present epoch determinations of the Mass–Temperature (M–T) relation [3, 113, 137, 138]. The M–T relation derived from 13 high redshift ($0.4 < z < 0.7$) clusters is consistent with self-similar evolution of the M–T relation for low redshift clusters [77], which confirms an important prediction from the theory of cluster formation [35].

Recently two methods using Chandra observations of distant clusters have been used to constrain cosmological parameters. In one study, Allen et al. measured the ratio of the mass in baryons (gas plus stars) to the total cluster mass (i.e. the "baryon mass fraction") for a sample of 26 clusters ranging in redshift from $z = 0.07$ to 0.9 [2]. Since the baryon mass fraction in clusters is expected to be constant with redshift (distance to the cluster), the cosmological parameters are constrained by determining when a constant baryon mass fraction is obtained [106]. Figure 25 shows the measured ratios for the 26 clusters for SCDM parameters (panel **a**) and for ΛCDM with $\Omega_m = 0.25$ and $\Omega_\Lambda = 0.96$ (panel **b**). The gas mass fraction appears to decrease with redshift in the SCDM cosmology, because this cosmology underestimates the distance to higher redshift clusters.

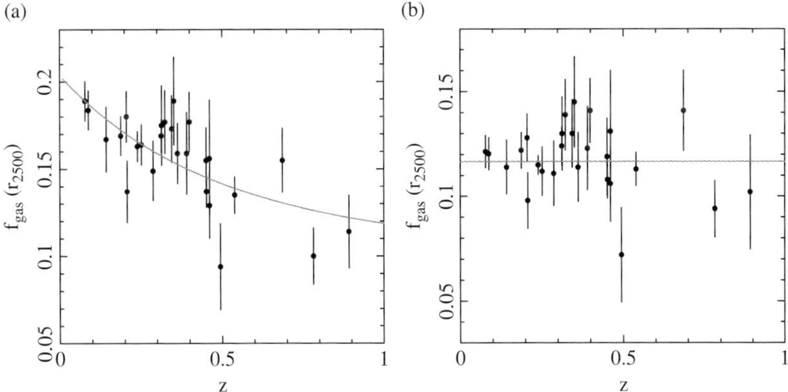

Fig. 25. (**a**) The gas mass fraction in 26 clusters plotted as a function of redshift, for the standard CDM cosmology. (**b**) The gas mass fraction plotted against redshift for the ΛCDM cosmology [2]

Like the cosmological tests that use SN Ia's as standard candles, the cluster baryon mass fraction is a "distance" measurement. A disadvantage of this technique is that it is difficult to accurately measure the total cluster mass at large radii, since the mass measurement at a particular radius is only as accurate as the measurement of the temperature and its gradient at that radius.

While WMAP observations of the CMB have greatly improved the precision with which many cosmological parameters are known, in the CMB data alone there are some degeneracies concerning dark energy and its equation of state (i.e. the ratio of pressure and energy density). Recently, Rapetti, Allen and Weller extended the work of Allen et al. and performed a combined analysis of the Cosmic Microwave Background data, the SN Ia observations and the cluster observations and, with the assumption that the Universe is flat, obtained constraints on the current equation of state for dark energy of $w = -1.05$ $(+0.10, -0.12)$ [115].

The second method, developed by Vikhlinin et al., relies on the assumption that the composition of clusters is typical of the Universe and thus that the baryon mass fraction at large radii in massive clusters should be the same as that in the Universe (i.e. Ω_b/Ω_m is constant) [136]. From the accurate temperature and density measurements for 13 clusters (see Fig. 5), Vikhlinin et al. find that the fraction of the total cluster mass in luminous baryons (gas plus stars) in hot relaxed clusters is consistent with the global baryon fraction measured from CMB observations [116, 127]. However instead of using the baryon fraction to constrain cosmological parameters (e.g. [106]), which would also require that the total cluster mass be measured, Vikhlinin et al. uses the cluster baryon mass function as a proxy for the cluster mass function.

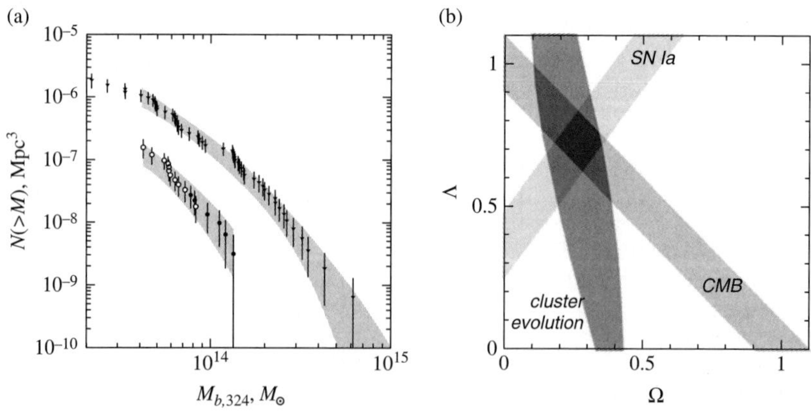

Fig. 26. (**a**) The upper set of points define the local baryon mass function [139], while the lower points are from clusters at $z > 0.4$ [135]. (**b**) A comparison of the constraints on Ω_m and Ω_Λ derived from SN 1a, CMB, and cluster observations [135]

Figure 26 (panel **a**) shows the change in the baryon mass function determined from a sample of local clusters compared with a sample of clusters at $z > 0.4$ [136]. Deriving constraints on cosmological parameters through different techniques is important both for confirming parameter values, but also because the combination of different techniques can often better define the parameters, since different methods yield different constraints, as illustrated in panel **b** of Fig. 26 [135].

Acknowledgements

With the launch of the Chandra observatory, we can now view the X-ray sky with a spatial resolution comparable to that of ground-based optical telecopes. For understanding clusters of galaxies, this increased resolution, as well as Chandra's sensitivity, has had a tremendous impact. We thank all the scientists, engineers, programmers, administrators and everyone who has been responsible for the success of Chandra. We also thank Omar López-Cruz, Manolis Plionis and David Hughes for organizing the 2005 Advanced School "Pan-Chromatic View of Clusters and Large Scale Structures" where the lectures that form the basis of this chapter were presented. We thank Ben Maughan, Ralph Kraft, Ming Sun and Marusa Bardac for providing figures. Lastly we acknowledge support by the Smithsonian Institution and the Chandra Science Center, as well as several NASA Grants.

References

1. Allen, S., Schmidt, R., Fabian, A., Ebeling, H.: MNRAS **342**, 287 (2003)
2. Allen, S., Schmidt, R., Ebeling, H., Fabian, A., vanSpeybroeck, L.: MNRAS **353**, 457 (2004)
3. Arnaud, M., Pointecouteau, E., Pratt, G.: A&A **441**, 893 (2005)
4. Ascasibar, Y., Markevitch, M.: ApJ, **650**, 102 (2006)
5. Belsole, E., et al.: A&A **365**, L188 (2001)
6. Bernabei, R., et al.: astro-ph/0307403 (2003)
7. Biller, B., Jones, C., Forman, W., Kraft, R., Ensslin, T.: ApJ **613**, 238 (2004)
8. Binney, J., Tabor, G.: MNRAS **276**, 663 (1995)
9. Birzan, L., Rafferty, D., McNamara, B., et al.: ApJ **607**, 800 (2004)
10. Böhringer, H., Morfill, G. ApJ **330**, 609 (1988)
11. Böhringer, H., et al.: MNRAS **264**, L25 (1993)
12. Böhringer, H., et al.: ApJS **129**, 435 (2000)
13. Böhringer, H., et al.: A&A **365**, L181 (2001)
14. Bradac, M., Clowe, D., Gonzales, A., Marshall, R., Jones, C., Forman, W., Markevitch, M., Randall, S.: ApJ, **652**, 937 (2006)
15. Butcher, H., Oemler, A.: ApJ **219**, 18 (1978)
16. Churazov, E., Forman, W., Jones, C., Böhringer, H.: A&A **356**, 788 (2000)
17. Churazov, E., Bruggen, M., Kaiser, C., Böhringer, H., Forman,W.: ApJ **554**, 261 (2001)

18. Churazov, E., Forman, W., Jones, C., Sunyaev, R., Bohringer, H.: ApJ **590**, 225 (2003)
19. Churazov, E., Sazonav, S., Sunyaev, R., Forman, W., Jones, C., Bohringer, H.: MNRAS **363**, L91 (2005)
20. Clowe, D., Gonzalez, A., Markevitch, M.: ApJ **604**, 596 (2004)
21. Clowe, D., et al.: ApJ, **648**, L109 (2006)
22. Cowie, L., Binney, J.: ApJ **215**, 723 (1978)
23. David, L., Hughes, J., Tucker, W.: ApJ **394**, 452 (1992)
24. David, L., Jones, C., Forman, W.: ApJ **445**, 578 (1995)
25. David, L., et al.: ApJ **557**, 546 (2001)
26. De Grandi, S., Molendi, S.: ApJ **567**, 163 (2002)
27. Donahue, M., Voit, M.: ApJ (Letters) **523**, L137 (1999)
28. Donahue, M., Stocke, J., Gioia, I.: ApJ **385**, 49 (1992)
29. Dressler, A.: ApJ **236**, 351 (1980)
30. Dressler, A., Smail, I., Poggianti, B., Butcher, H., Oemler, A.: ApJ (Suppl) **122**, 51 (1999)
31. Dunn, R., Fabian, A., Taylor, G.: MNRAS **364**, 1343 (2005)
32. Durret, F., Forman, W., Gerbal, D., Jones, C., Vikhlinin, A.: A&A **335**,41 (1998)
33. Durret, F., LimaNeto, G., Forman, W., Churazov, E.: A&A (Letters) **430**, 29 (2003)
34. Durret, F., LimaNeto, G., Forman, W.: A&A **432**, 809 (2005)
35. Eke, V., Cole, S., Frenk, C.: MNRAS **282**, 263 (1996)
36. Eke, V., Cole, S., Frenk, C., Henry, J.P.: MNRAS **298**, 1145 (1998)
37. Ensslin, T., Bruggen, M.: A&A **331**, 1011 (2002)
38. Ettori, S., Fabian, A.: MNRAS **305**, 834 (1999)
39. Ettori, S., Fabian, A.: MNRAS **317**, 57 (2000)
40. Evrard, A.: MNRAS **292**, 289 (1997)
41. Fabian, A., Nulsen, P.: MNRAS **180**, 479 (1977)
42. Fabian, A.: ARA&A **32**, 277 (1994)
43. Fabian, A., et al.: MNRAS **344**, L43 (2003)
44. Fabian, A., et al.: MNRAS **366**, 417 (2006)
45. Fabricant, D., Lecar, M., Gorenstein, P.: ApJ **241**, 552 (1980)
46. Feretti, L.: astro-ph/0406090 (2004)
47. Finoguenov, A., Pietsch, W., Aschenbach, B., Miniati, F.: A&A **415**, 415 (2004)
48. Finoguenov, A., Jones, C.: ApJ (Letters) **547**, L107 (2001)
49. Forman, W., Kellogg, E., Gursky, H., Tananbaum, H., Giacconi, R.: ApJ **178**, 309 (1972)
50. Forman, W., Schwarz, J., Jones, C., Liller, W., Fabian, A.: ApJ (Letters) **234**, 27 (1979)
51. Forman, W., Jones, C., Tucker, W.: ApJ **61**, 33 (1985)
52. Forman, W., Jones, C.: ARA&A **20**, 547 (1982)
53. Forman, W., et al.: ApJ **635**, 894 (2005)
54. Forman, W., et al.: ApJ, **665**, 1057 (2007)
55. Fugita, S., Suzuki, T.: ApJ (Letters) **630**, 1 (2005)
56. Gavazzi, G., et al.: A&A **304**, 325 (1995)
57. Govoni, G., Markevitch, M., Vikhlinin, A., VanSpeybroeck, L., Feretti, L., Giovannini, G.: ApJ **605**, 695 (2004)
58. Gunn, J., Gott. R.: ApJ **176**, 1 (1972)

59. Gursky, H., et al.: ApJL **173**, L99 (1972)
60. Heinz, S., Reynolds, C., Begelman, M.: ApJ **501**, 126 (1998)
61. Heinz, S., Churazov, E.: ApJ (Letters) **634**, L141 (2005)
62. Helsdon, S., Ponman, T.: MNRAS **319**, 933 (2000)
63. Henry, J.P., Arnaud, K.: ApJ **372**, 410 (1991)
64. Henry, J.P., Gioia, I., Maccacaro, T., Morris, S., Stocke, J., Wolter, A.: ApJ **386**, 408 (1992)
65. Henry, J.P.: ApJ (Letters) **489**, 1 (1997)
66. Henry, J.P.: ApJ **534**, 565 (2000)
67. Israel, F.P.: A&ARv **8**, 2371 (1998)
68. Jones, C., Forman, W.: ApJ **224**, 1 (1978)
69. Jones, C., Forman, W.: ApJ **276**, 38 (1984)
70. Jones, C., Forman, W.: ApJ **511**, 65 (1999)
71. Jones, C., et al.: ApJ (Letters) **567**, L115 (2002)
72. Jones, C., et al.: In preparation (2007)
73. Kaiser, N.: MNRAS **222**, 323 (1986)
74. Kaiser, N.: ApJ **383**, 104 (1991)
75. Kauffmann, G., White, S., Heckman, T., Menard, B., Brinchmann, J., Charlot, S., Tremonti, C., Brinkmann, J.: MNRAS **353**, 713 (2004)
76. Kellogg, E., Gursky, H., Tananbaum, H., Giacconi, R., Pounds, K.: ApJ (Letters) **174**, L65 (1972)
77. Kotov, O., Vikhlinin, A.: ApJ **633**, 781 (2005)
78. Kraft, R., Forman, W., Jones, C., Murray, S., Hardcastle, M., Worrall, D.: ApJ **569**, 54 (2002)
79. Kraft, R., Vazquez, S., Forman, W., et al.: ApJ **592**, 129 (2003)
80. Kraft, R., Forman, W., Churazov, E., et al.: ApJ **601**, 221 (2004)
81. Machacek, M., Jones, C., Forman, W.: ApJ **610**, 183 (2004)
82. Machacek, M., Dosaj, A., Forman, W., Jones, C., Markevitch, M., Vikhlinin, A., Warmflash, A., Kraft, R.: ApJ **621**, 663 (2005)
83. Machacek, M., Nulsen, P., Stirbat, L., Jones, C., Forman, W.: ApJ **630**, 280 (2005)
84. Machacek, M., et al.: ApJ, **644**, 155 (2006)
85. Markevitch, M., Forman, W., Sarazin, C., Vikhlinin, A.: ApJ **503**, 77 (1998)
86. Magorrian, J. et al.: AJ **115**, 2285 (1998)
87. Markevitch, M., Sarazin, C., Vikhlinin, A.: ApJ **521**, 526 (1999)
88. Markevitch, M. et al.: ApJ (Letters) **567**, 27 (2002)
89. Markevitch, M. et al.: ApJ **606**, 819 (2004)
90. Markevitch, M., Govoni, F., Brunetti, G., Jerius, D.: ApJ **627**, 733 (2005)
91. Markevitch, M., et al.: In preparation (2007)
92. Mathews, W., Faltenbacher, A., Brighenti, F.: ApJ **638**, 659 (2006)
93. McNamara, B. et al.: ApJ (Letters) **534**, L135 (2000)
94. McNamara, B., et al.: Nature **433**, 45 (2005)
95. Mullis, C., Vikhlinin, A., et al.: ApJ **607**, 175 (2004)
96. Mitchell, R., Culhane, J., Davison, P., Ives, J.: MNRAS **175**, 29 (1976)
97. Navarro, J., Frenk, C., White, S.: ApJ **462**, 563 (1996)
98. Navarro, J., Frenk, C., White, S.: ApJ **490**, 493 (1997)
99. Nulsen, P.: MNRAS **198**, 1007 (1982)
100. Nulsen, P., et al.: ApJ **628**, 629 (2005)
101. Nulsen, P., et al.: ApJ **625**, L9 (2005)

102. Oegerle, W., Cowie, L., Davidsen, A., Hu, E., Hutchings, J., Murphy, E., Sembach, K., Woodgate, B.: ApJ **560**, 1870 (2001)
103. Oemler, A.: ApJ **194**, 10 (1974)
104. Owen, F., Eilek, J., Kassim, N. ApJ **543**, 611 (2000)
105. Owen, F., Ledlow, M., Keel, W., Wang, Q.D., Morrison, G.: AJ **129**, 31 (2005)
106. Pen, U.: New Astron. **2**, 309 (1997)
107. Peterson, J. et al.: A&A **365**, L104 (2001)
108. Perlmutter, S., et al.: ApJ **517**, 565 (1999)
109. Peterson, J., et al.: ApJ **590**, 207 (2003)
110. Piffaretti, R., Jetzer, P., Kaastra, J., Tamura, T.: A&A **433**, 101 (2005)
111. Poggianti, B., et al.: ApJ **518**, 576 (1999)
112. Ponman, T., Cannon, D., Navarro, J.: Nature **297**, 135 (1999)
113. Pointecouteau, E., Arnaud, M., Pratt, G.: A&A **435**, 1 (2005)
114. Rangarajan, F., White, D., Ebeling, H., Fabian, A.: MNRAS **277**, 1047 (1995)
115. Rapetti, S., Allen, S., Weller, J.: MNRAS **360**, 555 (2005)
116. Readhead, A., et al.: ApJ **609**, 498 (2004)
117. Riess, A., et al.: ApJ **607**, 665 (2004)
118. Rines, K., Forman, W., Pen, U., Jones, C., Burg, R.: ApJ **517**, 70 (1999)
119. Rosner, R., Tucker, W.: ApJ **338**, 761 (1989)
120. Ruszkowski, M., Bruggen, M., Begelman, M.: ApJ **611**, 158 (2004)
121. Ruszkowski, M., Bruggen, M., Begelman, M.: ApJ **615**, 675 (2004)
122. Ryle, M., Windram, M.: MNRAS **138**, 1 (1968)
123. Sun, M., Vikhlinin, A., Forman, W., Jones, C., Murray S.: ApJ **619**, 169 (2005)
124. Sanderson, A., Ponman, T.: MNRAS **345**, 1241 (2003)
125. Sun, M., Vikhlinin, A.: ApJ **621**, 718 (2005)
126. Sun, M., Jones, C., Forman, W., Nulsen, P., Donahue, M., Voit, G.: ApJ (Letters) **637**, 81 (2006)
127. Spergel, D., et al.: ApJ (Suppl) **148**, 175 (2003)
128. Stewart, G., et al.: ApJ **285**, 1 (1984)
129. Thomas, P, Fabian, A., Arnaud, K., Forman, W., Jones, C.: MNRAS **222**, 655 (1986)
130. Tucker, W., David, L.: ApJ **484**, 602 (1997)
131. Vikhlinin, A., McNamara, B., Forman, W., Jones, C., Quintana, H., Hornstrup, A.: ApJ (Letters) **498**, 21 (1998)
132. Vikhlinin, A., Markevitch, M., Murray, S.: ApJ **551**, 160 (2001)
133. Vikhlinin, A., Markevitch, M., Murray, S.: ApJ (Letters) **549**, 47 (2001)
134. Vikhlinin, A., Markevitch, M., Forman, W., Jones, C.: ApJ (Letters) **555**, 87 (2001)
135. Vikhlinin, A., VanSpeybroeck, L., Markevitch, M., Forman, W., Grego, L.: ApJ (Letters) **578**, 107 (2002)
136. Vikhlinin, A., Voevodkin, A., Mullis, C., VanSpeybroeck, L., Quintana, H., McNamara, B., Gioia, I., Hornstrup, A., Henry, J.P., Forman, W., Jones, C.: ApJ **590**, 15 (2003)
137. Vikhlinin, A., Markevitch, M., Murray, S., Jones, C., Forman, W., VanSpeybroeck, L.: ApJ **628**, 655 (2005)
138. Vikhlinin, A., et al.: In preparation (2007)
139. Voevodkin, A., Vikhlinin, A.: ApJ **601**, 610 (2004)
140. Wang, Q.D., Owen, F., Ledlow, M.: ApJ **611**, 821 (2004)
141. White, D., Fabian, A., Forman, W., Jones, C., Stern, C.: ApJ **375**, 35 (1991)

142. White, S., Navarro, J., Evrard, A., Frenk, C.: Nature **366**, 429 (1993)
143. Yamasaki, N., Ohashi, T., Furusho, T.: ApJ **578**, 833 (2002)
144. Zakamska, N., Narayan, R.: ApJ **582**, 162 (2003)
145. Zwicky, F.: ApJ **86**, 217 (1937)

Dynamics of Galaxies and Clusters of Galaxies

L. A. Aguilar

Instituto de Astronomía, UNAM, Baja California México
Aguilar@astrosen.unam.mx

1 Introduction

The purpose of these lectures is to describe some of the dynamical phenomena that are important in the evolution of galaxies and clusters of galaxies.

Galactic Dynamics is usually perceived as an arid discipline, whose mathematical formalism make it hard to apply to real astronomical problems. This is very unfortunate, since gravity is behind all astronomical phenomena and its resulting dynamics has a role to play. This misconception of Galactic Dynamics stems from the fact that often students are introduced to the subject using a formal approach, full of mathematical rigor, that leads through a lengthy path, before reaching astronomical applications. Although mathematical correctness is gained, a physical grasp of the dynamics behind is sometimes lost.

It is our intention to bypass the usual formal introduction and to use an eclectic list of topics related to the dynamics of galaxies and clusters of galaxies, to illustrate the use of Galactic Dynamics. Our main goal is understanding, rather than rigor. We will attempt to show the student how, using some basic equations, we can extract useful dynamical information that can help in our understanding of the realm of galaxies.

In Sect. 2, Poisson's and Boltzmann's equations are presented as the basic equations of Galactic Dynamics. After briefly reviewing what makes a dynamical system collisionless, we introduce in Sect. 3 one of the Jeans equations in spherical coordinates as a quick tool to derive the dynamical properties of a model built from a density profile, without having to build a self-consistent dynamical model. The Navarro, Frenk and White profile is used as an example. A *Mathematica* Notebook version of this section is provided as well. Students with access to this program can interact with this version and change parameters in the examples provided. The full Notebook is available from the *Guillermo Haro* and the author's web pages.

In Sect. 4 we mention two additional density profiles that can be used as exercises for the reader. We also note that simulations of interacting galaxies

L. A. Aguilar: *Dynamics of Galaxies and Clusters of Galaxies*, Lect. Notes Phys. **740**, 71–118 (2008)
DOI 10.1007/978-1-4020-6941-3_3

tend to give density profiles that go as $\rho \propto r^{-4}$ at large radii, when simulated in isolation, as opposed to those of systems immersed in an expanding background, which tend to give the gentler $\rho \propto r^{-3}$ of the Navaro, Frenk & White (NFW) profile. We show why the former dependence is expected for an isolated finite mass system that has been perturbed.

In Sect. 5 we talk about the orbital structure of spherical potentials. Although idealized, spherical potentials allow us to introduce basic concepts about orbits. We describe the use of the Lindblad diagram as a tool to classify orbits in spherical potentials.

In Sect. 6 we tackle a sticky issue: Dynamical friction, one of the main culprits responsible for robbing center of mass motion to satellite systems, causing them to spiral toward the center of the host system. As we will see, this force is a case of bipolar personality, depending on the speed of the object being reduced, its behavior can change quite a bit.

In Sect. 7 we examine the tidal force, responsible for truncating small systems when they move within the gravitational influence of a larger one. We will see that, contrary to common opinion, tides do not necessarily stretch along the radial direction. It all depends on the mass distribution of the tide-producing object. We then discuss the concept of tidal radius at some length, going from a very simplified static model of two point particles, to the inclusion of effects due to the motion of the satellite and host systems, non-circular orbits and extended mass distributions.

In Sect. 8 we talk about what happens when two galaxies collide. The rapid variation of the perturbing force pumps orbital energy into the internal degrees of motion of the interacting galaxies and heats them up, promoting mass loss above the rate due to tidal radius truncation. We discuss the role of two very important timescales: the interaction time and the internal dynamical time. As we will see, the amount of damage the interacting systems suffer, depends to a great extent on the ratio of these two numbers.

In Sect. 9 we present a simple exercise, where the combined action of dynamical friction and tidal truncation, acting on a Plummer sphere that spirals, following a path close to circular within a flat rotation curve halo, is modeled. This wraps up much of what has been discussed in the three previous sections.

Finally, in Sect. 10 we suggest some references to those interested student that which to pursue, in more detail, the topics covered in these lectures. Finally, Sect. 11 provides some concluding remarks.

2 Basic Galactic Dynamics

The two most important equations of Stellar and Galactic Dynamics are the *Poisson's equation*, which relates the gravitational potential ϕ with its source (the mass density function ρ),

$$\nabla^2\phi = 4\pi G\rho\ , \tag{1}$$

and the *Boltzmann's equation*, which is a transport equation that describes the evolution of a dynamical system in phase-space,

$$\frac{\partial f}{\partial t} + \boldsymbol{v}\cdot\frac{\partial f}{\partial \boldsymbol{x}} - \nabla\phi\cdot\frac{\partial f}{\partial \boldsymbol{v}} = \left(\frac{\partial f}{\partial t}\right)_{\rm coll} \tag{2}$$

The phase-space distribution function, $f(\boldsymbol{x},\boldsymbol{v},t)$, contains all the dynamical information of the system. Its projection in configuration space gives the spatial density of the system,

$$\rho(\boldsymbol{x},t) = \int f(\boldsymbol{x},\boldsymbol{v},t){\rm d}^3v\ .$$

The right hand side of Boltzmann's equation contains the so-called *collisional term.* Its effect differs from the term in the left hand side, mainly by the timescale on which the collisions, it describes, operate. The left hand side describes the flow of particles in a given parcel of the system as it moves in phase-space, on a timescale determined essentially by its potential, whereas the right hand side describes the flow of particles in and out of this parcel due to 2-body collisions, which usually are so fast, with respect to the former timescale, that the collisional time derivative appears, to the rest of the equation, as an instantaneous source and sink term.

The natural timescale of the left hand side of Boltzmann's equation is the so-called *dynamical timescale*, which is essentially the orbital time for particles within the system. As a very rough approximation to a system-wide average of this timescale, we can use the ratio of the system size and the *rms* velocity,

$$t_{\rm dyn} \approx R/v_{\rm rms}\ .$$

The effect of the right hand side of Boltzmann's equation operates on the so-called *collisional timescale.* It can be shown that this time is of the form (e.g. see Sect. 3.2 of [46]),

$$t_{\rm coll} \approx (R/v_{\rm rms})N/\log(N)\ ,$$

where N is the number of particles within the system.

The ratio of $t_{\rm coll}$ to $t_{\rm dyn}$ is a measure of the degree of collisionality of a dynamical system:

$$t_{\rm coll}/t_{\rm dyn} \approx N/\log(N)\ . \tag{3}$$

Notice that the parameters that characterize the properties of the system in physical units cancel out, and the only dependence left is on the number of particles within the system: the more particles, the larger this ratio is, and so the less important collisions are. This is a bit counter-intuitive and must be explained further. One would have thought that the more particles there are in a system, the more collisions will be, and indeed this is the case, but what matters at the end is not the number of collisions but rather the effect they have in the system.

Let us take a system with N particles. If we split each particle in two, keeping everything else constant in the system, like the total mass, size and the *rms* velocity, we double the number of particles and thus the number of collisions. But since the strength of the gravitational interaction scales as the product of the masses, and each mass is now one half of what it was before, the result is a reduction in the effect of collisions. In reality, doubling the number of particles also reduces the mean inter-particle distance ($l \propto N^{-1/3}$), which has the outcome of increasing the collisional effect, but in 3D space this is dominated by the mass effect, even when taking into account the increased number of collisions per unit time.

So, the more particles the more collisions, but the less effect they will have. When the effect of collisions is negligible over the period of time we are interested in, we are lead to the collisionless Boltzmann equation:

$$\frac{\partial f}{\partial t} + \boldsymbol{v} \cdot \frac{\partial f}{\partial \boldsymbol{x}} - \nabla \phi \cdot \frac{\partial f}{\partial \boldsymbol{v}} = 0 \, . \tag{4}$$

This is a partial differential equation that, together with Poisson's equation and proper boundary conditions, presumably set by observations, can in principle be solved. In practice, there is not enough observational information and furthermore, the mathematical complexity of the task of solving it, makes the direct solution an impossibility for realistic cases.

Other alternative approaches have been used. In particular, a fruitful approach is obtained rewriting the previous equation as a total time derivative, using the fact that minus the gradient of the potential is just the acceleration,

$$\frac{\partial f}{\partial t} + \dot{\boldsymbol{x}} \cdot \frac{\partial f}{\partial \boldsymbol{x}} + \dot{\boldsymbol{v}} \cdot \frac{\partial f}{\partial \boldsymbol{v}} = \frac{\mathrm{D}f}{\mathrm{D}t} = 0 \, . \tag{5}$$

This means that, as we move along dynamical trajectories, the value of the distribution function does not change. This is a very strong restriction that can be exploited. If furthermore, the system is in steady state, the trajectories become invariant orbits and f must be a function of those quantities that are conserved along them:

$$f(\boldsymbol{x}, \boldsymbol{v}) = f(I_i), \quad \text{where } I_i(\boldsymbol{x}, \boldsymbol{v}) \text{ is such that } \mathrm{D}I_i/\mathrm{D}t = 0 \, , \tag{6}$$

with I_i's the so called *integrals of motion.* This results in the Jeans' theorem that can be used to find solutions [12]:

Theorem 1 (Jeans' Theorem). *Any steady-state solution of the collisionless Boltzmann equation depends on the phase-space coordinates only through integrals of motion in the galactic potential, and any function of the integrals yields a steady-state solution of the collisionless Boltzmann equation.*

This lead to a whole industry of model building based on classical integrals of motion: $f = f(E)$, $f = f(E, L)$, $f = f(E, L^2)$, $f = f(E_J)$, etc., where E, L and E_J are the energy, angular momentum and the Jacobi energy.

Before leaving this approach, we must note that Jeans' theorem is strictly valid only when the system is integrable (i.e., there is a canonical transformation to a coordinate system where the motion can be separated in each of its degrees of freedom). Such systems have orbits completely determined by isolating integrals of motion and are called regular. The appearance of irregular orbits invalidates Jeans' theorem [11]. However, steady-state spherical systems, like those we examine in these notes, do not have irregular orbits. We will come back to this subject in Sect. 5.

Another approach has been to extract useful dynamical information without necessarily building a full dynamical model. This method is based in the *Jeans' equations*, which are just the result of taking velocity moments of Boltzmann's equation. In spherical coordinates, a very useful equation that we will be using is the one obtained from the 2nd velocity moment:

$$\frac{\mathrm{d}}{\mathrm{d}r}(\rho\sigma_r^2) + \frac{\rho}{r}[2\sigma_r^2 - (\sigma_\theta^2 + \sigma_\phi^2)] = -\rho\frac{\mathrm{d}\phi}{\mathrm{d}r} \, , \tag{7}$$

where σ_r, σ_θ and σ_ϕ are the three spherical coordinate components of the velocity dispersion. Using this equation, we can extract information about these components for a system for which we know the density profile and the resulting potential, *without building a phase-space distribution function for it.* This is a very useful tool. But we must pay a price for this expedience, as finding a solution to Jeans' equations does not guarantee that a physical model may exist (i.e., f is positive everywhere).

In the next section we will learn to use this equation to explore some dynamical properties of the Navarro, Frenk and White density profile, without building a full self-consistent dynamical model for it.

3 A Case Study: The Navarro, Frenk and White profile

The Navarro, Frenk & White (NFW) profile was proposed as a universal density profile, produced by hierarchical clustering in cosmological simulations [40]. Here it is used as an example of how to extract information, using just a few basic equations.

3.1 Density Profile

The density profile is given by:

$$\rho(r) = \rho_o (r/r_o)^{-1}(1 + r/r_o)^{-2} \, . \tag{8}$$

It is convenient to cast it in a dimensionless form, by defining a dimensionless length:

$$\zeta \equiv r/r_o ,$$

and a dimensionless density:

$$\delta(\zeta) \equiv \rho(r)/\rho_o = \zeta^{-1}(1+\zeta)^{-2} . \tag{9}$$

Its limits at the center and infinity are,

$$\lim_{\zeta\to 0} \delta(\zeta) \to \zeta^{-1} \to \infty, \qquad \lim_{\zeta\to\infty} \delta(\zeta) \to \zeta^{-3} \to 0 .$$

We note that at $r = r_o$, the density has a value equal to $\rho(r_o) = \rho_o/4$, and $\rho(r) = 1 \implies \zeta(1+\zeta)^2 = 1 \implies \zeta = 0.465571...$

The NFW profile has a varying slope in the log–log plane, it diverges as ζ^{-1} at the center and goes as ζ^{-3} for $\zeta \to \infty$. The midpoint for this slope variation is at $\zeta \sim -1$ (Fig. 1).

3.2 Cumulative Mass

The mass enclosed within a sphere of radius r, is:

$$\begin{aligned} M_r &\equiv \int_0^r 4\pi\rho(r')r'^2 \mathrm{d}r' = 4\pi\rho_o \int_0^r \left(\frac{r'}{r_o}\right)^{-1} \left(1+\frac{r'}{r_o}\right)^{-2} r'^2 \mathrm{d}r' \\ &= 4\pi\rho_o r_o^3 \int_0^\zeta (1+\zeta')^{-2}\zeta' \mathrm{d}\zeta' = 4\pi\rho_o r_o^3 \left[\log(1+\zeta) + (1+\zeta)^{-1} - 1\right] . \end{aligned}$$

Again, it is convenient to define a dimensionless enclosed mass,

$$\mu(\zeta) \equiv (M_r/M^*) = \log(1+\zeta) + (1+\zeta)^{-1} - 1 , \tag{10}$$

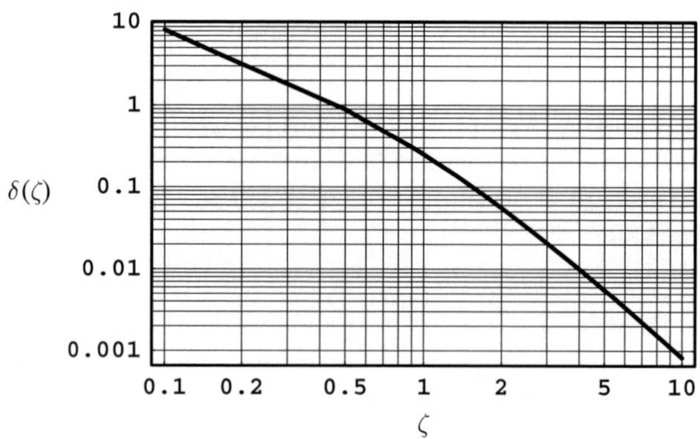

Fig. 1. NFW density profile

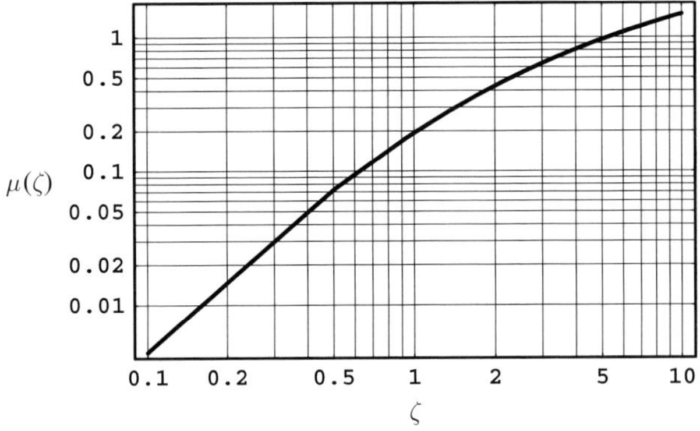

Fig. 2. NFW cumulative mass

where we have also defined a characteristic mass, $M^* \equiv 4\pi \rho_o r_o^3$.

We note the following values and limits:

$$\mu(\zeta) = 1 \Rightarrow \zeta = 5.3054..., \quad \mu(\zeta = 1) = 0.193147...$$

$$\lim_{\zeta \to 0} \mu = 0 \, , \quad \lim_{\zeta \to \infty} \mu = \log(\zeta) \to \infty \, ,$$

thus, the enclosed mass diverges, but only logarithmically (Fig. 2).

3.3 Potential

The potential of a spherical mass distribution can be calculated as,

$$\phi(r) = -4\pi G \left[\frac{1}{r} \int_0^r \rho(r') r'^2 \mathrm{d}r' + \int_r^\infty \rho(r') r' \mathrm{d}r' \right] .$$

In our case, the first integral is,

$$\begin{aligned} \phi_1(r) &= -\frac{4\pi G}{r} \int_0^r \rho(r') r'^2 \, \mathrm{d}r' = -\frac{G}{r} \int_0^r \rho(r') \, 4\pi r'^2 \, \mathrm{d}r' = -\frac{GM_r}{r} \\ &= -\frac{GM^*}{r_o} \frac{\mu(\zeta)}{\zeta} = -4\pi G \rho_o r_o^2 \frac{\mu(\zeta)}{\zeta} \, , \end{aligned}$$

and the second integral is given by,

$$\begin{aligned} \phi_2(r) &= -4\pi G \int_r^\infty \rho(r') r' \, \mathrm{d}r' \\ &= -4\pi G \rho_o r_o^2 \int_\zeta^\infty \frac{\mathrm{d}\zeta'}{(1+\zeta)^2} = -4\pi G \rho_o r_o^2 \, \zeta^{-1} (1+\zeta)^{-1} \, . \end{aligned}$$

Putting everything together, we get:

$$\phi(r) = -4\pi G\rho_o r_o^2 \zeta^{-1} \log(1+\zeta) \ .$$

It is natural then to define a dimensionless potential as:

$$\Psi(\zeta) = \zeta^{-1} \log(1+\zeta) \ , \tag{11}$$

where $\phi(r) = \phi_o \Psi(\zeta)$, and $\phi_o \equiv -4\pi G\rho_o r_o^2$.

Since $\lim_{\zeta\to 0} \Psi(\zeta) = 1$, it is clear that ϕ_o is the depth of the potential well. We also note that $\lim_{\zeta\to\infty} \Psi(\zeta) = 0$, so despite the divergent mass, the potential well has a finite depth. Since $\phi_o < 0$, our dimensionless potential is a positive function (Fig. 3).

3.4 Force

We now compute the magnitude of the force exerted by the NFW profile,

$$F(r) = -\frac{\mathrm{d}\phi}{\mathrm{d}r} = -\frac{\phi_o}{r_o}\left(\frac{\mathrm{d}\Psi}{\mathrm{d}\zeta}\right) = -\frac{\phi_o}{r_o}\mathcal{F}(\zeta) \ .$$

The dimensionless force is given by,

$$\mathcal{F}(\zeta) \equiv \frac{\mathrm{d}\Psi}{\mathrm{d}\zeta} = \frac{\zeta - (1+\zeta)\log(1+\zeta)}{\zeta^2(1+\zeta)} \ . \tag{12}$$

The limits of the dimensionless force are,

$$\lim_{\zeta\to 0} \mathcal{F}(\zeta) = -1/2 \ , \quad \lim_{\zeta\to\infty} \mathcal{F}(\zeta) = 0 \ .$$

Notice that the force is discontinuous at the origin. This is due to the central cusp of the profile (Fig. 4).

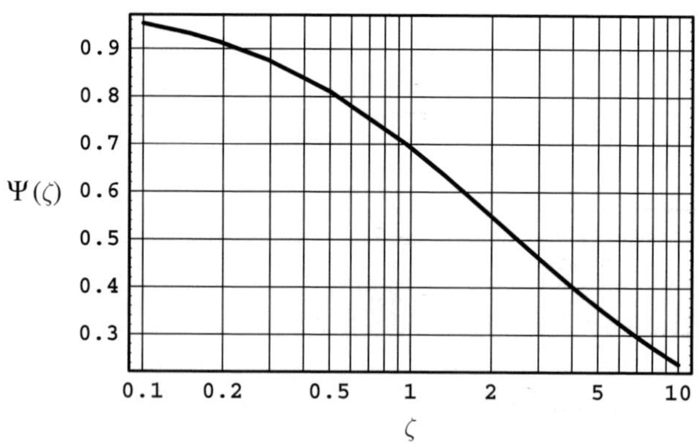

Fig. 3. NFW potential

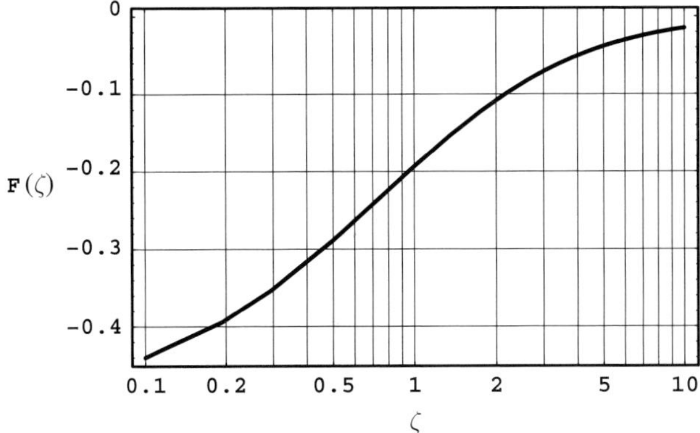

Fig. 4. NFW force

3.5 Escape and Circular Velocities

The escape velocity is easily obtained by the condition of null energy:

$$E = (1/2)v_{\rm esc}^2 + \phi(r) = 0 \Longrightarrow v_{\rm esc}^2 = -2\phi(r) = -2\phi_o \Psi(\zeta) \ .$$

It is clear that the natural unit of velocity is $\sqrt{\phi_o}$. We can then define a dimensionless escape velocity as:

$$\beta_{\rm esc}^2 \equiv v_{\rm esc}^2/\phi_o = 2\log(1+\zeta)/\zeta \tag{13}$$

The limits are,

$$\lim_{\zeta\to 0} \beta_{\rm esc}^2 = 2 \ , \quad \lim_{\zeta\to\infty} \beta_{\rm esc}^2 = 0 \ ,$$

so, despite the infinite mass of the model, the escape velocity is finite.

The circular velocity is obtained from the centrifugal equilibrium condition:

$$v_c^2/r = -F(r) \Longrightarrow \beta_c^2 = -\zeta\mathcal{F}(\zeta) = \frac{(1+\zeta)\log(1+\zeta) - \zeta}{\zeta(1+\zeta)} \ . \tag{14}$$

The limits in this case are,

$$\lim_{\zeta\to 0} \beta_c^2 \propto \zeta/2 \to 0 \ , \quad \lim_{\zeta\to\infty} \beta_c^2 = 0 \ .$$

Notice that the rotation curve of the NFW profile rises as $\sqrt{\zeta}$ from the center, reaches a maximum of $\beta_c^{\rm max} \approx 0.465...$ at $\zeta \approx 2.16258...$, and then goes down very gently, falling 10% of its peak value for $\zeta \approx 6.66...$

The ratio of escape velocity to circular velocity goes to infinity at the center, while at large radii, it goes to $\sqrt{2}$, which is the Keplerian value (Fig. 5).

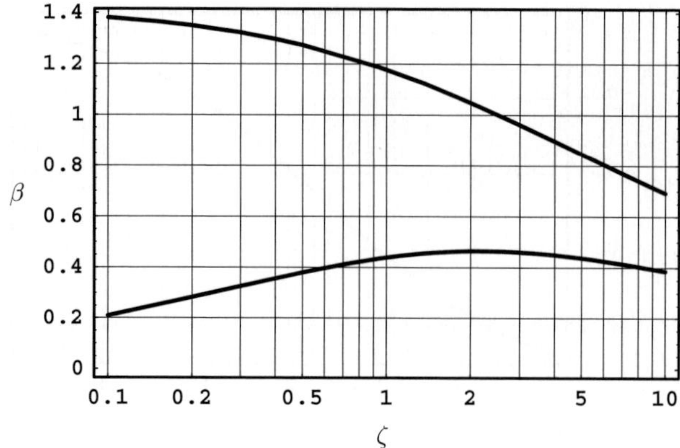

Fig. 5. NFW escape (*upper curve*) and circular (*lower curve*) velocities

3.6 Velocity Dispersion

Up to now, all the properties that we have derived from the NFW profile have not required information about the velocity distribution of the model. Even the escape and circular velocities that we have derived, are not diagnostics of the velocity distribution, but rather characterizations of the potential.

The question thus arises as to the range of variations that are possible in the velocity distribution as a function of position for the NFW profile. This is an important issue, because at least in the case of luminous elliptical galaxies, although there is some homogeneity in the surface brightness profiles and isophotal shapes, there is a wider range of variation in the velocity dispersion profiles. This can be interpreted as changes in the velocity distribution, or in the mass to light ratio. We will explore the first possibility.

One possible approach is to build appropriate dynamical models by finding, by whatever means may be available, the range of phase-space distributions $f(\boldsymbol{r}, \boldsymbol{v})$, which project onto the same $\rho(r)$ when integrated over velocity space.

Another, more limited but simpler approach which is quite useful, is to use Jeans' equations to impose restrictions, not in the velocity distribution but in its moments, in particular in the velocity dispersion. We will assume no net rotation and a velocity distribution that is invariant under rotations. The two components of the tangential velocity dispersion are then equal:

$$\sigma_\theta = \sigma_\phi \ ,$$

and the velocity ellipsoid everywhere can be characterized by its radial velocity dispersion, σ_r, and an anisotropy parameter:

$$\beta \equiv 1 - \sigma_\theta^2 / \sigma_r^2 \ . \tag{15}$$

Notice that β is negative when tangential motions dominate, it goes to 0 for the isotropic case, and can reach up to 1 for the purely radial motion case.

In spherical coordinates, the Jeans equation that corresponds to the 2nd moment of Boltzmann's equation is (7) and (15):

$$\frac{1}{\rho}\frac{\mathrm{d}}{\mathrm{d}r}(\rho\sigma_r^2) + 2\beta\frac{\sigma_r^2}{r} = -\frac{\mathrm{d}\phi}{\mathrm{d}r} \,. \tag{16}$$

Isotropic Case

The first case we will study is the model whose velocity distribution is isotropic everywhere. From the Jeans' equation in spherical coordinates for an isotropic (in velocity space) model, we can obtain the 1-dimensional velocity dispersion as a function of position ((16) with $\beta = 0$):

$$\frac{1}{\rho}\frac{\mathrm{d}}{\mathrm{d}r}(\rho\sigma^2) = -\frac{\mathrm{d}\phi}{\mathrm{d}r} \quad \Longrightarrow \quad \sigma^2(r) = -\frac{1}{\rho(r)}\int_r^\infty \rho(r')\left(\frac{\mathrm{d}\phi}{\mathrm{d}r'}\right)\mathrm{d}r' \,.$$

In dimensionless form, this equation is,

$$\chi_{\rm iso}^2(\zeta) = -\frac{1}{\delta(\zeta)}\int_\zeta^\infty \delta(\zeta')\left(\frac{\mathrm{d}\Psi}{\mathrm{d}\zeta'}\right)\mathrm{d}\zeta' \,,$$

where we have defined the dimensionless velocity dispersion as $\chi_{\rm iso} \equiv \sigma/\sqrt{\phi_o}$.

Using our previously defined dimensionless density and force functions (9) and (12), we can evaluate this expression:

$$\begin{aligned}\chi_{\rm iso}^2(\zeta) &= -\zeta(1+\zeta)^2\int_\zeta^\infty \frac{1}{\zeta'(1+\zeta')^2}\,\frac{\zeta' - (1+\zeta')\log(1+\zeta')}{\zeta'^2(1+\zeta')}\,\mathrm{d}\zeta' \\ &= -\zeta(1+\zeta)^2\int_\zeta^\infty \frac{\zeta' - (1+\zeta')\log(1+\zeta')}{\zeta'^3(1+\zeta')^3}\,\mathrm{d}\zeta' \,.\end{aligned} \tag{17}$$

The integrand is a function, positive everywhere, that diverges as ζ^{-1} at the origin and approaches zero for large radii. We can not integrate it analytically but we can do so numerically (Fig. 6).

The velocity dispersion of the isotropic model goes to zero at the origin and at large radii while it reaches a maximum of $\chi_{\rm iso} \approx 0.30707$ at $\zeta \approx 0.7625$.

The shrinking velocity dispersion at the center is a result of the mild divergence of the density cusp, diverging as $\rho \propto \zeta^{-1}$. We can see this as follows: the equation we used to obtain $\sigma(r)$ can be written as:

$$\rho(r)\,\sigma^2(r) = \int_r^\infty \rho(r')\,F(r')\,\mathrm{d}r' \,,$$

with the left hand side being the local amount of kinetic energy per unit volume, or local pressure. The right hand side is the force per unit volume, integrated on a radial column from the local position all the way to infinity; this is the force per unit area that the local element has to support. Now, in

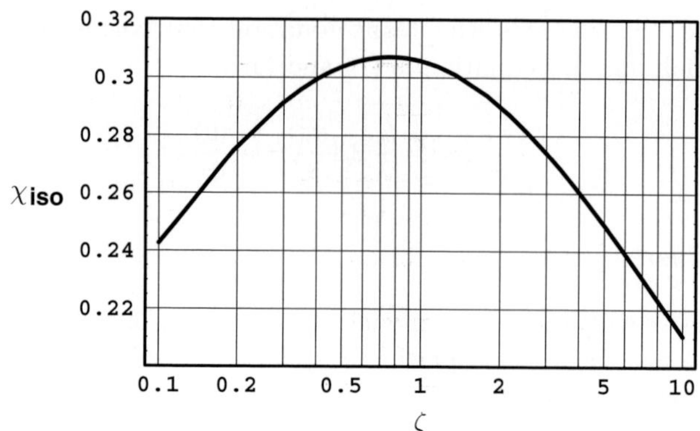

Fig. 6. NFW isotropic velocity dispersion

a spherical distribution of mass, the force goes as the enclosed mass divided by radius squared and the enclosed mass goes as the density times the radius cubed:

$$F(r) \propto \frac{M_r}{r^2} \propto \frac{\rho\, r^3}{r^2} \propto \rho\, r \quad \Longrightarrow \quad \int \rho F\, \mathrm{d}r \propto \rho^2 r^2\,,$$

so the local pressure $\rho\sigma^2$ has to go as $\rho^2 r^2$, or $\sigma^2 \propto \rho r^2$. If we assume that $\rho \propto r^\alpha$, it is clear that as $r \to 0$ we have:

$$\begin{aligned} \alpha > -2 &\implies \sigma^2 \to 0\,, \\ \alpha = -2 &\implies \sigma^2 \to \text{constant}\,, \\ \alpha < -2 &\implies \sigma^2 \to \infty\,, \end{aligned}$$

so a cusp steeper than r^{-2} is required to force a divergent central isotropic velocity dispersion. Another way of looking at this is that you need to pack a lot of mass at the center, so that the resulting gravitational force makes the local velocity dispersion to soar without bound.

Radial Case

As an extreme case, we will now explore the possibility of building an NFW model in which all orbits are radial. This would maximize the observed central velocity dispersion. In principle, one can solve the relevant Jeans' equation for the general case ($\beta \neq 0$) as follows. We begin multiplying both sides of (16) by $\rho\, r^2$:

$$r^2 \frac{\mathrm{d}}{\mathrm{d}r}(\rho\sigma_r^2) + 2\beta r\,(\rho\sigma_r^2) = -\rho\, r^2 \frac{\mathrm{d}\phi}{\mathrm{d}r}\,,$$

our next step is to realize that,

$$\frac{\mathrm{d}}{\mathrm{d}r}(r^2\rho\sigma_r^2) = r^2 \frac{\mathrm{d}}{\mathrm{d}r}(\rho\sigma_r^2) + 2r(\rho\sigma_r^2)\,.$$

The first equation can then be written as,

$$\frac{\mathrm{d}}{\mathrm{d}r}(r^2\rho\sigma_r^2) - 2r(\rho\sigma_r^2)\,(1-\beta) = -\rho r^2\,\frac{\mathrm{d}\phi}{\mathrm{d}r} \quad \Longrightarrow$$
$$\frac{\mathrm{d}}{\mathrm{d}r}(r^2\rho\sigma_r^2) = \rho r^2\left[\frac{2\sigma_r^2}{r}(1-\beta) - \frac{\mathrm{d}\phi}{\mathrm{d}r}\right] ,$$

from which we finally obtain ($\beta = 1$),

$$\sigma_r^2(r) = -\frac{1}{r^2\rho(r)}\int_r^\infty r'^2\rho(r')\frac{\mathrm{d}\phi}{\mathrm{d}r'}\,\mathrm{d}r' \,. \tag{18}$$

Our next step would be to write the dimensionless form of this equation and substitute the appropriate functions for the density and potential of the NFW model. This can be done and, indeed, it gives an answer that diverges strongly at the center. However, it is important to emphasize that, although the Jeans' equation can be solved, the implied solution may not be physical.

In the particular case of purely radial orbits, we should realize that we are putting a strong constraint on the central density: since all orbits are radial, all go through the center, and so the central region must accommodate all particles, although not at the same time. It can be proved that the solution obtained from the Jeans' equation for the purely radial orbit in the NFW case, implies a phase-space distribution function that becomes negative at the center, which something is clearly non-sensical. This is the mathematical way, of the formal solution, to accommodate all particles on radial orbits within a central cusp that does not diverge quickly enough.

We will examine in more detail this question and derive a general lower limit to the rate of divergence that a central cusp must have to accommodate a population of particles on radial orbits. The relation between the phase-space distribution function and the spatial density is:

$$\rho(r) = \int f(\boldsymbol{r},\boldsymbol{v})\,\mathrm{d}^3v \,.$$

In a system that is integrable, the distribution function $f(\boldsymbol{r},\boldsymbol{v})$ should be expressible as a function of the integrals of motion (Jeans' Theorem). Now, if the system is invariant with respect to spatial rotations, then we can use $f = f(E, L^2)$, since the energy E and the magnitude (squared) of the angular momentum L are invariant with respect to rotations. We can then write the density as,

$$\rho(r) = \int f(E,L^2)\,\mathrm{d}^3v = 2\pi\int_{-\infty}^{\infty}\mathrm{d}v_r\int_0^\infty v_t\,\mathrm{d}v_t\;f(v_r^2/2 + v_t^2/2 + \phi,\, r^2v_t^2) \,,$$

where we have separated the integration over velocity space in two parts, one over the radial direction (v_r) and the other over the tangential plane (v_t).

Now, because we are building a model with radial orbits only, the phase-space distribution function can be written as:

$$f(E, L^2) = g(E)\,\delta(L^2) = g(v_r^2/2 + \phi)\,\delta(r^2 v_t^2)\,,$$

where the radial velocity dependence is in the g function and that of the tangential velocity is in the Dirac δ function:

$$\delta(x) = 0\ \forall x \neq 0, \qquad \int \delta(x)\,\mathrm{d}x = 1\,.$$

The density integral can then be split in two factors:

$$\rho(r) = 2\pi \int_{-\infty}^{\infty} g(v_r^2/2 + \phi)\,\mathrm{d}v_r \int_0^{\infty} \delta(r^2 v_t^2) v_t\,\mathrm{d}v_t\,.$$

To integrate on the tangential velocity we change the integration variable: $x = r^2 v_t^2 \quad \Rightarrow \quad \mathrm{d}x = 2r^2 v_t\,\mathrm{d}v_t$, to obtain:

$$\int_0^{\infty} \delta(r^2 v_t^2) v_t\,\mathrm{d}v_t = \frac{1}{2r^2} \int_0^{\infty} \delta(x)\,\mathrm{d}x = \frac{1}{2r^2}\,.$$

Putting this result back in the density integral, we have:

$$\rho(r) = \frac{\pi}{r^2} \int_{-\infty}^{\infty} g(v_r^2/2 + \phi)\,\mathrm{d}v_r \qquad \Longrightarrow \qquad r^2 \rho(r) = \pi \int_{-\infty}^{\infty} g(v_r^2/2 + \phi)\,\mathrm{d}v_r\,.$$

Let's assume that at the center, the density profile behaves as $\rho \propto r^{\alpha}$, then it is clear that $r^2 \rho(r) \to 0$ for $\alpha > -2$, forcing the left hand side of our result to go to zero at the origin. However, the right hand side is an integral over $g(E)$, which is a positive function, and so the only way that this integral can be zero is if $g(E) = 0$, which gives no model! So, we conclude that the only way to build a dynamical model having only radial orbits, is to have a central density cusp that diverges at least as fast as $1/r^2$ at the center. The NFW profile does not satisfy this condition and so no radial orbit model is possible.

Tangential Case

The opposite extreme to a radial orbit model is one with tangential motion only. In a spherical model, this means that all orbits are circular and so no radial mixing exists. Such a model is always possible, since we are free to put as many stars as required by the density profile at each radius. It is easy to see that in this case, the phase-space distribution function is:

$$f(r, v_r, v_t) = \frac{1}{\pi}\,\rho(r)\,\delta(v_r)\,\delta(v_t^2 - v_c^2)\,, \tag{19}$$

where v_c is the local circular velocity (14).

In this case, (7) reduces to $\sigma_t^2 = r(\mathrm{d}\phi/\mathrm{d}r)$ and so, the tangential velocity dispersion is simply the circular velocity,

$$\chi_{\tan}^2(\zeta) = \beta_c^2 \tag{20}$$

A word of caution is appropriate here: just because we can find a solution, does not imply that it is stable. This is particularly critical for models built with circular orbits only.

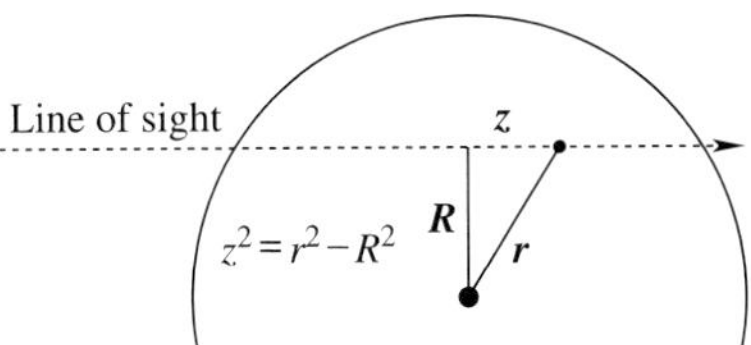

Fig. 7. Projection on the plane of the sky, with r the $3D$ radius, R the projected component on the plane of the sky radius and z the corresponding along the line of sight

3.7 Projected Properties

So far we have obtained $3D$ information about the NFW profile. What is observed, however, is projections on the plane of the sky. So we now proceed to obtain the projected versions of the density and velocity dispersion profiles.

Projected Density

Having computed the velocity dispersion for the isotropic and tangential versions of the NFW profile, we can now compute the line of sight velocity dispersion. Our first step is to obtain the projected density as an integral over the line of sight:

$$\Sigma(R) = \int_{-\infty}^{\infty} \rho(r)\,\mathrm{d}z = 2\int_{R}^{\infty} \rho(r)\frac{r\,\mathrm{d}r}{\sqrt{r^2 - R^2}} ,$$

where R is the projected distance on the plane of the sky and z is along the line of sight (see Fig. 7).

Defining a dimensionless projected radius as $\eta \equiv R/r_o$, and using our dimensionless functions, we cast the previous equation in dimensionless form:

$$\Sigma(r) = 2\int_{\eta}^{\infty} \frac{\rho_o\delta(\zeta)\,r_o\zeta}{r_o\sqrt{\zeta^2 - \eta^2}} r_o\,\mathrm{d}\zeta \;=\; 2\rho_o r_o \int_{\eta}^{\infty} \frac{\delta(\zeta)\zeta}{\sqrt{\zeta^2 - \eta^2}}\,\mathrm{d}\zeta \,.$$

We can now define a dimensionless projected density as,

$$\Gamma(\eta) \equiv \Sigma(R)/(2\rho_o r_o) \,.$$

In the case of the NFW profile, we get (Fig. 8):

$$\Gamma(\eta) = \int_{\eta}^{\infty} \frac{\mathrm{d}\zeta}{(1+\zeta)^2\sqrt{\zeta^2 - \eta^2}} \;=\; \frac{1}{\eta^2 - 1} - \frac{\operatorname{arcsec}(\eta)}{(\eta^2 - 1)^{3/2}} \,. \tag{21}$$

Caution should be used for $\eta < 1$, where numerator and denominator in the second term are imaginary, but the result is real and finite. In this range, it is preferable to use an equivalent form with only real values:[1]

[1] To get the second form we use $\operatorname{arcsec}(z) = -i\,\ln[(1 + \sqrt{1 - z^2})/z]$.

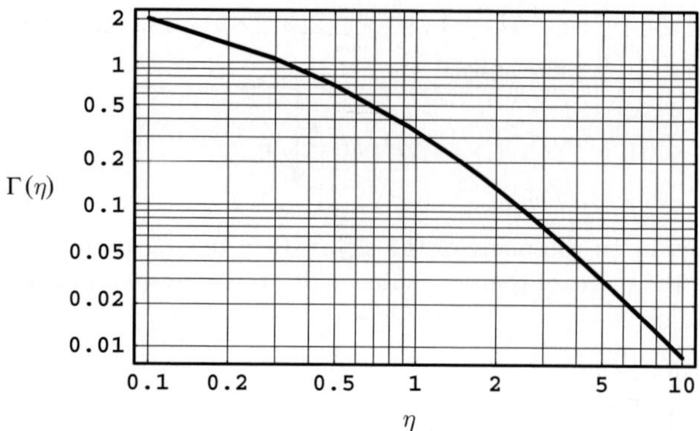

Fig. 8. NFW projected density

$$\Gamma(\eta) = \frac{1}{\eta^2 - 1} + \frac{1}{(1-\eta^2)^{3/2}} \ln\left(\frac{1+\sqrt{1-\eta^2}}{\eta}\right), \quad 0 \le \eta \le 1 . \qquad (22)$$

We also note that as $\eta \to 1$, $\Gamma \to 1/3$. The projected density has the following limits:

$$\lim_{\eta \to 0} \Gamma(\eta) = \infty, \qquad \lim_{\eta \to \infty} \Gamma(\eta) = 0 .$$

Projected Velocity Dispersion

We can now compute the projected velocity dispersion for the isotropic and the purely tangential orbits cases. This can be done using the following expression:

$$\sigma_p^2(R) = \frac{2}{\Sigma(R)} \int_R^\infty \frac{\rho(r)\, \sigma_{\rm los}^2(r,R)\, r}{\sqrt{r^2 - R^2}}\, {\rm d}r ,$$

where $\sigma_{\rm los}(r,R)$ is the line of sight velocity dispersion, on a volume element at distance r from the center along the line of sight, at projected distance R. Elementary geometry (Fig. 9) shows that it can be written as,

$$\sigma_{\rm los}^2(r,R) = \left(1 - \frac{R^2}{r^2}\right) \sigma_r^2(r) + \left(\frac{R^2}{2r^2}\right) \sigma_t^2(r) ,$$

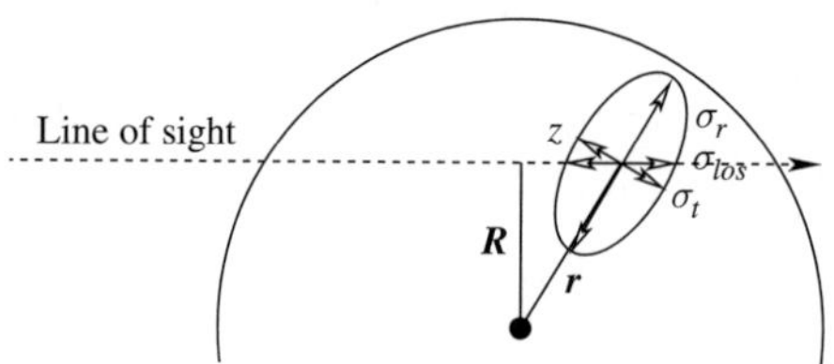

Fig. 9. Projection along the line of sight $\sigma_{\rm los}$ of the local radial σ_r and tangential σ_t velocity dispersions

where $\sigma_r(r)$ and $\sigma_t(r)$ are the radial and tangential velocity dispersions at r.

We see that in the isotropic case,

$$\sigma_t^2 = 2\sigma_r^2 \quad \Longrightarrow \quad \sigma_{\rm los}^2(r,R) = \sigma_r^2 \ .$$

The projected squared velocity dispersion in the isotropic case is then,

$$\sigma_{\rm p-iso}^2(R) = \frac{2}{\Sigma(R)} \int_R^\infty \frac{\rho(r)\,\sigma_r^2(r)\,r}{\sqrt{r^2-R^2}}\,{\rm d}r \ ,$$

or in dimensionless form,

$$\chi_{\rm p-iso}^2(\eta) = \frac{1}{\Gamma(\eta)} \int_\eta^\infty \frac{\delta(\zeta)\,\chi_{\rm iso}^2(\zeta)\,\zeta}{\sqrt{\zeta^2-\eta^2}}\,{\rm d}\zeta \ , \tag{23}$$

where $\chi_{\rm iso}^2$ is given by (17).

In the tangential case we have,

$$\sigma_{\rm p-tan}^2(R) = \frac{2}{\Sigma(R)} \int_R^\infty \frac{\rho(r)\,(R^2/2r^2)\,\sigma_t^2(r)\,r}{\sqrt{r^2-R^2}}\,{\rm d}r \ ,$$

and in dimensionless form,

$$\chi_{\rm p-tan}^2(\eta) = \frac{\eta^2}{2\Gamma(\eta)} \int_\eta^\infty \frac{\delta(\zeta)\,\chi_{\rm tan}^2(\zeta)\,\zeta}{\zeta\sqrt{\zeta^2-\eta^2}}\,{\rm d}\zeta \ , \tag{24}$$

where $\chi_{\rm tan}^2$ is given by (20).

In both cases the projected squared velocity dispersion rises from the center to a maximum at $\eta \sim 0.6$ (isotropic case) or $\eta \sim 2.5$ (tangential case) and then decreases steadily for larger distances (Fig. 10).

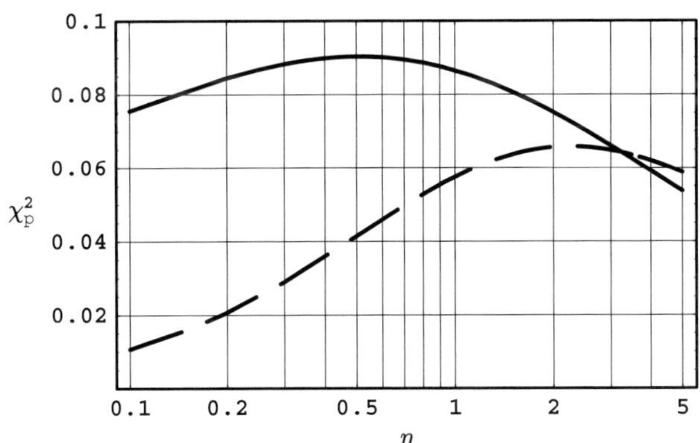

Fig. 10. NFW projected squared velocity dispersion for the isotropic (*solid*) and tangential (*dashed*) models

4 Other Interesting Profiles

There are two very useful, but still simple profiles, that have been used extensively. These are the Jaffe ([33, 37]) and the Hernquist ([6, 28]) profiles:

$$\rho_J(r) = (M/4\pi r_o^3)\,(r/r_o)^{-2}\,(1 + r/r_o)^{-2}\,, \tag{25}$$

$$\rho_H(r) = (M/2\pi r_o^3)\,(r/r_o)^{-1}\,(1 + r/r_o)^{-3}\,. \tag{26}$$

Here M is the total mass and r_o is a scale-length whose physical meaning is different in each case. Notice that these two profiles behave as $\rho \propto r^{-4}$ at large radii, as opposed to the shallower NFW profile. This behaviour has been found in N-body simulations of perturbed galaxies when they are not part of a cosmological expansion [4], and has also been observed in some real elliptical galaxies [35].

What is the reason for a $\rho \propto r^{-4}$ profile at large radii? It is clear that any finite mass, power-law profile, should be of the form $\rho \propto 1/r^{3+\epsilon}$, with $\epsilon > 0$ at large radius, however, why the particular -4 exponent?

The distribution function of an isolated, steady-state, galaxy is such that the number of stars $N(E)$ with energy between E and $E + \mathrm{d}E$ goes to zero at $E = 0$, where the energy boundary of the system lies. However, when a galaxy suffers an external perturbation, like the tidal force of a passing galaxy, the external layers of the perturbed galaxy heat up and a continuous non-zero distribution of stars at the zero energy boundary develops. We will show that this ensures a $\rho \propto r^{-4}$ tail at large galacto-centric distances (Fig. 11).

Proposition 1. *If a spherical galaxy with finite mass, no rotation and isotropic velocity distribution, develops a finite, non-zero population of particles at* $E = 0$*, then the tail of the density profile at large radii will exhibit a* $\rho \propto r^{-4}$ *behaviour.*

Proof. The energy distribution is given by,[2]

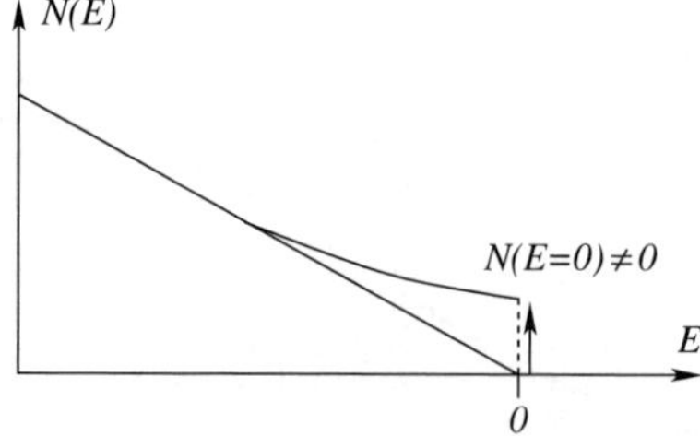

Fig. 11. Initial (*lower*) and perturbed (*upper*) energy distribution

[2] It is important to distinguish between $f(E)$ and $N(E)$. The former is the fraction of stars per unit phase-space volume, $\mathrm{d}^3r\mathrm{d}^3v$, while the latter is the fraction of stars per unit interval in E. f may be very large at some energy, but if the corresponding volume between E and $E + \mathrm{d}E$ is tiny, N will be small.

$$N(E)\mathrm{d}E = \int_{\Omega_E} f(\boldsymbol{r},\boldsymbol{v})\,\mathrm{d}^3r\,\mathrm{d}^3v \;=\; 16\pi^2 \int_{\Omega_E} f(E)\,r^2\mathrm{d}r\,v^2\mathrm{d}v\;, \tag{27}$$

where Ω_E is the volume in phase-space with energy between E and $E+\mathrm{d}E$ (Fig. 12), and we have used the isotropy of f in $\boldsymbol{r}$ and $\boldsymbol{v}$.

As f is a function of E only, it is convenient to change the integral to

$$N(E)\mathrm{d}E = 16\pi^2 \int_{\Omega_E} f(E)\,r^2v^2\,\frac{\partial(r,v)}{\partial(r,E)}\,\mathrm{d}r\mathrm{d}E\;.$$

Since the Jacobian is equal to $1/v$, we get

$$N(E)\mathrm{d}E = 16\pi^2\,f(E)\mathrm{d}E \int_0^{r_E} r^2v\,\mathrm{d}r \;=\; f(E)\,A(E)\,\mathrm{d}E\;,$$

where we have taken $f(E)$ out of the integral, since it is done at fixed energy, r_E is the largest radius that a particle of energy E can reach, and

$$A(E) \equiv 16\pi^2 \int_0^{r_E} r^2\,\sqrt{2[E-\phi(r)]}\,\mathrm{d}r\;,$$

where we have used $v=\sqrt{2[E-\phi(r)]}$. $A(E)$ is the "area" of the constant energy surface in phase-space [10].

Now, at sufficiently large radii, any spherical finite mass distribution has a potential that approaches the Keplerian limit $\phi\propto 1/r$. Using this asymptotic dependence we get,

$$A(E)\propto r^2\,r^{-1/2}\,r \;=\; r^{5/2} \;\propto\; E^{-5/2}\;.$$

Then, if $N(E\sim 0)\mathrm{d}E = f(E\sim 0)A(E\sim 0)\mathrm{d}E$, is non–zero and finite, we should have $f(E\sim 0)\propto E^{5/2}$. The density can be written as,

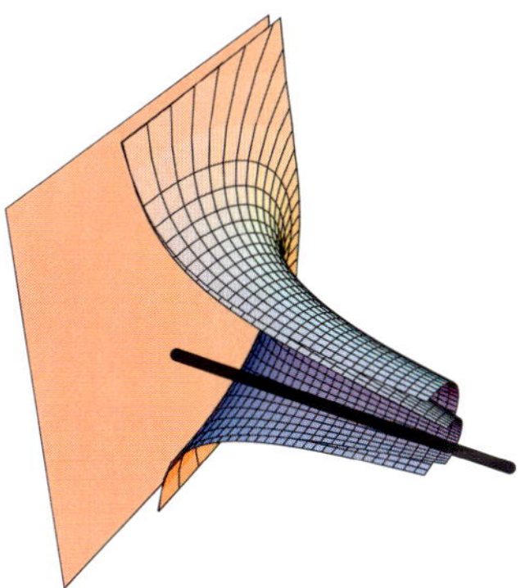

Fig. 12. Two constant energy surfaces in phase-space for the Kepler potential. The thick axis is the radial distance and the orthogonal plane is the velocity space. The volume in between the surfaces is Ω_E. They have been cut open for clarity

$$\rho(r) = 4\pi \int_{\phi(r)}^{0} f(E)\, \sqrt{2[E - \phi(r)]}\, \mathrm{d}E$$

$$\Longrightarrow \quad \rho \propto E^{5/2} E^{1/2} E \;=\; E^4 \;\propto\; r^{-4} \; \square$$

This result does not apply to the NFW profile, whose mass diverges.

5 The Orbital Structure of Spherical Potentials

We now study the orbital structure of spherical mass models. Why should we be interested in the orbital structure of a model? There are interesting problems where it is important to find out, for instance, the radial region spanned by individual orbits: we may be interested in the fraction of stars in a galactic bulge that plunge within the radius of influence of a central massive black hole, or the effect of radial mixing in galactic metallicity gradients.

There is another, more fundamental, reason for studying orbits. At the beginning of the XX century, the mathematician Emmy Noether proved a result that implies that, when a potential presents a symmetry (i.e., its functional form is unchanged by a spatial and/or temporal transformation), there is a corresponding physical quantity that is conserved, when moving along the orbits supported by the potential. For instance, energy is conserved for orbits in potentials that are time-invariant, linear momentum is conserved when the potential is invariant under a spatial translation and angular momentum is conserved when we have rotational invariance.

Now, in Sect. 2 we introduced the concept of integrals of motion and saw that the distribution function is a function of them. Clearly, the conserved physical quantities in Noether's result are the integrals of motion of the collisionless Boltzmann equation (in fact, energy, and linear and angular momentum are the so-called classical integrals of motion). Since orbits are the set of points in phase-space where the integrals keep a constant value, the distribution function must depend on the orbital structure of the potential. Indeed, for steady-state, collisionless systems, orbits are the basic bricks used to assemble them in phase-space.

5.1 A Phase-Space Portrait of Orbits

Each integral of motion defines a hyper-surface in phase-space, and orbits move along the intersection of all of them. A system with N degrees of freedom has a $2N$-dimensional phase-space. Each integral of motion lowers by one the dimensions of the allowed region, and thus a system with M integrals has orbits restricted to a region of dimension $2N$–M. Figure 13 shows this for the Kepler potential. Since the direction of the angular momentum is fixed, the motion is restricted to a plane in configuration space and we only need 2 spatial coordinates to describe the motion (say polar coordinates r, θ). Since we can

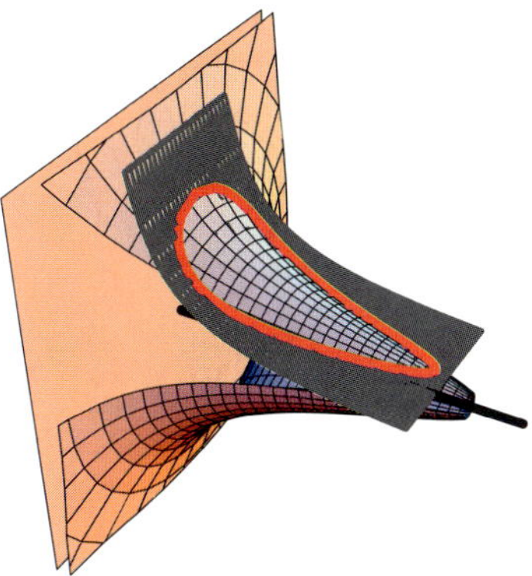

Fig. 13. Two integral of motion surfaces in phase-space. The view is the same as in Fig. 12, with the tangential velocity axis running vertically. The funnel is the constant energy surface (drawn open for clarity). The folded plane is the constant angular momentum surface. The orbit lies at the intersection (*thick line*)

not draw the 4-dimensional phase-space, one coordinate must be sacrificed. Since the potential is symmetric in θ, we drop it from the figure, knowing that whatever we get, must be wrapped around in θ to obtain the full picture.

The intersection of the funnel-like energy surface and the bent plane angular momentum surface defines a loop in Fig. 13. It is clear that at fixed energy, if we increase angular momentum, the bent plane moves upward and the loop shrinks to a point that corresponds to the maximum angular momentum orbit: the circular orbit.

Figure 13 describes orbits in any spherical potential, since E and L are always conserved in them (using a different potential only changes details). To get the orbit in configuration space, we must introduce the missing θ direction sacrificing one of the velocity axes. This is shown in Fig. 14, where the base plane is configuration space and the vertical axis is the radial velocity. Wrapping the loop in θ results in a torus in phase-space: the *invariant orbital torus*. Its projection in configuration space gives a rosette limited by two circles whose radii are the periapsis and apoapsis.[3] It can be shown that all regular orbits move on orbital torii in phase-space, although their form, and corresponding projection, can be very complicated.

5.2 The Lindblad Diagram

Since energy and angular momentum define the shape of orbits, in spherical potentials, we can use them as labels to catalogue them. This is precisely the idea behind a diagram first used by B. Lindblad in 1933.

[3] In Celestial Mechanics, the apsis of an orbit is the point of maximum or minimum distance from the center of attraction. Periapsis is the minimum distance point while apoapsis is the maximum distance point. In this and the next section, we will also use these terms as proxies for the actual distance at these points, as no generally accepted term exists for them.

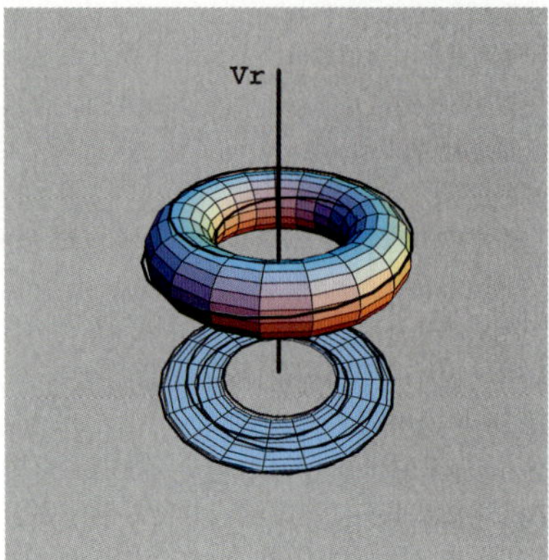

Fig. 14. Invariant orbital torus. The base plane is the configuration space while the vertical axis is the radial velocity. The orbit wraps around the torus while the projection in ordinary space traces a rosette

A Lindblad diagram is like a chart that allows us to see in one glance, the orbital make up of any spherical model. In this section we will illustrate its use by building the Lindblad diagram of a simple spherical model with a flat rotation curve: the singular, truncated logarithmic potential. Its density profile, potential function and circular velocity curve are given by,

$$\rho(r) = \begin{cases} v_o^2/4\pi G r^2, & r < r_t \\ 0, & r \geq r_t \end{cases} \tag{28}$$

$$\phi(r) = \begin{cases} -v_o^2\,[1 - \log(r/r_t)], & r < r_t \\ -v_o^2\,(r_t/r), & r \geq r_t \end{cases} \tag{29}$$

$$v_c^2(r) = \begin{cases} v_o^2, & r < r_t \\ v_o^2\,(r_t/r), & r \geq r_t \end{cases} \tag{30}$$

where v_o is the constant circular velocity within the truncation radius r_t.

Since at a fixed energy, orbits can range in angular momentum from the radial to the circular orbit (see 5.1) our first job is to find the locus of circular orbits in the Lindblad Diagram, where E will be plotted on the horizontal axis and L on the vertical one. All possible orbits will lie beneath this curve. The energy and angular momentum of a circular orbit of radius r_c is given by,

$$E_c = \phi(r_c) + v_c^2/2, \quad L_c = r_c v_c \ ,$$

where v_c is the local circular velocity. To find E_c *vs.* L_c, we eliminate r_c from these equations and substitute the potential function. This gives,

$$E_c = \begin{cases} -v_o^2\,[(1/2) - \log(L_c/r_t v_o)], & r < r_t \\ -(1/2)\,(r_t v_o^2/L_c)^2, & r \geq r_t \end{cases}$$

It is convenient to have the inverse relation as well,

$$L_c = \begin{cases} r_t v_o \exp[(E_c/v_o^2) + (1/2)], & r < r_t \\ (r_t v_o^2/2\sqrt{|E_c|}), & r \geq r_t \,. \end{cases}$$

We can now plot the locus of circular orbits in the Lindblad diagram (Fig. 15). The axes are shown in dimensionless units. The lower horizontal axis shows the radial position at which the potential function is equal to the energy in the upper axis. The envelope of circular orbits goes from $L_c \to 0$ at the infinitely deep center of the potential well, to $L_c \to \infty$ at infinite distance ($E \to 0$), denoted by the vertical dotted line.

We now introduce the concept of the *characteristic parabola*. Let us take a spherical shell of radius r_o and consider all orbits that touch it, but do not cross it (Fig. 16). The condition for this to happen is ($v_r = 0$ at $r = r_o$),

$$E_* = \phi(r_o) + v_t^2/2 = \phi(r_o) + (L_*^2/2r_o^2) \,, \tag{31}$$

where E_* and L_* are the energy and angular momentum of these orbits. This is the equation of a parabola in the Lindblad diagram that opens up to the right and crosses the E axis at $\phi(r_o)$.

Figure 15 shows one such characteristic parabola. Point A is the radial orbit that just reaches $r = r_o$, before plunging back to the center. Point B is the circular orbit at this radius and point C is the parabolic orbit that comes

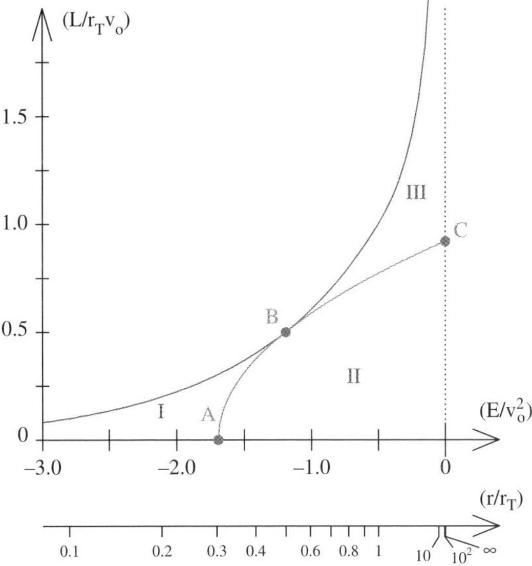

Fig. 15. Lindblad diagram for the singular, truncated, logarithmic potential. The thick upper envelope is the locus of circular orbits. The thin lower curve that goes through points A, B and C is a characteristic parabola

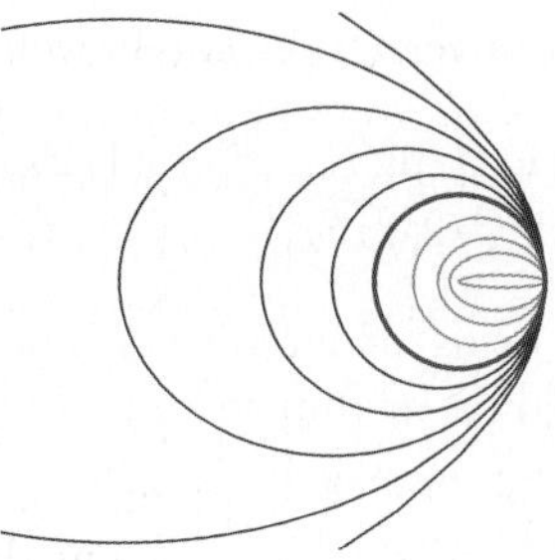

Fig. 16. Orbits in Kepler potential that lie along the characteristic parabola that corresponds to the thick red circle

from infinity and reaches the r_o radius before going back to infinity. Points along the characteristic parabola between A and B are orbits that share r_o as their apoapsis (green orbits in Fig. 16), while points between B and C are orbits that share r_o as their periapsis (blue orbits in same figure).

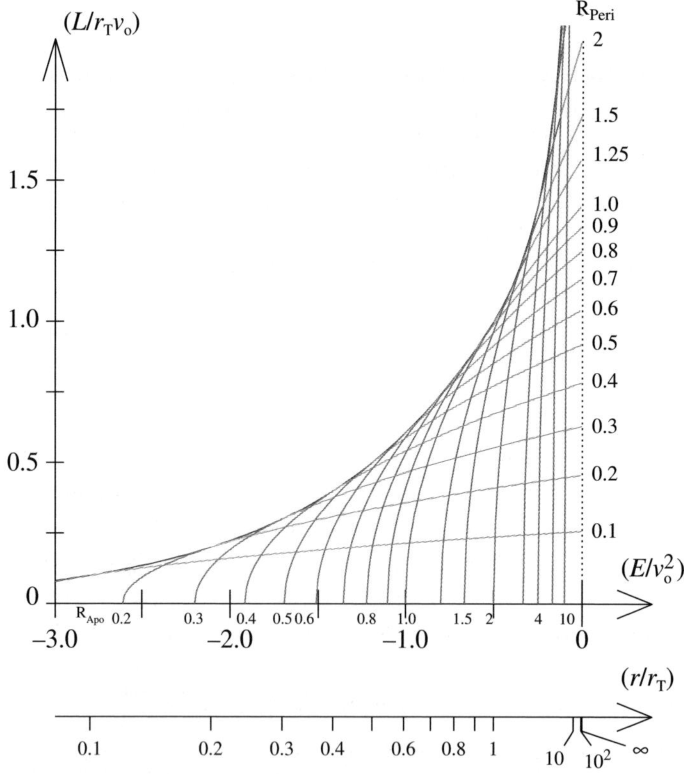

Fig. 17. Lindblad diagram with grid of iso-periapsis (*blue*) and iso-apoapsis (*green*) extrema

A characteristic parabola splits the allowed region of the Lindblad diagram in three regions (Fig. 15). Region I contains orbits enclosed completely by the spherical shell of radius r_o, while region III contains those that are always outside this shell. Region II then is where the orbits that cross this shell lie.

If we draw characteristic parabola for several radial distances in the Lindblad diagram (Fig. 17), we produce a grid of iso-apoapsis with the segments of the parabola between the E axis and the point where they touch the circular orbit envelope, and iso-periapsis for the segments beyond to the $E = 0$ boundary. Each point in the Lindblad diagram, below the circular orbit envelope and to the left of the zero energy boundary, represents a unique bound orbit.[4] The two unique characteristic parabola that go through it, define its radial extrema. The Lindblad diagram thus provides us with a unique and complete catalogue for all orbits in a spherical potential, arranged according to orbital characteristics. From it, we can easily figure out the radial range of individual orbits. Furthermore, if we actually have the form of the distribution function as a function of E and L, we can compute from this diagram the fraction of the model that shares some particular orbital characteristics (for this we need to get from $f(E, L)$ to $N(E, L)$ in a way analogous to (27).

6 A Sticky Story: Dynamical Friction

When a massive object moves in a sea of background particles, the gravitational force of the former stirs the latter; since the energy invested in producing the stirring must come from somewhere, the impinging object losses kinetic energy. This is a classic tale of pumping ordered motion energy into a thermal bath.[5] A classic cartoon model depicts an overdense wake in the background, trailing the moving object and decelerating it with its own gravity.

6.1 Chandrasekhar's Formula

S. Chandrasekhar [19] thought about this in 1942 and came with a very famous formula that describes the situation when a massive point of mass M, moves along a straight path with velocity $\boldsymbol{v}$, within a uniform background of non self-interacting and uncorrelated point particles, of individual mass m and density ρ_f, with zero mean motion and velocity distribution given by $f(v_f)$,

$$(\mathrm{d}\boldsymbol{v}/\mathrm{d}t) = -16\pi^2 G^2 \ln(\Lambda)\, \rho_f (m + M) \int_0^v f(v_f) v_f^2 \mathrm{d}v_f\, (\boldsymbol{v}/v^3)\,. \tag{32}$$

The Λ is the so-called *Coulomb term*[6] and it is given by,

---

[4] Except for the spatial orientation of the orbital plane, line of apses and sense of rotation.

[5] If Maxwell's demon worked here, we would talk about dynamical acceleration!

[6] The name comes from Plasma Physics.

$$\Lambda = p_{\max} v^2 / G(m + M) , \tag{33}$$

where $p_{\max}$ is formally the maximum impact parameter.

It is important to take a second look at the paragraph that precedes (32). There are a lot of caveats attached to Chandrasekhar's result. We emphasize this, because this equation has been used, and abused, under many different physical situations that clearly invalidate those caveats, and yet, it has been found to provide an adequate description on many different situations, provided some details are taken into account.

There is a rich literature in this subject, we will just mention a few examples. Tremaine and Weinberg [51], have reviewed the derivation of the dynamical friction formula in the context of a satellite in a bound orbit within a spherical system. The key question here is that, in contrast to the infinite background medium used by Chandrasekhar, background particles in resonant orbits with the satellite play a very prominent role in slowing down the satellite and can lead to unusual effects. If the satellite's orbit decays rapidly enough, however, Chandrasekhar's formula remains approximately valid. Bontekoe and Van Albada [14], as well as Zaristky and White [55], debated about the global response that a satellite in a bound orbit produces in the halo of the host galaxy and its consequences upon dynamical friction. They concluded that the purely local description of the effect, as given by (32), is adequate. Cora et al. [17] worried about the effect of chaotic orbits for the background particles, and came to a similar conclusion.

Since dynamical friction is so important in the accretion process of a small system into a larger one, we will examine it in some detail. There are important issues here: How fast is the orbital decay? Does the orbit becomes circularized? Can the accreting system make it all the way to the center of the host?

We begin with the Coulomb term. What exactly is $p_{\max}$? In deriving (32), an integration over impact parameters must be performed. Unfortunately, however, such an integration diverges at the upper end. Chandrasekhar [18] argued that an upper cut-off should be used, since distant collisions are not isolated binary encounters, as modeled in his derivation. He used the mean particle separation, while other authors have advocated "the distance where the average density significantly drops", some others, the radial distance of the spiraling system, and yet others have used the size of the host system. What should one do? Fortunately Λ enters into the dynamical friction formula through its logarithm and this waters down our ignorance about it. In fact, a fractional error δ in Λ, translates into a fractional error $\delta/\ln(\Lambda)$ in the computed deceleration. Let us compute a specific example: (33) in astronomical units can be written as,

$$\Lambda = 2.32\,(p_{\max}/\mathrm{kpc})\,(v/10^2\,\mathrm{km/s})^2\,/\,(M/10^9\,M_\odot)\ .$$

For a $10^9\,M_\odot$ satellite moving at 300 km/s, the logarithm of the Coulomb factor varies only from 5.3 to 7.6, if we take 10 or 100 kpc for $p_{\max}$, respectively. So you can plug your favorite value for $p_{\max}$ and compute the dynamical friction deceleration without worrying much for Λ (some authors have turned the problem around and measured Λ from N-body simulations [14]).

The next thing to notice in (32) is the $\rho_f(m + M)$ dependence. Although both, the mass of the object being slowed down and that of the individual background particles appear, it is usually the case the the former is much bigger than the latter. In this limit, we can forget about m and just say that the deceleration is directly proportional to the mass of the massive object and the background density. If a globular cluster decays in the galactic halo, it really does not matter whether the halo is made up of subatomic particles, brown dwarfs, or stellar remnants.

The final actor, in the dynamical friction play, is the integral over the velocity distribution of the background particles. There are a couple of things that must be noted from the outset: Firstly, dynamical friction tries to bring massive objects to a stand-still *with respect to the velocity centroid of the background particles.* This can have important consequences. Secondly, the integral limit only goes up to the massive object speed: It is only the slower background particles that contribute to dynamical friction (faster particles can overtake the massive object and do not necessarily contribute to the overdense wake).

The shape of the velocity distribution function depends on the details of the background model, however, the overall shape is pretty much like that of a Gaussian distribution: it peaks at the origin and then drops smoothly for increasing velocity. Assuming a Gaussian of dispersion σ, we get Fig. 18 for the velocity part of (32). And here is where the split personality of dynamical friction arises: in the low velocity regime ($v \ll \sigma$), f is approximately constant and the integral goes as v^3; plugging this into Chandrasekhar's formula we get a deceleration that goes as v, just like in Stoke's formula for the drag on a solid object moving inside a viscous fluid. However, in the high velocity regime ($v \gg \sigma$), f drops precipitously and the integral is approximately constant. This leads to a deceleration that goes as $1/v^2$, opposite to your run-of-the-mill

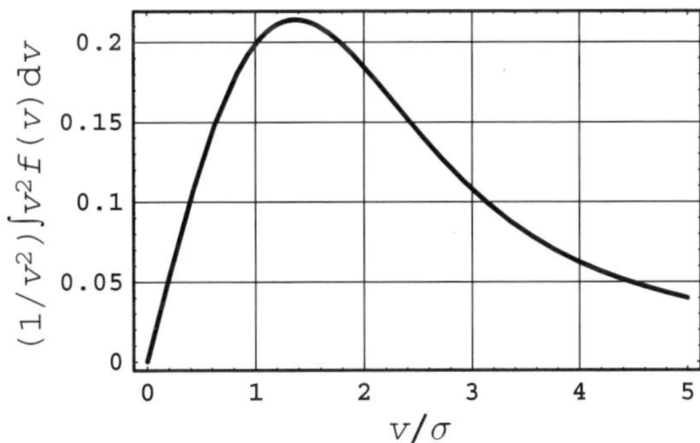

Fig. 18. Velocity dependence of dynamical friction assuming a Gaussian distribution of velocity dispersion σ for the background particles

frictional force. So dynamical friction has bipolar disorder, which makes it work at low speeds and it becomes more efficient the faster you try to go; but if you are going too fast, the faster you go, the less efficient it becomes.[7] This is a characteristic that is important to consider, as we examine now.

6.2 Dynamical Friction in an NFW Profile

An elongated orbit goes fastest near its periapsis and slowest near its apoapsis. Indeed, it needs to go faster that the local circular velocity at periapsis to climb up from that point. Similarly, it goes slow with respect to the local circular velocity at apoapsis, to fall down from this point. If the whole span in velocity of an elongated orbit is such that it remains to the left of the peak in Fig. 18, dynamical friction will be more efficient at periapsis and the orbital eccentricity will increase. If on the contrary, it remains to the right, the eccentricity will decrease. Orbits that straddle the peak can go either way.

Let us examine the situation for the NFW profile that we explored in Sect. 3. Figure 19 shows the escape and circular velocities, normalized to the velocity dispersion for the isotropic model. If a satellite spirals, following an orbit close to circular, it will move from right to left along the circular velocity curve. We can see that as its orbit shrinks from far away, dynamical friction behaves inversely with velocity and its efficiency increases as v_c/σ steadily approaches peak efficiency at $\zeta \sim 0.771...$; past this point, dynamical friction enters the Stoke's regime, but its efficiency shrinks to zero as v_c/σ goes to zero at the center.

If we now have a satellite that plunges to its periapsis along a parabolic orbit, it will follow the escape velocity line in Fig. 19. In this case, dynamical friction will remain in the regime that scales inversely with velocity. Since v/σ

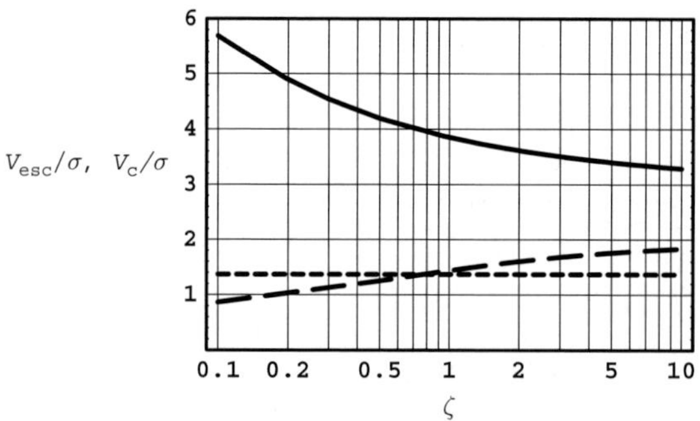

Fig. 19. Escape (*solid line*) and circular (*long dashed line*) velocities, normalized to the local isotropic velocity dispersion, for a spherical NFW model. The constant short dashed line at $v/\sigma = 1.3688$, is where the peak of Fig. 18 occurs

[7] If you are going very fast, the overdense wake trails far behind. The faster you go, the more distant it becomes and thus its retarding effect diminishes.

is very high at all times, dynamical friction is very inefficient. Furthermore, the deeper the satellite penetrates on its first pass, the less drag it will experience.

A satellite on a bound, but elongated orbit, will move between the escape and the circular velocity curves in Fig. 19. For $\zeta > 0.771$, dynamical friction will work in the inverse velocity regime, where orbits are circularized. Past this point, the situation gets murky.

There is, however, a big caveat we must make before leaving this subject. In this section we have only considered the effect of velocity in Chandrasekhar's dynamical friction formula. The variation of the other parameters that enter must be accounted for, before any conclusion is reached. The variation in background density, in particular, is important. We also have that as the satellite moves in, it gets peeled layer by layer by the tidal force and so its mass also changes. We will say more about this latter effect in the next section.

Before leaving the subject of dynamical friction, we want to stress the very important role that (32) has had, and continues to have, in our understanding of stellar and galactic dynamics. The original Chandrasekhar article [19] was chosen to be part of a collection of fundamental papers published during the XX century, and reprinted as the centennial volume of *ApJ* [1]. We encourage the reader to take a look at F. Shu's commentary on Chandrasekhar's contribution [45].

7 The Effect of Tides

We will now take a look at a subject very familiar to sailors and careless galaxies that venture too close to others: the tidal force. The idea behind it, is very simple: an extended object moving within an external gravitational field, will experience a changing force across its body. If we sit at its center of mass, we will see a differential force acting on the rest of the body, and it is up to the self-gravity of the object to keep itself together. From this cartoon model it is apparent that the tidal force depends on two factors: the rate of change of the underlying force and the size of the object:[8] $F_{\mathrm{tid}} \propto (\mathrm{d}F/\mathrm{d}r)\delta r$.

In a Kepler potential the force decays with distance, this produces tidal stretching along the radial direction. It is usually assumed that tides always produce radial stretching, however, it all depends on the sign of the force gradient. Let us assume a spherical mass model with a power-law density profile: $\rho \propto r^{\alpha}$. In this case, the enclosed mass within radius r goes as $M_r \propto r^{3+\alpha}$. The gravitational force then goes as $F \propto M_r/r^2 \propto r^{1+\alpha}$. It is clear that for $\alpha < -1$ we have the usual radial stretching. However, $\alpha > -1$ produces radial squeezing and $\alpha = -1$ produces no radial deformation!

[8] The gravitational force is the vector field that results from applying the $-\partial/\partial x_i$ operator to the potential scalar field. The tidal force is described by the 2nd rank tensor that results from applying the $-\partial^2/\partial x_i \partial x_j$ operator to the potential. This results in a force that may produce distortions in all directions. For simplicity, we focus here on the radial direction only.

Can this happen in astronomical objects? Well, a constant density core, like that of a King model, produces radial squeezing (can this help star formation at the center of flat-core galaxies?). The NFW spherical model has a cusp ($\alpha = -1$) that produces no radial deformation, while its outer part ($\alpha = -3$) produces the usual radial stretching.

7.1 The Tidal Radius

An important concept is the so called *tidal radius*, usually defined as the radius beyond which, a test particle in orbit around a satellite becomes unbound to it and flies apart within the gravitational force field of the larger system. Direct evidence for the existence of such radius is provided by the sharp cut-off in the density profiles of globular clusters in our galaxy and of satellite galaxies, like M32, trapped deep within the potential well of a larger system.

How do we compute the tidal radius? One possibility is to compute the point at which the tidal force and the self-gravity of the object are equal. If we assume point masses for the tide-producing object (M) and tide-distorted object (m), we get (Fig. 20):

$$\left.\begin{array}{l} F_g = -\frac{Gm}{r^2} \\ F_t = \frac{\mathrm{d}}{\mathrm{d}R}\left(-\frac{GM}{R^2}\right) r = \left(\frac{GM}{R^3}\right) r \end{array}\right\} F_t = -F_g \implies 2\,(r_R/R)^3 = m/M\,. \quad (34)$$

r_R is called the *Roche limit*[9] and is the largest radius that a self-gravitating object can have when immersed within an external gravitational force.

Now, is this the tidal radius? We said that it defines the region beyond which it is not possible to have bound orbits around m; isn't this r_R?, well no. The Roche limit is a stationary concept, no information whatsoever about the motion of M and m is used in its derivation.

If we consider m and M to be in circular orbit around their center of mass, besides their combined forces, we must include a centrifugal term that arises in the co-rotating frame. The energy defined in the this frame, using the combined potential functions of both objects, together with that of the centrifugal term, is conserved. This is the *Jacobi energy*. The combined potential produces a series of spherical equipotentials that surround each object

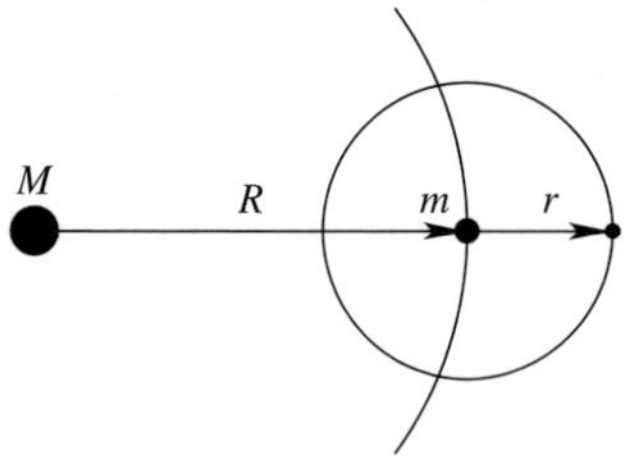

Fig. 20. Geometry for computation of tidal radius

[9] First derived by Éduard Roche in 1848.

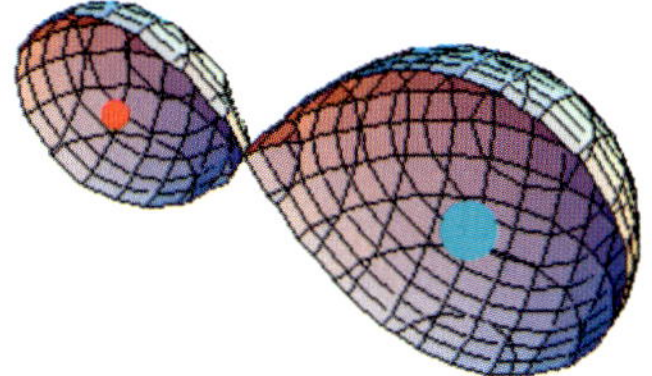

Fig. 21. Roche lobes for two point masses in circular orbit

close to them. As we move away, the equipotentials become elongated until we finally arrive at the particular equipotential that goes through a saddle point, where both lobes touch each other[10] (Fig. 21). Any higher energy equipotential surrounds both masses. Since this particular equipotential is the least bound that encloses separately either object, no particle with Jacobi energy less than it can escape.

The distance from either point mass to the saddle point is its so-called *Roche lobe radius* (it is also called the *Hill radius*, since George Hill discovered it independently of Roche). Its magnitude is given by,

$$3\,(r_H/R)^3 = m/M\ . \tag{35}$$

Comparing with (34), we see that the Roche limit and Hill radius have exactly the same functional form and differ only by a factor of 2/3. The important thing to consider is that both can be cast as an argument that relates densities: the average density of M, within the orbit of m, and the average density of m within its limiting radius:[11] $\langle\rho_M(R)\rangle \propto \langle\rho_m(r_{R,H})\rangle$.

Which one should we use? Well, the Hill radius has additional dynamical information and we may be tempted to use it as a bona fide tidal radius. Unfortunately, there are additional complications to consider. First of all, the criteria that define the Roche limit and the Hill radius lead to non-spherical regions, so strictly speaking, these radii are direction dependent additionally, the dynamics in the co-rotating frame is complicated by the appearance of the Coriolis term, which can not be expressed as the gradient of a potential function. As a result of this, an orbit whose Jacobi energy exceeds that of the Roche lobe does not necessarily escapes. In fact, the region just beyond the Roche Lobe becomes so complex, that some authors have characterized it as "fractal" [39] (see Fig. 2 in this reference for an example of a hovering orbit that ends up escaping). In fact, it was a study of the three-body problem, that lead H. Poincaré to discover what we now call "chaos".

---

[10] This is the L_1 Lagrange point. In the combined potential of a two body system there are five points where the forces cancel. These are the Lagrangian points first determined by Lagrange in 1772.

[11] Although we are considering point masses, we define an average density within r simply as $\rho \propto m/r^3$.

7.2 The Coriolis Effect

The effect of the Coriolis term gives rise to a distinction between prograde and retrograde orbits for particles orbiting the satellite. The distinction is defined by the relative orientation of the orbital angular momentum of the satellite and the particle. If they are parallel, it is a prograde orbit, if they are anti-parallel, it is a retrograde orbit (see Fig. 22).

Now, the Coriolis acceleration is of the form $a_c \propto -(\boldsymbol{\omega} \times \boldsymbol{v})$, where $\boldsymbol{\omega}$ is the satellite's angular velocity and $\boldsymbol{v}$ is the instantaneous velocity of the particle, in the co-rotating frame. It is easy to see, using your right hand, that this acceleration opposes the self-gravity of the satellite for prograde orbits, while it reinforces it for retrograde orbits.

This effect makes prograde orbits more fragile and leads to net retrograde rotation in the outer parts of a tidally pruned satellite [34], from which it becomes necessary to define separate tidal radii for prograde and retrograde orbits [44].

7.3 Non-Circular Orbits and Extended Mass Distributions

Even if we take into account all that we said before, our job is not finished. As soon as the satellite moves in a non-circular orbit, the problem becomes time-dependent and away goes conservation of energy for the orbiting particle. The so-called elliptical restricted three-body problem, where a test particle moves in the same plane of two point masses in an elliptical orbit around each other, is very rich in history and has been investigated very thoroughly [49]. Although the situation may seem hopeless, in reality it is not. It turns out that, except at the very edge of the system, lingering orbits that end up escaping, occupy a vanishingly small volume in phase-space and, although a bit fuzzy, a boundary between bound and unbound orbits can be defined. A working definition of the tidal radius can be the instantaneous Hill radius, or its value at peripasis, where the tidal effect is largest and mass loss peaks [38]. A tidal radius valid for point masses in an elliptical orbit around each other is [32],

$$\frac{r_t}{R_{\rm Peri}} = \frac{2}{3}\left(\frac{m}{(3+e)M}\right)^{1/3}, \tag{36}$$

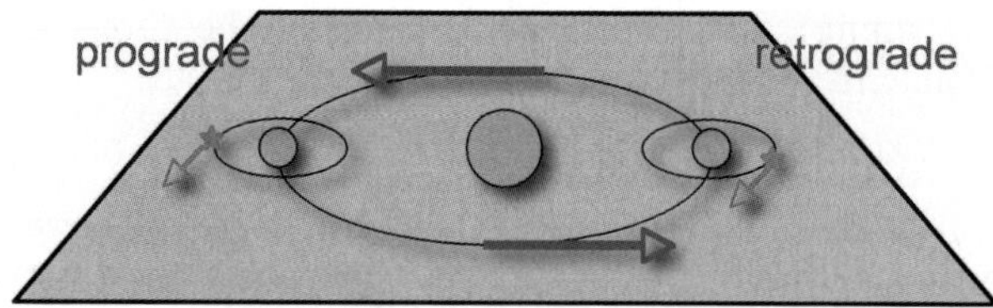

Fig. 22. Prograde and retrograde orbits around a satellite

where $R_{\rm Peri}$ is the distance between M and m at periapsis and e is the orbital eccentricity. Comparing with (35), we see that in the limit of circular orbits, r_t is a factor 2/3 smaller that r_H. This is because our calculation of r_H was done along the radial direction, whereas r_t represents an average value.

Another complication is that in our discussion of the tidal radius we have assumed point masses, and this is not the case when a globular cluster, or satellite galaxy, move inside the halo of a large galaxy. Fortunately, this problem is easier to remedy. The potential produced by the extended mass distributions should be used when calculating the tidal radius. An expression valid for a point mass satellite, moving within a flat rotation curve galaxy is [32],

$$\frac{r_t}{R_{\rm Peri}} = \frac{2}{3}\left[1 - \ln\left(\frac{2R_{\rm Peri}}{R_{\rm Peri} + R_{\rm Apo}}\right)\right]^{-1/3}\left(\frac{m}{2M_P}\right)^{1/3} . \qquad (37)$$

$R_{\rm Apo}$ is the separation between the centers of the galaxy and the satellite at apoapsis and M_P is the galaxy's mass enclosed within $R_{\rm Peri}$. An even more general formula, valid for galaxies with power-law density profiles, can be derived [44].

Before leaving this section, we mention that for very elongated orbits, the tidal force varies abruptly near periapsis. Besides the mass shed due to the much diminished tidal radius, a new effect appears: tidal shocking. We will examine this phenomenon in the next section.

8 Tidal Encounters

Anyone who has been unfortunate enough to have suffered a collision in the highway will tell you that it is a shocking experience. It is the same for galaxies. However, a key difference between collisions in the highway and collisions in your galactic backyard, is that in the former, we have physical contact and the larger the relative velocity, the bigger the damage. In the astronomical case, we have long range interactions, where the larger the relative velocity, the shorter the interaction time and less damage will occur. This is an important lesson: if you are a galaxy, zip by very fast when traversing crowded environments, like rich clusters of galaxies, if you want to survive the experience.

Both, dynamical friction and the tidal force dissipate encounter orbital energy and dump it within the thermal reservoir of internal motions of the hapless galaxies. The action of the tidal force, in this context, is a bit different from what we saw in the last section. In that case, we examined the static effect of the tidal force, which basically leads to the imposition of a boundary. When we have an encounter, the time-varying part of the tidal force heats up the interacting systems (this is often called tidal shocking [41]). In any case, at this point the distinction between these forces becomes a semantic issue and we must look for a more holistic description.

A gravitational encounter is a very complicated phenomenon [8]. It is illuminating just listing the reasons for this. Imagine a perturber passing close to a target system (it doesn't matter which one is which). As the perturber passes nearby, its time-varying gravity perturbs the motion of the stars in the target, which in turn perturb the perturber in a manner different to what you would have guessed, if you had not taken into account this back reaction. Since the situation is symmetric, the same goes when you place yourself in the other galaxy. It is this coupling of perturbations that makes any analytical treatment hopeless (except for very specific settings) and we must turn to N-body simulations.

With simulations the problem is availability of computer time (as always!). Even with the impressive advances in computational technology, predicated by Moore's law, we lack computer cycles to tackle some basic issues in a comprehensive manner that may give us a proper statistical description. You may have seen a lot of computer simulations of galaxy collisions out there, but do you know what is the range of parameters that lead to mergers when two galaxies interact (the capture cross section)? If you search in the literature you will find that a lot of parameter space remains to be explored, particularly in the intermediate mass ratio regime [15]. The basic problem is the vastness of the parameter space that uniquely defines a galactic encounter [7], and so it is difficult to undertake a proper exploration that could give us a good statistical description of the problem.[12] So, fortunately for us, a lot of work remains to be done.

8.1 A Simplified Picture: The Impulse Approximation

Even if a full N-body experiment may be needed to study galactic encounters, some basic physical insight can be gained by using an approximation, with a rich history in stellar and galactic dynamics since it was introduced: the impulse approximation [47].

In this approximation, we assume that a perturber of mass M_p moves at constant speed $v_{\rm col}$ along a straight path, so fast, that we can neglect the motion of stars in the target system. Figure 23 gives the basic geometry.

The impulse in the velocity of the target star is obtained by integrating the component of the gravitational force of the perturber that is orthogonal to the path, along the entire path:

$$\Delta v_* = \int_{-\infty}^{\infty} a_\perp \, \mathrm{d}t \; = \; 2 \int_{o}^{\infty} \frac{G M_P}{r^2} \frac{p}{r} \, \mathrm{d}t \, .$$

Simple geometry allows us to compute this integral,

$$\Delta v_* = \left(\frac{2 G M_P}{p v_{\rm col}} \right) \, . \tag{38}$$

[12] The dynamics of clusters of galaxies may benefit from the approach used to understand the dynamics of globular clusters, where all the complicated dynamics of binary scattering was synthesized in a few interaction cross sections [30].

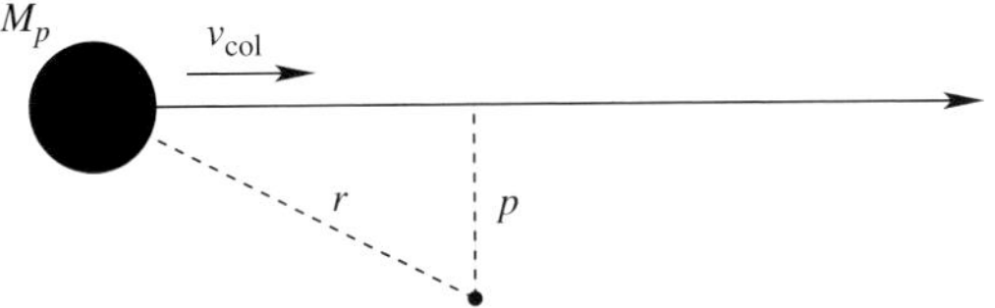

Fig. 23. Encounter geometry for impulse approximation

An easy way to remember this result, and give physical meaning to it, is splitting it as the product of two factors, whose physical units are acceleration and time, respectively:

$$\Delta v_* = \left(\frac{GM_P}{p^2}\right)\left(\frac{2p}{v_{\rm col}}\right) . \tag{39}$$

The first factor is just the acceleration at the point of closest distance, the second is the time it takes the perturber to travel a distance of twice the impact parameter (Fig. 24), we will call it, the *interaction time*. So, here you have it: the longer the interaction time, the larger the impulse.

There is, however, an undesired quirk in (39): what happens in the case of a head-on collision ($p = 0$)? The undesired behavior arises because we assumed a point mass for the perturber. This is easy to solve, and doing a proper calculation for an extended mass, spherical perturber, gives the same result, but multiplied by a correction factor given by [3],

$$f(p) = \int_1^\infty \frac{\mu_P(p\xi)}{\xi^2\sqrt{\xi^2-1}}\,\mathrm{d}\xi , \tag{40}$$

where $\mu_{\rm P}(r)$ is the mass fraction of the perturber within a radius r. It is easy to see that f has the right asymptotic behavior (approaches zero for an infinitely spread perturber, and goes to unity for a point mass).

Now, to estimate the tidal heating we must obtain the change in energy suffered by the star within the target. Since the motion of the star is neglected in the impulse approximation, the only change is due to Δv_*,

$$\Delta E_* = \Delta(m_* v_*^2/2) = m_*(\boldsymbol{v}_* \cdot \boldsymbol{\Delta v}_*) + m_* \Delta v_*^2/2$$

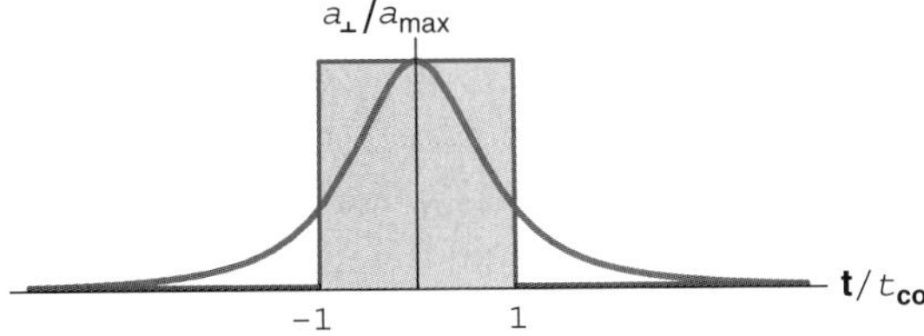

Fig. 24. Impulse acceleration as a function of time (*thick line*). The equivalent step function (*thin line*), has the same area under it

The first term averages to zero when integrating over all stars in the target.[13] so we are left with the quadratic term. Integrating it, we get,

$$\Delta E = \frac{1}{3} \left(\frac{2GM_P f(p)}{p^2 v_{\rm col}} \right)^2 \langle r^2 \rangle \,, \tag{41}$$

where $\langle r^2 \rangle$ is the mean of the squared radial distance of all stars in the target.

8.2 A Tale of Two Timescales: Adiabatic and Impulsive Regimes

Spitzer did not introduce (41), he went a bit beyond. He knew that the motion of stars in the target system should be considered, even if very crudely. He modeled these stars as harmonic oscillators and casted the encounter as a standard problem in Classical Mechanics: the forced harmonic oscillator [48].

He found that the efficiency for energy transfer to the oscillator varies dramatically as a function of an *adiabatic parameter* given by,

$$\beta = \frac{2p}{v_{\rm col}} \left(\frac{GM_s}{r^3} \right)^{1/2} \,, \tag{42}$$

where M_s is the target mass. This parameter is proportional to the ratio of the interaction time to the oscillator period (local orbital time). And here is where our tale of two timescales begins: When $\beta < 1$, the encounter is fast with respect to the natural period of the oscillator, while for $\beta > 1$, the encounter is slow. Spitzer found that the energy transfer efficiency could be described by a multiplicative factor η that decays exponentially with β [48]. Weinberg has reviewed this problem, removing the 1-dimensional treatment used by Spitzer and taking into account the effect of resonances. He found a function that decays as a power-law [52]. Both corrections approach unity for $\beta \to 0$, this is the *impulsive regime* that we saw in Sect. 8.1. For $\beta >$ 1, both correction factors shrink rather quickly, this is the *adiabatic regime* (Fig. 25).

Equation (41) should then be multiplied by a net shock efficiency, obtained as a mass-weighted average of η over the whole target [5],

$$\eta^* = \frac{4\pi}{\langle r^2 \rangle M_s} \int_0^{r_t} \rho_s(r) \eta(\beta) r^4 \, {\rm d}r \tag{43}$$

Why is that the amount of energy transfer diminishes so rapidly as we enter the adiabatic regime? This has to do with *adiabatic invariants*, a subject discussed by Einstein at the first Solvay Conference in 1911. The basic idea is that when a dynamical system undergoes a change that is slow with respect to its internal timescale, there are properties of the motion that will remain invariant, these are the actions (see [25], Sect. 12.5). Here is a cartoon answer:

[13] For this argument to work, we must include escaping stars.

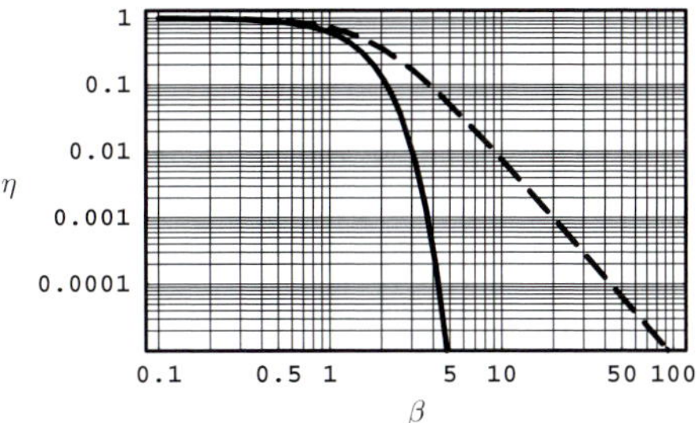

Fig. 25. Adiabatic correction of Spitzer (*solid line*) and Weinberg (*dashed line*) [24]

when a perturber encounters a slow star, the star sweeps only a fraction of its orbit while the encounter takes place, which can then be distorted. On the other hand, a fast star covers many periods and so it appears to the perturber, not as a point mass, but as a mass spread over the whole orbit. The perturber then can only push the orbit as a whole (Fig. 26).

This effect can clearly be seen in Fig. 27, which shows an old N-body simulation of the effect of a hyperbolic encounter on a de Vaucouleurs galaxy. The perturber was launched directly across the frame but is deflected downward during the encounter. The halo of the target reacts impulsively and absorbs energy that leads to its heating up and expanding without any appreciably overall displacement. The inner core of the target, on the contrary, hardly expands but is displaced as a whole from its original position, leaving behind the halo that becomes unbound. This is called *tidal stripping*.

At the end, we must get a copy of an N-body code and run our own simulations. However, a close examination of the impulse approximation and

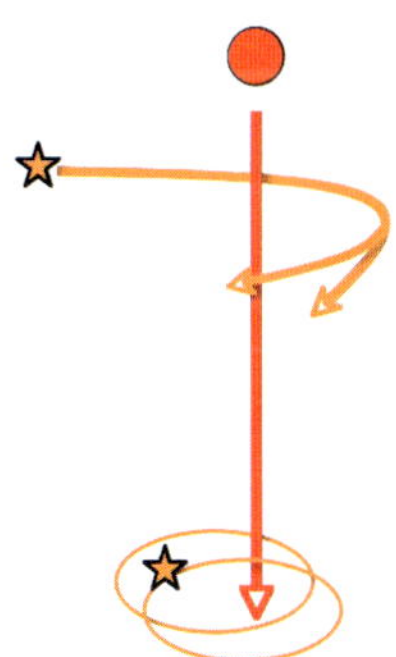

Fig. 26. The same encounter can be impulsive for a star in the outskirts of a target galaxy and adiabatic for one in its central region. In the first case, the orbit of the star gets distorted, in the second, it gets pushed as a whole

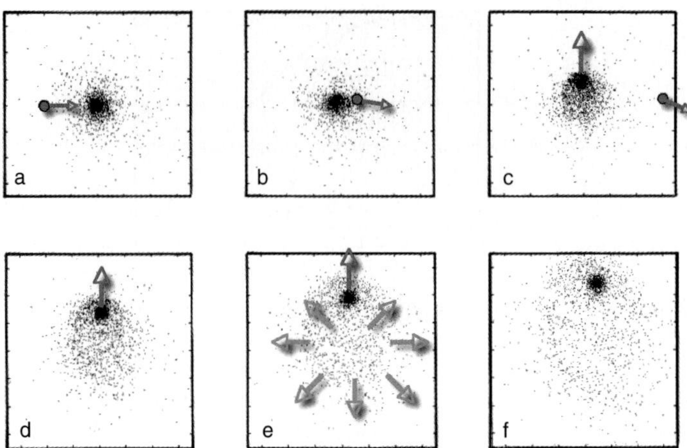

Fig. 27. N-body simulation of the hyperbolic encounter of two de Vaucouleur galaxies. The perturber is represented by the filled circle of radius equal to its effective radius. The frames are set 2 crossing times (at the effective radius) apart [3]

knowledge about what to expect in the impulsive and adiabatic regimes, can help us understand the results we obtain.[14]

9 Putting Things Together: The Orbital Decay of a Satellite within an Extended Halo

In this section we put together a bit of what we have learned about dynamical friction and tidal truncation, to follow a satellite whose orbit decays as it moves within an extended halo. This problem has been treated by Binney and Tremaine [12] (see their Sect. 7.1.1.a) for the case of a point mass satellite of fixed mass, that spirals down a singular isothermal halo. A similar calculation has also been made elsewhere for a constant density halo [27]. The treatment we present here is inspired by that of Binney and Tremaine, but with the distinction that we allow for an extended satellite and describe the shrinking tidal radius, as it plunges into deeper layers of the host halo. Ours is a simple model that wraps up what we have learned in Sects. 6 and 7.

9.1 The Cast of Characters

We need a model for the satellite and the host halo. Due to its simplicity, we choose a Plummer [43] model for the satellite:

$$\rho_s(x) = \left(\frac{3M_s}{4\pi r_o^3}\right) (1 + x^2)^{-5/2} , \tag{44}$$

[14] "Computers are useless, they can only give you answers". Pablo Picasso.

$$\phi_s(x) = -\frac{\mathrm{GM_s}}{r_o}\,(1+x^2)^{-1/2}\,, \tag{45}$$

$$\mathcal{M}_s(x) = M_s\,x^3(1+x^2)^{-3/2}\,, \tag{46}$$

Here ρ_s, ϕ_s and $\mathcal{M}_s$ are the density profile, potential function and cumulative mass. M_s is the total initial mass of the satellite and $x \equiv r/r_o$, where r_o is the nuclear radius (it contains 35% of the mass). This is a model with an approximately flat density core within r_o. As we will soon see, this is a very important feature.

For the host halo, we use a mass model that results in a flat rotation curve (see (28), (29) and (30)). This is the same halo used by Binney and Tremaine. Its cumulative mass, within a galacto-centric distance R, is given by:

$$M_R = (v_o^2/G)\,R\,, \tag{47}$$

with v_o being the constant circular velocity of the halo. Notice that both of our models are spherical. We will use r to denote radial distances within the satellite and R for the radial position of the satellite within the halo.

9.2 Tidal Truncation

We can now use the tidal truncation condition to find the tidal radius of the satellite r_t as a function of position within the halo. We will make the simplifying assumption that, as the satellite travels inward within the halo and its tidal radius shrinks, the satellite inside r_t remains unaffected and it is just the layers outside it, that are lost.

We saw in Sect. 7 that the tidal radius can be written as a condition that relates the average density of the satellite inside its tidal radius, with the average density of the underlying halo within the satellite orbit:

$$r_t/R = \alpha\,[\mathcal{M}_s(r_t)/M_R]^{1/3}\,, \tag{48}$$

where $\alpha = 2^{2/3}/3 \approx 0.529$, for the tidal radius imposed by a singular isothermal halo (37).[15] We will assume that the satellite decays following a spiraling orbit very close to circular, and so we only consider the case ($e = 0$).

Substituting the mass profile of the satellite and halo (46) and (47) in the tidal radius condition, we get:

$$\frac{r_t}{R} = \alpha\left(\frac{GM_s}{Rv_o^2}\right)^{1/3}\frac{x_t}{\sqrt{1+x_t^2}}\,. \tag{49}$$

The left hand side term can be written as,

$$\frac{r_t}{R} = \frac{r_t/r_o}{R/r_o} = \frac{x_t}{R/r_o}\,,$$

[15] This equation assumes a point mass satellite. We will have to live with this.

where $x_t \equiv r_t/r_o$ is the dimensionless tidal radius.

Substituting this back in (49), we get,

$$x_t = \alpha \frac{R}{r_o} \left(\frac{GM_s}{Rv_o^2} \right)^{1/3} \frac{x_t}{\sqrt{1+x_t^2}} .$$

This can be easily manipulated to yield,

$$x_t^2 = \alpha^2 \left(\frac{GM_s R^2}{r_o^3 v_o^2} \right)^{2/3} - 1 . \tag{50}$$

A bit more algebra leads to r_t,

$$r_t = \sqrt{\alpha^2 \left(\frac{GM_s^2}{v_o^2} \right)^{2/3} - r_o^2} . \tag{51}$$

We have thus obtained the satellite size as a function of its position within the halo. Notice that the size shrinks to zero at a particular position R_t, which occurs when the radical vanishes. This condition can be cast as a tidal condition:

$$r_o/R_t = \alpha \left[M_s/M_{R_t} \right]^{1/3} , \tag{52}$$

comparing with (48), we see that a Plummer model, dropped within a flat rotation curve halo, will survive until its Roche lobe size (computed using its original mass) is equal to its core radius. What is behind the existence of a terminal galacto-centric position? As we mentioned at the beginning of this section, the tidal radius criterion can be interpreted as a condition that relates densities: If the satellite does not have a central density higher than the corresponding value for the halo, it will be destroyed before reaching the center of the halo. The density of the Plummer model does not vary much within its core radius. Once the Roche lobe reaches it, the satellite will be gone very soon.

The terminal galacto-centric position can be obtained from the previous equation. It is equal to,

$$R_t = v_o \sqrt{\frac{r_o^3}{\alpha^3 GM_s}} . \tag{53}$$

Finally, the remaining satellite mass can be computed using (50) and (46) to evaluate $\mathcal{M}_s(x_t)$.

In astronomical units, appropriate for our problem, (51) and (53) are,

$$(r_r/\text{pc}) = \sqrt{1,595 \left[\frac{(M_s/10^6\, M_\odot)(R/\text{kpc})^2}{(v_o/10^2\ \text{kms}^{-1})^2} \right]^{2/3} - \left(\frac{r_0}{\text{pc}} \right)^2} , \tag{54}$$

$$(R_t/\text{kpc}) = 5.868 \times 10^{-4}\, (v_o/10^2\ \text{kms}^{-1}) \sqrt{\frac{(r_o/\text{pc})^3}{(M_s/10^6\, M_\odot)}} . \tag{55}$$

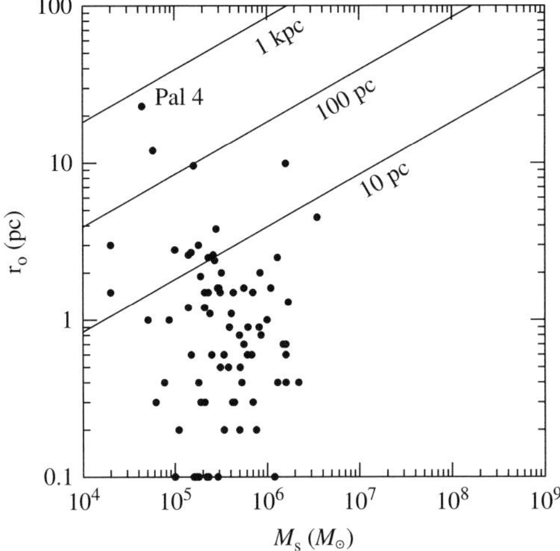

Fig. 28. Terminal galactocentric distance for Plummer models as a function of core radius r_o and satellite mass. Three lines of constant terminal distance are shown. The dots correspond to globular clusters in our Galaxy. Clusters that lie on the lower horizontal axis are core-collapsed clusters

We present in Fig. 28 the terminal galacto-centric distance as a function of satellite mass and core radius. The halo has a characteristic velocity of $v_o = 220$ km s^{-1}. Data for globular clusters in our galaxy are shown too [23]. It is clear from this figure that the vast majority of clusters are dense enough at their centers, to survive all the way to the central region of our Galaxy, the exception being a group of three, or maybe five clusters, of which the most extreme example is Palomar 4.[16]

In Fig. 29 we show the cluster tidal radius as a function of galactocentric position, for satellites of mass 10^6 and $10^9\,M_\odot$, and in each case, for core radii of 1, 10 and 100 pc. It is clear that all clusters of a given mass shrink steadily in size as $\propto R^{2/3}$ until they reach close to their terminal galactocentric distance, at which point, they are very quickly destroyed, this terminal distance depending on the core radius: the smaller it is, the deeper the satellite survives.

Finally, Fig. 30 shows the satellite mass as a function of galactocentric position. The same three core radii of Fig. 29 are used for a satellite mass of $10^6\,M_\odot$. We note that the $r_o = 1$ and 10 pc curves coincide with those of a $10^9\,M_\odot$ satellite with $r_o = 10$ and 100 pc, respectively.

Given the form of the cumulative mass distribution of the Plummer model, it is clear that our satellites loose little mass until they approach their terminal position, at which point they are quickly destroyed. It is then a reasonable

[16] We must remember that we have modeled the dark halo only. The bulge and galactic disk have not been taken into account.

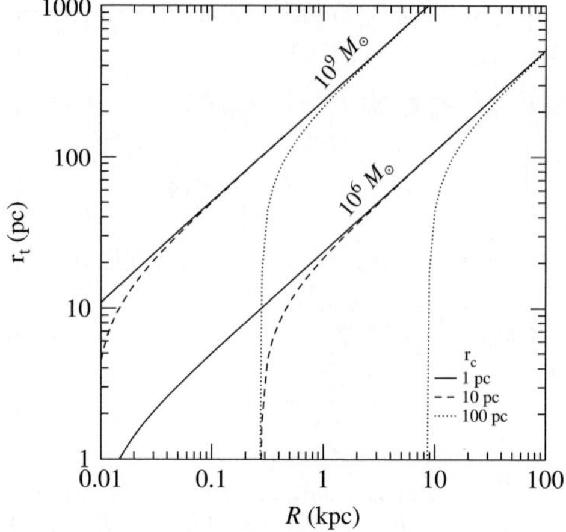

Fig. 29. Satellite tidal radius as a function of galactocentric position. Two sets of curves are shown for satellite masses of 10^6 and $10^9\ M_\odot$. In each set we show the results for three core radii (1, 10 and 100 pc). Notice the abrupt destruction when the satellite reaches its terminal galactocentric distance

approximation to assume that the satellites evolve at fixed mass. We will use this approximation in the next section.

9.3 Orbital Decay

We now compute the rate of orbital decay for the satellites. This decay is driven by dynamical friction, which produces a force that opposes the motion of the satellite (32). To apply this formula we need the halo density, which is

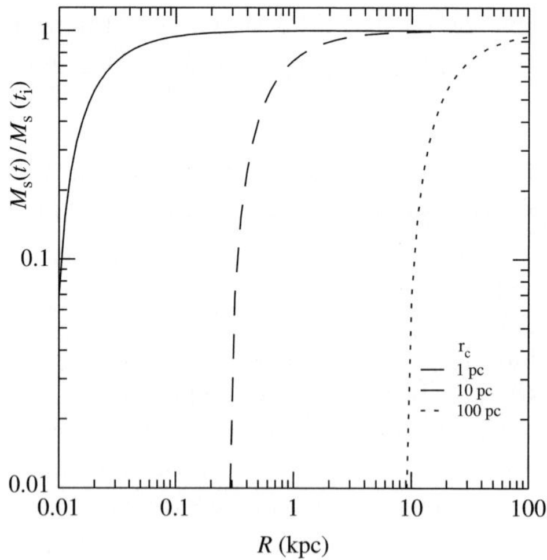

Fig. 30. Fraction of satellite mass as a function of galactocentric position. The same core radii as in Fig. 29 are shown, for a $10^6\ M_\odot$ satellite

given by (28), and the velocity distribution at each radial distance, here using a Maxwellian (see Fig. 18). These assumptions are identical to the model of Binney and Tremaine, so we refer to their (7.23) for the retarding force:

$$F_{\mathrm{DF}} \approx -0.428 \ln(\Lambda) \frac{\mathrm{GM}_{s^2}}{R^2} ,$$

where Λ is the Coulomb term (33) and R is the galactocentric position of the satellite.

This force produces a torque $\tau = F_{\mathrm{DF}} R$, for our case of a circular orbit. The change in orbital angular momentum is then,

$$\frac{\mathrm{d}}{\mathrm{d}t}(R M_s v_o) = -0.428 \ln(\Lambda) \frac{\mathrm{GM}_s^2}{R} .$$

Strictly speaking, we should now take into account the variation in M_s due to tidal truncation that we saw in Sect. 9.2. This, however, would greatly complicate our model, which we want to keep simple. Furthermore, as we saw in Fig. 30, assuming a constant mass for the satellite is not too bad, so we will do so. Solving the resulting equation we get the galactocentric position of the satellite, as a function of time,

$$R(t) = \sqrt{R_i^2 - 0.856 \ln(\Lambda) \frac{G M_s}{v_o} t} . \tag{56}$$

In astronomical units, this can be written as:

$$\frac{R(t)}{\mathrm{kpc}} = \sqrt{\left(\frac{R_i}{\mathrm{kpc}}\right)^2 - 0.0171 \ln(\Lambda) \left(\frac{M_s}{10^6 M_\odot}\right) \left(\frac{v_o}{220\,\mathrm{kms}^{-1}}\right)^{-1} \left(\frac{t}{\mathrm{Gyrs}}\right)} . \tag{57}$$

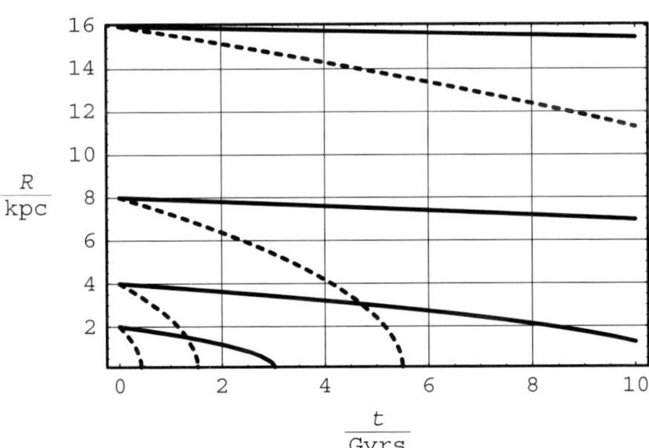

Fig. 31. Satellite galactocentric position as a function of time. The satellite masses are 10^7 (*solid lines*) and 10^8 $M_\odot$ (*dashed lines*). Initial positions of 2, 4, 8 and 16 *kpc*, have been considered

Fig. 31 shows the evolution of 10^7 and 10^8 $M_\odot$ satellites launched from several initial galactocentric distances.

Our model is very simple but fulfills our pedagogical goal of introducing the main actors that play a role in bringing down and trimming the unfortunate accreting satellites. Given the observational evidence and prevailing cosmological scenario, this is an important problem to study. In particular, a proper coupling of mass loss due to tidal truncation and dynamical friction, must be modeled carefully. We refer the interested reader to other, more sophisticated models, that have been introduced recently [50, 56].

10 Further Reading

For those interested in digging up some more, here is a list of recommended readings. The literature is very rich and we have chosen just a few references to get somebody interested started.

For Sects. 3 and 4 we recommend the following:

- Fully self-consistent anisotropic models in phase-space have been obtained by [6] for the Hernquist profile and by [37] for the Jaffe profile.
- A very general class of models that includes the Jaffe and Hernquist models, as special cases, is presented by [20]. Most of these models can be expressed as combinations of elementary functions.
- Many properties of spherical models with NFW profiles, including phase-space distribution functions with various kinematics is given by [36].
- A class of self-consistent models for disk galaxies that consist of an NFW dark halo, an exponential disk, a Hernquist bulge and even a massive central black hole, has been introduced by [54].

For Sect. 5, we suggest the following references,

- An approximation to orbits in the Hernquist model based in epicycloids has been obtained by [2]
- Sections 1 and 2 of [16] contain a good introduction to orbits, integrals of motion and orbital torii.
- A good discussion of the Lindblad diagram and its application to generate dynamical models of various degrees of eccentricities for their orbits can be found in [37].

For Sect. 6 there is a very extensive literature. Our list of suggested references is as follows,

- For an illuminating discussion about the subtleties and perils involved in simulating dynamical friction with N-body codes, and whether the global response of the background invalidates the classical Chandrasekhar formula, take a look at [14, 53] and [55].

- An examination of the effect on dynamical friction of velocity anisotropy is found in [9] and [42].
- The effect of chaotic orbits on dynamical friction is studied by [17].
- An investigation of whether dynamical friction produces orbit circularization can be found in [26].

For Sect. 7 there is also a very extensive literature. There is not, unfortunately, a good reference for the calculation of tidal forces produced by extended objects.

- Descriptions of the tidal force and computation of the tidal radius can be found in Chap. 5 of [48] and Sect. 7.3 of [12].
- A basic introduction to the three-body problem can be found in almost any book on Celestial Mechanics [13]. For a lighter description check [25].
- To gain the historical perspective in the three-body problem look at [22]. This reference presents a very nice account of Poincaré's musings about the three-body problem and the discovery of chaos.
- A gallery of orbits that illustrates the intricacies of the three-body problem can be found in Fig. 20.7 of [31]
- A description of the three-body problem and the tidal force, although in the context of planetary dynamics, can be found in Sects. 2.2 and 2.6, respectively, of [21]
- An illustrative reference that plots the magnitude of the Coriolis, centrifugal and gravitational terms in the co-rotating frame of reference for the restricted three-body problem is [34]
- For an up-to-date re-examination of the tidal radius take a look at [44]

For Sect. 8, we suggest the following,

- The first N-body simulation of an encounter between galaxies was done by E. Holmberg in 1941. It is interesting to see how this author was able to do the simulation at a time when no computers were available to him [29].
- A review of the field of N-body simulations of galactic encounters can be found in [8]
- A detailed description of the impulse approximation together with Spitzer correction for adiabaticity can be found in Sect. 5.2 of [48]
- An example of the use of the impulse approximation, with Weinberg correction for adiabaticity for the problem of globular cluster survival in our Galaxy, can be found in [24]

For Sect. 9, we recommend these references,

- Our model for orbital decay has followed [12]. A similar model, but for a flat density halo motivated by dwarf galaxies, is given by [27]
- More sophisticated, but still analytical models of the orbital decay of a satellite, have been presented by [50] and [56]

Finally, [12] is the standard graduate level reference for Galactic dynamics. It is very thorough and it may contain far more than what you may need for just a quick computation, but if you really want to dig up what is behind the topics we have been discussing, you will find it here.

11 Some Final Words

It has been my job to give you a theoretical view of clusters of galaxies, or rather a glimpse. I guess that if we stretch the electromagnetic spectrum a bit, my talk fits within the title of this school: "A Panchromatic View ...". The organizers certainly think so, and asked me to do this job. We live in the golden era of Astronomy. I am sure that you have heard this so many times that by now it may sound like a cliché, however, it is true. Those of us old enough to remember the time when an *IR* image meant laboriously taking readings with your bolometer on a 10×10 sampling array on the sky, will tell you so. However, with the current deluge of observational information comes the task of making sense of all of it, and for this we need theoretical understanding. It is appropriate here to quote a Philosopher of old times,

Οι αδαείς, ελαφρόμυαλοι άνθρωποι,
πού θαρρούν πως η Αστρονομία μαθαίνεται
μονάχα κοιτάζοντας τ' αστέρια
δίχως τή βαθιά γνώση τών μαθηματικών,
θά καταλήξουν στή μέλλουσα ζωή πτηνά ...

Πλάτων, Τίμαιος

Innocent light minded men,
who think that Astronomy can be learnt
by looking at the stars without knowledge
of Mathematics will, in the next life, be birds ...

Plato (Timaeus)

Dynamics is one of the pillars of Classical Physics. Given the spread of Gravity in the Universe, its application in Astronomy is only natural. In fact, a "good dynamical intuition" ought to be second nature to Astronomers. I hope that these lectures help students a bit to move in this direction, motivating them to explore dynamical aspects of Galaxies, even if, and specially if, they are not theoreticians.

I want to thank Omar López, Manolis Plionis and David Hughes for inviting me and organizing a wonderful "Gullermo Haro" school. INAOE for providing a beautiful environment to ponder these matters about far away galaxies, within a setting rich in history and cultural heritage. I also want to thank the students who had to suffer my lectures and then wait for me to produce these notes.

References

1. Abt, H., ed. 1999, “The Astrophysical Journal American Astronomical Society Centennial Issue”, Univ. of Chicago Press
2. Adams, F., Bloch, A.: ApJ **629**, 204 (2005)
3. Aguilar, L.A., White, S.D.M.: ApJ **295**, 374 (1985)
4. Aguilar, L.A., White, S.D.M.: ApJ **307**, 97 (1986)
5. Aguilar, L.A., Hut, P., Ostriker, J.P.: ApJ **335**, 720 (1988)
6. Baes, M., Dejonghe, H.: AA **393**, 485 (2002)
7. Barnes, J.E.: ApJ **393**, 484 (1992)
8. Barnes, J.E.: Galaxies: Interactions and Induced Star Formation, Saas-Fee Advanced Course 26. Swiss Society for Astrophysics and Astronomy XIV, In: Kennicutt, R.C., Schweizer, F., Barnes, J., Friedli, D., Martinet, L., Pfenniger, D. (eds.) p. 275. Springer-Verlag, New York (1998)
9. Binney, J.: MNRAS **181**, 735 (1977)
10. Binney, J.: MNRAS **200**, 951 (1982)
11. Binney, J.: MNRAS **201**, 15 (1982)
12. Binney, J., Tremaine, S.: Galactic Dynamics. Princeton University Press, Princeton (1987)
13. Boccaletti, D., Pucacco, G.: Theory of Orbits, Vol. 1, Ch. 4, Springer-Verlag, New York, Astronomy & Astrophysics Library (1996)
14. Bontekoe, T., van Albada, T.S.: MNRAS **224**, 349 (1987)
15. Bournaud, F., Jog, C.J., Combes, F.: AA **437**, 69 (2005)
16. Carpintero, D.D., Aguilar, L.A.: MNRAS **298**, 1 (1998)
17. Cora, S., Vergne, M., Muzzio, J.C.: ApJ **546**, 165 (2001)
18. Chandrasekhar, S.: Principles of Stellar Dynamics, p. 56. University of Chicago Press, Chicago (1942)
19. Chandrasekhar, S.: ApJ **97**, 255 (1943)
20. Dehnen, W.: MNRAS **255**, 250 (1993)
21. de Pater, I., Lissauer, J.: Planetary Sciences. Cambridge University Press, Cambridge (2001)
22. Diacau, F., Holmes, P.: Celestial Encounters, Ch. 1, Princeton University Press, Princeton (1996)
23. Djorgovski, S., Meylan, G. (ed.): Structure and Dynamics of Globular Clusters, ASPCS, Vol. 50 (Astron. Soc. Pacific, Provo) (1993)
24. Gnedin, O.Y., Ostriker, J.P.: ApJ **474**, 223 (1997)
25. Goldstein, H., Poole, C., Safko, J.: Classical Mechanics, 3rd ed., Sec. 3.12, Addison-Wesley (2002)
26. Hashimoto, Y., Funato, Y., Makino, J.: ApJ **582**, 196 (2003)
27. Hernández, X., Gilmore, G.: MNRAS **297**, 517 (1998)
28. Hernquist, L.: ApJ **356**, 359 (1990)
29. Holmberg, E.: ApJ **94**, 385 (1941)
30. Hut, P., Inagaki, S.: ApJ **298**, 502 (1985)
31. Hut, P., Heggie, D.: The Gravitational Million-Body Problem. Cambridge University Press, Cambridge (2003)
32. Innanen, K., Harris, W., Webbink, R.F.: AJ **88**, 338 (1983)
33. Jaffe, W.: MNRAS **202**, 995 (1983)
34. Keenan, D.W., Innanen, K.A.: AJ **80**, 290 (1975)
35. Kormendy, J.: ApJ **218**, 333 (1977)

36. Lokas, E., Mamon, G.: MNRAS **321**, 155 (2001)
37. Merritt, D.: MNRAS **214**, 25 (1985)
38. Meza, A., Navarro, J.F., Abadi, M., Steinmetz, M.: MNRAS **359**, 93 (2005)
39. Murison, M.: AJ **98**, 2346 (1989)
40. Navarro, J., Frenk, C.S., White, S.D.M.: ApJ **490**, 493 (1997)
41. Ostriker, J.P., Spitzer, L., Chevalier, R.: ApJ **176**, L51 (1972)
42. Peñarrubia, J., Andrew, J., Kroupa, P.: MNRAS **349**, 747 (2004)
43. Plummer, H.: MNRAS **76**, 107 (1905)
44. Read, J.I., Wilkinson, M.I., Evans, N.W., Gilmore, G., Kleyna, J.T.: MNRAS **366**, 429 (2006)
45. Shu, F.: ApJ **525**, 347 (2000)
46. Sparke, L.S., Gallagher, J.S.: Galaxies in the Universe. Cambridge University Press, Cambridge (2000)
47. Spitzer, L.: ApJ **127**, 17 (1958)
48. Spitzer, L.: Dynamical Evolution of Globular Clusters, Princeton University Press, Princeton (1987)
49. Szebehely, V., Giacaglia, G.: AJ **69**, 230 (1964)
50. Taylor, J.E., Babul, A.: ApJ **559**, 716 (2001)
51. Tremaine, S., Weinberg, M.: MNRAS **209**, 729 (1984)
52. Weinberg, M.D.: AJ **108**, 1403 (1994)
53. White, S.D.M.: ApJ **274**, 53 (1983)
54. Widrow, L.M., Dubinski, J.: ApJ **631**, 838 (2005)
55. Zaritsky, D., White, S.D.M.: MNRAS **235**, 289 (1988)
56. Zhao, H.: MNRAS **351**, 891 (2004)

Optical Detection of Clusters of Galaxies

R. R. Gal

University of Hawaii, Institute for Astronomy, 2680 Woodlawn Dr., Honolulu, HI 96822, USA
rgal@ifa.hawaii.edu

1 Introduction

Taken literally, galaxy clusters must be comprised of an overdensity of galaxies. Almost as soon as the debate was settled on whether or not the "nebulae" were extragalactic systems, it became clear that their distribution was not random, with regions of very high over- and under-densities. Thus, from a historical perspective, it is important to discuss the detection of galaxy clusters through their galactic components. Today we recognize that galaxies constitute a very small fraction of the total mass of a cluster, but they are nevertheless some of the clearest signposts for detection of these massive systems. Furthermore, the extensive evidence for differential evolution between galaxies in clusters and the field means that it is imperative to quantify the galactic content of clusters.

Perhaps even more importantly, optical detection of galaxy clusters is now inexpensive both financially and observationally. Large arrays of CCD detectors on moderate sized telescopes can be utilized to perform all-sky surveys with which we can detect clusters to $z \sim 0.5$. Using some of the efficient techniques discussed later in this section, we can now survey hundreds of square degrees for rich clusters at redshifts of order unity with 4-m class telescopes, and similar surveys, over smaller areas but with larger telescopes are finding group-mass systems to similar distances.

Looking to the future, ever larger and deeper surveys will permit the characterization of the cluster population to lower masses and higher redshifts. Projects such as the Large Synoptic Survey Telescope (LSST) will map thousands of square degrees to very faint limits (29th magnitude per square arcsecond) in at least five filters, allowing the detection of clusters through their weak lensing signal (i.e., mass) as well as the visible galaxies. Ever more efficient cluster-finding algorithms are also being developed, in an effort to produce catalogs with low contamination by line-of-sight projections, high completeness, and well-understood selection functions.

R. R. Gal: *Optical Detection of Clusters of Galaxies*, Lect. Notes Phys. **740**, 119–142 (2008)
DOI 10.1007/978-1-4020-6941-3_4

This chapter provides an overview of past and present techniques for optical detection of galaxy clusters. It follows the progression of cluster detection techniques through time, allowing readers to understand the development of the field while explaining the variety of data and methodologies applied. Within each section (Sect. 2) we describe the datasets and algorithms used, pointing out their strengths and important limitations, especially with respect to the characterizability of the resulting catalogs. The next section provides a historical overview of pre-digital, photographic surveys that formed the basis for most cluster studies until the start of the 21st century. Section 3 describes the hybrid photo-digital surveys that created the largest current cluster catalogs. Section 4 is devoted to fully digital surveys, most specifically the Sloan Digital Sky Survey and the variety of methods used for cluster detection. We also describe smaller surveys, mostly for higher redshift systems. In Sect. 5 we give an overview of the different algorithms used by these surveys, with an eye towards future improvements. The concluding (Sect. 6) discusses various tests that remain to be done to fully understand any of the catalogs produced by these surveys, so that they can be compared to simulations.

2 Photographic Cluster Catalogs

Even before astronomers had a full grasp of the distances to other galaxies, the creators of the earliest catalogs of nebulae recognized that they were sometimes in spectacular groups. Messier and the Herschels observed the companions of Andromeda and what we today know as the Pisces-Perseus supercluster. With the invention of the wide-field Schmidt telescope, astronomers undertook imaging surveys covering significant portions of the sky. These quickly revealed some of the most famous clusters, including Virgo, Coma, and Hydra. The earliest surveys relied on visual inspection of vast numbers of photographic plates, usually by a single astronomer. As early as 1938, Zwicky [64] discussed such a survey based on plates from the 18″ Schmidt telescope at Palomar. In 1942, Zwicky [65] and Katz & Mulders [30] published a pair of papers presenting the first algorithmic analyses of galaxy clustering from the Shapley-Ames catalog, using galaxies brighter than 12.7^m. Examining counts in cells, cluster morphologies, and clustering by galaxy type, these surveys laid the foundation for decades of galaxy cluster studies, but were severely limited by the very bright magnitude limit of the source material. Nevertheless, many fundamental properties of galaxy clusters were discovered. Zwicky, with his typical prescience, noted that elliptical galaxies are much more strongly clustered than late-type galaxies (Fig. 1), and attempted to use the structure and velocity dispersions of clusters to constrain the age of the universe as well as galaxy masses.

However, the true pioneering work in this field did not come until 1957, upon the publication of a catalog of galaxy clusters produced by George Abell as his Caltech Ph.D. thesis, which appeared in the literature the following

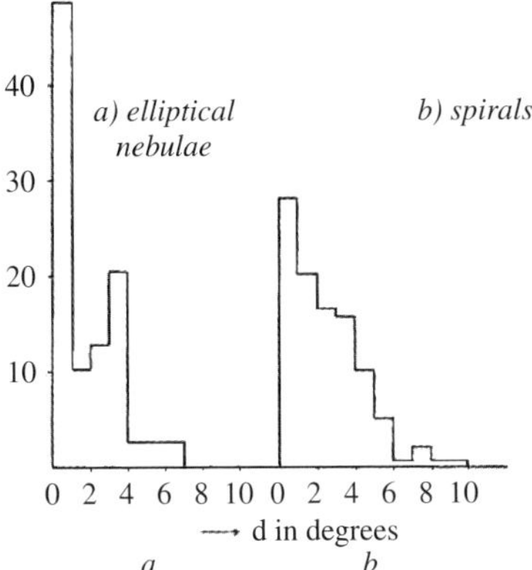

Fig. 1. The radial distribution of elliptical and spiral "nebulae" in the Virgo cluster. The enhanced clustering of elliptical galaxies is apparent, and is used to construct many modern cluster catalogs

year [1]. Zwicky followed suit a decade later, with his voluminous Catalogue of Galaxies and of Clusters of Galaxies [66]. However, Abell's catalog remained the most cited and utilized resource for both galaxy population and cosmological studies with clusters for over 40 years. Abell used the red plates of the first National Geographic-Palomar Observatory Sky Survey. These plates, each spanning $\sim 6°$ on a side, covered the entire Northern sky, to a magnitude limit of $m_{\rm r} \sim 20$. His extraordinary work required the visual measurement and cataloging of hundreds of thousands of galaxies. To select clusters, Abell applied a number of criteria in an attempt to produce a fairly homogeneous catalog. He required a minimum number of galaxies within two magnitudes of the third brightest galaxy in a cluster ($m_3 + 2$), a fixed physical size within which galaxies were to be counted, a maximum and minimum distance to the clusters, and a minimum galactic latitude to avoid obscuration by interstellar dust. The resulting catalog, consisting of 1,682 clusters in the statistical sample, remained the only such resource until 1989. In that year, Abell et al. [2] (hereafter ACO) published an improved and expanded catalog, now including the Southern sky. These catalogs have been the foundation for many cosmological studies over the last four decades, even with serious questions about their reliability. Despite the numerical criteria laid out to define clusters in the Abell and ACO catalogs, their reliance on the human eye and use of older technology and a single filter led to various biases. These include a bias towards centrally-concentrated clusters (especially those with cD galaxies), a relatively low redshift cutoff ($z \sim 0.15$; [4]), and strong plate-to-plate sensitivity variations. Photometric errors and other inhomogeneities in the Abell catalog [15, 59], as well as projection effects are a serious and difficult-to-quantify issue [29, 39]. These resulted in early findings of excess large-scale power in the angular correlation function [4], and later attempts to disentangle these issues relied on models to decontaminate the catalog ([46, 59]. The extent of these effects is also surprisingly unknown; measures of completeness and contamination in the Abell catalog disagree by factors of a few. For instance, Miller et al. [42] claim that under- or over-estimation of richness is not a significant problem, whereas van Haarlem et al. [60] suggest that one-third of Abell clusters have incorrect richnesses, and that one-third of rich ($R \geq 1$) clusters are missed. Unfortunately, some of these problems will plague any optically selected cluster sample, but objective selection criteria and a strong statistical understanding of the catalog can mitigate their effects.

In addition to the Zwicky and Abell catalogs, a few others based on plate material have also been produced [54], from the galaxy counts of Shane and Wirtanen [53], and a search for more distant clusters carried out on plates from the Palomar 200″ by Gunn et al. ([25]; hereafter GHO). None of these achieved the level of popularity of the Abell catalog, although the GHO survey was one of the first to detect a significant number of clusters at moderate to high redshifts ($0.15 < z < 0.9$), and remains in use to this day.

3 Hybrid Photo-Digital Surveys

Only in the past 10 years has it become possible to utilize the objectivity of computational algorithms in the search for galaxy clusters. These more modern studies required that plates be digitized, so that the data are in machine readable form. Alternatively, the data had to be digital in origin, coming from CCD cameras. Unfortunately, this latter option provided only small area coverage, so the hybrid technology of digitized plate surveys blossomed into a cottage industry, with numerous catalogs being produced in the past decade. All such catalogs relied on two fundamental data sets: the Southern Sky Survey plates, scanned with the Automatic Plate Measuring (APM) machine [41] or COSMOS scanner (to produce the Edinburgh/Durham Southern Galaxy Catalog/EDSGC, [27]), and the POSS-I, scanned by the APS group [47]. The first objective catalog produced was the Edinburgh/Durham Cluster Catalog (EDCC, [40]), which covered 0.5 sr ($\sim 1,600$ square degrees) around the South Galactic Pole (SGP). Later, the APM cluster catalog was created by applying Abell-like criteria to select overdensities from the galaxy catalogs, and is discussed in detail in [12]. More recent surveys, such as the EDCCII [9] did not achieve the large area coverage of DPOSS (see below), and perhaps more importantly, are not nearly as deep. For instance, the EDCCII's limiting magnitude is $b_J = 20.5$. For an L_* elliptical this corresponds to a limiting redshift of $z \sim 0.23$. The work by Odewahn and Aldering [44], based on the POSS-I, provided a Northern sky example of such a catalog, while utilizing additional information (namely galaxy morphology). Some initial work on this problem, using higher quality POSS-II data, was performed by Picard [48] in his thesis.

The largest, most recent, and likely the last photo-digital cluster survey is the Northern Sky Optical Survey (NoSOCS; [16, 17, 21, 36]). This survey relies on galaxy catalogs created from scans of the second generation Palomar Sky Survey plates. The POSS-II [52] covers the entire northern sky ($\delta > -3°$) with 897 overlapping fields (each 6.5° square, with 5° spacings), and, unlike the old POSS-I, has no gaps in the coverage. Approximately half of the survey area is covered at least twice in each band, due to plate overlaps. Plates are taken in three bands: blue-green, IIIa-J + GG395, $\lambda_{\rm eff} \sim 480$ nm; red, IIIa-F + RG610, $\lambda_{\rm eff} \sim 650$ nm; and very near-IR, IV-N + RG9, $\lambda_{\rm eff} \sim 850$ nm. Typical limiting magnitudes reached are $B_J \sim 22.5$, $R_F \sim 20.8$, and $I_N \sim 19.5$, i.e., $\sim 1^m - 1.5^m$ deeper than POSS-I. The image quality is improved relative to POSS-I, and is comparable to the southern photographic sky surveys. The original survey plates are digitized at STScI, using modified PDS scanners [33]. The plates are scanned with 15 μ (1.0″) pixels, in rasters of 23,040 square, giving ~1 GB/plate, or ~3 TB of pixel data total for the entire digital survey. The digital scans are processed, calibrated, and cataloged, with detection of all objects down to the survey limit, and star/galaxy classifications accurate to 90% or better down to $\sim 1^m$ above the detection limit [45]. They are photometrically calibrated using extensive CCD observations of Abell clusters [18].

The resulting galaxy catalogs are used as an input to an adaptive kernel galaxy density mapping routine (discussed in Sect. 5), and photometric redshifts based on galaxy colors are calculated, along with cluster richnesses in a fixed absolute luminosity interval. The NoSOCS survey utilizes F (red) plates, with a limiting magnitude of $m_r = 20$. This corresponds to a limiting redshift of 0.33 for an L_* elliptical galaxy. Because of the increase in g–r color with redshift, the APM would have to go as deep as $b_J = 22.0$ to reach the same redshift from their data for early type galaxies. Similarly, even at lower redshift, this implies that DPOSS can see $\sim 0.5^m - 1^m$ deeper in the cluster luminosity functions. Additionally, NoSOCS uses at least one color (two filters), and a significantly increased amount of CCD photometric calibration data. The final catalog covers 11,733 square degrees, with nearly 16,000 candidate clusters (Fig. 2), extending to $z \sim 0.3$, making it the largest such resource in existence. However, new CCD surveys, discussed in the next section, are about to surpass even this benchmark.

4 Digital CCD Surveys

With the advent of charge-coupled devices (CCDs), fully digital imaging in astronomy became a reality. These detectors provided an order-of-magnitude increase in sensitivity, linear response to light, small pixel size, stability, and much easier calibration. The main drawback relative to photographic plates was (and remains) their small physical size, which permits only a small area (of order 10′) to be imaged by a typical 2048^2 pixel detector. As detector sizes grew, and it became possible to build multi-detector arrays covering large

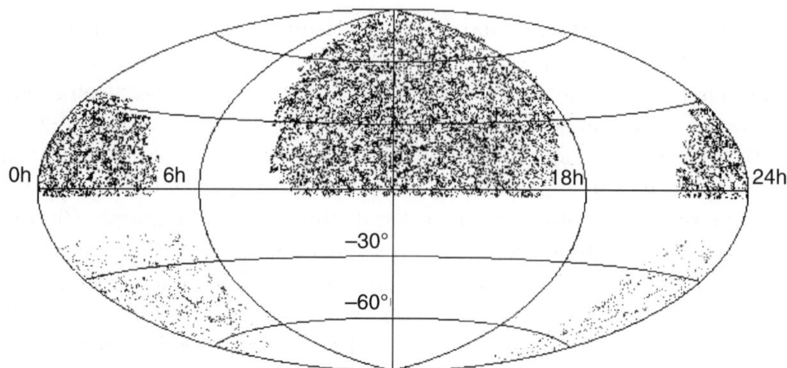

Fig. 2. The sky distribution of NoSOCS (*northern sky*) and APM (*southern sky*) candidate clusters in equatorial coordinates. The much higher density of NoSOCS is due to its deeper photometry and lower richness limit

areas, it became apparent that new sky surveys with this modern technology could be created, far surpassing their photographic precursors. Unfortunately, in the 1990s most modern telescopes did not provide large enough fields-of-view, and building a sufficiently large detector array to efficiently map thousands of square degrees was still challenging.

Nevertheless, realizing the vast scientific potential of such a survey, an international collaboration embarked on the Sloan Digital Sky Survey (SDSS, [61]), which included construction of a specialized 2.5 m telescope, a camera with a mosaic of 30 CCDs, a 640-fiber multi-object spectrograph, a novel observing strategy, and automated pipelines for survey operations and data processing. Main survey operations were completed in the fall of 2005, with over 8,000 square degrees of the northern sky image in five filters to a depth of $r' \sim 22.2$ with calibration accurate to $\sim 2\% - 3\%$, as well as spectroscopy of nearly one million objects.

With such a rich dataset, many groups both internal and external to the SDSS collaboration have generated a variety of cluster catalogs, from both the photometric and the spectroscopic catalogs, using techniques including:

1. Voronoi Tessellation [32]
2. Overdensities in both spatial and color space (maxBCG, [3])
3. Subdividing by color and making density maps (Cut-and-Enhance, [24])
4. The Matched Filter and its variants [32]
5. Surface brightness enhancements [6, 62, 63]
6. Overdensities in position and color spaces, including redshifts (C4; [43])

These techniques are described in more detail in Sect. 5. Each method generates a different catalog, and early attempts to compare them have shown not only that the catalogs are quite distinct, but also that comparison of two photometrically-derived catalogs, even from the same galaxy catalogs, is not straightforward [5].

In addition to the SDSS, smaller areas, to much higher redshift, have been covered by numerous deep CCD imaging surveys. Notable examples include the Palomar Distant Cluster Survey (PDCS, [49]), the ESO Imaging Survey (EIS, [35, 62]), and many others. None of these surveys provide the angular coverage necessary for large-scale structure and cosmology studies, and are specifically designed to find rich clusters at high redshift. The largest such survey to date is the Red Sequence Cluster Survey (RCS, [23], based on moderately deep two-band imaging using the CFH12K mosaic camera on the CFHT 3.6 m telescope, covers $\sim$100 square degrees. This area coverage makes it comparable to or larger than X-ray surveys designed to detect clusters at $z \sim 1$. The use of the red sequence of early-type galaxies makes this a very efficient survey, and the methodology is described is Sect. 5.

5 Algorithms

From our earlier discussion, it is obvious that many different mathematical and methodological choices must be made when embarking on an optical cluster survey. Regardless of the dataset and algorithms used, a few simple rules should be followed to produce a catalog that is useful for statistical studies of galaxy populations and for cosmological tests:

1. Cluster detection should be performed by an objective, automated algorithm to minimize human biases and fatigue.
2. The algorithm utilized should impose minimal constraints on the physical properties of the clusters, to avoid selection biases. If not, these biases must be properly characterized.
3. The sample selection function must be well-understood, in terms of both completeness and contamination, as a function of both redshift and richness. The effects of varying the cluster model on the determination of these functions must also be known.
4. The catalog should provide basic physical properties for all the detected clusters, including estimates of their distances and some mass-related quantity (richness, luminosity, overdensity) such that specific subsamples can be selected for future study.

This section describes many of the algorithms used to detect clusters in modern cluster surveys. No single one of these generates an "optimal" cluster catalog, if such a thing can even be said to exist. Therefore, I provide some of the strengths and weaknesses of each technique. In addition to the methods discussed here, many other variants are possible, and in the future, joint detection at multiple wavelengths (i.e. optical and X-ray, [56]) may yield more complete samples to higher redshifts and lower mass limits, with less contamination.

5.1 Counts in Cells

The earliest cluster catalogs, like those of Abell, utilized a simple technique of counting galaxies in a fixed magnitude interval, in cells of a fixed physical or angular size. Indeed, Abell simply used visual recognition of galaxy overdensities, whose properties were then measured ex post facto in fixed physical cells. This technique was used by Couch et al. [11] and Lidman and Peterson [34] to detect clusters at moderate redshifts ($z \sim 0.5$), by requiring a specified enhancement, above the mean background, of the galaxy surface density in a given area. This enhancement, called the contrast, is defined as

$$\sigma_{\mathrm{cl}} = \frac{N_{\mathrm{cluster}} - N_{\mathrm{field}}}{\sigma_{\mathrm{field}}} \tag{1}$$

where N_{cluster} is the number of galaxies in the cell corresponding to the cluster, N_{field} is the mean background counts and σ_{field} is the variance of the field counts for the same area. The magnitude range and cell size used are parameters that must be set based on the photometric survey material and the type or distance of clusters to be found. For instance, Lidman and Peterson [34] chose these parameters to maximize the contrast above background for a cluster at $z = 0.5$. Using the distribution of cell counts, one can analytically determine the detection likelihood of a cluster with a given redshift and richness (assuming a fixed luminosity function), given a detection threshold. The false detection rate is harder, if not impossible, to quantify, without running the algorithm on a catalog with extensive spectroscopy. This is true for most of the techniques that rely on photometry alone. It is also possible to increase the contrast of clusters with the background by weighting galaxies based on their luminosities and positions. Galaxies closer to the cluster center are up-weighted, while the luminosity weighting depends on both the cluster and field luminosity functions, as well as the cluster redshift. This scheme is similar to that used by the matched filter algorithm, detailed later.

This technique, although straightforward, has numerous drawbacks. First, it relies on initial visual detection of overdensities, which are then quantified objectively. Since simple counts-in-cells methods use the galaxy distribution projected along the entire line of sight, chance alignments of poorer systems become more common, increasing the contamination. Optimizing the magnitude range and cell size for a given redshift reduces the efficiency of detecting clusters at other redshifts, especially closer ones since their core radii are much larger. Setting the magnitude range typically assumes that the cluster galaxy luminosity function at the redshift of interest is the same as it is today, which is not true. Furthermore, single band surveys observe different portions of the rest frame spectrum of galaxies at different redshifts, altering the relative sensitivity to clusters over the range probed. Finally, the selection function can only be determined analytically for circular clusters with fixed luminosity functions. Given these issues, this technique is inappropriate for modern, deep surveys.

5.2 Percolation Algorithms

A majority of current cluster surveys rely on a smoothed map of projected galaxy density from which peaks are selected (see below). However, smoothing invariable reduces the amount of information being used, leading some authors to employ percolation (or friends-of-friends, FOF) algorithms. In their simplest form, these techniques link pairs of galaxies that are separated by a distance less than some threshold (typically related to the mean galaxy separations). Galaxies that have links in common are then assigned to the same group; once a group contains more than a specified number of members, it becomes a candidate cluster. This technique was used to construct a cluster catalog from APM data [12]. However, it is not typically used on two dimensional data, because the results of this method are very sensitive to the linking length, and can easily combine multiple clusters into long, filamentary structures. On the other hand, FOF algorithms are very commonly used for structure finding in three-dimensional data, especially N-body simulations [13, 14] and redshift surveys [28, 51]. A variant of this technique utilizing photometric redshifts has been recently proposed [8].

5.3 Simple Smoothing Kernels

Another objective and automated approach to cluster detection is the use of a smoothing kernel to generate a continuous density field from the discrete positions of galaxies in a catalog. For instance, Shectman [54] used the galaxy counts of Shane and Wirtanen in $10'$ bins, smoothed with a very simple weighting kernel. A minimum number of galaxies within this smoothed region (in this case, 20) were then required to detect a cluster. This type of kernel is fixed in angular size and thus does not smooth clusters at different redshifts with consistent *physical* radii, making its sensitivity highly redshift dependent. Similarly, it uses the full projected galaxy distribution (much as Abell did), and is thus insensitive to the different parts of the LF sampled at different redshifts.

5.4 The Adaptive Kernel

A slightly more sophisticated technique is to use an adaptive smoothing kernel [57]. This technique uses a two-stage process to produce a density map. First, at each point t, it produces a pilot estimate $f(t)$ of the galaxy density at each point in the map. Based on this pilot estimate, it then applies a smoothing kernel whose size changes as a function of the local density, with a smaller kernel at higher density. This is achieved by defining a *local bandwidth factor*:

$$\lambda_i = [f(t)/g]^{-\alpha}, \tag{2}$$

where g is the geometric mean of $f(t)$ and α is a sensitivity parameter that sets the variation of kernel size with density. NoSOCS uses a sensitivity parameter

$\alpha = 0.5$, which results in a minimally biased final density estimate, and is simultaneously more sensitive to local density variations than a fixed-width kernel [57]. This is then used to construct the adaptive kernel estimate:

$$\hat{f}(t) = n^{-1} \sum_{i=1}^{n} h^{-2} \lambda_i^{-2} K\{h^{-2} \lambda_i^{-2} (t - X_i)\} \tag{3}$$

where h is the bandwidth, which is a parameter that must be set based on the survey properties.

The adaptive kernel was used to generate the Northern Sky Optical Cluster Survey [16, 17, 21]. The smoothing size (in their case, 500″ radius) is set to prevent over-smoothing the cores of higher redshift ($z \sim 0.3$) clusters, while avoiding fragmentation of most low redshift ($z \sim 0.08$) clusters. Because the input galaxy catalog is relatively shallow, and the redshift range probed is not very large, it is possible to do this. For deeper surveys, this is not practical, and therefore this technique cannot be used in its simplest form. Figure 3 demonstrates example density maps, showing the effect of varying the initial smoothing window. In this figure, four simulated clusters are placed into a simulated background, representing the expected range of detectability in the NoSOCS survey. There are two clusters at low z (0.08), and two at high z

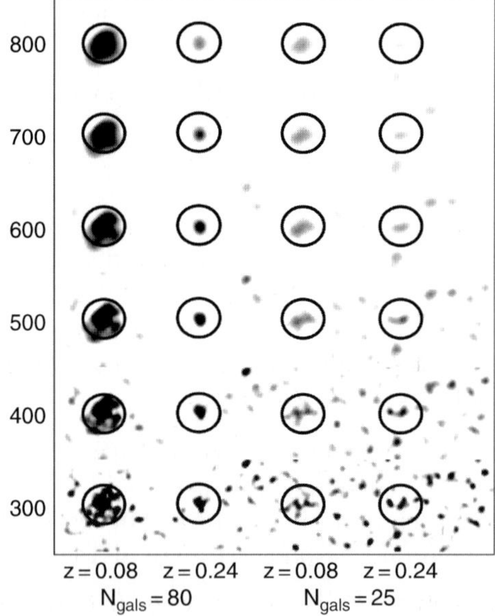

Fig. 3. The effect of varying the initial smoothing window for the adaptive kernel on cluster appearance. Each panel contains a simulated background with four simulated clusters, as described in the text. The smoothing kernel ranges in size from 300″ to 800″ in 100″ increments. Taken from [17]

(0.24), with one poor and one rich cluster at each redshift (100 and 333 total members, $N_{\mathrm{gals}} = 25$ and 80 respectively).

After a smooth density map is generated, cluster detection can be performed analogously to object detection in standard astronomical images. In NoSOCS, Gal et al. used SExtractor [7] to detect density peaks. The tuning of parameters in the detection step is fundamentally important in such surveys, and can be accomplished using simulated clusters placed in the observed density field, from which the completeness and false detection rates can be determined. Even so, this method involves many adjustable parameters (the smoothing kernel size, sensitivity parameter, and all the source detection parameters) such that it must be optimized with care for the data being used. Given an end-to-end cluster detection methodology, one can use simulations to determine the selection function's dependence on redshift, richness, and other cluster properties (see [17] for details). However, the measurement of cluster richness and redshift are done in a step separate from detection, using the input galaxy catalogs, further complicating this technique. The adaptive kernel is very fast and simple to implement, making it suitable for all-sky surveys, but is only truly useful in situations where the photometry is poor, and the survey is not very deep, as is the case for NoSOCS.

5.5 Surface Brightness Enhancements

It is not necessary to have photometry for individual galaxies to detect clusters. A novel but difficult approach is to detect the localized cumulative surface brightness enhancement due to unresolved light from galaxies in distant clusters. This method was pioneered by Zaritsky et al. [62, 63], who showed that distant clusters could be detected using short integration times on small 1-m class telescopes. However, this method requires extremely accurate flat-fielding, object subtraction, masking of bright stars. and excellent data homogeneity. Once all detected objects are removed from a frame, and nuisance sources such as bright stars masked, the remaining data is smoothed with a kernel comparable to the size of clusters at the desired redshift. The completeness and contamination rates of such a catalog are extremely difficult to model. Thus, this technique is not necessarily appropriate for generating statistical catalogs for cosmological tests, but is an excellent, cost-effective means to find interesting objects for other studies.

5.6 The Matched Filter

With accurate photometry, and deeper surveys, one can use more sophisticated tools for cluster detection. As we will discuss later, color information is very powerful, but is not always available. However, even with single-band data, it is possible to simultaneously use the locations and magnitudes of galaxies. One such method is the matched filter [49], which models the spatial and luminosity distribution of galaxies in a cluster, and tests how well galaxies in a given sky region match this model for various redshifts. As a result, it outputs an estimate of the redshift and total luminosity of each detected cluster as an integral part of the detection scheme. Following Postman et al. we can describe, at any location, the distribution of galaxies per unit area and magnitude $D(r, m)$ as a sum of the background and possible cluster contributions:

$$D(r, m) = b(m) + \Lambda_{cl} P(r/r_c) \phi(m - m^*) \tag{4}$$

Here, D is the number of galaxies per magnitude per arcsec^2 at magnitude m and distance r from a putative cluster center. The background density is $b(m)$, and the cluster contribution is defined by a parameter Λ_{cl} proportional to its total richness, its differential luminosity function $\phi(m - m^*)$, and its projected radial profile $P(r/r_c)$. The parameter r_c is the characteristic cluster radius, and m^* is the characteristic galaxy luminosity. One can then construct a likelihood for the data given this model, which is a function of the parameters r_c, m^*, and $\Lambda_{\rm cl}$. Because two of these parameters, especially m^*, are sensitive to the redshift, one obtains an estimated redshift when maximizing the likelihood relative to this parameter. The algorithm outputs the richness $\Lambda_{\rm cl}$ at each redshift tested, and thus provides an integrated estimator of the total cluster richness. The luminosity function used by Postman et al. is a Schechter [55] function with a power law cutoff applied to the faint end, while they use a circularly symmetric radial profile with core and cutoff radii (see their 19).

Like the adaptive kernel, this method produces density maps on which source detection must still be run. These maps have a grid size set by the user, typically of order half the core radius at each redshift used, with numerous maps for each field, one for each redshift tested. The goal of the matched filter is to improve the contrast of clusters above the background, by convolving with an "optimal" filter, and also to output redshift and richness estimates. Given a set of density maps, one can use a variety of detection algorithms to select peaks. A given cluster is likely to be detected in multiple maps (at different redshifts) of the same region; its redshift is estimated by finding the filter redshift at which the peak signal is maximized. By using multiple photometric bands, one can run this algorithm separately on each band and improve the reliability of the catalogs. The richness of a cluster is measured from the density map corresponding to the cluster redshift, and represents approximately the equivalent number of L_* galaxies in the cluster.

The matched filter is a very powerful cluster detection technique. It can handle deep surveys spanning a large redshift range, and provides redshift and richness measures as an innate part of the procedure. The selection function can be estimated using simulated clusters, as was done in significant detail by Postman et al. However, the technique relies on fixed analytic luminosity functions and radial profiles for the likelihood estimates. Thus, clusters which have properties inconsistent with these input functions will be detected at lower likelihood, if at all. While this is not likely to be an issue at low to moderate redshifts, as the population of clusters becomes increasingly merger dominated at $z \sim 0.8$ [10], these simple representations will fail. Similarly, the cluster and field LF both evolve with redshift, which can effect the estimated redshift. Also, as the redshifts and k-corrections become large, one samples a very different region of the LF than at low redshift. Nevertheless, this remains one of the best cluster detection techniques for cluster detection in moderately deep surveys.

5.7 Hybrid and Adaptive Matched Filter

The matched filter can be extended to include estimated (photometric) or measured (spectroscopic) redshifts. This extension has been called the adaptive matched filter (AMF, [31]). The adaptive here refers to this method's ability to accept 2-dimensional (positions and magnitudes), 2.5-dimensional (positions, magnitudes, and estimated redshifts), and 3-dimensional (positions, magnitudes, and redshifts) data, adapting to the redshift errors. In implementation, this technique uses a two-stage method, first maximizing the cluster likelihood on a coarse grid of locations and redshifts, and then refining the redshift and richness on a finer grid. Unlike the standard matched filter, the AMF evaluates the likelihood function at each galaxy position, and not on a fixed grid for each redshift interval. Thus, for each galaxy, the output includes a likelihood that there is a cluster centered on this galaxy, and the estimated redshift.

The inclusion of photometric redshifts should substantially improve detection of poor clusters, which is very important since most galaxies live in poor systems, and these are suspected to be sites for significant galaxy evolution. However, [32], using SDSS data, found that the simple matched filter is more efficient at detecting faint clusters, while the AMF estimated cluster properties more accurately. The matched filter performs better for detection because the significance threshold for finding candidates is redshift dependent, determined separately for each map produced in different redshift intervals. The AMF, on the other hand, finds peaks first in redshift space, and then selects candidates using a universal threshold. Thus, they propose a hybrid system, using the matched filter to detect candidate clusters, and the AMF to obtain its properties.

5.8 Cut-and-Enhance

Despite the popularity of matched filter algorithms for cluster detection, their assumption of a radial profile and luminosity function are cause for concern. Thus, development of semi-parametric detection methods remains a vibrant area of research. While the adaptive kernel described earlier is such a technique, more sophisticated algorithms are possible, especially with the inclusion of color information. One such technique is the Cut-and-Enhance method [24], which has been applied to SDSS data. This method relies on the presence of the red sequence in clusters, applying a variety of color and color-color cuts to generate galaxy subsamples which should span different redshift ranges. Within each cut, pairs of galaxies with separations less than $5'$ are replaced by Gaussian clouds, which are then summed to generate density maps. In this technique, the presence of many close pairs (as in a high redshift cluster) yields a more compact cloud, making it easier to detect, and thus possibly biasing the catalog against low-z clusters (see Fig. 4). As with the AK technique, this method yields a density map on which object detection must be performed; Goto et al. [24] utilize SExtractor. Once potential clusters are detected in the maps made using the various color cuts, these catalogs must be merged to produce a single list of candidates. Redshift and richness estimates are performed a posteriori, as they are with the AK. Similar to the AK, there are many tunable parameters which make this method difficult to optimize.

5.9 Voronoi Tessellation

Considering a distribution of particles it is possible to define a characteristic volume associated with each particle. This is known as the Voronoi volume, whose radius is of the order of the mean particle separation. The complete division of a region into these volumes is known as Voronoi Tessellation (VT), and it has been applied to a variety of astronomical problems, and in particular to cluster detection ([32, 50]). As pointed out by the latter, one of the main advantages of employing VT to look for galaxy clusters is that this technique does not distribute the data in bins, nor does it assume a particular source geometry intrinsic to the detection process. The algorithm is thus sensitive to irregular and elongated structures.

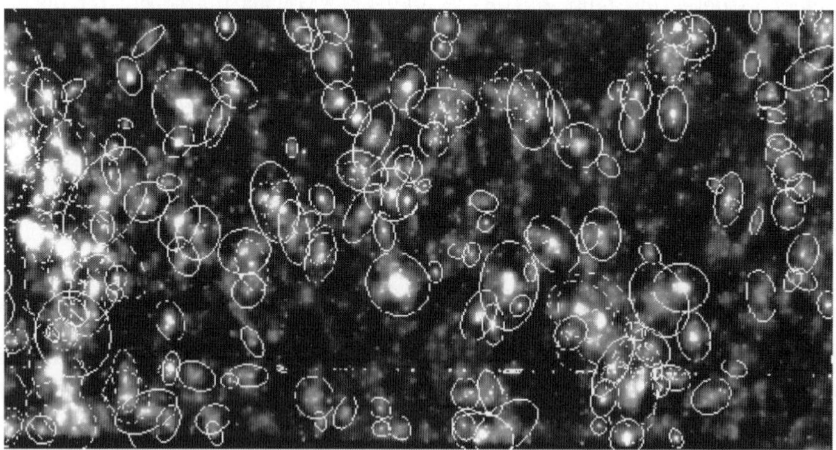

Fig. 4. An enhanced map of the galaxy distribution in the SDSS Early Data Release, after applying the g^*–r^*–i^i color-color cut. Detected clusters are circled. Taken from [24]

The parameter of interest in this case is the galaxy density. When applying VT to a galaxy catalog, each galaxy is considered as a seed and has a Voronoi cell associated to it. The area of this cell is interpreted as the effective area a galaxy occupies in the plane. The inverse of this area gives the local density at that point. Galaxy clusters are identified by high density regions, composed of small adjacent cells, i.e., cells small enough to give a density value higher than the chosen density threshold. An example of Voronoi Tessellation applied to a galaxy catalog for one DPOSS field is presented in Fig. 5. For clarity, we show only galaxies with $17.0 \leq m_{\rm r} \leq 18.5$.

Once such a tessellation is created, candidate clusters are identified based on two criteria. The first is the density threshold, which is used to identify fluctuations as significant overdensities over the background distribution, and is termed the search confidence level. The second criterion rejects candidates from the preliminary list using statistics of Voronoi Tessellation for a poissonian distribution of particles, by computing the probability that an overdensity is a random fluctuation. This is called the rejection confidence level. Kim et al. [32] used the color-magnitude relation for cluster ellipticals to divide the galaxy catalog into separate redshift bins, and ran the VT code on each bin. Candidates in each slice are identified by requiring a minimum

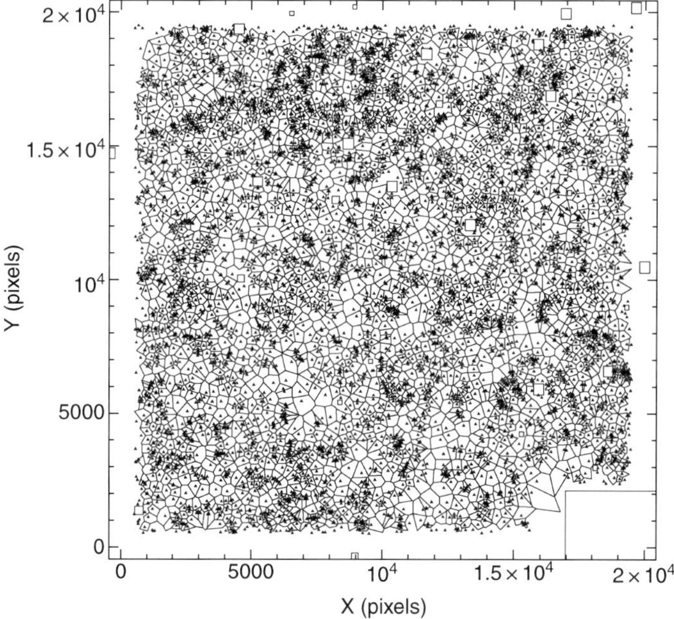

Fig. 5. Voronoi Tessellation of galaxies with $17.0 \leq m_r \leq 18.5$ in a DPOSS field. Each triangle represents a galaxy surrounded by its associated Voronoi cell (indicated by the polyhedrals). Excised areas (due to bright objects) are shown as rectangles. Taken from [36]

number $N_{\rm hdg}$ of galaxies having overdensities δ greater than some threshold $\delta_{\rm c}$, within a radius of 0.7 $\rm h^{-1}$ Mpc. The candidates originating in different bins are then cross-correlated to filter out significant overlaps and produce the final catalog. Ramella et al. [50] and Lopes et al. [36] follow a different approach, as they do not have color information. Instead, they use the object magnitudes to minimize background/foreground contamination and enhance the cluster contrast, as follows:

1. The galaxy catalog is divided into different magnitude bins, starting at the bright limit of the sample and shifting to progressively fainter bins. The step size adopted is derived from the photometric errors of the catalog.
2. The VT code is run using the galaxy catalog for each bin, resulting in a catalog of cluster candidates associated with each magnitude slice.
3. The centroid of a cluster candidate detected in different bins will change due to the statistical noise of the foreground/background galaxy distribution. Thus, the cluster catalogs from all bins are cross-matched, and overdensities are merged according to a set criterion, producing a combined catalog.
4. A minimum number ($\rm N_{min}$) of detections in different bins is required in order to consider a given fluctuation as a cluster candidate. $\rm N_{min}$ acts as

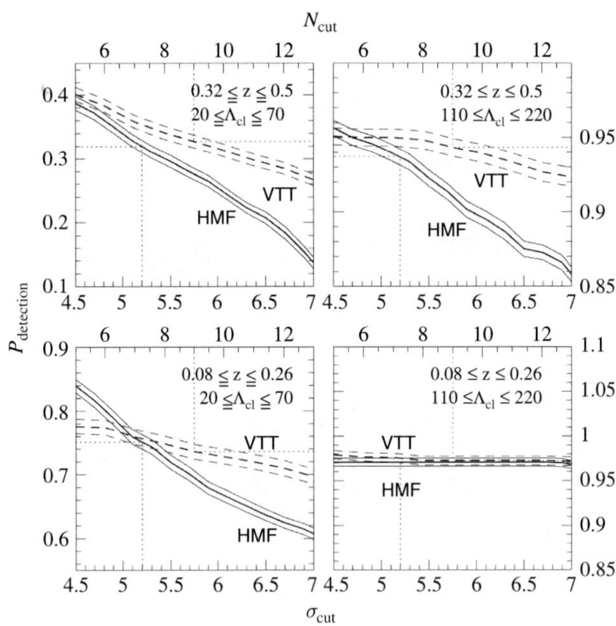

Fig. 6. The absolute recovery rates of clusters from the SDSS in four different ranges of cluster parameters for the HMF (*solid line*) and the VTT (*dashed line*). Taken from [32]

a final threshold for the whole procedure. After this step, the final cluster catalog is complete.

Kim et al. [32] and Lopes et al. [36] compare the performance of their VT algorithms with the HMF and adaptive kernel, respectively. Figure 6 (taken from [32]) shows the absolute recovery rates of clusters in four different ranges of cluster parameters for the HMF (solid line) and the VT (dashed line). Both algorithms agree very well for clusters with the highest signals (rich, low redshift), but the VT does slightly better for the thresholds determined from the uniform background case. Similarly, Lopes et al. [36] find that the VT algorithm performs better for poor, nearby clusters, while the adaptive kernel goes deeper when detecting rich systems, as seen in Fig. 7, where the VT-only detections are preferentially poor and low redshift, and the AK-only detections are richer and at high redshift.

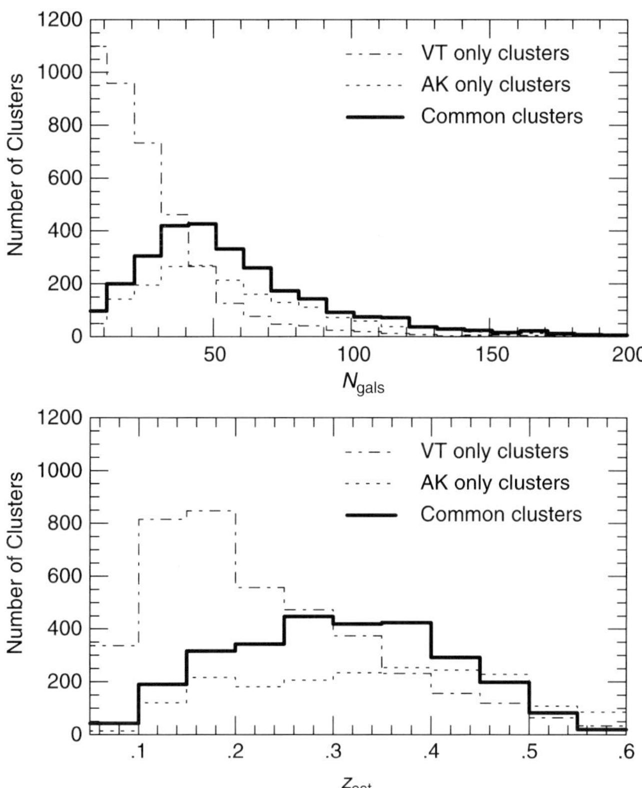

Fig. 7. Richness (**top**) and estimated redshift (**bottom**) distributions for clusters detected in DPOSS by only the VT code (*dash-dotted line*), only the AK code (*dotted line*), and by both methods (*heavy solid line*). Taken from [36]

5.10 MaxBCG

The maxBCG algorithm, developed for use on SDSS data [3, 26], is another technique that relies on the small color dispersion of early-type cluster galaxies. The brightest of the cluster galaxies (BCGs) have predictable colors and magnitudes out to redshifts of order unity. Unlike many of the other techniques discussed above, maxBCG does not generate density maps. Instead, it calculates a likelihood as a function of redshift for *each* galaxy that it is a BCG, based on its colors and the presence of a red sequence from the surrounding objects (see also Fig. 8). This is calculated as

$$\mathcal{L}_{\mathrm{max}} = \mathrm{max}\mathcal{L}(z); \mathcal{L}(z) = \mathcal{L}_{\mathrm{BCG}} + \log N_{\mathrm{gal}} \tag{5}$$

where $\mathcal{L}_{\mathrm{BCG}}$ is the likelihood, at redshift z, that a galaxy is a BCG, based on its colors and luminosity, and N_{gal} is the number of galaxies within 1 h^{-1} Mpc with colors and magnitudes consistent with the red sequence (i.e. within 0.1 mag of the mean BCG color at the redshift being tested). This procedure results in a maximum likelihood and redshift for each galaxy in the catalog. The peaks in the $\mathcal{L}_{\mathrm{max}}$ distribution are then selected as the candidate clusters.

This algorithm appears to be extremely powerful for selecting clusters in the SDSS. Simulations suggest that maxBCG recovers and correctly estimates the richness for greater than 90% of clusters and groups present with $N_{\mathrm{gal}} \geq 15$ out to $z = 0.3$, with an estimated redshift dispersion of $\delta z = 0.02$. As long as one can obtain a sufficiently deep photometric catalog, with the appropriate colors to map the red sequence, this technique can be used to very efficiently detect clusters. Like all methods that rely on the presence of a red sequence, it will eventually fail at sufficiently high redshifts, where the cluster galaxy population becomes more heterogeneous. However, clusters detected out to

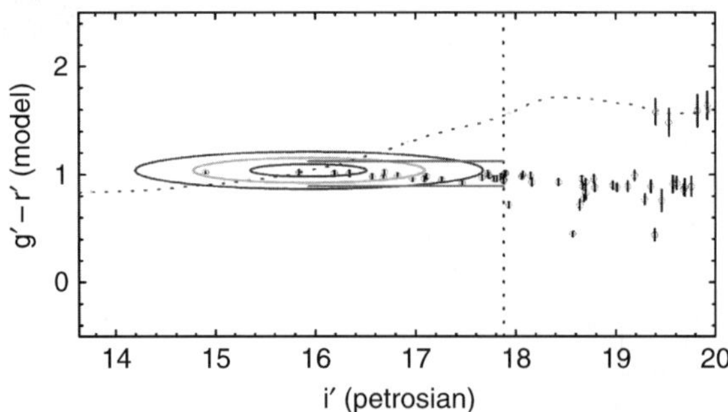

Fig. 8. SDSS color-magnitude diagram of observed $g-r$ vs. apparent i band for galaxies near a rich cluster at $z = 0.15$. Ellipses represent 1, 2, and 3 σ contours around the mean BCG color and magnitude at that redshift. The dotted line indicates the track of BCG color and magnitude as a function of redshift. The horizontal lines and vertical dashed line show the region of inclusion for N_{gal} determination. Taken from [26]

$z \sim$1–1.5, even using non-optical techniques, still show a red sequence, albeit with larger scatter, which will reduce the efficiency of this method. Additionally, the definition of $N_{\rm gals}$ as the number of red sequence galaxies may introduce a bias, as poorer, less concentrated, or more distant clusters have less well defined color-magnitude relations, and the luminosity functions for clusters vary with richness as well (Fig. 10 of [26]).

5.11 The Cluster Red Sequence Method

As we have discussed already, the existence of a tight color-magnitude relation for cluster galaxies provides a mechanism for reducing fore- and background contamination, enhancing cluster contrast, and estimating redshifts in cluster surveys. Because the red sequence is such a strong indicator of a cluster's presence, and is especially tight for the brighter cluster members, it can be used to detect clusters to high redshifts ($z \sim 1$) with comparatively shallow imaging, if an optimal set of photometric bands is chosen. This is the idea behind the Cluster Red Sequence (CRS; [22]) method, utilized by the Red Sequence Cluster Survey (RCS; [23]). Figure 9a shows model color-magnitude tracks for different galaxy types for $0.1 \leq z \leq 1.0$. The cluster ellipticals are the reddest objects at all redshifts. Even more importantly, if the filters used

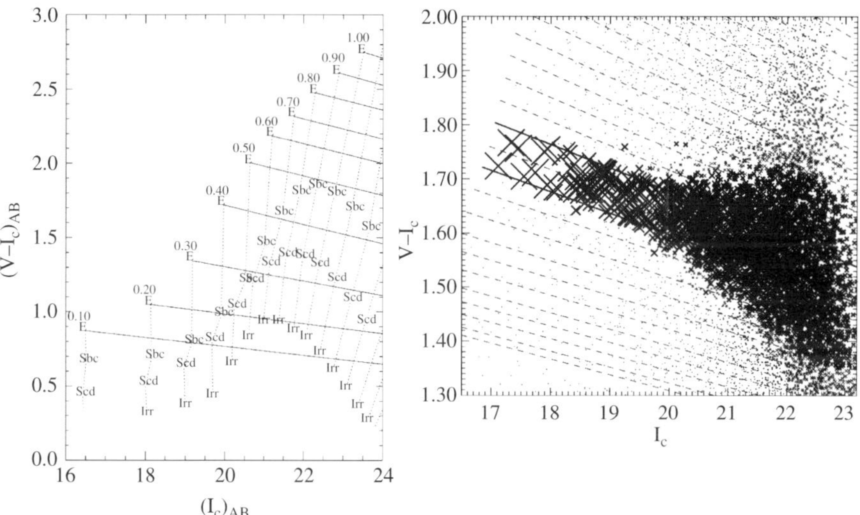

Fig. 9. Left: Simulated $(V{-}I_{\rm c})_{\rm AB}$ vs. $(I_{\rm c})_{\rm AB}$ color-magnitude diagram. Model apparent magnitudes and colors at various redshifts for several types of galaxies at a fixed M_I of –22 are shown. The dotted lines connect galaxies at the same redshift. Solid near-horizontal lines show the expected slope of the red sequence at each redshift. **Right**: CMD of a CNOC2 Redshift Survey Patch, with dashed lines showing various color CRS slices. The galaxy symbols are sized by the probability that they belong to the color slice defined by the solid lines. Taken from [22]

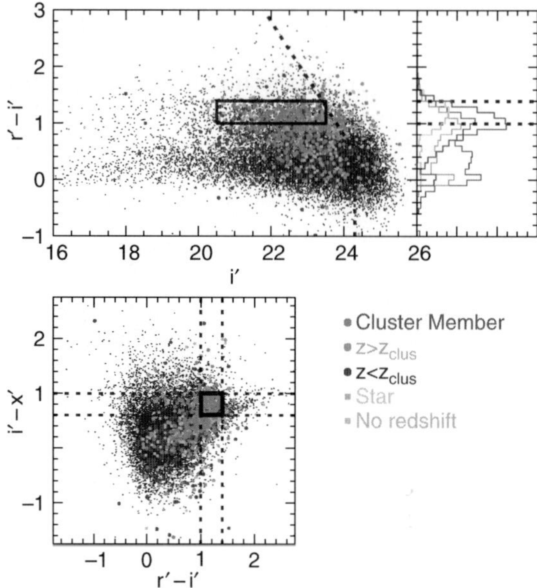

Fig. 10. r–i vs. i color–magnitude and r–i vs. i–z color–color diagrams for objects in the Cl1604 field

straddle the 4000 Åbreak at a given redshift, the cluster ellipticals at that redshift are redder than all galaxies at all lower redshifts. The only contaminants are more distant, bluer galaxies, eliminating most of the foreground contamination found in imaging surveys. The change of the red sequence color with redshift at a fixed apparent magnitude also makes it a very useful redshift estimator [37].

Gladders and Yee generate a set of overlapping color slices based on models of the red sequence. A subset of galaxies is selected that belong to each slice, based on their magnitudes, colors, color errors, and the models. A weight for each chosen galaxy is computed, based on the galaxy magnitude and the likelihood that the galaxy belongs to the color slice in question (Fig. 9b). A surface density map is then constructed for each slice using a fixed smoothing kernel, with a scale radius of 0.33 h^{-1} Mpc. All the slices taken together form a volume density in position and redshift. Peaks are then selected from this volume. Gladders et al. [23] present the results of this technique applied to the first two RCS patches.

In a similar vein, the High Redshift Large Scale Structure Survey [19, 20, 38], uses deep multicolor photometry around known clusters at $z > 0.7$ to search for additional large scale structure. They apply color and color–color cuts to select galaxies with the colors of spectroscopically confirmed members in the original clusters. The selected galaxies are used to make adaptive kernel

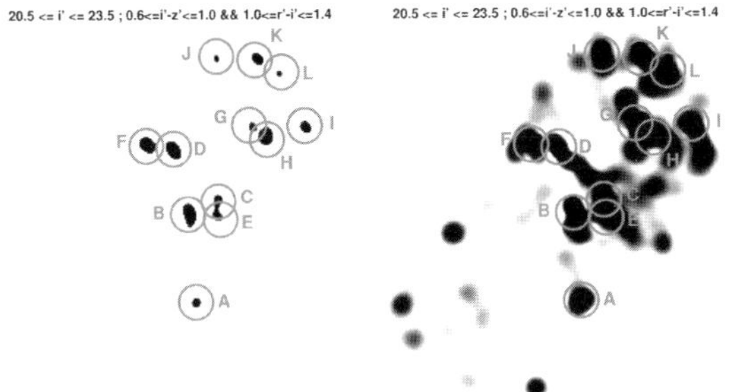

Fig. 11. Density maps of galaxies meeting the $z \sim 0.9$ red galaxy criteria in the Cl1604 field

density maps from which peaks are selected. This technique was applied to the Cl1604 supercluster at $z \sim 0.9$. Starting with two known clusters with approximately 20 spectroscopic members, there are now a dozen structures with 360 confirmed members known in this supercluster. These galaxies typically follow the red sequence, but as can be seen in Fig. 10, the scatter is very large, and many cluster or supercluster members are actually bluer than the red sequence at this redshift. Figure 10 shows the $r-i$ vs. i color-magnitude and $r-i$ vs. $i-z$ color–color diagrams for objects in a $\sim 30'$ square region around the Cl1604 supercluster, with all known cluster members shown in red. and the color selection boxes marked. Figure 11 shows the density map for this region, with two different significance thresholds, and the clusters comprising the supercluster marked. Clearly, in regions such as this, traditional cluster detection techniques will yield incorrect results, combining multiple clusters, and measuring incorrect redshifts and richnesses. Figure 12 shows a 3-d map of

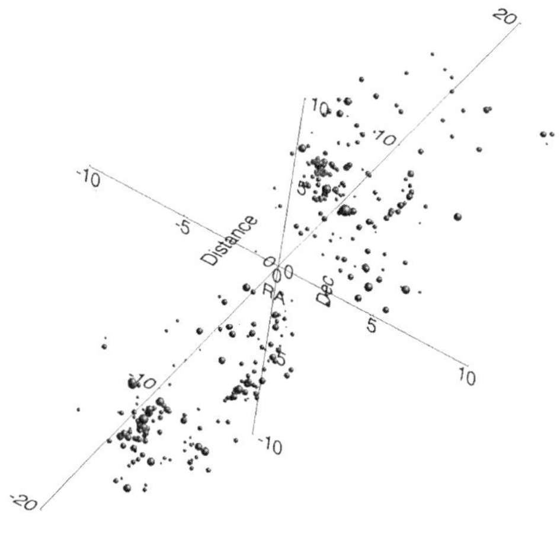

Fig. 12. Three dimensional spatial distribution of the spectroscopically confirmed Cl1604 supercluster members. Dots are scaled by galaxy luminosity

the spectroscopically confirmed supercluster members, revealing the complex nature of this structure. Dots are scaled with galaxy luminosity. While only ~ 10 Mpc across on the sky, the apparent depth of this structure is nearly 10 times greater, making it comparable to the largest local superclusters.

6 Conclusions

It is clear that there exist many methods for detecting clusters in optical imaging surveys. Some of these are designed to work on very simple, single-band data (AK, Matched Filter, VT), but will work on multicolor data as well. Others, such as maxBCG and the CRS method, rely on galaxy colors and the red sequence to potentially improve cluster detection and reduce contamination by projections and spurious objects. Very little work has been done to compare these techniques, with some exceptions ([5, 32, 36], each of whom compared the results of only two or three algorithms. Even from these tests it is clear that no single technique is perfect, although some (notably those that use colors) are clearly more robust. Certainly any program to find clusters in imaging data must consider the input photometry when deciding which, if any, of these methods to use.

One of the most vexing issues facing cluster surveys is our inability to compare directly to large scale cosmological simulations. Most such simulations are N-body only, but have perfect knowledge of object masses and positions. Thus, it is possible to construct algorithms to detect overdensities based purely on mass, but it is *not* possible to obtain the photometric properties of these objects! Recent work, such as the Millennium Simulation [58], is approaching this goal. It is necessary to extract from these simulations the magnitudes of galaxies in filters used for actual surveys, and run the various cluster detection algorithms on these simulated galaxy catalogs. The results can then be compared to that of pure mass selection, and the redshift-, structure- and mass-dependent biases understood. Ideally, this should be done for many large simulations using different cosmologies, since the galaxy evolution and selection effects will vary. Such work is fundamental if we are to use the evolution of the mass function of galaxy clusters for cosmology. As deeper and larger optical surveys, such as LSST, and other techniques such as X-ray and Sunyaev–Z'eldovich effect observations become available, the need for these simulations becomes ever greater.

References

1. Abell, G.O.: ApJS **3**, 211 (1958)
2. Abell, G.O., Corwin, H.G., Olowin, R.P.: ApJS **70**, 1 (1989)
3. Annis, J., et al.: AAS **31**, 1391 (1999)
4. Bahcall, N.A., Soneira, R.M.: ApJ **270**, 20 (1983)

5. Bahcall, N.A., et al.: ApJS **148**, 243 (2003)
6. Bartelmann, M., White, S.D.M.: A&A **388**, 732 (2002)
7. Bertin, E., Arnouts, S.: A&AS **117**, 393 (1996)
8. Botzler, C.S., et al.: MNRAS **349**, 425 (2004)
9. Bramel, D.A., Nichol, R.C., Pope, A.C.: ApJ **533**, 601 (2000)
10. Cohn, J.D., White, M.: APh **24**, 316 (2005)
11. Couch, W.J., et al.: MNRAS **249**, 606 (1991)
12. Dalton, G.B., et al.: MNRAS **289**, 263 (1997)
13. Davis, M., et al.: ApJ **292**, 371 (1985)
14. Efstathiou, G., et al.: MNRAS **235**, 715 (1988)
15. Efstathiou, G., et al.: MNRAS **257**, 125 (1992)
16. Gal, R.R., et al.: AJ **119**, 12 (2000)
17. Gal, R.R., et al.: AJ **125**, 2064 (2003)
18. Gal, R.R., et al.: AJ **128**, 3082 (2004a)
19. Gal, R.R., Lubin, L.M.: ApJ **607**, L1 (2004b)
20. Gal, R.R., Lubin, L.M., Squires, G.K.: AJ **129**, 1827 (2005)
21. Gal, R.R., et al.: In preparation (2007)
22. Gladders, M.D., Yee, H.K.C.: AJ **120**, 2148 (2000)
23. Gladders, M.D., Yee, H.K.C.: ApJS **157**, 1 (2005)
24. Goto, T., et al.: AJ **123**, 1807 (2002)
25. Gunn, J.E., Hoessel, J.G., Oke, J.B.: ApJ **306**, 30 (1986)
26. Hansen, S.M., et al.: ApJ **633**, 122 (2005)
27. Heydon-Dumbleton, N.H., Collins, C.A., MacGillivray, H.T.: MNRAS **238**, 379 (1989)
28. Huchra, J.P., Geller, M.J.: ApJ **257**, 423 (1982)
29. Katgert, P., et al.: A&A **310**, 8 (1996)
30. Katz, L., Mulders, G.F.W.: ApJ **95**, 565 (1942)
31. Kepner, J., et al.: ApJ **517**, 78 (1999)
32. Kim, R.S.J., et al.: AJ **123**, 20 (2002)
33. Lasker, B.M., et al.: ASPC **101**, 88 (1996)
34. Lidman, C.E., Peterson, B.A.: AJ **112**, 2454 (1996)
35. Lobo, C., et al.: A&A **360**, 896 (2000)
36. Lopes, P.A.A., et al.: AJ **128**, 1017 (2004)
37. López-Cruz, O., Barkhouse, W.A., Yee, H.K.C.: ApJ **614**, 679 (2004)
38. Lubin, L.M., Gal, R.R.: In preparation (2007)
39. Lucey, J.R.: MNRAS **204**, 33 (1983)
40. Lumsden, S.L., et al.: MNRAS **258**, 1 (1992)
41. Maddox, S.L., et al.: MNRAS **243**, 692 (1990)
42. Miller, C.J., et al.: ApJ **523**, 492 (1999)
43. Miller, C.J., et al.: AJ **130**, 968 (2005)
44. Odewahn, S.C., Aldering, G.: AJ **110**, 2009 (1995)
45. Odewahn, S.C., et al.: AJ **128**, 3092 (2004)
46. Olivier, S., et al.: ApJ **356**, 1 (1990)
47. Pennington, R.L., et al.: PASP **105**, 521 (1993)
48. Picard, A.: AJ **102**, 445 (1991)
49. Postman, M., et al.: AJ **111**, 615 (1996)
50. Ramella, M., et al.: A&A **368**, 776 (2001)
51. Ramella, M., et al.: AJ **123**, 2976 (2002)
52. Reid, I.N., et al.: PASP **103**, 661 (1991)

53. Shane, C.D., Wirtanen, C.A.: AJ **59**, 285 (1954)
54. Shectman, S.A.: ApJS **57**, 77 (1985)
55. Schechter, P.: ApJ **203**, 297 (1976)
56. Schuecker, P., Böhringer, H., Voges, W.: A&A **420**, 61 (2004)
57. Silverman, B.W.: Density Estimation for Statistics and Data Analysis. Chapman and Hall, London (1986)
58. Springel, V., et al.: Nature **435**, 629 (2005)
59. Sutherland, W.: MNRAS **234**, 159 (1988)
60. van Haarlem, M.P., Frenk, C.S., White, S.D.M.: MNRAS **287**, 817 (1997)
61. York, D.G., et al.: AJ **120**, 1579 (2000)
62. Zaritsky, D., et al.: ApJ **480**, L91 (1997)
63. Zaritsky, D., et al.: ASPC **257**, 133 (2002)
64. Zwicky, F.: PASP **50**, 218 (1938)
65. Zwicky, F.: PASP **54**, 185 (1942)
66. Zwicky, F., Herzog, E., Wild, P.: Catalogue of Galaxies and of Clusters of Galaxies. Caltech, Pasadena (1968)

Clusters of Galaxies in the Radio: Relativistic Plasma and ICM/Radio Galaxy Interaction Processes

L. Feretti[1] and G. Giovannini[1,2]

[1] Istituto di Radioastronomia INAF, Via P. Gobetti 101, 40129 Bologna, Italy
lferetti@ira.inaf.it
[2] Dipartimento di Astronomia, Universitá di Bologna, Via Ranzani 1, 40127 Bologna, Italy
gabriele.giovannini@unibo.it

1 Introduction

Studies at radio wavelengths allow the investigation of important components of clusters of galaxies. The most spectacular aspect of cluster radio emission is represented by the large-scale diffuse radio sources, which cannot be obviously associated with any individual galaxy. These sources indicate the existence of relativistic particles and magnetic fields in the cluster volume, thus the presence of non-thermal processes in the hot intracluster medium (ICM). The knowledge of the properties of these sources has increased significantly in recent years, due to higher sensitivity radio images and to the development of theoretical models. The importance of these sources is that they are large scale features, which are related to other cluster properties in the optical and X-ray domain, and are thus directly connected to the cluster history and evolution.

The radio emission in clusters can also originate from individual galaxies, which have been imaged over the last decades with sensitive radio telescopes. The emission from radio galaxies often extends well beyond their optical boundaries, out to hundreds of kiloparsec, and hence it is expected that the ICM would affect their structure. This interaction is indeed observed in extreme examples: the existence of radio galaxies showing distorted structures (tailed radio sources), and radio sources filling X-ray cavities at the centre of cooling core clusters. Finally, the cluster environment may play a role in the statistical radio properties of galaxies, i.e. their probability of forming radio sources.

The organization of this paper is as follows: The basic formulae used to derive the age of synchrotron sources and the equipartition parameters are presented in Sect. 2, while the observational properties of diffuse radio sources are presented in Sect. 3. Then in Sect. 4 we give a general outline of the models

L. Feretti and G. Giovannini: *Clusters of Galaxies in the Radio: Relativistic Plasma and ICM/Radio Galaxy Interaction Processes*, Lect. Notes Phys. **740**, 143–176 (2008)
DOI 10.1007/978-1-4020-6941-3_5

of the relativistic particle origin and re-acceleration; while the current results on cluster magnetic fields are described in Sect. 5. Finally, Sect. 6 reports the properties of cluster radio emitting galaxies.

The intrinsic parameters quoted in this paper are computed for a ΛCDM cosmology with $H_0 = 70$ km s^{-1}Mpc^{-1}, Ω_m=0.3 and Ω_Λ=0.7.

2 Basic Formulas from the Synchrotron Theory

2.1 Synchrotron Radiation

The synchrotron emission is produced by the spiralling motion of relativistic electrons in a magnetic field. An electron with energy $E = \gamma m_e c^2$ (where γ is the Lorentz factor) in a magnetic field $\boldsymbol{B}$, experiences a $\boldsymbol{v}\times\boldsymbol{B}$ force that causes it to follow a helical path along the field lines, emitting radiation into a cone of half-angle $\simeq \gamma^{-1}$ about its instantaneous velocity. To the observer, the radiation is essentially a continuum with a fairly peaked spectrum concentrated near the frequency

$$\nu_{\rm syn} = \frac{3e}{4\pi m_e^3 c^5}(B\sin\theta)\mathrm{e}^2 \ , \qquad (1)$$

where θ is the pitch angle between the electron velocity and the magnetic field direction. The synchrotron power emitted by a relativistic electron is

$$-\frac{\mathrm{d}E}{\mathrm{d}t} = \frac{2e^4}{3m_e^4c^7}(B\sin\theta)^2E^2 \ . \qquad (2)$$

In c.g.s units:

$$\begin{aligned}\nu_{\rm syn} &\simeq 6.27\times10^{18}(B\sin\theta)E^2 \qquad (3)\\ &\simeq 4.2\times10^{6}(B\sin\theta)\gamma^2,\end{aligned}$$

$$\begin{aligned}-\frac{\mathrm{d}E}{\mathrm{d}t} &\simeq 2.37\times10^{-3}(B\sin\theta)^2E^2 \qquad (4)\\ &\simeq 1.6\times10^{-15}(B\sin\theta)^2\gamma^2.\end{aligned}$$

From (3), it is easily derived that electrons of $\gamma \simeq 10^3-10^4$ in magnetic fields of $B \simeq 1\mu$ G radiate in the radio domain.

The case of astrophysical interest is that of a homogeneous and isotropic population of electrons with a power-law energy distribution, i.e., with the particle density between E and E+dE given by:

$$N(E)\mathrm{d}E = N_0E^{-\delta}\mathrm{d}E \ . \qquad (5)$$

To obtain the total monochromatic emissivity $J(\nu)$, one must integrate over the contributions of all electrons. In regions which are optically thin to their own radiation (i.e. without any internal absorption), the total intensity spectrum varies as [14]:

$$J(\nu) \propto N_0 (B \sin\theta)^{1+\alpha} \nu^{-\alpha} , \tag{6}$$

therefore it follows a power-law with spectral index related to the index of the electron energy distribution $\alpha = (\delta - 1)/2$.

2.2 Time Evolution of the Synchrotron Spectrum

By integrating the expression of the electron energy loss (2) it is found that the particle energy decreases with time, as:

$$E = \frac{E_0}{1 + b(B \sin\theta)^2 E_0 t} , \tag{7}$$

where E_0 is the initial energy at $t = 0$, and $b = 2e^4/(3m_e^4 c^7) = 2.37 \times 10^{-3}$ c.g.s units (see 4). Therefore, the particle energy halves after a time t^* $= [b(B \sin\theta)^2 E_0]^{-1}$. This is a characteristic time which can be identified as the particle lifetime. Similarly, we can define a characteristic energy $E^* = [b(B \sin\theta)^2 t]^{-1}$, such that a particle with energy $E_0 > E^*$ will lose most of its energy in a time t^*.

In an ensemble of particles, the energy losses of each particle affect the overall particle energy distribution, and consequently the resulting synchrotron spectrum undergoes a modification. Indeed, after a time t^* the particles with $E > E^*$ will lose most of their energy. This produces a critical frequency ν^* in the radio spectrum, such that for $\nu < \nu^*$ the spectrum is unchanged, whereas for $\nu > \nu^*$ the spectrum steepens. If particles were produced in a single event with power law energy distribution, $N(E,0)\mathrm{d}E = N_0 E^{-\delta} \mathrm{d}E$, the radio spectrum would fall rapidly to zero for $\nu > \nu^*$. In the case that new particles were injected in the source, the spectrum beyond ν^* steepens by 0.5. These various cases are illustrated in Fig. 1. Any radio spectrum showing a cutoff is evidence of ageing of the radio emitting particles. In addition, any spectrum showing no cutoff but having a steep spectral index is also indicative

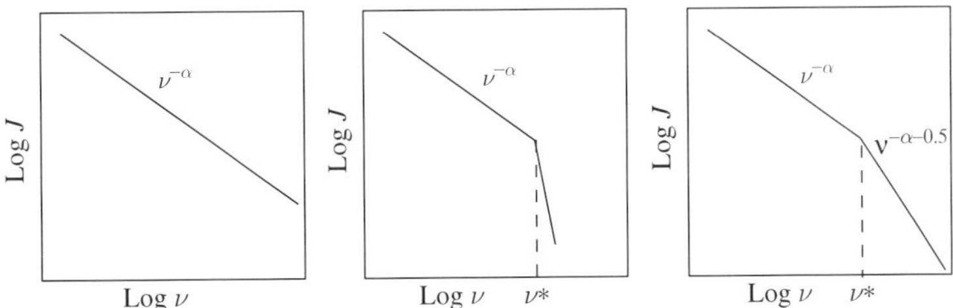

Fig. 1. Sketch of synchrotron spectra. The left panel shows a standard spectrum, the central panel shows an aged spectrum produced in a source with a single event of particle production, the right panel shows an aged spectrum with particle injection. The critical frequency ν^* is related to the particle lifetime

of ageing, since it naturally refers to a range of frequencies higher than the critical frequency. For a rigorous treatment of the evolution of synchrotron spectra we refer to [75] and [117].

From the critical frequency ν^*, it is possible to derive the radiating electron lifetime, which represents the time since the particle production (or the time since the last injection event, depending on the shape of spectral steepening). Since the synchrotron emission depends on $\sin\theta$ (1), one has to take into account the distribution of electron pitch angles. Moreover, for a correct evaluation, also the electron energy losses, due to the inverse Compton process, must be considered.

The electron lifetime (in Myr), assuming an *anisotropic* pitch angle distribution is given by:

$$t^* = 1060 \frac{B^{0.5}}{B^2 + \frac{2}{3}B^2_{\rm CMB}} \left[(1+z)\nu^*\right]^{-0.5} , \tag{8}$$

where the magnetic field B is in μG, the frequency ν is in GHz and $B_{\rm CMB}(= 3.25\,(1+z)^2\ \mu{\rm G})$ is the equivalent magnetic field of the Cosmic Microwave Background. If the distribution of electron pitch angles is *isotropic*, the above formula becomes:

$$t^* = 1590 \frac{B^{0.5}}{B^2 + B^2_{\rm CMB}} \left[(1+z)\nu^*\right]^{-0.5} . \tag{9}$$

A derivation of the expressions in (8) and (9) can be found in [111].

2.3 Energy Content and Equipartition Magnetic Fields

The total energy of a synchrotron source is due to the energy in relativistic particles ($U_{\rm el}$ in electrons and $U_{\rm pr}$ in protons) plus the energy in magnetic fields (U_B):

$$U_{\rm tot} = U_{\rm el} + U_{\rm pr} + U_B . \tag{10}$$

The magnetic field energy contained in the source volume V is given by

$$U_B = \frac{B^2}{8\pi}\Phi V , \tag{11}$$

where Φ is the fraction of the source volume occupied by the magnetic field (filling factor). The electron total energy in the range E_1–E_2,

$$U_{\rm el} = V \times \int_{E_1}^{E_2} N(E) E \,{\rm d}E = V N_0 \int_{E_1}^{E_2} E^{-\delta+1}\,{\rm d}E , \tag{12}$$

can be expressed as a function of the synchrotron luminosity, $L_{\rm syn}$, observed between two frequencies ν_1 and ν_2, i.e.,

$$U_{\rm el} = L_{\rm syn}(B\sin\theta)^{-\frac{3}{2}} f(\delta, \nu_1, \nu_2) , \tag{13}$$

where $f(\delta, \nu_1, \nu_2)$ is a function of the index of the electron energy distribution and of the observing frequencies (see [96] for a rigorous derivation). The energy contained in the heavy particles, $U_{\rm pr}$, can be related to $U_{\rm el}$ assuming:

$$U_{\rm pr} = kU_{\rm el} \,. \tag{14}$$

Finally, taking $\sin\theta$=1, the total energy is:

$$U_{\rm tot} = (1+k)L_{\rm syn}B^{-\frac{3}{2}}f(\delta, \nu_1, \nu_2) + \frac{B^2}{8\pi}\Phi V \,. \tag{15}$$

The trend of the radio source energy content is shown in Fig. 2. The condition of minimum energy, $U_{\rm min}$, computed by equating to zero the first derivative of the expression of $U_{\rm tot}$ (15), is obtained when the contributions of the magnetic field and the relativistic particles are approximately equal:

$$U_B = \frac{3}{4}(1+k)U_{\rm el} \,. \tag{16}$$

For this reason the minimum energy is known also as equipartition value.

The total minimum energy density $u_{\rm min} = U_{\rm min}/V\Phi$, assuming same volume in particles and magnetic field (Φ=1), and applying the K-correction, can be expressed in terms of observable parameters, as:

$$u_{\rm min} = 1.23 \times 10^{-12}(1+k)^{\frac{4}{7}}(\nu_0)^{\frac{4\alpha}{7}}(1+z)^{\frac{(12+4\alpha)}{7}}I_0^{\frac{4}{7}}d^{\frac{4}{7}} \,, \tag{17}$$

where I_0 is the source brightness which is directly observed at the frequency ν_0, d is the source depth along the line of sight, z is the source redshift and α

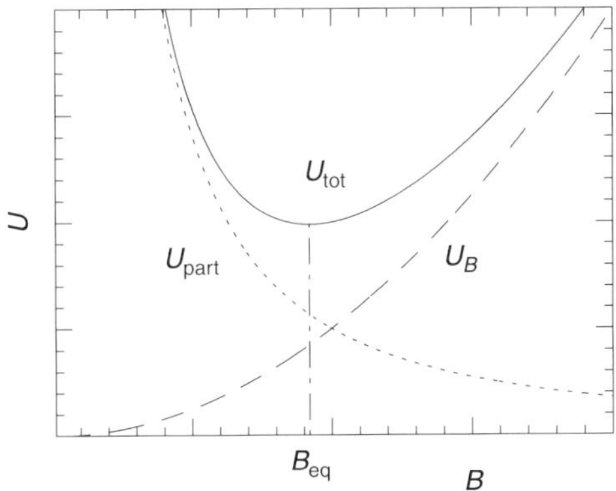

Fig. 2. Trend of the energy content in a radio source (in arbitrary units): the energy in magnetic fields is $U_B \propto B^2$, the energy in relativistic particles is $U_{\rm part} = U_{\rm el}+U_{\rm pr} \propto {\rm B}^{-3/2}$. The total energy content $U_{\rm tot}$ is minimum when the contributions of magnetic fields and relativistic particles are approximately equal (equipartition condition). The corresponding magnetic field is commonly referred to as equipartition value $\rm B_{eq}$

is the spectral index of the radio emission. The energy density is in erg cm^{-3}, ν_0 in MHz, I_0 in mJy arcsec^{-2} and d in kpc. I_0 can be measured from the contour levels of a radio image (for significantly extended sources) or can be obtained by dividing the source total flux by the source solid angle, while d can be inferred from geometrical arguments. The constant has been computed for $\alpha = 0.7$, $\nu_1 = 10$ MHz and $\nu_2 = 100$ GHz (tabulated in [69], for other values of these parameters).

The magnetic field for which the total energy content is minimum is referred to as the equipartition value and is derived as follows:

$$B_{\rm eq} = \left(\frac{24\pi}{7} u_{\rm min}\right)^{\frac{1}{2}} . \tag{18}$$

One must be aware of the uncertainties inherent to the determination of the minimum energy density and equipartition magnetic field strength. The value of k, the ratio of the energy in relativistic protons to that in electrons (14), depends on the mechanism of generation of relativistic electrons, which, so far, is poorly known. Values usually assumed in literature for clusters are $k = 1$ (or $k = 0$). Uncertainties are also related to the volume filling factor Φ.

In the standard approach presented above, the equipartition parameters are obtained from the synchrotron radio luminosity observed between the two fixed frequencies ν_1 and ν_2. Brunetti et al. [18] demonstrated that it is more appropriate to calculate the radio source energy by integrating the synchrotron luminosity over a range of electron energies. This avoids the problem

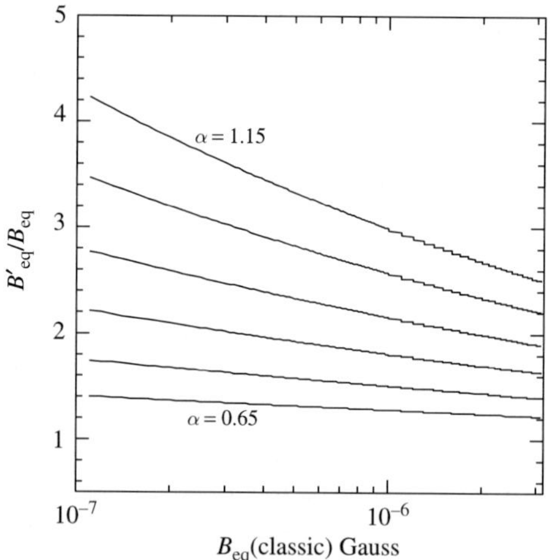

Fig. 3. Values of the ratio $B'_{\rm eq}/B_{\rm eq}$ (see text) as a function of the equipartition magnetic field obtained with the classical approach, assuming an electron minimum Lorentz factor $\gamma_{\rm min} = 50$. Different lines refer to different values of the initial spectral index (i.e. not affected by ageing), from $\alpha = 1.15$ (top line) to $\alpha = 0.65$ (bottom line) in steps of $\alpha = 0.1$

that electron energies corresponding to frequencies ν_1 and ν_2 depend on the magnetic field value (see 1), thus the integration over a range of fixed frequencies is equivalent to considering radiating electrons over a variable range of energies. Moreover, it has the advantage that electrons of very low energy are also taken into account. The equipartition quantities obtained by following this approach are presented by [18] and [6]. Representing the electron energy by its Lorentz factor γ, and assuming that $\gamma_{\rm min} \ll \gamma_{\rm max}$, the new expression for the equipartition magnetic field $B'_{\rm eq}$ in Gauss (for $\alpha > 0.5$) is:

$$B'_{\rm eq} \sim 1.1\, \gamma_{\rm min}^{\frac{1-2\alpha}{3+\alpha}}\, B_{\rm eq}^{\frac{7}{2(3+\alpha)}}\,, \qquad (19)$$

where $B_{\rm eq}$ is the value of the equipartition magnetic field obtained with the standard formulae by integrating the radio spectrum between 10 MHz and 100 GHz. It should be noticed that $B'_{\rm eq}$ is larger than $B_{\rm eq}$ for $B_{\rm eq} < \gamma_{\rm min}^{-2}$ (see Fig. 3).

3 Radio Emission from the ICM: Diffuse Radio Sources

In recent years, there has been growing evidence for the existence of cluster large-scale diffuse radio sources, which have no optical counterpart and no obvious connection to cluster galaxies, and are therefore associated with the ICM. These sources are typically grouped in 3 classes: halos, relics and mini-halos. The number of clusters with halos and relics is presently around 50, and whose properties have been recently reviewed by Giovannini & Feretti [60] and Feretti [48]. The synchrotron nature of this radio emission indicates the presence of cluster-wide magnetic fields of the order of $\sim$ 0.1–1 μG, and of a population of relativistic electrons with Lorentz factor $\gamma \gg 1000$. The understanding of these non-thermal components is important for a comprehensive physical description of the ICM.

3.1 Radio Halos

Radio halos are diffuse radio sources of low surface brightness ($\sim$ µJy arcsec^{-2} at 20 cm) permeating the central volume of a cluster. They are typically extended with sizes $\gtrsim$ 1 Mpc and are unpolarized down to a few percent level. The prototype of this class is the diffuse source Coma C at the centre of the Coma cluster ([57] and Fig. 4), first classified by Willson [122]. The halo in A2163, shown in left panel of Fig. 5, is one of the most extended and powerful halos. Other well studied giant radio halos are present in A665 [59], A2219 [2], A2255 [42], A2319 [43], A2744 (Fig. 7, left panel), 1E0657-56 [84], and in the distant cluster CL 0016+16 [59] at redshift $z = 0.555$. All these clusters show recent merging processes, and no cooling core.

Radio halos of small size, i.e. $\ll$ 1 Mpc, have also been revealed in the central regions of clusters. Some examples are in A401 [59], A1300 [99], A2218

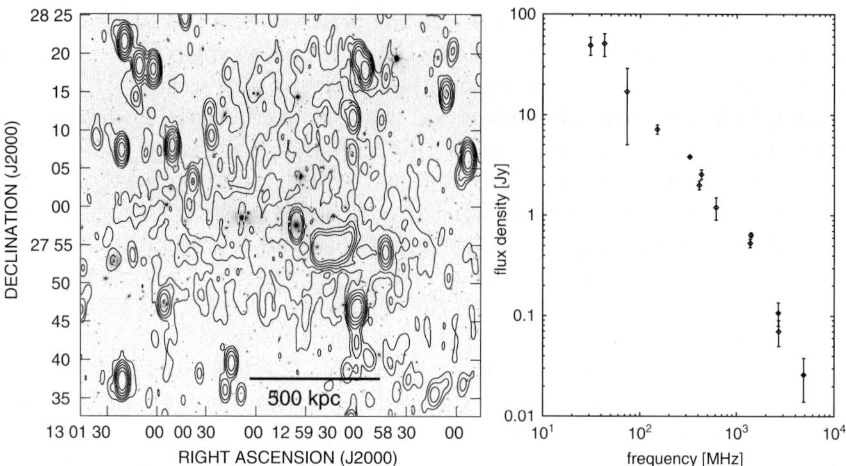

Fig. 4. Left panel: Diffuse radio halo Coma C in the Coma cluster ($z = 0.023$) at 0.3 GHz, superimposed onto the optical image from the DSS1. The resolution of the radio image is $55'' \times 125''$(FWHM, RA $\times$ DEC); contour levels are: 3, 6, 12, 25, 50, 100 mJy/beam. **Right panel:** Total radio spectrum of the radio halo Coma C (from [115])

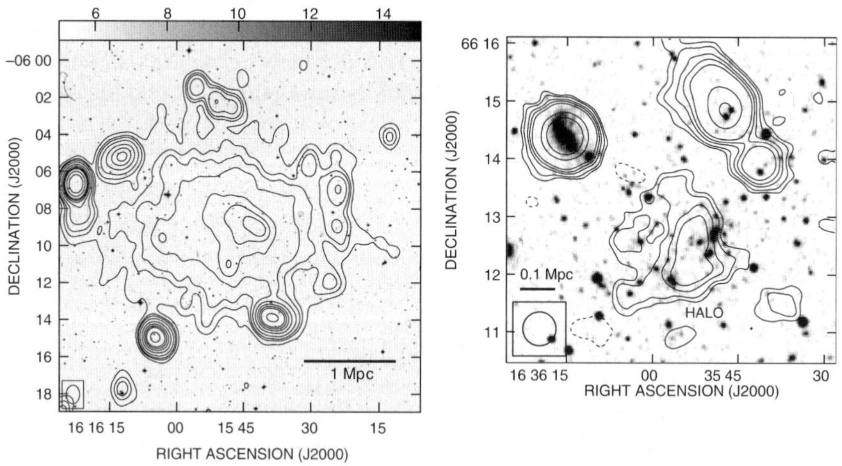

Fig. 5. Left panel: Radio emission in A2163 ($z = 0.203$) at 20 cm [45]. The radio halo is one of the most powerful and extended halos known so far. **Right panel:** Radio emission of the cluster A2218 ($z = 0.171$) at 20 cm [59]. In both clusters the radio contours are overlayed onto the grey-scale optical image

(Fig. 5, right panel) and A3562 [55]. All these clusters, as well as those hosting giant radio halos, are characterized by recent merger processes and no cooling core.

Unlike the presence of thermal X-ray emission, the presence of diffuse radio emission is not common in clusters of galaxies: the detection rate of radio halos, at the detection limit of the NRAO VLA Sky Survey (NVSS) is $\sim 5\%$ in a complete cluster sample [58]. However, the probability is much larger, if clusters with high X-ray luminosity are considered. Indeed, $\sim 35\%$ of clusters with X-ray luminosity larger than 10^{45} erg s^{-1} X-ray (in the ROSAT band 0.1–2.4 keV, computed assuming $H_0 = 50$ km s^{-1}Mpc^{-1} and $q_0 = 0.5$) show a giant radio halo [60].

The physical parameters in radio halos can be estimated assuming equipartition conditions, and further assuming equal energy in relativistic protons and electrons, a volume filling factor of 1, a low frequency cut-off of 10 MHz, and a high frequency cut-off of 10 GHz. The derived minimum energy densities in halos and relics are of the order of 10^{-14}–10^{-13} erg cm^{-3}, i.e. much lower than the energy density in the thermal gas. The corresponding equipartition magnetic field strengths range from 0.1 to 1 μG.

The total radio spectra of halos are steep ($\alpha \gtrsim 1$),[1] as typically found in aged radio sources. Only a few halos have good multi-frequency observations that allow an accurate determination of their integrated spectrum. Among them, the spectrum of the Coma cluster halo is characterized by a steepening at high frequencies, which has been recently confirmed by single dish data (Fig. 4, right panel). The spectrum of the radio halo in A1914 is very steep, with an overall slope of $\alpha \sim 1.8$. A possible high frequency curvature is discussed by Komissarov & Gubanov [79]. In A754, Bacchi et al. [2] estimate $\alpha_{0.07\,\mathrm{GHz}}^{0.3\,\mathrm{GHz}} \sim 1.1$, and $\alpha_{0.3\,\mathrm{GHz}}^{1.4\,\mathrm{GHz}} \sim 1.5$, and infer the presence of a possible spectral cutoff. Indication of a high frequency spectral steepening is also obtained in the halo of A2319, where Feretti et al. [43] report $\alpha_{0.4\,\mathrm{GHz}}^{0.6\,\mathrm{GHz}} \sim 0.9$ and $\alpha_{0.6\,\mathrm{GHz}}^{1.4\,\mathrm{GHz}} \sim 2.2$. In the few clusters where maps of the spectral index are available (Coma C, [57]; A665 and A2163, [49], the radio spectrum steepens radially with the distance from the cluster centre. In addition, it is found that the spectrum in A665 and A2163 is flatter in the regions influenced by merger processes (see Sect. 4.1).

In general, from the spectra of halos, it is derived that the radiative lifetime of the relativistic electrons, considering synchrotron and inverse Compton energy losses, is of the order of $\sim 10^8$ yr [107]. Since the expected diffusion velocity of the electron population is of the order of the Alfvén speed (~ 100 km s^{-1}), the radiative electron lifetime is too short to allow the particle diffusion throughout the cluster volume. Thus, the radiating electrons cannot have been produced at some localized point of the cluster, but they must undergo *in situ* energization, acting with an efficiency comparable to the energy loss

[1] $S(\nu) \propto \nu^{-\alpha}$ as in (6).

processes [97]. We will show in Sect. 4 that recent cluster mergers are likely to supply energy to the halos and relics.

The radio and X-ray properties of halo clusters are related. The most powerful radio halos are detected in the clusters with the highest X-ray luminosity. This follows from the correlation shown in Fig. 6 between the monochromatic radio power of a halo at 20 cm and the bolometric X-ray luminosity of the parent cluster [60, 84]. The right panel of Fig. 6 shows the correlation between the average surface brightness of the radio halo and the cluster X-ray luminosity. Since the brightness is an observable, this correlation can be used to set upper limits to the radio emission to those clusters in which a radio halo is not detected. It is worth reminding the reader that the radio power versus X-ray luminosity correlation is valid for merging clusters with radio halos, and therefore cannot be generalized to all clusters. Among the clusters with high X-ray luminosity and no radio halo, there are A478, A576, A2204, A1795, A2029, all well known relaxed clusters with a massive cooling flow. An extrapolation of the above correlation to low radio and X-ray luminosities indicates that clusters with $L_X \lesssim 10^{45}$ erg s^{-1} would host halos of power of a few 10^{23} W Hz^{-1}. With a typical size of 1 Mpc, they would have a radio surface brightness (easily derived from the right panel of Fig. 6) lower than current limits obtained in the literature and in the NVSS. On the other hand, it is possible that giant halos are only present in the most X-ray luminous clusters, i.e. above a threshold of X-ray luminosity (see [2]). Future radio data with next generation instruments (LOFAR, LWA, SKA) will allow the detection of low brightness/low power large halos, in order to clarify if halos are present in all merging clusters or only in the most massive ones.

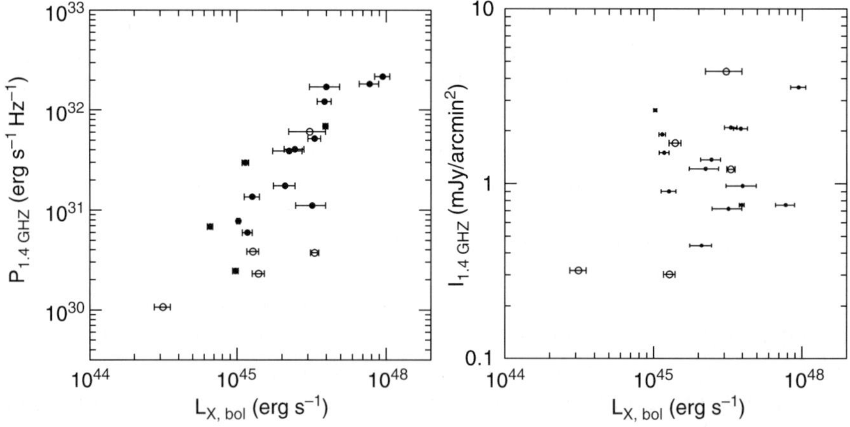

Fig. 6. Left panel: Monochromatic radio power at 20 cm versus cluster bolometric X-ray luminosity. **Right panel:** Average surface brightness of the radio halos versus cluster X-ray luminosity. In both panels, filled and open circles refer to halos of size > and < 1 Mpc, respectively

Since cluster X-ray luminosity and mass are correlated [100], the correlation between radio power ($P_{1.4\ \mathrm{GHz}}$) and X-ray luminosity could reflect a dependence of the radio power on the cluster mass. A correlation of the type $P_{1.4\ \mathrm{GHz}} \propto M^{2.3}$ has been derived [48, 66], where M is the total gravitational mass within a radius of $3h_{50}^{-1}$ Mpc. Using the cluster mass within the virial radius, the correlation is steeper (Cassano et al. in preparation). A correlation of radio power vs cluster mass could indicate that the cluster mass may be a crucial parameter in the formation of radio halos, as also suggested by [23]. Since it is likely that massive clusters are the result of several major mergers, it is concluded that both past mergers and current mergers are the necessary ingredients for the formation and evolution of radio halos. This scenario may provide a further explanation of the fact that not all clusters showing recent mergers host radio halos, which is expected from the recent modeling of Cassano & Brunetti [24].

3.2 Radio Relics

Relic sources are diffuse extended sources, similar to the radio halos in their low surface brightness, large size ($\gtrsim$ 1 Mpc) and steep spectrum ($\alpha \gtrsim 1$), but they are generally detected in the cluster peripheral regions. They typically show an elongated radio structure with the major axis roughly perpendicular to the direction of the cluster radius, and they are strongly polarized ($\sim$ 20–30%). The most extended and powerful sources of this class are detected in clusters with central radio halos: in the Coma cluster (the prototype relic source 1253+275, [56], A2163 [45], A2255 [42], A2256 [103] and A2744 (Fig. 7, left panel). A spectacular example of two giant almost symmetric relics in the same cluster is found in A3667 (Fig. 7, right panel). There are presently only a few cases of double opposite relics in clusters.

Other morphologies have been found to be associated with relics (see [61] for a review). In the cluster A1664 (Fig. 8, left panel), the structure is approximately circular and regular. In A115 (Fig. 8, right panel), the elongated relic extends from the cluster center to the periphery. This could be due to projection effects, however this is the only relic showing such behaviour.

There are diffuse radio sources which are naturally classified as relics, because of their non-central cluster location, but their characteristics are quite different from those of giant relics. Examples of these sources are in A13, A85 (Fig. 9), A133, A4038 [111]: they show a much smaller size than relics ($\lesssim$ 300 kpc down to $\sim$ 50 kpc), are generally closer to the cluster center, and show extremely steep radio spectra ($\alpha \gtrsim 2$). They are strongly polarized ($\gtrsim$ 30%), and often quite filamentary when observed with sufficient resolution. The relic in A133 was suggested to be related to past activity of a nearby galaxy [50].

The detection rate of radio relics in a complete sample of clusters is $\sim$6% at the detection limit of the NVSS [60]. Relics are found in clusters both with and without a cooling core, suggesting that they may be related to minor or off-axis

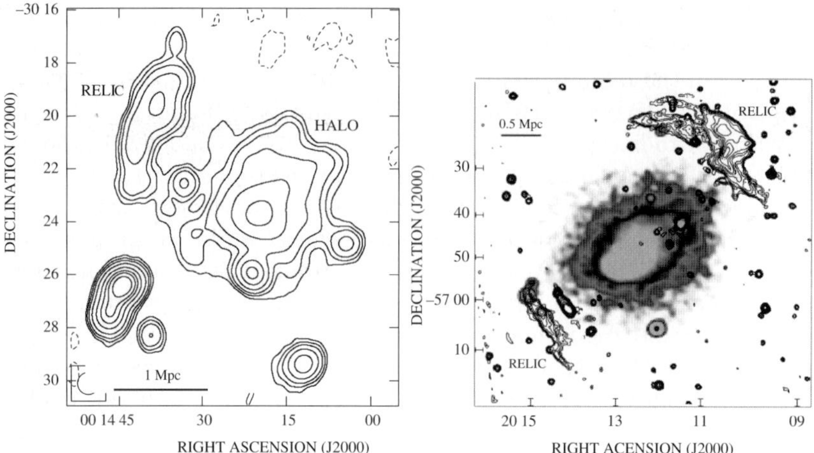

Fig. 7. Left panel: Radio emission of A2744 ($z = 0.308$) showing a peripheral elongated relic, and a central radio halo [66]. **Right panel:** A3667 ($z = 0.055$): contours of the radio emission at 36 cm [104] overlayed onto the grey-scale ROSAT X-ray image. Two radio relics are located on opposite sides of the cluster along the axis of the merger, with the individual radio structures elongated perpendicular to this axis

mergers, as well as to major mergers. Theoretical models propose that they are tracers of shock waves in merger events (see Sect. 4.3). This is consistent with their elongated structure, almost perpendicular to the merger axis. The radio power of relics correlates with the cluster X-ray luminosity [46, 61], as also found for halos (see Sect. 3.1 and Fig. 6), although with a larger dispersion.

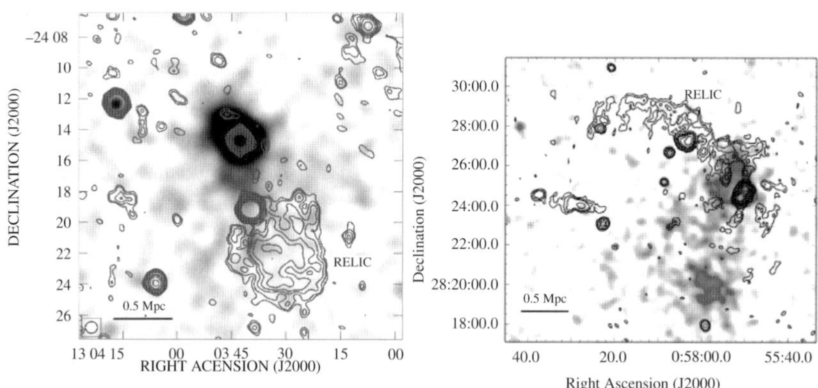

Fig. 8. Radio emission at 20 cm (contours) of the clusters: **Left panel:** A1664 ($z = 0.128$), **Right panel:** A115 ($z = 0.197$), superimposed onto the grey-scale cluster X-ray emission detected from ROSAT PSPC [66]

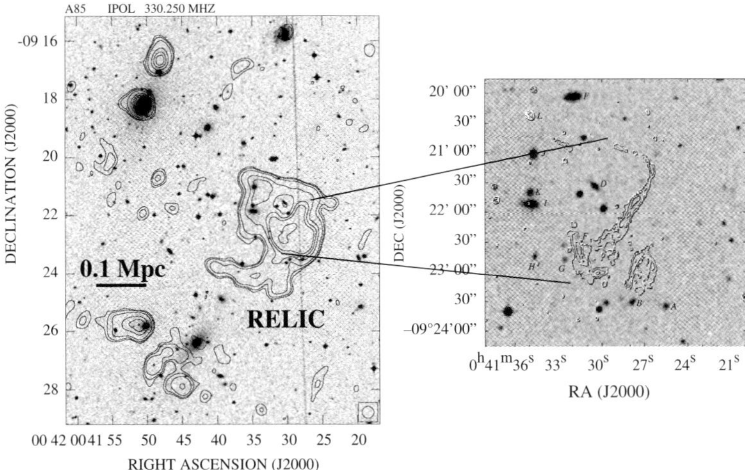

Fig. 9. Radio emission at 90 cm (contours) in A85 ($z = 0.056$), superimposed onto the optical image [59]. The zoom to the right shows the filamentary structure detected at high resolution by Slee et al. [111] at 20 cm

The existence of this correlation indicates a link between the thermal and relativistic plasma also in peripheral cluster regions.

3.3 Mini-Halos

Mini-halos are small size ($\sim$ 500 kpc) diffuse radio sources at the center of cooling core clusters, usually surrounding a powerful radio galaxy, as in the Perseus cluster (Fig. 10, left panel), Virgo cluster [95], PKS 0745-191 [4], A2626 [63]. Since there is an anticorrelation between the presence of a cooling core and that of a major merger event, mini-halos are the only diffuse sources which are not associated with cluster mergers. A peculiar example is represented by the cluster A2142, which contains a cooling core but also shows a cold front and thus merging activity [87]. The mini-halo in this cluster is about 200 kpc in size and does not surround any powerful radio galaxy (Fig. 10, right panel). For the latter reason, it could be also considered as a small halo.

The radio spectra of mini-halos are steep, as those of halos and relics. In the Perseus mini-halo, the integrated spectrum steepens at high frequency and the spectral index distribution shows a radial steepening [110].

Gitti et al. [62] argued that the radio emitting particles in mini-halos cannot be connected to the central radio galaxy in terms of particle diffusion or buoyancy, but they are likely associated with the ICM in the cooling flow region (see Sect. 4.4). This is supported by the correlation observed between the mini-halo radio power and the cooling flow power [63]; however, the number of objects is still low and the parameters are affected by large errors.

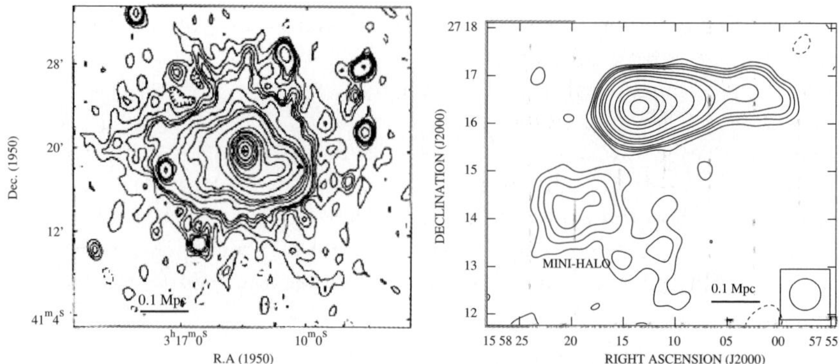

Fig. 10. Left panel: Radio contour map of the mini-halo in the Perseus cluster ($z = 0.018$) at 92 cm [110], **Right panel:** The mini-halo in A2142 ($z = 0.089$), superimposed onto the optical image [59]

4 Radio Emitting Particles

From the diffuse radio emission described in the previous sections, it is determined that highly energetic relativistic electrons ($\gamma \sim 10^4$) are present in clusters, either in the central or in the peripheral regions. They are found both in merging (halos and relics) and relaxed (mini-halos) clusters, thus under different cluster conditions. These radio features are currently not known to be present in all clusters. They show steep radio spectra, thus the radiating particles have short lifetimes ($\sim 10^8$ yr). Given the large size of the radio emitting regions, the relativistic particles need to be reaccelerated by some mechanism, acting with an efficiency comparable to the energy loss processes. Several possibilities have been suggested for the origin of relativistic electrons and for the mechanisms transferring energy into the relativistic electron population.

4.1 Connection Between Halos/Relics and Cluster Merger Processes

Evidence favour the hypothesis that clusters with halos and relics are characterized by strong dynamical activity, related to merging processes. These clusters indeed show: (i) substructures and distortions in the X-ray brightness distribution [109]; (ii) temperature gradients [86] and gas shocks [90]; (iii) absence of a strong cooling flow [109]; (iv) values of the spectroscopic β parameter which are on average larger than 1 [46]; (v) core radii significantly larger than those of clusters classified as single/primary [46]; (vi) larger distance from the nearest neighbours, compared to clusters with similar X-ray luminosity [108]. The fact that they appear more isolated supports the idea that recent merger events lead to a depletion of the nearest neighbours.

Buote [23] derived a correlation between the radio power of halos and relics and the dipole power ratio of the cluster two-dimensional gravitational potential. Since power ratios are closely related to the dynamical state of a cluster, this correlation represents the first attempt to quantify the link between diffuse sources and cluster mergers.

Maps of the radio spectral index between 0.3 and 1.4 GHz of the halos in the two clusters A665 and A2163 show that the regions influenced by the merger, as deduced from X-ray maps, show flatter spectra [49]. This is the first direct confirmation that the cluster merger supplies energy to the radio halo. Finally, we point out that we are not presently aware of any radio halo or relic in a cluster where the presence of a merger has been clearly excluded.

4.2 Relativistic Electrons in Radio Halos

Origin

The relativistic electrons present in the cluster volume, which are responsible for the diffuse radio emission, can be either *primary* or *secondary electrons*. Primary electrons were injected into the cluster volume by AGN activity (quasars, radio galaxies, etc.), or by star formation in normal galaxies (supernovae, galactic winds, etc.) during the cluster dynamical history. This population of electrons suffers strong radiation losses mainly because of synchrotron and inverse Compton emission, thus reacceleration is needed to maintain their energy to the level necessary to produce radio emission. For this reason, primary electron models can also be referred to as reacceleration models. These models predict that the accelerated electrons have a maximum energy at $\gamma < 10^5$ which produces a high frequency cut-off in the resulting synchrotron spectrum [20]. Thus a high frequency steepening of the integrated spectrum is expected, as well as a radial steepening and/or a complex spatial distribution of the spectral index between two frequencies, the latter due to different reacceleration processes in different cluster regions. Moreover, in these models, a tight connection between radio halos and cluster mergers is expected.

Secondary electrons are produced from inelastic nuclear collisions between the relativistic protons and the thermal ions of the ambient intracluster medium. The protons diffuse on large scales because their energy losses are negligible. They can continuously produce *in situ* electrons, distributed throughout the cluster volume [10]. Secondary electron models can reproduce the basic properties of the radio halos provided that the strength of the magnetic field, averaged over the emitting volume, is larger than a few μG. They predict synchrotron power-law spectra which are independent on cluster location, i.e., do not show any features and/or radial steepening, and the spectral index values are flatter than $\alpha \sim 1.5$ [20]. The profiles of the radio emission should be steeper than those of the X-ray gas (e.g. [67]). Since the radio emitting electrons originate from protons accumulated during the cluster formation history, no correlation to recent mergers is expected, but halos should be present in virtually all clusters. Moreover, emission of gamma-rays and of neutrinos is predicted.

Present observational results, i.e., the behaviour of radio spectra (see Sect. 3.1), the association between radio halos and cluster mergers (Sect. 4.1), and the fact that halos are not common in galaxy clusters [81], are in favour of electron reacceleration models. A two-phase scenario including the first phase of particle injection, followed by a second phase during which the aged electrons are reaccelerated by recent merging processes was successfully applied by Brunetti et al. [19] to the radio halo Coma C, reproducing its observational properties.

Reacceleration

In the framework of primary electron models, a cluster merger plays a crucial role in the energetics of radio halos. Energy can be transferred from the ICM thermal component to the non-thermal component through two possible basic mechanisms: (1) acceleration at shock waves [77, 107]; (2) resonant or non-resonant interaction of electrons with magneto-hydrodynamic (MHD) turbulence [19, 21, 51, 97].

Shock acceleration is a first-order Fermi process of great importance in radio astronomy, since it is recognized as the mechanism responsible for particle acceleration in the supernova remnants. The acceleration occurs diffusively, in that particles scatter back and forth across the shock, gaining at each crossing and recrossing an amount of energy proportional to the energy itself. The acceleration efficiency is mostly determined by the shock Mach number. In the case of radio halos, however, the following arguments do not favour a connection to merger shocks: (i) the shocks detected so far with Chandra at the center of several clusters (e.g. A2744, [76]; A665, [88]; 1E0657-56, [89]) have inferred Mach numbers in the range of $\sim$ 1–2.5, which seem too low to accelerate the radio halo electrons [53]; (ii) the radio emission of halos can be very extended up to large scales, thus it is hardly associable with localized shocks; (iii) the comparison between radio data and high resolution Chandra X-ray data, performed by Govoni et al. [70], shows that some clusters exhibit a spatial correlation between the radio halo emission and the hot gas regions. This is not a general feature, however, and in some cases the hottest gas regions do not exhibit radio emission; (iv) the radio spectral index distribution in A665 [49] shows no evidence of spectral flattening at the location of the hot shock detected by Chandra [88].

Although it cannot be excluded that shock acceleration may be efficient in some particular regions of a halo (e.g. in A520, [92]), current observations globally favour the scenario that cluster turbulence might be the major mechanism responsible for the supply of energy to the electrons radiating in radio halos. Numerical simulations indicate that mergers can generate strong fluid turbulence on scales of 0.1–1 Mpc. Turbulence acceleration is similar to a second-order Fermi process and is therefore rather inefficient compared with shock acceleration. The time during which the process is effective is only a few

10^8 years, so that the emission is expected to correlate with the most recent or ongoing merger event. The mechanism involves the following steps [12, 21]:

(1) the fluid turbulence which is injected into the ICM must be converted to MHD turbulence; the mechanism for this process is not fully established (although the Lighthill mechanism is mostly used in the recent literature);
(2) several types of MHD turbulence modes can be activated (Alfvèn waves, slow and fast magnetosonic modes, etc.) and each of them has a different channel of wave-particle interaction;
(3) the cascade process due to wave-wave interaction, i.e., the decay of the MHD scale size to smaller values, must be efficient to produce the MHD scale relevant for the wave-particle interaction, i.e., for the particle reacceleration process;
(4) the MHD waves are damped because of wave-particle interaction, so the reacceleration process could be eventually reduced.

The particle reacceleration through Alfvèn waves has the following limitations: (i) the scale relevant for wave-particle interaction is $\sim$ 1 pc, thus the reacceleration process is efficient only after a significant cascade process; (ii) Alfvèn waves are strongly damped through interaction with protons. It follows that if protons are too abundant in the ICM, they suppress the MHD turbulence and consequently the reacceleration of electrons. Brunetti et al. [21] derived that the energy in relativistic protons should be $< 5\%$–10% than the cluster thermal energy to generate radio halos. In the case of fast magnetosonic (MS) waves, the difficulty of wave cascade to small scales is alleviated by the fact that their scale of interaction with particles is of the order of a few kpc. Moreover, the MS wave damping is due to thermal electrons, and thus hadrons do not significantly affect the electron reacceleration process [24]. Therefore, fast MS waves represent a promising channel for the MHD turbulence reacceleration of particles.

The emerging scenario is that turbulence reacceleration is the likely mechanism to supply energy to the radio halos. All the different aspects discussed above need to be further investigated in time-dependent regimes, considering all types of charged particles [22], and the contribution of different mechanisms.

4.3 Relativistic Electrons in Radio Relics

Peripheral cluster regions do not host a sufficiently dense thermal proton population which is required as the target for the efficient production of secondary electrons, and therefore secondary electron models cannot operate in the case of relics. There is increasing evidence that the radio emitting particles in relics are powered by the energy dissipated in shock waves produced in the ICM by the flows of cosmological large-scale structure formation. The production of outgoing shock waves at the cluster periphery is indeed observed in numerical simulations of cluster merger events [106]. Because of the electron short radiative lifetimes, radio emission is produced close to the location of the shock waves. This is consistent with the almost perpendicular to the merger axis elongated structure of relics. The electron acceleration required to produce the relic emission could result from Fermi-I diffusive shock acceleration of thermal ICM electrons [33], or by adiabatic energization of relativistic electrons confined in fossil radio plasma, released by a former active radio galaxy [34, 35, 73]. These models predict that the magnetic field within the relic is aligned with the shock front, and that the radio spectrum is flatter at the shock edge, where the radio brightness is expected to decline sharply.

The detection of shocks in the cluster outskirts is presently very difficult because of the very low X-ray brightness of these regions. The X-ray data for radio relics are indeed very scarce. The Chandra data of A754 [91] indicate that the easternmost boundary of the relic coincides with a region of hotter gas. From XMM data of the same cluster, Henry et al. [72] show that the diffuse radio sources (halo + relic) appear to be associated with high pressure regions.

4.4 Relativistic Electrons in Mini-Halos

Current models for mini-halos involve primary or secondary electrons, similar to halos. Gitti et al. [62] suggest that the relativistic primary electrons are continuously undergoing reacceleration due to the MHD turbulence associated with the cooling flow region. Pfrommer & Enßlin [98], on the other hand, discuss the possibility that relativistic electrons in mini-halos are of secondary origin and thus are produced by the interaction of cosmic ray protons with the ambient thermal protons. Predictions of these models are similar to those of the halo models. The electron reacceleration model is favoured by the spectral behaviour of the Perseus mini-halo, i.e. high frequency steepening and radial spectral steepening [110], and by the observed correlation between the mini-halo radio power and the cooling flow power [62]. Data on this class of diffuse radio sources, however, are too poor to draw conclusions.

5 Cluster Magnetic Fields

The presence of magnetic fields in clusters is directly demonstrated by the existence of large-scale diffuse synchrotron sources, which have been discussed in Sect. 3. In this section, we present an independent way of obtaining indirect information about the cluster magnetic field strength and geometry, using data at radio wavelengths. This is the analysis of the Faraday rotation of radio sources in the background of clusters or in the galaxy clusters themselves.

Measurements of the ICM magnetic fields can also be obtained through X-ray data from the studies of cold fronts (e.g. [119]) and from the detection of non-thermal X-ray emission of inverse Compton origin, due to scattering of the cosmic microwave background photons by the synchrotron electrons. The latter emission can be detected in the hard X-ray domain (e.g. [52]), where the cluster thermal emission becomes negligible. The studies in the radio band are, however, the most relevant and provide the most detailed field estimates.

5.1 Rotation Measure

The synchrotron radiation from cosmic radio sources is well known to be linearly polarized. A linearly polarized wave of wavelength λ, traveling from a radio source through a magnetized medium, experiences a phase shift of the left versus right circularly polarized components of the wavefront, leading to a rotation $\Delta\chi$ of the position angle of the polarization, according to the law: $\Delta\chi = \mathrm{RM}\ \lambda^2$, where RM is the Faraday rotation measure. The RM is obtained as:

$$\mathrm{RM} = \frac{e^3}{2\pi m_e^2 c^4} \int_0^L n_e \mathbf{B} \cdot d\mathbf{l}\ . \tag{20}$$

In practical units, RM is related to the electron density n_e, in units of cm^{-3}, and to the magnetic field along the line of sight $B_\parallel$, in units of μG, through the relation:

$$\mathrm{RM} = 812 \int_0^L n_\mathrm{e} B_\parallel dl \qquad \mathrm{rad\ m}^{-2}\ , \tag{21}$$

where the path length l is in kpc. By convention, RM is positive (negative) for a magnetic field directed toward (away from) the observer.

The RM values can be derived from multi-frequency polarimetric observations of sources within or behind the clusters, by measuring the position angle of the polarized radiation as a function of wavelength. In general, the position angle must be measured at three or more wavelengths in order to determine RM accurately and remove the position angle ambiguity: $\chi_\mathrm{true} = \chi_\mathrm{obs} \pm n\pi$. Once the contribution of our Galaxy is subtracted, the RM should be dominated by the contribution of the ICM, and therefore it can be combined with measurements of n_e to estimate the cluster magnetic field along the line of sight. This approach can be followed analytically only for simple distributions of n_e and B.

A recent technique to analyse and interpret the RM data is the RM Synthesis, developed by Brentjens & De Bruyn [17], which uses the RM transfer function to solve the $n\pi$ ambiguity related to the RM computation, and allows one to distinguish the emission as a function of Faraday depth.

Below we present some simple cases, where the strength of the magnetic field can be derived by RM measurements:

Uniform Screen

In the simplest approximation of an external screen with uniform magnetic field, no depolarization is produced and the rotation measure follows directly from (21):

$$\mathrm{RM} = 812\, n_e B_{\|} L, \tag{22}$$

where n_e is in cm^{-3}, $B_{\|}$ is in μG, and L is the depth of the screen in kpc.

Screen with Tangled Magnetic Field

The effect of a Faraday screen with a tangled magnetic field has been analyzed by Lawler and Dennison [82] and by Tribble [116] in the ideal case that the screen is made of cells of uniform size, with the same electron density and the same magnetic field strength, but with field orientation at random angles in each cell. The observed RM along any given line of sight will be generated by a random walk process, which results in a gaussian RM distribution with mean and variance given by:

$$\langle \mathrm{RM} \rangle = 0\,, \quad \sigma^2_{\mathrm{RM}} = \langle \mathrm{RM}^2 \rangle = 812^2\, \Lambda_c \int (n_e B_{\|})^2 \mathrm{d}l\,, \tag{23}$$

where n_e is in cm^{-3}, B is in μG, and Λ_c is the size of each cell in kpc. A tangled magnetic field also produces depolarization (see [116]).

Screen with Tangled Magnetic Field and Radial Gas Density Distribution

The case of a screen with tangled magnetic field can be treated analytically if a realistic cluster gas density distribution is considered, given that the cells have uniform size, the same magnetic field strength and random field orientation. If the gas density follows a hydrostatic isothermal beta model [25], i.e.,

$$n_e(r) = n_0 (1 + r^2/r_c^2)^{-\frac{3\beta}{2}}\,, \tag{24}$$

where n_0 is the central electron density, and r_c is the core radius of the gas distribution, the value of the RM variance is given by:

$$\sigma_{\mathrm{RM}}(r_\perp) = \frac{K B n_0 r_c^{\frac{1}{2}} \Lambda_c^{\frac{1}{2}}}{(1 + r_\perp^2/r_c^2)^{\frac{(6\beta-1)}{4}}} \sqrt{\frac{\Gamma(3\beta - 0.5)}{\Gamma(3\beta)}}\,, \tag{25}$$

where $r_\perp$ is the projected distance from the cluster centre and Γ indicates the Gamma function. The constant K depends on the integration path over the gas density distribution: $K = 624$, if the source lies completely beyond the cluster, and $K = 441$ if the source is halfway through the cluster.
For β=0.7 the previous formula becomes:

$$\sigma_{\rm RM} \approx \frac{575B}{(1+r^2/r_c^2)^{0.8}} n_0 M^{\frac{1}{2}} l \,. \quad (26)$$

Note that depolarization is also produced, due to the fact that the magnetic field is tangled.

5.2 Current Results from RM Studies

Cluster surveys of the Faraday rotation measures of polarized radio sources both within and behind clusters provide an important probe of the existence of intracluster magnetic fields. The RM values derived in background or embedded cluster sources are of the order of tens to thousands rad m^{-2} (an example is shown in Fig. 11). The observing strategy to derive information on the magnetic field intensity and structure is twofold: (i) obtain the average value of the RM of sources located at different impact parameters of the cluster, (ii) derive maps of the RM of extended radio sources, to evaluate the σ of the RM distribution.

Studies have been carried out on both statistical samples and individual clusters (see e.g. the review by Govoni & Feretti [69] and references therein). Kim et al. [78] analyzed the RM of 53 radio sources in and behind clusters and 99 sources in a control sample. This study, which contains the largest

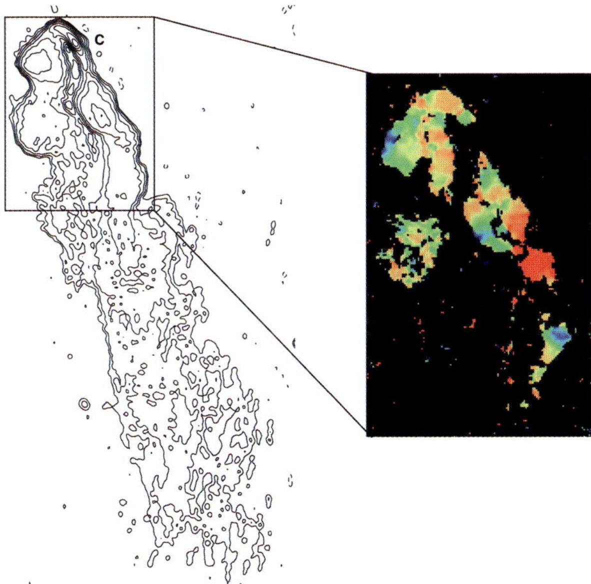

Fig. 11. VLA contour plot of the tailed radio galaxy 0053-015 in A119 at 1.4 GHz (**left**), and RM image (**right**). The values of RM range between –350 and +450 rad m^{-2}, with $\langle \mathrm{RM} \rangle = +\,28$ rad m^{-2}, and a dispersion of $\sigma_{\rm RM} = 152$ rad m^{-2}. They show fluctuations on scales of $\sim$ 3.5 arcsec [44]

cluster sample to date, demonstrated that μG level fields are widespread in the ICM. In a more recent statistical study, Clarke et al. [27] analyzed RMs for a representative sample of 16 cluster sources, plus a control sample, and found a statistically significant broadening of the RM distribution in the cluster sample, and a clear increase in the width of the RM distribution toward smaller impact parameters (see Fig. 12). They derived that the ICM is permeated with a high filling factor of magnetic fields at levels of 4–8 μG and with a correlation length of ∼15 kpc, up to ∼0.75 Mpc from the cluster centre. The results are confirmed by new data on an expanded sample [28].

The first detailed studies of RM within individual clusters have been performed on cooling core clusters, owing to the extremely high RMs of the powerful radio galaxies at their centres (e.g., Hydra A, [113]; 3C295, [1]). High values of the magnetic fields, up to tens of μG, have been obtained, but they only refer to the innermost cluster regions. Studies on larger areas of clusters have been carried out e.g. for Coma [41], A119 [44], A514 [68], 3C129 [114].

Overall, the data are consistent with cluster atmospheres containing magnetic fields in the range of 1–5 μG, regardless of the presence or not of diffuse radio emission. At the centre of cooling core clusters, magnetic field strengths can be larger by more than a factor of 2. The RM distributions are generally patchy, indicating that large-scale magnetic fields are not regularly ordered on cluster scales, but have coherence scales between 1 and 10 kpc. In most clusters the magnetic fields are not dynamically important, with magnetic pressures much lower than the thermal pressures, but the fields may play a

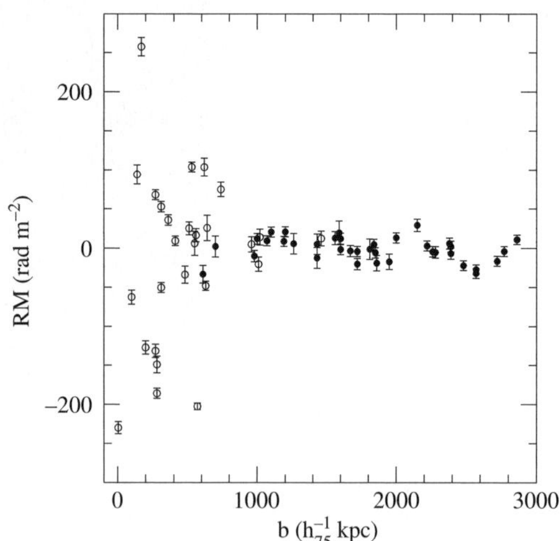

Fig. 12. Galaxy-corrected rotation measure plotted as a function of source impact parameter in kiloparsecs for the sample of sources from Clarke et al. [27]. Open dots refer to cluster sources, closed dots to control sources

fundamental role in the suppression of the particle thermal conduction [26] and in the energy budget of the ICM.

5.3 Magnetic Field Structure

The simplest model is a uniform field throughout the cluster. However, this is not realistic: if the field values detected at the cluster centres extend over several core radii, up to distances of the order of $\sim$ Mpc, then the magnetic pressure would exceed the thermal pressure in the outer parts of the clusters. The magnetic field intensity is likely to decrease with the distance from the cluster centre, as derived in Coma [19]. This is also predicted as a result of compression of the thermal plasma during the cluster gravitational collapse, where the magnetic field-lines are frozen into the plasma, and compression of the plasma results in compression of flux lines. As a consequence of magnetic flux conservation, the expected growth of the magnetic field is proportional to the gas density as B $\propto \rho^{2/3}$.

Dolag et al. [31] showed that in the framework of hierarchical cluster formation, the correlation between two observable parameters, the RM and the X-ray surface brightness, is expected to reflect the correlation between the magnetic field and gas density. Therefore, from the analysis of the RM versus X-ray brightness it is possible to infer the trend of magnetic field versus gas density. The application of this approach has been possible so far only in A119, giving the radial profile of the magnetic field as $B \propto n_e^{0.9}$ [31]. The magnetic field decline with radius is confirmed in this case.

Another important aspect to consider is the structure in the cluster magnetic field, i.e. the existence of filaments and flux ropes [32]. The magnetic field structure can be investigated by deriving the power spectrum of the field fluctuations, defined as: $|B_\kappa|^2 \propto \kappa^{-n}$, where κ represents the wave number of the fluctuation scale. By using a semi-analytic technique, Enßlin & Vogt [36] and Vogt & Enßlin [120] showed that the magnetic field power spectrum can be estimated by Fourier transforming RM maps, if very detailed RM images are available. Alternatively, a numerical approach using Monte Carlo simulations has been developed by Murgia et al. [94] to reproduce the rotation measure and the depolarization produced by magnetic field with different power spectra.

5.4 Reconciling Values Derived with Different Approaches

The cluster magnetic field values obtained from RM arguments are about an order of magnitude higher than those derived from both the synchrotron diffuse radio emission (Sect. 3.1) and the inverse Compton (IC) hard X-ray emission (e.g. [52]). The discrepancy can be alleviated by taking into account that:

- estimates of equipartition fields rely on several assumptions (see Sect. 2.3);
- Goldsmith & Rephaeli [64] suggested that the IC estimate is typically expected to be lower than the Faraday rotation estimate, because of the spatial profiles of the magnetic field and gas density. For example, if the magnetic field strength has a radial decrease, most of the IC emission will come from the weak field regions in the outer parts of the cluster, while most of the Faraday rotation and synchrotron emission occurs in the strong field regions in the inner parts of the cluster;
- it has been shown that IC models which include the effects of aged electron spectra, combined with the expected radial profile of the magnetic field, and anisotropies in the pitch angle distribution of the electrons, allow higher values of the ICM magnetic field in better agreement with the Faraday rotation measurements [19, 97];
- the magnetic field may show complex structure, as filamentation and/or substructure with a range of coherence scales (power spectrum). Therefore, the RM data should be interpreted using realistic models of the cluster magnetic fields (see Sect. 5.3);
- Beck et al. [5] pointed out that field estimates derived from RM may be too large in the case of a turbulent medium where small-scale fluctuations in the magnetic field and the electron density are highly correlated ;
- it has been recently pointed out that in some cases a radio source could compress the gas and fields in the ICM to produce local RM enhancements, thus leading to overestimates of the derived ICM magnetic field strength [105];
- evidence suggests that the magnetic field strength will vary depending on the dynamical history and location within the cluster. A striking example of the variation of magnetic field strength estimates for various methods and in various locations throughout the cluster is given in [74].

Future studies are needed to shed light on these issues and improve our current knowledge on the strength and structure of the magnetic fields.

5.5 Origin of Cluster Magnetic Fields

The field strengths that we observe in clusters greatly exceed the amplitude of the seed fields produced in the early universe, or fields injected by some mechanism by high redshift objects. There are two basic possibilities for their origin:

(1) ejection from galactic winds of normal galaxies or from active and starburst galaxies [80, 121];
(2) amplification of seed fields during the cluster formation process.

Support for a galactic injection in the ICM comes from the evidence that a large fraction of the ICM is of galactic origin, since it contains a significant concentration of metals. However, fields in clusters have strengths and coherence size comparable to, and in some cases larger than, galactic fields [71]. Therefore, it seems quite difficult that the magnetic fields in the ICM derive purely from ejection of the galactic fields, without invoking other amplification mechanisms [29, 101].

Magnetic field amplification is likely to occur during the cluster collapse, simply by compression of an intergalactic field. Clusters have present day overdensities $\rho \sim 10^3$ and in order to get $B_{\rm ICM} > 10^{-6}$ G by adiabatic compression ($B \propto \rho^{2/3}$) requires intergalactic (seed) fields of at least 10^{-8} G. These are somewhat higher than current limits derived in the literature [3, 11]. A possible way to obtain a larger field amplification is through cluster mergers. Mergers generate shocks, bulk flows and turbulence within the ICM. The first two of these processes can result in some field amplification simply through compression. However, it is the turbulence which is the most promising source of non-linear amplification. MHD calculations have been performed [30, 102, 112] to investigate the evolution of magnetic fields. The results of these simulations show that cluster mergers can dramatically alter the local strength and structure of cluster-wide magnetic fields, with a strong amplification of the magnetic field intensity. Shear flows are extremely important for the amplification of the magnetic field, while the compression of the gas is of minor importance. The initial field distribution at the beginning of the simulations at high redshift is irrelevant for the final structure of the magnetic field. The final structure is dominated only by the cluster collapse. Fields can be amplified from initial values of $\sim 10^{-9}$ G at $z = 15$ to $\sim 10^{-6}$ G at the present epoch [30]. Roettiger et al. [102] found a significant evolution of the structure and strength of the magnetic fields during two distinct epochs of the merger evolution. In the first, the field becomes quite filamentary as a result of stretching and compression caused by shocks and bulk flows during infall, but only minimal amplification occurs. In the second, amplification of the field occurs more rapidly, particularly in localized regions, as the bulk flow is replaced by turbulent motions. Mergers change the local magnetic field strength drastically, but also the structure of the cluster-wide field is influenced. At early stages of the merger the filamentary structures prevail. This structure breaks down later ($\sim$ 2–3 Gyr) and leaves a stochastically ordered magnetic field. Subramanian et al [112] argue that the dynamo action of turbulent motions in the intracluster gas can amplify a random magnetic field by a net factor of 10^4 in 5 Gyr. The field is amplified by random shear, and has an intermittent spatial distribution, possibly producing filaments.

6 Radio Emission from Cluster Radio Galaxies

Recent results on the thermal gas in clusters of galaxies has revealed a significant amount of spatial and temperature structure, indicating that clusters are dynamically evolving by accreting gas and galaxies and by merging with other clusters/groups (roughly every few Gyrs). Simulations suggest that the ICM within clusters is violent, filled with shocks, high winds and turbulence. This gas can interact with a radio source in different ways: modifying its morphology via ram pressure, confining the radio lobes, possibly feeding the active nucleus. We discuss below some of the recent results on these topics (see also the review of Feretti & Venturi [47]).

6.1 Interaction Between the Radio Galaxies and the ICM

Tailed Radio Galaxies

A dramatic example of the interaction of the radio galaxies with the ICM is represented by the tailed radio galaxies, i.e. low-power radio sources (FR I type, [38]) where the large scale low-brightness emission is bent towards the same direction, forming features similar to tails. These radio galaxies were originally distinguished in two classes: narrow-angle tailed sources (NAT), which are "U" shaped with a small angle between the tails, and wide-angle tailed sources (WAT), which are "V" shaped with a larger angle between the tails (see Fig. 13). We note that distortions in powerful radio galaxies (FR II type, [38]) are marginal and only present in weak structures.

The standard interpretation of the tailed radio morphology is that the jets are curved by ram pressure from the high-velocity host galaxy moving through the dense ICM, whereas the low brightness tails are material left

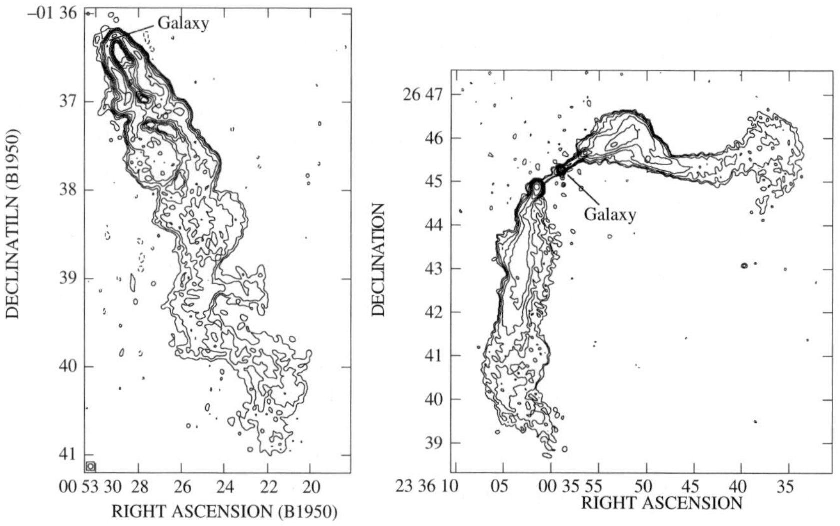

Fig. 13. Examples of tailed radio galaxies: the NAT 0053-016 in the cluster A119 (**left panel**) and the WAT 3C465 in the cluster A2634 (**right panel**). The location of the optical galaxy is indicated

behind by the galaxy motion. The ram-pressure model was first developed by Begelman et al. [7]. Following dynamical arguments, the bending is described by the Euler equation:

$$R \sim h \left(\frac{\rho_j}{\rho_e}\right)\left(\frac{v_j}{v_g}\right)^2 , \tag{27}$$

where R is the radius of curvature, ρ is density, v is velocity (the subscript j refers to the jet, e to the external medium, g to the galaxy) and h is the scale height over which the ram pressure is transmitted to the jets. Thus, from the jet bending, important constraints on both the jet dynamics and the ICM can be placed. In some cases there is evidence that the radio jets travel first through the galactic atmosphere and then are sharply bent at the transition between the galactic atmosphere and the ICM. Bends can occur very close to the nucleus, as in NGC 4869 in the Coma cluster [40], indicating that the bulk of interstellar medium has been stripped by the galaxy during its motion.

In general, the ram-pressure model can explain the radio jet deflection when the galaxy velocity with respect to the ICM is of the order of $\sim$ 1000 km s^{-1}. Therefore, it can successfully explain the structure of NAT sources, which are indeed identified with cluster galaxies located at any distance from the cluster centre and thus characterized by significant motion. However, Bliton et al. [13] derived that NATs are preferentially found in clusters with X-ray substructure. Additionally, NAT galaxies tend to have, on average, velocities similar to those of typical cluster members, instead of high peculiar motions expected if NATs were bent only from ram pressure. Thus, they suggested a new model for the NAT formation, in which NATs are associated with dynamically complex clusters with possible recent or ongoing cluster-subcluster mergers. The U-shaped morphology is then suggested to be produced, at least in part, by the merger-induced bulk motion of the ICM bending the jets. This is supported, in some clusters, by the existence of NAT radio galaxies with their tails oriented in the same direction (e.g., A2163, Fig. 14; A119, [44]), since it seems unlikely that their parent galaxies are all moving towards the same direction.

The interpretation of WAT sources may be problematic in the framework of the ram-pressure model, since these sources are generally associated with dominant cluster galaxies moving very slowly ($\lesssim$ 100 km s^{-1}) relative to the cluster velocity centroid. Such slow motion is insufficient to bend the jets/tails of WATs to their observed curvature by ram pressure. It has therefore been suggested that WATs must be shaped mostly by other ram-pressure gradients not arising from the motion of the host galaxy, but produced by mergers between clusters [65, 85]. Numerical simulations lead support to this idea: peak gas velocities are found well in excess of 1000 km s^{-1} at various stages of the cluster merger evolution, which generally do not decay below 1000 km s^{-1} for nearly 2 Gyr after the core passage. This is consistent with the observations, as modelled in the cluster A562 (Fig. 15).

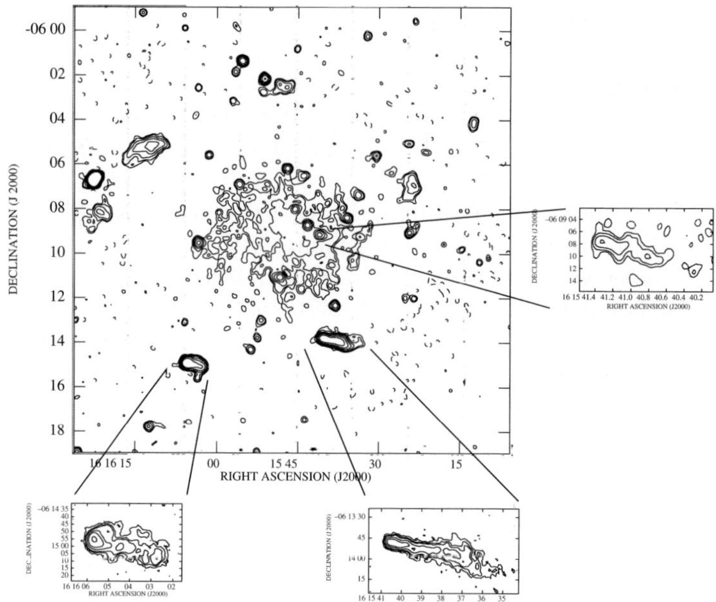

Fig. 14. Radio image of the cluster A2163 at 1.4 GHz, with angular resolution of 15″ [45]. The structure of tailed radio galaxies as detected at higher resolution is shown in the insets. The tails are all oriented in the same direction

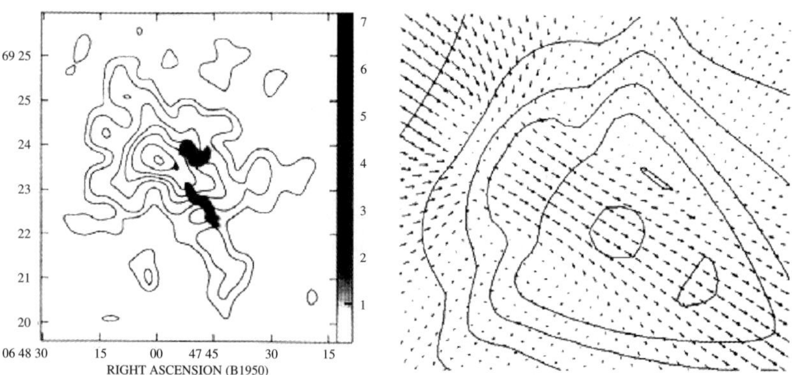

Fig. 15. Left panel: Overlay of the grey-scale radio image of the WAT source in A562 onto the ROSAT X-ray surface brightness contour image of the same cluster. **Right panel**: Overlay of a synthetic X-ray image of a cluster merger onto a velocity vector field that represents the gas velocity. Note that the X-ray contours in the left panel look very similar to the synthetic X-ray image and that the radio tails are in the direction of the gas velocity (from [65])

Radio Emission in X-ray Cavities

A clear example of the interaction between the radio plasma and the hot intracluster medium was found in the ROSAT image of the Perseus cluster [15], where X-ray cavities associated with the inner radio lobes to the north and south of the bright central radio galaxy 3C84 have been first detected. The high spatial resolution of the Chandra X-ray Observatory has confirmed the presence of such X-ray holes [37], coinciding with the radio lobes and showing rims cooler than the surrounding gas. Chandra has permitted the detection of X-ray deficient bubbles in the inner region of many cooling flow clusters, e.g., Hydra A, A2052, A496, A2199, RBS797. These features are discussed by C. Jones et al. in this volume.

6.2 Trigger of Radio Emission

An important issue is to understand whether and how the cluster environment plays any role in the statistical radio properties of galaxies, in particular their probability of forming radio sources. The high density of galaxies within clusters, especially in the innermost cluster regions, and the peculiar velocities of galaxies, most extreme in merging clusters, enhance the probability of galaxy–galaxy interactions. These special conditions raise the questions whether cluster galaxies have enhanced probability of developing a radio source, and whether they tend to have more powerful and long lived radio emission.

A powerful statistical tool to address the above questions is the radio luminosity function (hereafter RLF). The fractional RLF is defined as:

$$f_i(P,z) = \frac{\rho_i(P,z)}{\phi_i(z)} \,, \tag{28}$$

where $\phi_i(z)$ is the density of objects of a particular class i at the epoch z, and $\rho_i(P,z)$ is the density of the same class objects showing a radio emission of power P. The fractional RLF, $f(P)$, thus represents the probability that a galaxy in a defined sample at a given epoch emits with radio power in the interval $P \pm \mathrm{d}P$. From an operational point of view, the RLF can be expressed as:

$$f(P) = \frac{n(\Delta P_i)}{N(\Delta P_i)} \,, \tag{29}$$

where n and N are respectively the number of detected radio galaxies in the power interval ΔP_i and the total number of optical galaxies which could have been detected in the same power bin. The integral form of the RLF $F(>P)$ can be obtained simply summing over all radio power intervals up to the power P. In order to take into account the correlation between the optical and radio properties of galaxies, it is useful to introduce the bivariate luminosity function $f(P,M)$, which gives the probability that a galaxy with absolute magnitude in the range $M \pm \mathrm{d}M$ is radio emitting in the radio power range $P \pm \mathrm{d}P$.

The RLF of galaxies in clusters has been first investigated by Fanti [39], and latter by Ledlow and Owen [83]. The most striking result is that statistical properties of radio galaxies are surprisingly similar for sources both inside and outside rich clusters. For both cluster and non-cluster galaxies, the only parameter relevant for the radio emission seems to be the optical magnitude, i.e. brighter galaxies have a higher probability of developing a radio galaxy. Furthermore, the radio luminosity function is independent on richness class, Bautz-Morgan or Rood-Sastry cluster class. Recently, Best et al. [9] demonstrated that, while the radio power of a radio galaxy does not correlate to its mass, the probability of a galaxy to become a radio source is a very strong function of both stellar mass and central black hole mass.

It is still under debate whether the universality of the local RLF for early type galaxies can be applied also to merging clusters. According to some authors (e.g. [54, 118]) the enhanced probability of galaxy interaction in merging clusters has no effect on the probability of galaxies to develop a radio active galactic nucleus in their centres.

In the cluster A2255, instead, Miller & Owen [93] found an excess of powerful radio galaxies, which is interpreted as due to the dynamical state of the cluster. Best [8] showed that the fraction of radio loud AGN appears to be strongly dependent upon the large scale environment of a galaxy. This supports the argument that a merger process may affect the AGN activity, since infalling galaxies or galaxy groups more likely produce galaxy interactions or galaxy–galaxy mergers which can trigger the AGN activity. The effect of cluster merger processes on the trigger of radio emission would imply an enhanced number of radio source in cluster at high redshift, i.e. at the earlier epochs when the clusters are being assembled. These issues are under investigation. The result of Branchesi et al. [16] points to a higher number of radio galaxies in distant clusters, although with poor statistics. In conclusion, whereas the ICM in clusters has strong effect on the structures of radio galaxies, the probability of forming radio sources is likely unaffected by the cluster environment, but may be affected by cluster mergers.

Other effects of the interaction between galaxies and ICM, as the trigger of star formation, the gas stripping, HI deficiency, etc., are discussed by other authors in this volume.

Acknowledgments

LF is grateful to the organizers David Hughes, Omar López-Cruz and Manolis Plionis for their invitation to this stimulating and very interesting school. We acknowledge Gianfranco Brunetti for illuminating discussions on the models of relativistic particle origin and re-acceleration.

References

1. Allen, S.W., Taylor, G.B., P.Nulsen, E.J., et al.: MNRAS **324**, 842 (2001)
2. Bacchi, M., Feretti, L., Giovannini, G., Govoni, F.: A&A **400**, 465 (2003)
3. Barrow, J.D., Ferreira, P.G., Silk, J.: Phys. Rev. Let. **78**, 3610 (1997)
4. Baum, S.A., O'Dea, C.P.: MNRAS **250**, 737 (1991)
5. Beck, R., Sukurov, A., Sokoloff, D., Wielebinski, R.: A&A **411**, 99 (2003)
6. Beck, R., Krause, M.: Astron. Nachr. **326**, 414 (2005)
7. Begelman, M.C., Rees, M.J., Blandford, R.D.: Nature **279**, 770 (1979)
8. Best, P.N.: MNRAS **351**, 70 (2004)
9. Best, P.N., Kauffman, G., Heckman, T.M., et al.: MNRAS **362**, 25 (2005)
10. Blasi, P., Colafrancesco, S.: Astrop. Phys. **12**, 169 (1999)
11. Blasi, P., Burles, S., Olinto, A.V.: ApJ **514**, L79 (1999)
12. Blasi, P.: Proceedings of the International Conference on Cosmic Rays and Magnetic Fields in Large Scale Structure, Kang H., Ryu, D. (eds.) J. Korean. Astron. Soc. **37**, 483 (2004)
13. Bliton, M., Rizza, E., Burns, J.O., Owen, F.N., Ledlow, M.J.: MNRAS **301**, 609 (1998)
14. Blumenthal, G.R., Gould, R.J.: Rev. Modern Phys. **42**, 237 (1970)
15. Böhringer, H., Voges, W., Fabian, A.C., Edge, A.C., Neumann, D.M.: MNRAS **264**, L25 (1993)
16. Branchesi, M., Gioia, I.M., Fanti, C., Fanti, R., Perley, R.: A&A **446**, 97 (2006)
17. Brentjens, M.A., de Bruyn, A.G.: A&A **441**, 1217 (2005)
18. Brunetti, G., Setti, G., Comastri, A.: A&A **325**, 898 (1997)
19. Brunetti, G., Setti, G., Feretti, L., Giovannini, G.: MNRAS **320**, 365 (2001)
20. Brunetti G.: Proceedings of the International Conference on Cosmic Rays and Magnetic Fields in Large Scale Structure, Kang, H., Ryu, D. (eds.) J. Korean. Astron. Soc. **37**, 493 (2004)
21. Brunetti, G., Blasi, P., Cassano, R., Gabici, S.: MNRAS **350**, 1174 (2004)
22. Brunetti, G., Blasi, P.: MNRAS **363**, 1173 (2005)
23. Buote, D.A.: ApJ **553**, L15 (2001)
24. Cassano, R., Brunetti, G.: MNRAS **357**, 1313 (2005)
25. Cavaliere, A., Fusco-Femiano, R.: A&A **100**, 194 (1981)
26. Chandran, B.D.G., Cowley, S.C., Ivanushkina, M., Sydora, R.: ApJ **525**, 638 (1999)
27. Clarke, T.E., Kronberg, P.P., Böhringer, H.: ApJ **547**, L111 (2001)
28. Clarke, T.E.: Proceedings of the International Conference on Cosmic Rays and Magnetic Fields in Large Scale Structure, Kang, H., Ryu, D. (eds.) J. Korean. Astron. Soc. **37**, 337 (2004)
29. De Young, D.S.: ApJ **386**, 464 (1992)
30. Dolag, K., Bartelmann, M., Lesch, H.: A&A **348**, 351 (1999)
31. Dolag, K., Schindler, S., Govoni, F., Feretti, L.: A&A **378**, 777 (2001)
32. Eilek, J.: Diffuse Thermal and Relativistic Plasma in Galaxy Clusters, Böhringer, H., Feretti, L., Schuecker, P. (eds.) MPE Report vol. 271, p. 71 (1999)
33. Enßlin, T.A., Biermann, P.L., Klein, U., Kohle, S.: A&A **332**, 395 (1998)
34. Enßlin, T.A., Gopal-Krishna: A&A **366**, 26 (2001)
35. Enßlin, T.A., Brüggen, M.: MNRAS **331**, 1011 (2002)
36. Enßlin, T.A., Vogt, C.: A&A **401**, 835 (2003)

37. Fabian, A.C., Sanders, J.S., Ettori, S.: MNRAS **318**, L65 (2000)
38. Fanaroff, B.L., Riley, J.M.: MNRAS **167**, 31P (1974)
39. Fanti, R.: Clusters and Groups of Galaxies, Mardirossian, F., Giuricin G., Mezzetti, M., (eds.) Publ. Reidel, Dordrecht, p. 185 (1984)
40. Feretti, L., Dallacasa, D., Giovannini, G., Venturi, T.: A&A **232**, 337 (1990)
41. Feretti, L., Dallacasa, D., Giovannini, G., Tagliani, A.: A&A **302**, 680 (1995)
42. Feretti, L., Böhringer, H., Giovannini, G., Neumann, D.: A&A **317**, 432 (1997a)
43. Feretti, L., Giovannini, G., Böhringer, H.: New Astron. **2**, 501 (1997b)
44. Feretti, L., Dallacasa, D., Govoni, F., Giovannini, G., Taylor, G.B., Klein, U.: A&A **344**, 472 (1999)
45. Feretti, L., Fusco-Femiano, R., Giovannini, G., Govoni, F.: A&A **373**, 106 (2001)
46. Feretti, L.: The Universe at Low Radio Frequencies, IAU Symp. 199, Pramesh Rao, Swarup, G., Gopal-Krishna (eds.) (San Francisco: ASP), p. 133 (2002)
47. Feretti, L., Venturi, T.: Merging Processes of Galaxy Clusters, Feretti, L., Gioia, I.M., Giovannini, G. (eds.) ASSL, Kluwer Academic Publishers, p. 163 (2002)
48. Feretti, L.: Matter and energy in clusters of galaxies, Bowyer, S., Hwang, C.-Y. (eds.) ASP Conference Series Vol. 301, p. 143 (2003)
49. Feretti, L., Orrú, E., Brunetti, G., Giovannini, G., Kassim, N., Setti, G.: A&A **423**, 111 (2004)
50. Fujita, Y., Sarazin, C.L., Kempner, J.C., et al.: ApJ **575**, 764 (2002)
51. Fujita, Y., Takizawa, M., Sarazin, C.L.: ApJ **584**, 190 (2003)
52. Fusco-Femiano, R., Dal Fiume, D., Orlandini, M., et al.: Matter and Energy in Clusters of Galaxies, Bowyer, S., Hwang, C.Y. (eds.) ASP Conference Series Vol. 301, p. 109, (2003)
53. Gabici, S., Blasi, P.: ApJ **583**, 695 (2003)
54. Giacintucci, S., Venturi, T., Bardelli, S., Dallacasa, D., Zucca, E.: A&A **419**, 71 (2004)
55. Giacintucci, S., Venturi, T., Brunetti, G., et al.: A&A **440**, 867 (2005)
56. Giovannini, G., Feretti, L., Stanghellini, C.: A&A **252**, 528 (1991)
57. Giovannini, G., Feretti, L., Venturi, T., Kim, K.-T., Kronberg, P.P.: ApJ **406**, 399 (1993)
58. Giovannini, G., Tordi, M., Feretti, L.: New Astron. **4**, 141 (1999)
59. Giovannini, G., Feretti, L.: New Astron. **5**, 335 (2000)
60. Giovannini, G., Feretti, L.: Merging Processes of Galaxy Clusters, Feretti, L., Gioia, I.M., Giovannini, G., (eds.) ASSL, Kluwer Academic Publishers, p. 197 (2002)
61. Giovannini, G., Feretti, L.: Proceedings of the International Conference on Cosmic Rays and Magnetic Fields in Large Scale Structure, Kang, H., Ryu, D. (eds.) J. Korean. Astron. Soc. **37**, 323 (2004)
62. Gitti, M., Brunetti, G., Setti, G.: A&A **386**, 456 (2002)
63. Gitti, M., Brunetti, G., Feretti, L., Setti, G.: A&A **417**, 1 (2004)
64. Goldshmidt, O., Rephaeli, Y.: ApJ **411**, 518 (1993)
65. Gómez, P.L., Pinkney, J., Burns, J.O., et al.: ApJ **474**, 580 (1997)
66. Govoni, F., Feretti, L., Giovannini, G., Böhringer, H., Reiprich, T.H., Murgia, M.: A&A **376**, 803 (2001a)
67. Govoni, F., Ensslin, T., Feretti, L., Giovannini, G.: A&A **369**, 441 (2001b)

68. Govoni, F., Taylor, G.B., Dallacasa, D., Feretti, L., Giovannini, G.: A&A **379**, 807 (2001c)
69. Govoni, F., Feretti, L.: Int. J. Mod. Phys. D, **13**(8) 1549 (2004)
70. Govoni, F., Markevitch, M., Vikhlinin, A., VanSpeybroeck, L., Feretti, L., Giovannini, G.: ApJ **605**, 695 (2004)
71. Grasso, D., Rubinstein, H.R.: Phys. Rep. **348**, 163 (2001)
72. Henry, P.J., Finoguenov, A., Briel, U.G.: ApJ **615**, 181 (2004)
73. Hoeft, M., Brüggen, M., Yepes, G.: MNRAS **347**, 389 (2004)
74. Johnston-Hollitt, M.: The Riddle of Cooling Flows in Galaxies and Clusters of Galaxies, Reiprich, T., Kempner, J., Soker, N. (eds.) published electronically at http://www.astro.virginia.edu/coolflow/ (2004)
75. Kardashev, N.S.: Soviet Astron. **6**, 317 (1962)
76. Kempner, J.C., David, L.P.: MNRAS **349**, 385 (2004)
77. Keshet, U., Waxman, E., Loeb, A.: ApJ **617**, 281 (2004)
78. Kim, K.T., Tribble, P.C., Kronberg, P.P.: ApJ **379**, 80 (1991)
79. Komissarov, S.S., Gubanov, A.G.: A&A **285**, 27 (1994)
80. Kronberg, P.P., Lesch, H., Hopp, U.: ApJ **511,** 56 (1999)
81. Kuo, P.-H., Hwang, C.-Y., Ip, W.-H.: ApJ **604,** 108 (2004)
82. Lawler, J.M., Dennison, B.: ApJ **252**, 81 (1982)
83. Ledlow, M.J., Owen, F.N., AJ, **112**, 9 (1996)
84. Liang, H., Hunstead, R.W., Birkinshaw, M., Andreani, P.: ApJ **544**, 686 (2000)
85. Loken, C., Roettiger, K., Burns, J.O., Norman, M.: ApJ, **445**, 80 (1995)
86. Markevitch, M., Forman, W.R., Sarazin, C.L., Vikhlinin, A.: ApJ, **503**, 77 (1998)
87. Markevitch, M., Ponman, T.J., Nulsen, P.E.J., et al.: ApJ, **541**, 542 (2000)
88. Markevitch, M., Vikhlinin, A.: ApJ, **563**, 95 (2001)
89. Markevitch, M., Gonzalez, A.H., David, L., et al.: ApJ **567**, L27 (2002)
90. Markevitch, M., Vikhlinin, A., Forman, W.R., Matter and energy in clusters of galaxies, Bowyer, S., Hwang, C.-Y. (eds.) ASP Conference Series Vol. 301, p. 37 (2003a)
91. Markevitch, M., Mazzotta, P., Vikhlinin, A.: ApJ **586**, L19 (2003b)
92. Markevitch, M., Govoni, F., Brunetti, G., Jerius, D.: ApJ **627**, 733 (2005)
93. Miller, N.A., Owen, F.N., AJ **125**, 2427 (2003)
94. Murgia, M., Govoni, F., Feretti, L., et al.: A&A **424**, 429 (2004)
95. Owen, F.N., Eilek, J.A., Kassim, N.E.: ApJ **543**, 611 (2000)
96. Pacholczyk, A.G.: Radio Astrophysics, San Francisco: Freeman (1970)
97. Petrosian, V.: ApJ **557**, 560 (2001)
98. Pfrommer, C., Enßlin, T.A.: A&A **413**, 17 (2004)
99. Reid, A.D., Hunstead, R.W., Lemonon, L., Pierre, M.M.: MNRAS **302**, 571 (1999)
100. Reiprich, T.H., Böhringer, H.: ApJ **567**, 716 (2002)
101. Rephaeli, Y., Comm. Mod. Phys., **12**, part C, 265 (1988)
102. Roettiger, K., Stone, J.M., Burns, J.O.: ApJ **518**, 594 (1999)
103. Röttgering, H., Snellen, I., Miley, G., et al.: ApJ, **436**, 654 (1994)
104. Röttgering, H.J.A., Wieringa, M.H., Hunstead, R.W., Ekers, R.D.: MNRAS **290**, 577 (1997)
105. Rudnick, L., Blundell, K.M.: ApJ **588**, 143 (2003)
106. Ryu, D., Kang, H., Hallman, E., Jones, T.W.: ApJ **593**, 599 (2003)
107. Sarazin, C.L.: ApJ **520**, 529 (1999)

108. Schuecker, P., Böhringer, H.: Diffuse thermal and relativistic plasma in galaxy clusters, Böhringer, H., Feretti, L., Schuecker, P. (eds.) MPE Report 271, p. 43 (1999)
109. Schuecker, P., Böhringer, H., Reiprich, T.H., Feretti, L.: A&A **378**, 408 (2001)
110. Sijbring, L.G.: PhD Thesis, University of Groningen, (1993)
111. Slee, O.B., Roy, A.L., Murgia, M., Andernach, H., Ehle, M.: AJ, **122**, 1172 (2001)
112. Subramanian, K., Shukurov, A., Haugen, N.E.L.: MNRAS **366**, 1437 (2006)
113. Taylor, G.B., Perley, R.A.: ApJ, **416**, 554 (1993)
114. Taylor, G.B., Govoni, F., Allen, S.A., Fabian, A.C.: MNRAS **326**, 2 (2001)
115. Thierbach, M., Klein, U., Wielebinski, R.: A&A **397**, 53 (2003)
116. Tribble, P.C.: MNRAS **250**, 726 (1991)
117. Van der Laan, H., Perola, G.C.: A&A **3**, 468 (1969)
118. Venturi, T., Bardelli, S., Zambelli, G., Morganti, R., Hunstead, R.W.: MNRAS **324**, 1131 (2001)
119. Vikhlinin, A., Markevitch, M., Murray, S.: ApJ **551**, 160 (2001)
120. Vogt, C., Enßlin, T.A.: A&A **412**, 373 (2003)
121. Völk, H.J., Atoyan, A.M.: Astrop. Phys. **11**, 73 (1999)
122. Willson, M.A.G.: MNRAS **151**, 1 (1970)

Metal Content and Production in Clusters of Galaxies

A. Renzini[1,2]

[1] European Southern Observatory, Garching bei Munchen, Germany
arenzini@eso.org
[2] INAF - Osservatorio Astronomico di Padova, Padova, Italy
arenzini@pd.astro.it

1 Introduction

X-ray observations allow the abundance of several elements in the intracluster medium (ICM) to be measured and optical observations allow similar studies for galaxies. Taken together, we have at hand the possibility of studying chemical evolution on the large scale of clusters of galaxies, and in doing so we have the opportunity to learn several aspects of how both galaxies and clusters work and evolve.

The topics to be covered by these lectures include:

- Metals in the ICM and galaxies: how much of them and how they got there
- Metal Production: Supernovae of Type Ia, their progenitors and rate
- Metal Production: Supernovae of Type II and the IMF
- The main epoch of metal production in clusters
- Clusters as anempirical nucleosynthesis template
- The chemical evolution of the universe
- The global chemical evolution of the Milky Way galaxy.

2 Metals in the ICM

In one of the rare cases in which theory anticipates observations, the existence of large amounts of heavy elements in the intracluster medium (ICM) was predicted shortly before it was actually observed [64]. This came from (now old-fashioned) so-called *monolithic* models of elliptical galaxy formation, in which the observed color-magnitude relation is reproduced in terms of a metallicity trend. In turn, this trend is established by supernova-driven galactic winds being more effective in less massive galaxies with shallower potential wells, compared to more massive galaxies, harbored in deep potential

A. Renzini: *Metal Content and Production in Clusters of Galaxies*, Lect. Notes Phys. **740**, 177–211 (2008)
DOI 10.1007/978-1-4020-6941-3_6

wells. While these models are now inadequate in many respects, their prediction was confirmed the following year by the discovery of the strong Iron-K line in the X-ray spectrum of galaxy clusters [73].

The iron and other heavy element production, and circulation on the galaxy cluster scale, has been widely discussed since their early discovery [1, 6, 21, 23, 57, 68, 71, 84, 91, 93, 99, 119].

Figure 1 shows the ASCA X-ray spectrum of the inner and outer regions of the cluster A496 [34], which best-fit temperatures are respectively ~ 3.4 and ~ 4.2 keV. Emission lines of several elements are clearly visible, with the most prominent feature being the Iron-K line at ~ 7 keV. Figure 2 shows the XMM-Newton X-ray spectrum of the core of the Virgo cluster, with the individual contribution of the various elements from a synthetic spectrum [46]. As fully detailed in the lectures by Craig Sarazin, the continuum X-ray emission is due to electron bremsstrahlung, while the lines come from decays from higher energy levels which are populated by collisional excitations. Thus, for example, the Fe K line comes from the transitions down to the K shell of H-like and He-like iron ions, while the Fe L complex comes from transitions down to the L shell of iron ions with 3 or more bound electrons. Obtaining abundances from such spectra is universally achieved using packages [90], which are based on theoretically calculated collisional excitation probabilities. Therefore, the derived abundances of simple ions (e.g., H-like and He-like) should be regarded as more reliable than those of ions still with many bound electrons.

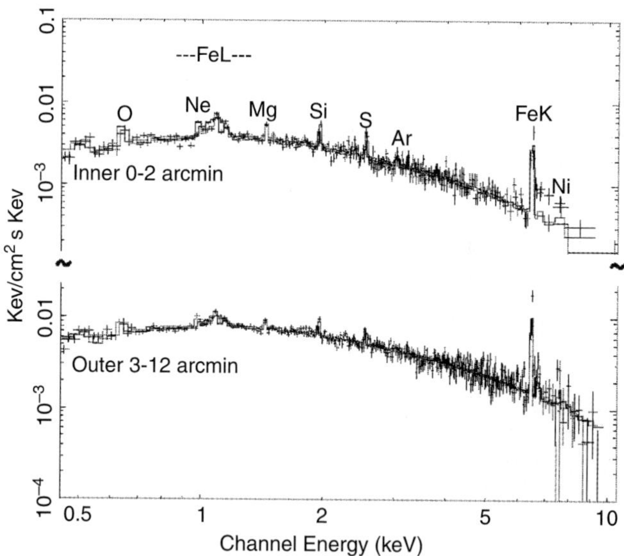

Fig. 1. The X-ray spectrum of the cluster A496 with the main emission lines identified (from [34])

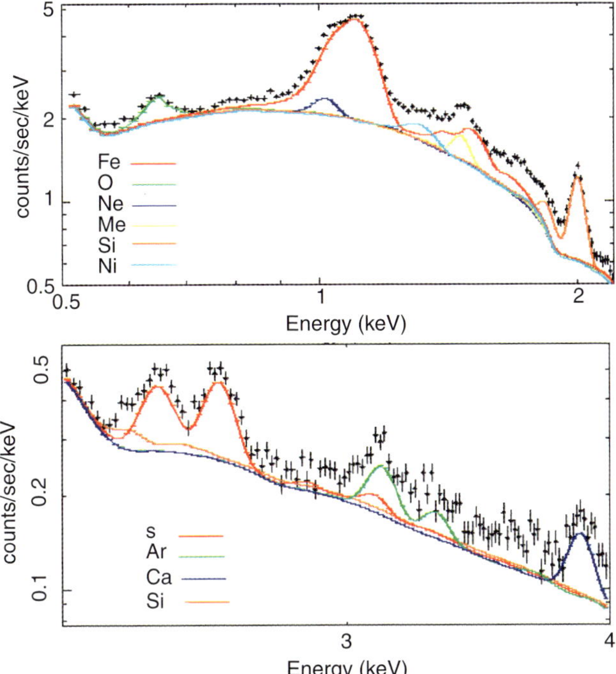

Fig. 2. The X-ray spectrum of the core of the Virgo cluster A496 and the superposition of the contribution of individual ions from a synthetic spectrum [46]

2.1 Iron

Iron is the best studied element in clusters of galaxies since ICM iron emission lines are present in all clusters and groups, either warm or hot. Figure 3 shows the iron abundance in the ICM of clusters and groups as a function of ICM temperature from an earlier compilation [97]. For $kT \gtrsim 3$ keV the ICM iron abundance is constant at $Z^{\rm Fe} \simeq 0.3 Z^{\odot}_{\rm Fe}$, independent of cluster temperature [29]. Abundances for clusters in this *horizontal* sequence come from the **Iron-K** complex at ~ 7 keV. At lower temperatures the situation is much less straightforward. Figure 4 shows data from [17], with the iron abundance having been derived with both one-temperature and two-temperature fits. The one-temperature fits give iron abundances for these cool groups which are more or less in line with those of the hotter clusters. The two-temperature-fit abundances, instead, form an almost vertical sequence, with a great deal of dispersion around a mean ~ 0.75 solar. Earlier estimates gave extremely low values for cooler groups, $kT \lesssim 1$ keV [75]. Compiling values from the literature a strong dependence of the abundance on ICM temperature is apparent, being very low at low temperatures, steeply increasing to a maximum around $kT \sim 1.5 - 2$ keV, then decreasing to reach ~ 0.3 solar by $kT \sim 3$ keV [78, 93].

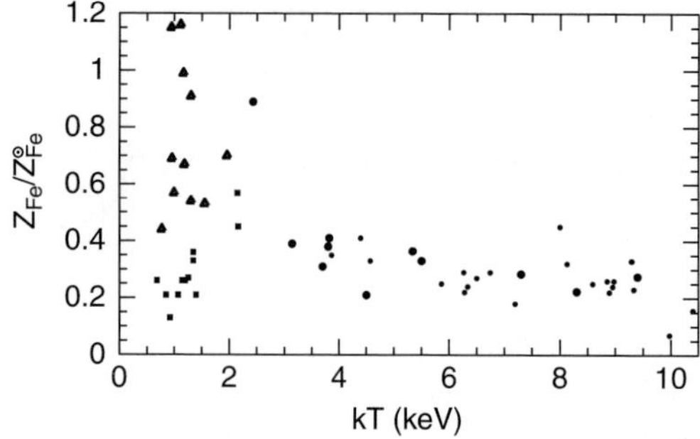

Fig. 3. A compilation of the iron abundances in the ICM as a function of the ICM temperature for a sample of clusters and groups [97]. Clusters at moderately high redshift ($\langle z\rangle \simeq 0.35$) are represented by small filled circles. For $kT \lesssim 2$ keV 11 groups are shown [17], whose temperatures and abundances are determined from one- and two-temperature fits (filled squares and open triangles, respectively)

Is this strong temperature dependence real? Perhaps some caution is in order? Besides the ambiguity as to whether one- or two-temperature fits are preferable, additional uncertainties on the iron abundances at $kT \lesssim 2$ keV are due to them being derived from the **Iron-L** complex at ~ 1 keV, where the emission lines are due to transitions to the L level of iron ions with three or more electrons. In these cooler groups/clusters most of the iron is indeed in such lower ionization stages, and the iron-K emission disappears. The atomic configurations of these more complex ions are not as simple as those giving rise to the iron-K emission, and their (calculated) collisional excitation probabilities may be more uncertain. In summary, iron abundances derived from the iron-L emission should be regarded with a little more caution, compared to those from the iron-K emission.

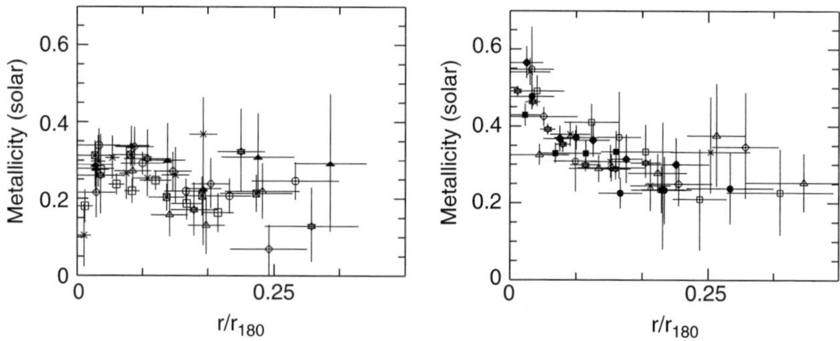

Fig. 4. Projected metallicity distributions (mostly iron) for non cool-core clusters (**left panel**) and cold-core clusters (**right panel**), from Beppo-SAX data [30]. The radial coordinate is normalized to the radius with overdensity factor 180

Abundances shown in Fig. 3 refer to the cluster central regions. However, radial gradients in the iron abundance have been reported for several clusters, starting with ASCA and then ROSAT data [34, 38, 40, 43, 122]. From Beppo-SAX data a systematic study of the radial distribution of iron in many clusters has been conducted [30], and Fig. 4 shows that clusters break up in two distinct groups. The so-called *cool core* clusters (CC, formerly known as *cooling flow* (CF) clusters before the failure of the CF model was generally acknowledged) are characterized by a steep iron gradient in the core, reaching ~ 0.6 solar near the center. Instead, in non-CC clusters (where no temperature gradient is found) no metallicity gradient appears either. The origin of the dichotomy remains to be fully understood and will be discussed later. It appears that silicon follows iron in the CC clusters, while instead there appears to be no abundance gradient in the lighter elements, such as oxygen, sulfur, etc. [70]. The fact that metallicity gradients are found in association with large temperature gradients in the central regions may look suspicious, as noted for the strong dependence of Z^{Fe} on ICM temperature, but the existence of such gradients appears to be well established.

Figure 5 shows the the iron abundance in clusters at high redshifts, including some at $z > 1$ [111]. Clearly, there is no appreciable departure from the ~ 0.3 solar abundance typical of the nearby clusters, indicating that the bulk of iron was produced and distributed in the ICM well before $z = 1$.

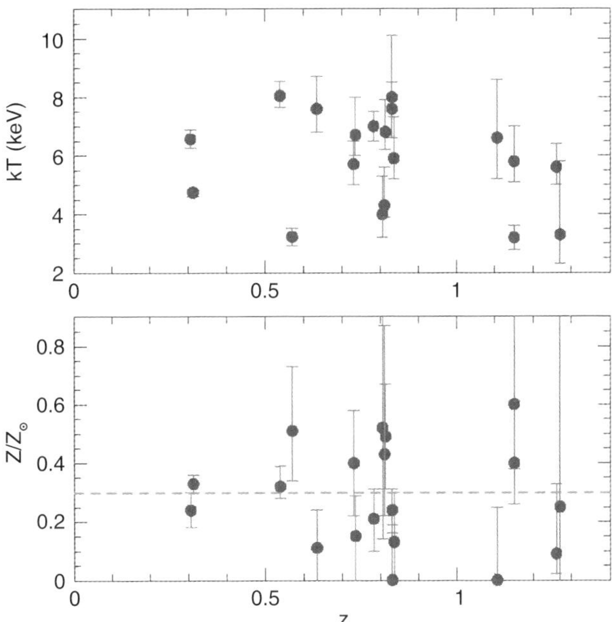

Fig. 5. The iron abundance in the ICM (**lower panel**) and the ICM temperature (**upper panel**) of a sample of high redshift clusters as a function of redshift [111]. The dashed line shows the typical abundance of $z \sim 0$ clusters

2.2 Elemental Ratios

X-ray observatories (especially ASCA, Beppo-SAX and XMM-Newton) have such high spectral resolution that, besides those of iron, the emission lines of many other elements can be detected and measured. These include oxygen, neon, magnesium, calcium, silicon, sulfur, argon, and nickel. Most of these are α elements (i.e., made by an integer number of α particles), which are predominantly synthesized in massive stars exploding as Type II supernovae. As it is well known, iron-peak elements are also produced by Type Ia supernovae, and 50–75% of iron in the sun may come from them.

Early estimates from ASCA suggested a sizeable α-element enhancement, with $\langle[\alpha/\mathrm{Fe}]\rangle \simeq +0.4$ [77], later reduced to $+0.2$ [79] and eventually found to be consistent with solar proportions $\langle[\alpha/\mathrm{Fe}]\rangle \simeq 0.0$ [57]. More recently, Finoguenov et al. (2000) report near solar Ne/Fe, slightly enhanced Si/Fe, and slightly depleted S/Fe, although with rather large error bars. From a systematic re-analysis of the ASCA archival data, a systematic increase of silicon and nickel and a decrease of sulfur with ICM temperature was reported [7] (see Fig. 6). Note that both silicon and sulfur are α elements, and apparently they do not follow the same trend. Note that the iron abundance has a pronounced

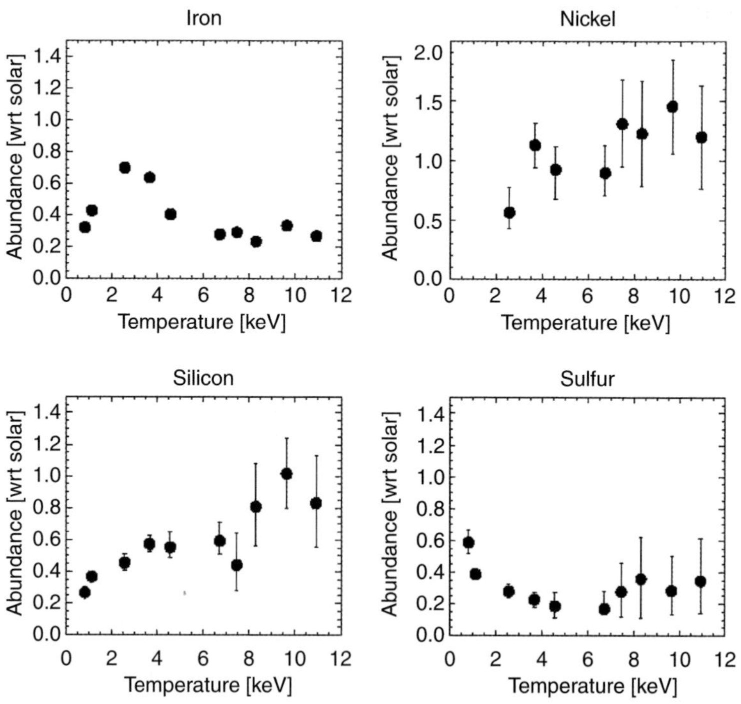

Fig. 6. The iron. nickel, silicon and sulfur abundances in the ICM (relative to solar abundances), as a function of ICM temperature, from archival ASCA data [7]

peak at $T \sim 2$ keV, which in conjunction with the increase of silicon would imply a very strong increase of the Si/Fe ratio with cluster temperature. If taken at face value, such a trend implies a progressive dominance of SN II's over SN Ia's in hotter and hotter clusters. No simple interpretation has so far emerged of why the SN mix should be sensitive to the cluster environment, and some caution is in order before accepting the reality of the trend. Indeed, one can note that Fe is most accurately determined from Fe K, i.e., in clusters hotter than ~ 2.5 keV, while Si is best determined in cooler clusters, because this element becomes virtually fully ionized in hot clusters. As a consequence the Si lines become very weak, hence with low S/N (note the increasing size of the Si error bars in Fig. 6). As a result, the Si/Fe ratio is poorly determined in cool as well as hot clusters.

Figure 7 shows more recent data from XMM-Newton, in which no apparent trend with cluster temperature is present in either iron or α-elements [108]. Therefore, I would conclude that no compelling evidence has so far emerged for other than near solar $[\alpha/\mathrm{Fe}]$ ratios **globally** in the ICM, especially when all α elements are lumped together. This leaves the possibility open for stellar nucleosynthesis having proceeded in much the same way in the solar neighborhood as well as at the galaxy cluster scale. In turn, this demands a similar ratio of the number of Type Ia to Type II SNs, as well as a similar IMF, suggesting that the star formation process (IMF, binary fraction, etc.) is universal, with little or no dependence on the global characteristics of the parent galaxies

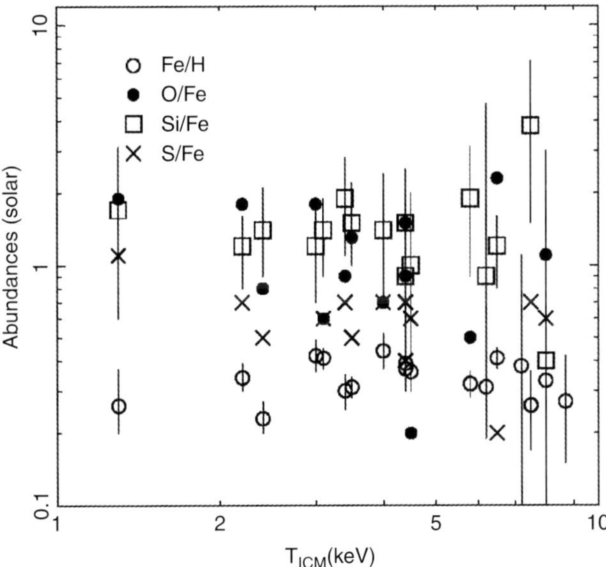

Fig. 7. The abundance of iron and the α-elements oxygen, sulfur, and silicon as a function of cluster temperature, as from XMM-Newton data [108]

(and their large scale structure environment) in which molecular clouds are turned into stars.

Alternatively, one can take at face value the variations of the abundance ratios with cluster temperature, as well as the overabundance of some α elements and the underabundance of others. One can then be forced to accept rather contrived conclusions, such as the mix of the two SN types, and perhaps even the nucleosynthesis of massive stars, depends on what the temperature of the ICM will become billions of years *after* star formation has ceased. On the other hand, one may argue that rich galaxy clusters are "special" places in many senses, and that ICM abundances reflect not only supernova nucleosynthesis yields, but also how efficiently these are ejected, mixed into, and retained by the ICM. However, no simple understanding of the apparent empirical trends has yet emerged [41, 47, 67]. Still, the story may be different for the cool-core clusters, which are characterized by a strong gradient in the iron abundance within the cool core. I shall return to cool-core clusters later.

In summary, in the following I will assume that clusters, on a global scale, have solar elemental ratios and the total heavy element abundance is 0.3 solar, or 0.006 by mass.

2.3 The Iron Mass-to-Light Ratio

One useful quantity is the iron-mass-to-light ratio (FeM/L) of the ICM, i.e. the ratio $M_{\rm Fe}^{\rm ICM}/L_{\rm B}$ of the total iron mass in the ICM over the total B-band luminosity of the galaxies in the cluster. In turn, the total iron mass in the ICM is given by the product of the iron abundance times the mass of the ICM, i.e., $M_{\rm Fe}^{\rm ICM} = M_{\rm ICM}Z_{\rm ICM}^{\rm Fe}$. Figure 8 shows the resulting FeM/L from an earlier compilation [97]. The drop of the FeM/L in poor clusters and groups (i.e. for $kT \lesssim 2$ keV) can be traced back to a drop in both the iron abundance (which however may not be real, see above) and in the ICM mass. Such groups appear to be gas poor compared to clusters, which suggests that they may have been subject to baryon and metal losses due to strong galactic winds driving much of the ICM out of them [28, 91, 93], or such winds having *preheated* and inflated the gas distribution around galaxies, thus prevents it to *fall* inside groups. In one way or another, the *break* seen in Fig. 8 is likely to be related to the break of self-similarity in the X-ray luminosity-temperature relation (see later).

In these lectures I will mainly deal with clusters with $kT \gtrsim$ 2–3 keV, for which the interpretation of the data appears more secure, but several cautionary remarks are in order, even concerning these hotter clusters. The first is that the iron abundances used to construct Fig. 8 did not take into account that some clusters have sizable iron gradients. In principle, X-ray observations can provide both the run of gas density and abundance with radius, making it possible to integrate their product over the cluster volume and get $M_{\rm Fe}^{\rm ICM}$. To my knowledge, so far this has been completed only for one cluster [89]. However, the extra-iron contained within the iron gradient core

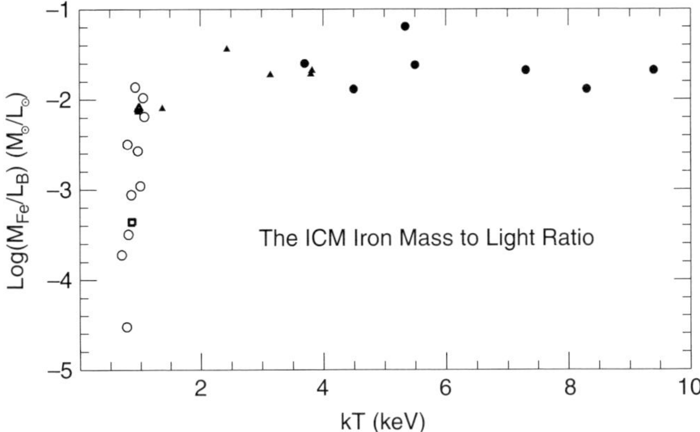

Fig. 8. The iron mass to light ratio of the ICM of clusters and groups as a function of the ICM temperature from an earlier compilation [93]. Data are taken from the following sources: filled circles [6]; filled triangles [113]; open triangle [27]; open square [75]; filled square [85]; open circles [76]

seems to be the product of the central cD galaxy, and may represent only a small fraction of the whole $M_{\rm Fe}^{\rm ICM}$ [31].

Another concern is that two of the three ingredients entering in the calculation of the FeM/L values shown in Fig. 8 (namely $M_{\rm ICM}$ and $L_{\rm B}$) may not be measured precisely in the same way in the various sources used for the compilation. Both quantities come from a radial integration up to an ill-defined cluster boundary, e.g., the Abell radius, the virial radius, or to a radius of some fixed overdensity. Sometimes it is quite difficult to ascertain what definition has been used by one author or another, with the complication that X-ray and optical data have generally been collected by different groups using different assumptions. There is certainly room for improvement here, and a new compilation paying attention to analyse all clusters in a homogeneous way would be highly desirable. Finally, estimated total luminosities $L_{\rm B}$ refer to the sum over all cluster galaxies, and do not include the population of stars which are diffused through the cluster, and which may account for at least $\sim 10\%$ of the total cluster light, and perhaps more [4, 37]. In any event, Fig. 9 shows a more recent compilation of FeM/L values [29], and compare them to the old ones finding fair agreement.

While keeping these cautions in mind, we see from Figs. 8 and 9 that the FeM/L runs remarkably flat with increasing cluster temperature, for $kT \gtrsim 2-3$ keV. This constancy of the FeM/L comes from both $Z_{\rm ICM}^{\rm Fe}$ and $M_{\rm ICM}/L_{\rm B}$ showing very little trend with cluster temperature, see Fig. 3 in this paper, and Fig. 4 in [93], where $M_{\rm ICM}/L_{\rm B} \simeq 25h_{70}^{-1/2}$ $(M_\odot/L_\odot)$. The resulting FeM/L is therefore

$$({\rm Fe}M/L)_{\rm ICM} = Z_{\rm ICM}^{\rm Fe}\frac{M_{\rm ICM}}{L_{\rm B}} \simeq 0.3Z_{\rm Fe}^{\odot}25h_{70}^{-1/2} \simeq 0.01\,h_{70}^{-1/2}(M_\odot/L_\odot)\,, \quad (1)$$

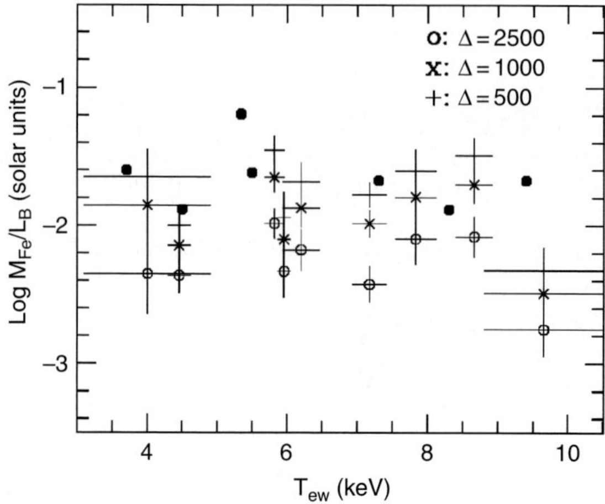

Fig. 9. The iron mass to light ratio of the ICM of clusters and groups as a function of the ICM temperature from a more recent compilation [29]. The Δ values refer to the overdensity within which the ratio is calculated. Note the good agreement with an old compilation [93], shown here as filled circles, when virtually all the cluster ICM mass is included (i.e., for the lowest value of Δ)

i.e., in the ICM there are about 0.01 solar masses of iron for each solar luminosity of the cluster galaxies. This value is $\sim$ 30% lower than adopted in [93] and shown in Fig. 8, having consistently adopted here for $Z^{\odot}_{\rm Fe}$ the recommended *meteoritic* iron abundance [2], i.e., $Z^{\odot}_{\rm Fe} = 0.0013$. Assuming solar elemental proportions for the ICM, the ICM *metal mass to light ratio* is therefore $\sim 0.3 \times 0.02 \times 25 h_{70}^{-1/2} = 0.15\ (M_{\odot}/L_{\odot})$, having adopted $Z_{\odot} = 0.02$ and for $h_{70} = 1$.

A very accurate analysis was performed recently for the A1983 cluster [89], paying attention to measure $M_{\rm ICM}$ and $L_{\rm B}$ within the same radius. The result is FeM/L= $(7.5 \pm 1.5) \times 10^{-3} h_{70}^{-1/2} (M_{\odot}/L_{\odot})$, in fair agreement with the estimate above.

The most straightforward interpretation of the constant FeM/L is that clusters did not lose iron (hence baryons), nor differentially acquired pristine baryonic material, and that the conversion of baryonic gas into stars and galaxies has proceeded with the same efficiency and the same stellar IMF in all clusters [93]. Otherwise, there should be cluster to cluster variations of $Z^{\rm Fe}_{\rm ICM}$ and FeM/L. All this is true in so far as the baryon to dark matter ratio is the same in all $kT \gtrsim 2$ keV clusters [123], and the ICM mass-to-light ratio and the gas fraction are constant. Nevertheless, there may be hints for some of these quantities showing (small) cluster to cluster variations [5, 74, 89], but no firm conclusion has yet been reached.

2.4 The Iron Share Between ICM and Cluster Galaxies

The metal abundance of the stellar component of cluster galaxies is derived from integrated spectra coupled to synthetic stellar populations. Much of the stellar mass in clusters is confined to passively evolving spheroids (ellipticals and bulges) for which the iron abundance $Z_*^{\rm Fe}$ may range from $\sim 1/3$ solar to a few times solar. For example, among ellipticals metal-sensitive spectral features such as the *magnesium index* $\rm Mg_2$ range from values slightly lower than those in the most metal rich globular clusters of the Milky Way bulge (which are nearly solar), to values for which models indicate a metallicity a few times solar [69]. The FeM/L of cluster galaxies is then given by:

$$({\rm Fe}M/L)_{\rm gal} = Z_*^{\rm Fe}\frac{M_*}{L_{\rm B}} \simeq 0.0046\,h_{70} \quad (M_\odot/L_\odot)\;, \tag{2}$$

where $Z_*^{\rm Fe} = Z_{\rm Fe}^\odot$ and one has adopted $M_*/L_{\rm B} = 3.5\,h_{70}$. This estimate comes from the $M/L_{\rm B}$ determinations for a sample of ellipticals [116], which have been used to derive the an average $\langle M/L_{\rm B}\rangle = 4.2\,h_{70}$ when adopting the Coma cluster luminosity function [123]. This value is finally reduced to $3.5\,h_{70}$ taking into account a likely $\sim 30\%$ dark matter contribution to the total mass within the galaxy effective radius [18]. The *total* cluster FeM/L (ICM+galaxies) is therefore $\sim 0.015 \quad (M_\odot/L_\odot)$, for $h_{70} = 1$. The ratio of the iron mass in the ICM to the iron mass locked into stars and galaxies is then

$$\frac{Z_{\rm ICM}^{\rm Fe}M_{\rm ICM}}{Z_*^{\rm Fe}M_*} \simeq 2.2h_{70}^{-3/2}\;, \tag{3}$$

having adopted $Z_{\rm ICM}^{\rm Fe} = 0.3$ solar, $Z_*^{\rm Fe} = 1$ solar, and $M_{\rm ICM}/M_* = 9.3h_{70}^{-3/2}$ as for the Coma cluster [123]. So, it appears that there is ~ 2 times more iron mass in the ICM than locked into cluster stars (galaxies), perhaps even more if $Z_*^{\rm Fe}$ is subsolar due to an abundance gradient within individual galaxies [3]. In turn, this empirical iron share (ICM vs. galaxies) sets a strong constraint to models of the chemical evolution of galaxies. Under the same assumptions as above, the total *metal mass to light ratio* (ICM + galaxies) is therefore $\sim 0.15h_{70}^{-1/2} + 0.07h_{70} \simeq 0.2 \;(M_\odot/L_\odot)$. This can be regarded as a fully empirical determination of the metal yield of (now) old stellar populations.

I would like to emphasize that the values of the total, cluster FeM/L and of the iron share derived in this section strictly depend on the adopted values of $M_*/L_{\rm B}$ and $M_{\rm ICM}/L_{\rm B}$, which may be subject to change as better estimates become available.

3 Metal Production: the Parent Stellar Population

The constant FeM/L of clusters means that the total mass of iron in the ICM is proportional to the total optical luminosity of the cluster galaxies [6, 23, 91, 104]. The simplest interpretation is that the iron and all the metals now in the ICM have been produced by supernovae belonging to the same stellar generation whose surviving low-mass stars now radiate the bulk of the cluster optical light. As much of the cluster light comes from old spheroids (ellipticals and bulges), one can conclude that *the bulk of cluster metals were produced by the same stellar generations that make up the old spheroids that we see today in clusters.*

It is also interesting to ask which galaxies have produced the bulk of the iron and the other heavy elements, i.e. the relative contribution as a function of the present-day luminosity of cluster galaxies. From their luminosity function it is easy to realize that the bright galaxies (those with $L \gtrsim L^*$) produce the bulk of the cluster light, while the dwarfs contribute a negligible amount of light in spite of dominating the galaxy counts by a large margin [109]. In practice, most galaxies don't do much, while only the brightest $\sim 3\%$ of all galaxies contribute $\sim 97\%$ of the whole cluster light. Giants dominate the scene while dwarfs don't count much. Following the simplest interpretation, according to which the metals were produced by the same stellar population that now shines, one can conclude that also the bulk of the cluster metals have been produced by the giant galaxies that contain most of the stellar mass. The relative contribution of dwarfs to ICM metals may have been somewhat larger than their small relative contribution to the cluster light, since metals can more easily escape from their shallower potential wells [109]. This is, however, unlikely to alter the conclusion that the giants dominate metal production by a very large margin.

Up to about 3/4 of the whole mass in stars in the local universe is now in spheroids, $\sim 1/4$ in disks, and less than 1% in irregular galaxies [10, 33, 44]. In clusters the dominance of spheroids is likely to be even stronger than in the general field. The prevalence of spheroids offers an opportunity to estimate the epoch (redshift) at which (most) metals were produces and disseminated, since we now know quite well when most stars in cluster spheroids were formed.

Following the first step in this direction [16], I believe that the most precise estimates of the age (redshift of formation) of stellar populations in cluster elliptical galaxies come from the tightness of several correlations, such as the color-magnitude, fundamental plane, and the $\mathrm{Mg}_2-\sigma$ relations, and especially by such relations remaining tight all the way to $z \sim 1$ [96, 105, 117]. This has taught us that the best way of breaking the age-metallicity degeneracy in the global proprties of stellar populations is to look back at high redshift galaxies. The collective evidence indicates that most stars in cluster ellipticals formed at $z \gtrsim 3$, while only minor episodes of star formation may have occurred later.

With most of the star formation having taken place at such high redshift, the major fraction of cluster metals should also have been produced and disseminated at $z \gtrsim 3$. Little evolution of the ICM composition is then expected all the way to high redshifts, with the possible exception of iron from SNIa's, for which the rate of release does not closely follow the star formation rate (SFR), as does the SNII rate, but for which the rate of release of iron is modulated by the distribution of the delay times between formation of the precursor and explosion time. As illustrated in the next section, one expects that the SNIa rate peaks shortly after a burst of star formation and then rapidly declines, with most events taking place within 1–2 Gyr after formation. If this is the case, no appreciable evolution of the iron abundance in clusters should be detectable from $z = 0$ to $z \sim 1$ – an argument supported by observational evidence [111]. Note, however, that *late winds* will keep enriching the ICM at a decreasing rate [23].

4 Metal Production: Type Ia vs. Type II Supernovae

Most heavy elements (metals) are produced by supernovae (SN), of which there are two main types: supernovae of Type II (SNII) result from the core collapse of massive stars ($M \gtrsim 8\,M_\odot$), while supernovae of Type Ia (SNIa) result from the thermonuclear explosion of a degenerate star, i.e., a white dwarf. Their relevance to the metal enrichment in clusters is discussed next.

4.1 Iron from SNIa's, SNIa Progenitors and SNIa Rate

As it is well known, clusters are now dominated by E/S0 galaxies, which produce only Type Ia SNs at a rate of $\sim (0.16 \pm 0.06) h_{70}^2$ SNU [19], with 1 SNU corresponding to 10^{-12} SNs yr$^{-1} L_{\mathrm{B}\odot}^{-1}$. Assuming such rate to have been constant through cosmological times ($\sim$13 Gyr), the number of SNIa's exploded in a cluster of present-day luminosity L_B is therefore $\sim 1.6 \times 10^{-13} \times 1.3 \times 10^{10} L_\mathrm{B} h_{70}^2 \simeq 2 \times 10^{-3} L_\mathrm{B}$. With each SNIa producing $\sim 0.7\,M_\odot$ of iron, the resulting FeM/L of clusters would be:

$$\left(\frac{M_\mathrm{Fe}}{L_\mathrm{B}}\right)_\mathrm{SNIa} \simeq (1.4 \pm 0.5) \times 10^{-3} h_{70}^2, \quad (4)$$

which falls short by a factor ~ 10 compared to the observed cluster FeM/L (0.015 for $h_{70} = 1$). The straightforward conclusion is that either SNIa's did not play any significant role in manufacturing iron in clusters, or their rate in what are now E/S0 galaxies had to be much higher in the past. This argues for a strong evolution of the SNIa rate in E/S0 galaxies and bulges, with the past average being $\sim 5 - 10$ times higher than the present rate [23].

In the case of SNIa's we believe we know fairly precisely the amount of iron released by each event, while we still don't know for sure what are the progenitors producing the events. There is universal agreement that SNIa's originate from the thermonuclear explosion of white dwarfs (WD) made of carbon and oxygen (CO), once they reach the Chandrasekhar limit ($\sim 1.4\,M_\odot$) by having accreted mass from a donor binary companion. Carbon is then ignited explosively and the star disrupts completely. The event produces $\sim 0.7\,M_\odot$ of iron-peak elements, mostly ^{56}Ni, which decays into ^{56}Co and finally into ^{56}Fe, powering the SN light curve. Along with iron, $\sim 0.15-0.28\,M_\odot$ of silicon are also synthetized [58].

Several different evolutionary paths, however, may lead to the SN explosion, and each of them would be characterized by a different evolution of the past SNIa rate. Two main channels are currently considered as viable, the so-called single-degenerate (SD) and the double-degenerate (DD) channel. In the SD case, a CO-WD accretes hydrogen-rich material from a companion-star, and processes it through H and He burning, increasing the CO mass until the WD exceeds the Chandrasekhar limit [121]. The case of helium being ignited explosively, even before the Chandrasekhar limit is reached, has also been considered (SD/Sub-Chandra exploders), but the resulting synthetic spectra differ markedly compared to those of observed SNIa's [66]. In the DD option, the secondary star in the binary is also a WD, and the two stars spiral in towards eachother, due to angular momentum loss via gravitational wave radiation (GWR), until the less massive star fills its Roche lobe and the two WDs merge together [56]. Once again, a SNIa may result if the combined mass exceeds the Chandrasekhar limit.

In the SD channel, the time of explosion, i.e. the delay between the star formation event and the supernova, is set by the time it takes the secondary (less massive) star to evolve off the main sequence and fill its Roche lobe. Hence the delay strongly increases with decreasing mass and, in principle, delay times of order the Huble time are possible, provided $\sim 1M_\odot$ donors are able to transfer enough mass to grow the WD beyond the threshold mass for ignition. In the DD channel, the delay time is further augmented by the time it takes the secondary WD to spiral in, due thanks to the GWR, which is given by:

$$\tau_{\rm GWR} = \frac{0.15A^4}{(M_1^{\rm WD} + M_2^{\rm WD})M_1^{\rm WD}M_2^{\rm WD}} \quad {\rm Gyr}\,, \tag{5}$$

where A is the intial separation of the DD system in units of the solar radius, and the two WD masses are also in solar units. Also in this case delay times can easily exceed the Hubble time. In both cases, the minimum delay time is set by the lifetime of the maximum initial mass that produces a WD remnant, i.e. ~ 35 Myr, the lifetime of $8\,M_\odot$ stars [55].

The run of the past SNIa rate is therefore proportional to the distribution function of the delay times, and hence depends on the distributions of the initial binary parameters (i.e., masses M_1 and M_2, and separations), since a wide range of them can lead to a successful explosion, as well as additional parameters describing the mass transfer phases that take place in the course of the binary evolution and their outcome.

The distributions of the delay times $f_{\rm Ia}(\tau)$ for both the SD and DD channels have been recently calculated in a fashion that allows one for an effective exploration of the parameter space [49], and the results are shown in Figs. 10 and 11, respectively. The distributions can differ widely, but have several characteristics in common: (a) $f_{\rm Ia}(\tau) = 0$ for $\tau < 35$ Myr, (b) from

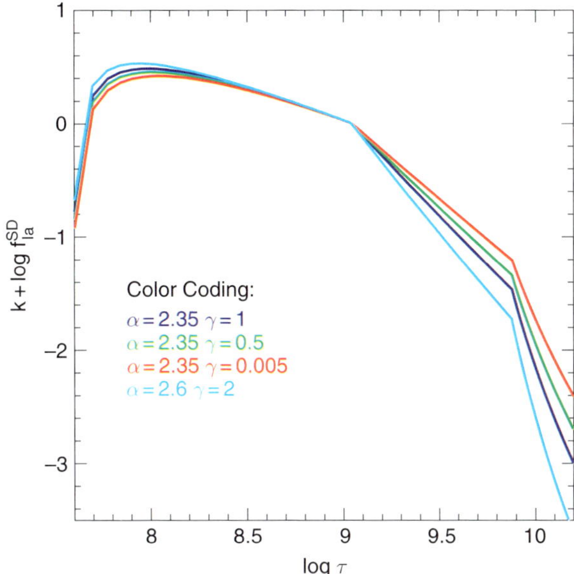

Fig. 10. The distribution function of the delay times for the SD model for the production of Type Ia supernovae, and for various values of the IMF slope (IMF$\propto$ $M^{-\alpha}$) and and the slope γ of the distribution function of the binary initial mass ratios $q = M_2/M_1$ (with $f(q) \propto q^{\gamma}$). The distributions present two cusps, one at $\tau \sim 1$ Gyr, which is due to the requirement that the primary produces a CO-WD, and the second at $\tau \sim 8$ Gyr is due to the requirement that the mass of the WD plus the mass of the envelope of the secondary star exceeds the Chandrasekhar limit (from [49])

zero, $f_{\rm Ia}(\tau)$ steeply increases, reaching is maximum by $\tau \lesssim 10^8$ yr, (c) the maximum is followed by a plateau phase with a duration $\lesssim 1$ Gyr and is not model-dependent, (d) the plateau is followed by a decline with a rate which is extremely model-dependent, especially in the SD scenario, and (e) in the SD case the late decline is much steeper than in the DD case. This is caused by the fact that at late times (e.g., 10 Gyr after the burst of star formation) the secondary components have quite low mass $\sim 1\,M_{\odot}$, their envelope mass available for transfer to the WD is just a fraction of this, and therefore successful exploders are restricted only to those few systems with very massive WDs. In general, the late decline is primarily controlled by the distributions of the binary masses that lead to a SNIa event in the SD case, and by the distribution of the initial separations of the WD+WD systems in the DD case, which are both difficult to predict. Nevertheless, properties (a), (b), (c) and (e) are *generic*, i.e., common to all combinations of model assumptions and parameters.

At first sight the various curves in Figs. 10 and 11 look very similar to each other, but it is only the use of a log-log plot that gives this impression.

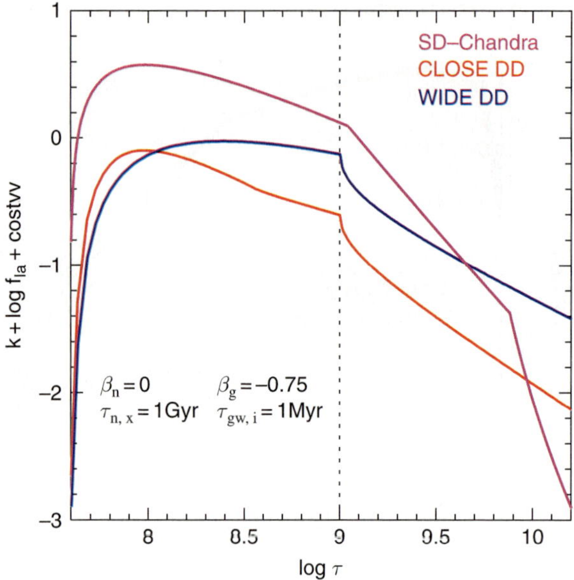

Fig. 11. The distribution function of the delay times for the DD case in the productiob of SNIa. Two different assumptions are considered which concern the first common envelope event (labelled CLOSE DD and WIDE DD respectively), compared to one rendition of the delay time for the SD case. The relative normalizations are arbitrary. (from [49])

Figure 12 shows as function of time (not log-time) the distributions of delay times for various model parameters, both for the SD and the DD cases. The curves are all normalized to give the same SNIa rate at $t = 10$ Gyr, equivalent to force all models to account for the present SNIa rate in ellipticals. Clearly, the time integrals of the delay times (which are proportional to the total number of SNIa events) differ dramatically from one case to another, and therefore so does the amount of iron from SNIa's predicted by the various models. This is shown in Fig. 13, where the FeM/L predicted by various models is plotted as a function of the SNIa rate at $t = 12$ Gyr divided by the average rate in the past [48]. Clearly some models dramatically overpredict the FeM/L compared to the cluster value, while others dramatically underpredict it. By and large, SD models appear to be excluded, because the late decline of their rate is too fast: if forced to account for the present rate in ellipticals their past rate would have been too high. Therefore, this comparison favors DD models, actually some particular version of them [48].

This conclusion, however, rests on the assumption that nature has chosen only one path to make SNIa's (either DD or SD), but by no means can we exclude that nature is able to make SNIa's from both SD and DD precursors. If so, the theoretical delay times shown in these figures suggest that SDs would

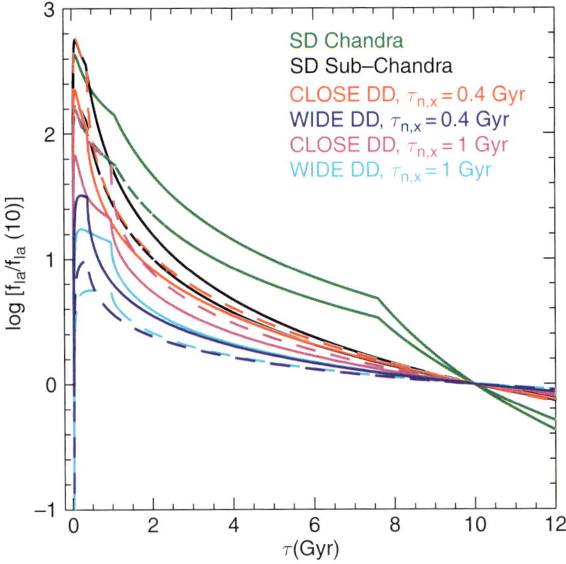

Fig. 12. The distribution function of the delay times for the DD (from [49])

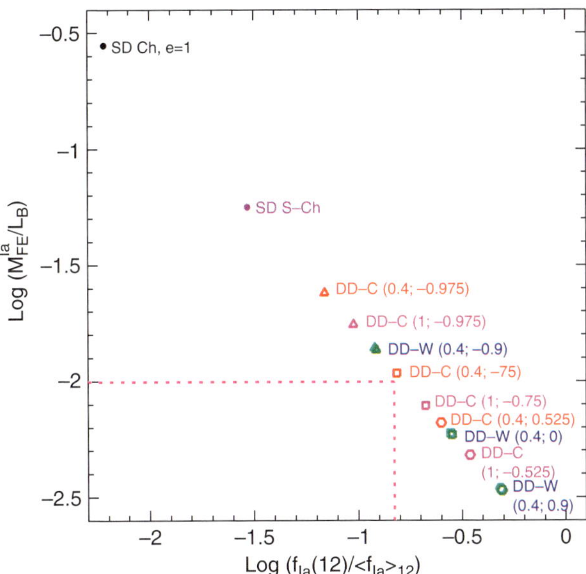

Fig. 13. The distribution function of the delay times for the DD (from [49])

dominate at early times, and DDs would dominate at late times, as indicated in Fig. 11. In any event, the bottom line is that these theoretical SNIa rates predict that the bulk of iron should be produced within 1–2 Gyr from the major phase of star formation phase in ellipticals.

The distributions of delay times derived for the SD and DD channels are sharply at variance with those derived from the SNIa statistics at high redshift by the GOODS "Piggyback" transient survey [107]. From the counts of SNIa's up to $z = 1.55$ in the GOODS fields, and from the cosmic history of star formation, it was claimed that such data would require the minimum delay time to be of the order of ~ 2 Gyr, with an average delay time of ~ 4 Gyr. Indeed, their best-fit solution is a Gaussian distribution fluctuation centered at $\tau = 4$ Gyr, with $\sigma = 0.8$ Gyr [107]. With only one possible exception, a distribution such as this bears no resemblance whatsoever to any of the theoretical distributions shown here or in [49], or ever considered for SNIa's existing scenarios (e.g., [126]). The exception would be a DD model in which the donor is a helium WD, which typically requires ~ 1 Gyr to appear after the formation of a stellar population. The explosion, however, would need to be a helium detonation that could ignite the underlying C–O core, but which would give a spectrum totally at variance with the observed SNIa spectra. I conclude that either we have so far completely missed the identification of the nature of the SNIa progenitors, or the mentioned estimate of the rate at $z \sim 1.5$ is severely biased. Being based on only 2 events at $z > 1.4$. I'm inclined to favor this latter option.

4.2 Iron and Metals from SNII's and the IMF

In the case of SNIa's, we currently believe to have a fairly precise knowledge of the amount of iron produced by each event, while the nature of the progenitors and the evolution of the SN rate still remain as open issues. The case of Type II SN's is quite the opposite: we believe to have unambiguously identified the progenitors (stars more massive than $\sim 8M_\odot$), while a great uncertainty affects the amount of iron $M_{\rm Fe}^{\rm II}(M)$ produced by each SNII event as a function of the progenitor's mass. This is due to the fundamental difficulty for core-collapse SN models to precisely locate the *mass cut* between the collapsing core that forms the neutron star remnant, and the ejected envelope. This cut is often within the iron-peak layer. On the other hand, the SN luminosity at late times can be used to infer the amount of radioactive Ni-Co (and hence eventually iron) that was ejected. An early study indicated small variations from one event to another ($0.04-0.10\,M_\odot$) [81]. This led to the assumption that $M_{\rm Fe}^{\rm II}$ is a weak function of initial mass, with an average yield of $0.07\,M_\odot$ of iron per SN event (as in SN 1987A) [93]. More recent studies, based on a larger sample of SNII events, have actually detected very large differences from one event to another (ranging from $\sim 0.002\,M_\odot$ to $\sim 0.3\,M_\odot$ [114]). Figure 14 shows the mass of ejected ^{56}Ni as a function of the velocity of the ejecta [51],

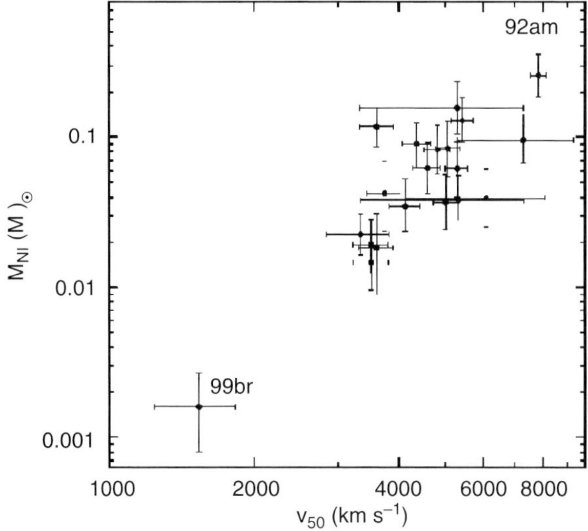

Fig. 14. The mass of ^{56}Ni ejected by a well studied sample of Type II supernovae, as a function of the velocity of the ejecta; from [51]

and averaging over the 22 SNII's in the sample one gets $\langle M_{\rm Ni}\rangle = 0.073\,M_\odot$, so close to the adopted value.

The total number of SNIIs $-N_{\rm SNII}-$ is obtained by integrating the stellar IMF from 8 to 40 $M_\odot$ for example, with the IMF being expressed as [95]:

$$\phi(M) = a(t,Z)L_{\rm B}M^{-s}\;, \tag{6}$$

where $a(t,Z)$ is a (slow) function of the SSP age and metallicity. For example, for $t = 12$ Gyr, $a(Z) = 2.22$ and 3.12, respectively for $Z = Z_\odot$ and $Z = 2Z_\odot$ [69], with $L_{\rm B}$ and M expressed in solar units.

Clearly, the flatter the IMF slope, the larger the number of massive stars per unit present luminosity, the larger the number of SNII's, and the larger the implied FeM/L. Thus, adopting $\langle M_{\rm Fe}^{\rm II}\rangle = 0.07M_\odot$ and $a = 3$ and integrating over the IMF one gets:

$$\left(\frac{M_{\rm Fe}}{L_{\rm B}}\right)_{\rm SNII} = \frac{M_{\rm Fe}^{\rm II}N_{\rm SNII}}{L_{\rm B}} \simeq \begin{cases} 0.003 & \text{for } s = 2.7 \\ 0.009 & \text{for } s = 2.35 \\ 0.035 & \text{for } s = 1.90 \end{cases} \tag{7}$$

Hence, if the Salpeter IMF ($s = 2.35$) applies also to clusters ellipticals, then SNII's underproduce iron by less than factor of ~ 2.

Constraints on the IMF slope in cluster ellipticals can be derived from the evolution of their M/L ratio with redshift, as inferred from the shift of the fundamental plane in clusters at increasingly high redshifts. This is illustrated in Fig. 15, showing the evolution of the $M/L_{\rm B}$ ratio all the way to $z = 1.27$, which indeed favors a Salpeter IMF [100].

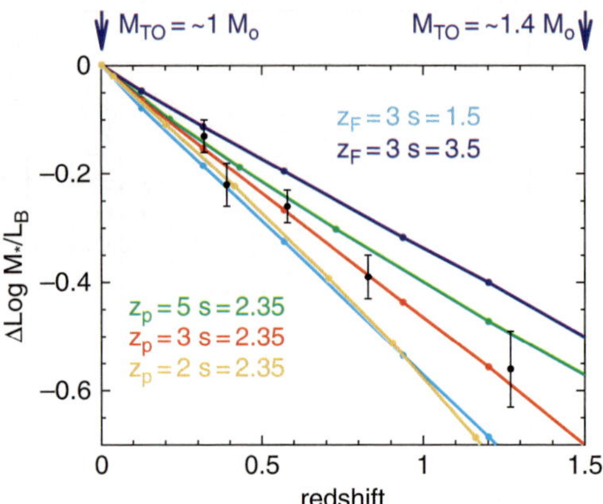

Fig. 15. The evolution with redshift of the M_*/L_B ratio of simple stellar populations (SSP) of solar metallicity and various IMF slopes ($\mathrm{d}N \propto M^{-s}\mathrm{d}M$), normalized to its value at $z = 0$ [100]. Convergence cosmology ($\Omega_\mathrm{m} = 0.3$, $\Omega_\Lambda = 0.7$, $H_\circ = 70\ \mathrm{kms}^{-1}\mathrm{Mpc}^{-1}$) and SSP formation redshifts as indicated. The data points [117] refer to the shifts in the fundamental plane locations for clusters of galaxies at various redshifts. Note that for such high formation redshifts the stellar mass at the main sequence turnoff is $\sim 1.4\,M_\odot$ at $z = 1.5$ and $\sim M_\odot$ at $z = 0$, as indicated by the arrows

At $z = 0$, however, elliptical galaxies harbour ~ 12 Gyr old stellar populations, with stars of $\sim 1\,M_\odot$ at the main sequence turnoff (MSTO). By $z = 1.5$ the precursors of such populations have an age of only ~ 3 Gyr, and correspondingly a higher mass at the MSTO, but not by much so. Specifically, the MSTO mass at an age of ~ 3 Gyr is ~ 1.4–$1.5\,M_\odot$, and therefore by following the evolution of the FP with redshift up to $z \sim 1.5$ (or equivalently of the mass-to-light ratio) we explore the IMF slope within the rather narrow mass interval $1 \lesssim M \lesssim 1.4\,M_\odot$. In practice, we measure the slope of the IMF only near $M = M_\odot$.

Given that iron is produced by both types of supernovae, iron is not the best element to constrain the IMF slope in the high mass range. Instead, α elements are produced almost exclusively by SNII's, and therefore the IMF slope can better be constrained by them, and in particular by oxygen and silicon, which abundance in the ICM is affected by relatively small errors. Therefore, in a similar way to the case for iron, the metal-mass-to-light ratio for the "X" element can be calculated in a straightforward manner from:

$$\frac{M_\mathrm{X}}{L_\mathrm{B}} = a(t, Z) \int_8^{40} m_\mathrm{X}(M) M^{-s} \mathrm{d}M\ , \tag{8}$$

where $m_\mathrm{X}(M)$ is the mass of the element "X" which is produced by a star of mass M. Adopting $a(t, Z) = 3$, $m_\mathrm{X}(M)$ for oxygen and silicon from [124], and integrating (8), one obtains the *oxygen-* and the *silicon-mass-to-light ratios*

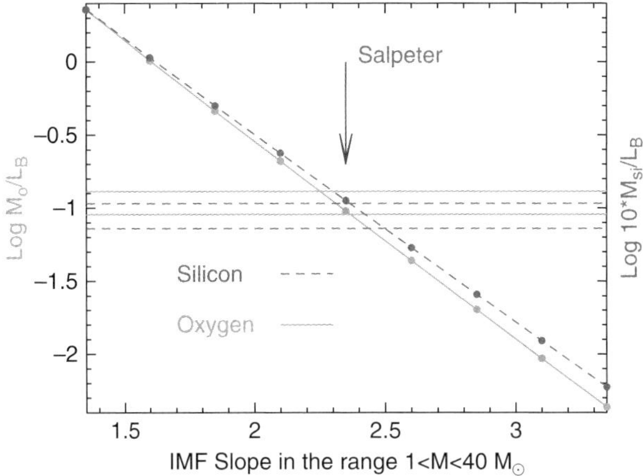

Fig. 16. The Oxygen- and the Silicon-Mass-to-Light Ratios as a function of the IMF slope calculated from (8) with $a(t, Z) = 3$ and using standard nucleosynthesis prescriptions [124]. The horizontal lines show the observed average values of these ratios in clusters of galaxies, and their range of uncertainty

which are shown in Fig. 16 as a function of the IMF slope [100]. As expected, the $M_\mathrm{O}/L_\mathrm{B}$ and $M_\mathrm{Si}/L_\mathrm{B}$ ratios are extremely sensitve to the IMF slope. The values observed in clusters of galaxies (ICM plus galaxies)[1] are ~ 0.1 and $\sim 0.01\, M_\odot/L_\odot$ respectively for oxygen and silicon [39, 88]. These empirical values are reported in Fig. 16 showing that with the *Salpeter* IMF slope ($s = 2.35$) the standard explosive nucleosynthesis from Type II supernovae produces just the right amount of oxygen and silicon to account for the observed $M_\mathrm{O}/L_\mathrm{B}$ and $M_\mathrm{Si}/L_\mathrm{B}$ ratios in cluster of galaxies, having assumed that most of the cluster B-band light comes from $\gtrsim 12$ Gyr old stellar populations.

Figure 16 also shows that with $s = 1.35$ such a *top heavy* IMF (in various circumstances invoked to ease discrepancies between theories and observations) would overproduce metals by more than a factor of 20. This is indeed the change one expects in $M_\mathrm{O}/L_\mathrm{B}$, $M_\mathrm{Si}/L_\mathrm{B}$, etc. for a change in the IMF slope $\Delta s = 1$ when considering that the light L_B is provided by $\sim M_\odot$ stars and the metals by $\sim 25\, M_\odot$ stars.

In summary, it appears that, with a Salpeter IMF and standard nucleosynthesis prescriptions, massive stars can produce the observed amonts of oxygen and silicon which are present in clusters of galaxies, while perhaps falling short by a factor ~ 2 to produce the observed iron. Yet, with an IMF just slightly shallower than Salpeter, SNII's could make also all the iron. There should then however be clearer evidence for an α-element overabundance than currently indicated by the observations(cf. Fig. 7). Hence, nucleosynthesis may

[1] These values result from averaging over the reported values for individual clusters with different ICM temperature, and take into account that ~10–30% of the stellar mass in clusters is not bound to individual galaxies [4, 45].

well have proceeded in clusters not unlike in the Milky Way, where we currently believe that about $\sim 1/2$ of iron has been produced by core-collapse Type II supernovae, and the other half by thermonuclear Type Ia supernovae. I shall discuss later the chemical evolution of clusters and the Milky Way. With $2.35 \lesssim s \lesssim 2.7$ and a past average rate of SNIa's in ellipticals $\gtrsim 5$ times the present rate, the iron content of clusters and the global ICM $[\alpha/\mathrm{Fe}]$ ratio are grossly accounted for, with SNIa's then having produced $\sim 1/2$ of the total cluster iron, not unlike in standard chemical models for the Milky Way. This is not to say that this has been firmly proved, but it seems to me to be premature to abandon the attractive simplicity of a universal nucleosynthesis process (i.e., IMF and SNIa/SNII ratio) before embarking towards more complex, multi-parametric scenarios.

5 Metals from Galaxies to the ICM: Ejection vs. Extraction

Having established that most metals in clusters are in the ICM and not in the ISM of their parent galaxies, it remains to be understood how they were transfered from galaxies to the ICM. There are two main possibilities: extraction by ram pressure stripping as galaxies plow through the ICM, and ejection by galactic winds powered from within the galaxies themselves. In the latter case the power can be supplied by supernovae (the so-called star formation feedback) and/or by AGN activity.

Ram pressure stripping certainly exists in clusters [102], as clusters are assembled by growing group and isolated galaxies which, by entering a dense ICM, are stripped of their gas and then become quiescent. Several arguments, however, favor ejection over extraction [99]:

⋆ There appears to be no trend of either $Z_{\mathrm{ICM}}^{\mathrm{Fe}}$ or the FeM/L with cluster temperature or cluster velocity dispersion (σ_{v}), while the efficiency of ram pressure stripping increases steeply with increasing σ_{v}.

⋆ Field ellipticals appear to be virtually identical to cluster ellipticals. They follow basically the same $\mathrm{Mg}_2 - \sigma$ and fundamental plane relations [11, 12], which does not show any appreciable trend with the local density of galaxies, or at most a very weak one. If stripping was responsible for extracting metals from galaxies one would expect galaxies in low density environments to have retained more metals, hence showing higher metal indices for given σ, which is not seen.

⋆ Non-gravitational energy injection of the ICM seems to be required to account for the break of the self-similar X-ray luminosity-temperature relation for groups and clusters [86]. While galactic winds are an obvious vehicle for such *pre-heating*, no pre-heating is associated to metal transfer by ram pressure.

⋆ Strong galactic (super)winds are actually observed in starburst galaxies at low and high redshift, which are thought to be the progenitors of local ellipticals [52, 83, 106, 118].

⋆ Most metals have been produced at very early times ($z \gtrsim 3$), probably well before the clusters were assembled, hence before the dense ICM was in place, hence when there was not much ram pressure exerted on the galaxies.

One can quite safely conclude that metals in the ICM have been *ejected* from galaxies by supernova (or AGN) driven winds, rather than stripped by ram pressure [34, 91]. Two kinds of galactic winds are likely to operate: *early winds* driven by the starburst forming much of the galaxy's stellar mass itself, and *late winds* or outflows where the gas comes from the cumulative stellar mass loss as the stellar populations passively age. Late winds are also likely to operate, as the stellar mass loss from the aging population flows out of spheroids, being either continuously driven by a declining SNIa rate [23], or intermittently by recurrent AGN activity [24].

5.1 The Metallicity Gradients in Cool-Core Clusters

Besides showing a radial gradient in the iron abundance (see Fig. 4), virtually all CC clusters host a cD galaxy at their center [30]. Hence, it is quite natural to associate the two phenomena, and attribute to the cD galaxy the responsibility to have further enriched the central regions of the clusters. This possibility has been recently explored in some detail [15], and Fig. 17 shows

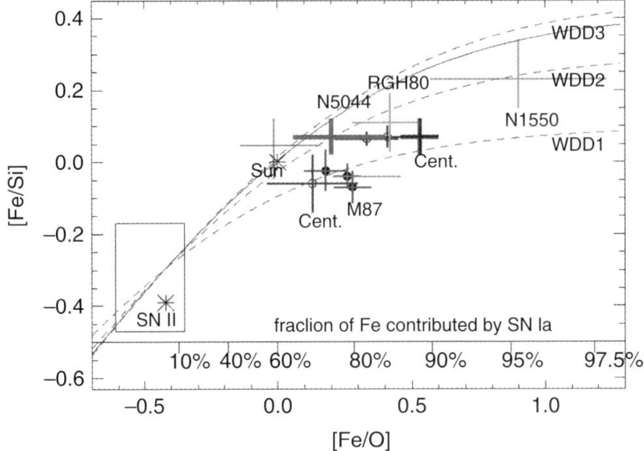

Fig. 17. The [Fe/Si] and [Fe/O] ratios for various proportions of SNIa contribution to the total iron abundance, an indicated in the inner scale [15]. The lines refer to different SNIa models, while the data refer to the innermost ($r \lesssim 100$ kpc) of some cool-core clusters. An asterisk shows the pure SNII elemental retios, and another asterisk refer to the solar proportions

the main results of this analysis. Both ratios [Fe/Si] and [Fe/O] depend on the relative proportion of the two supernova types that contribute to the nucleosynthesis, with the asterisk showing the ratio from SNII's only. By increasing the proportion of iron contributed by SNIa's, the ratios move along the trajectories shown by the various lines in Fig. 17, which refer to different SNIa models. The data points refer to the central regions (typically within 100 kpc) of the indicated (CC) clusters. It appears that these regions are exceptionally rich in SNIa products, which can be interpreted as due to the late winds for the cD's, powered by SNIa's themselves [23], and/or recurrent AGN activity [24].

By the way, sitting at the bottom of the cluster potential well, cD galaxies experience no ram pressure, yet, they appear to have further enriched the "Cool Core" of CC clusters. This is a further argument for a dominant role of winds in transfering metals from galaxies to the ICM.

6 Metals as Tracers of ICM Pre-heating

The total amount of iron (and metals) in clusters represents a record of the overall past supernova activity as well as the history of the mass and energy ejected from cluster galaxies. The empirical values FeM/L can be used to set a constraint on the energy injection into the ICM by supernova-driven galactic winds [92]. The total SN heating is given by the kinetic energy released by one SN ($\sim 10^{51}$ erg) times the number of SNs that have exploded. It is convenient to express this energy per unit present optical light $L_{\rm B}$, i.e.:

$$\frac{E_{\rm SN}}{L_{\rm B}} = 10^{51}\,\frac{N_{\rm SN}}{L_{\rm B}} = 10^{51}\left(\frac{M_{\rm Fe}}{L_{\rm B}}\right)^{\rm TOT}\frac{1}{\langle M_{\rm Fe}\rangle} \simeq 10^{50} \quad ({\rm erg}/L_\odot)\,, \tag{9}$$

where the total (ICM+galaxies) FeM/L=0.015 $M_\odot/L_\odot$ is adopted, and the average iron release per SN event is assumed to be 0.15 $M_\odot$ (appropriate if SNIa's and SNII's contribute equally to the iron production). This estimate should be accurate to within a factor ~ 2.

The kinetic energy injected into the ICM by galactic winds, again per unit cluster light, is given by 1/2 the ejected mass ($M_{\rm Fe}^{\rm ICM}/Z_{\rm w}^{\rm Fe}$) times the typical wind velocity squared, i.e.:

$$\frac{E_{\rm w}}{L_{\rm B}} = \frac{1}{2}\frac{M_{\rm Fe}^{\rm ICM}}{L_{\rm B}}\left\langle\frac{v_{\rm w}^2}{Z_{\rm w}^{\rm Fe}}\right\rangle \simeq 1.5\times 10^{49}\,\frac{Z_\odot^{\rm Fe}}{Z_{\rm w}^{\rm Fe}}\cdot\left(\frac{v_w}{500\,{\rm km\,s^{-1}}}\right)^2 \simeq 10^{49} \quad ({\rm erg}/L_\odot)\,, \tag{10}$$

where the empirical FeM/L for the ICM has been used, and the average metallicity of the winds $Z_{\rm w}^{\rm Fe}$ is assumed to be two times solar. As usual in the case of thermal winds, the wind velocity $v_{\rm w}$ is of the order of the escape velocity from individual galaxies. Again, this estimate may be regarded as accurate to within a factor of 2, or so.

A first inference is that of order of $\sim$ 5%–20% of the kinetic energy released by SNs is likely to survive as kinetic energy in galactic winds, thus contributing to the heating of the ICM. A roughly similar amount goes into work to extract the gas from the potential well of individual galaxies, while the rest of the SN energy has to be radiated away locally and does not contribute to the feedback. This estimated energy injection represents a small fraction of the thermal energy of the ICM of rich (hot) clusters and so it has only a minor impact on the history of the ICM. However, in groups it represents a non-negligible fraction of the thermal energy of the ICM, thus affecting its evolution and present structure. The necessity of some non-gravitational heating (or *pre-heating*) was recognized from the break of the self-similarity demanded by the observed X-ray luminosity-temperature relation, especially when groups are included [86].

The estimated $\sim 10^{49}$ erg/$L_{\odot}$ correspond to a pre-heating of ~ 0.1 keV per particle, for a typical cluster $M_{\rm ICM}/L_{\rm B} \simeq 25\ M_{\odot}/L_{\odot}$. This is $\gtrsim 10$ times lower than the ~ 1 keV/particle pre-heating that some models require to fit the cluster $L_{\rm X} - T$ relation [13, 39, 84, 110, 125]. This estimate depends somewhat on the gas density (hence environment and redshift) where/when the energy is injected, because what matters is the entropy change induced by the pre-heating, $\Delta S = k\Delta T/n_{\rm e}^{2/3}$ [13, 20, 60]. Hence the required energy decreases if it is injected at a lower gas density. Nevertheless, this extreme (1 keV/particle) requirement would be met only if virtually all the SN energy were to go into increasing the thermal energy of the ICM. Such extreme pre-heating requirement points toward an additional energy (entropy) source, such as AGN energy injection [115, 125]. Note however that in powerful starbursts most SNs explode inside hot bubbles made by previous SNs, thus reducing radiative losses, and the feedback efficiency may approach unity [53]. More recently it has been suggested that pre-heating requirements may be relaxed somewhat if the energy injection takes place at relatively low density, so as to boost the entropy increase with less energy deposition [87]. For example, pre-heating could take place within the filaments, prior to the time when they coalesce to form clusters. Indeed, if much of the star formation in cluster ellipticals took place at $z \gtrsim 3$, it likely predates by a long time the assembly of clusters. Further exploration of the metal enrichment connections to pre-heating are found in [14, 38].

7 Clusters vs. Field at $z = 0$ and the Overall Metallicity of the Universe

To what extent are clusters fair samples of the $z \sim 0$ universe as a whole? In many respects clusters look much different from the field, e.g., in the morphological mix of galaxies, or in the star formation activity, which in clusters has almost completely ceased while it is still going on in the field. Yet, when we restrict ourselves to some global properties, clusters and field are not so different. For example, the baryon fraction of the universe is $\Omega_{\rm b}/\Omega_{\rm m} \simeq 0.16 \pm 0.02$ [9], which compares ~ 0.15 as estimated for clusters [123] adopting $h_{70} = 1$. This tells us that no appreciable baryon vs. dark matter segregation has taken place at a cluster scale [123], a prediction that X-ray observations should be able to check.

Even more interesting may be the case of the stellar mass over baryonic mass in clusters and in the field. For the field, i.e., the local universe, the contribution of stars to Ω is estimated as $\Omega_* = 0.0035h_{70}^{-1}$ [44], or $\Omega_* = 0.0041h_{70}^{-1}$ from the 2dF K-band luminosity function [25], with a $\sim 15\%$ uncertainty (adopting a Salpeter IMF). The total baryon density is $\Omega_{\rm b} = 0.039h_{70}^{-2}$, as derived from the Standard Big Bang nucleosynthesis (and confirmed by WMAP [9]). This gives a global baryon to star conversion efficiency $\Omega_*/\Omega_{\rm b} \simeq 0.10h_{70}$, i.e., over the whole cosmic time $\sim$ 10% of the baryons have been converted and locked into stars. At the galaxy cluster level, the same efficiency can be measured directly, and following [123] one gets:

$$\frac{M_*}{M_{\rm ICM} + M_*} \simeq \frac{1}{9.3h_{70}^{-3/2} + 1} \simeq 0.1. \tag{11}$$

One can safely conclude that **the efficiency of baryon to galaxies/stars conversion has been $\sim$ 10%, quite the same in the "field" as well as within rich clusters of galaxies.** At this very basic level, the environment seems to be irrelevant! Note that we may be living in a rather special time, as the friction of baryons locked in stars must have evolved at a different rate in clusters and in the field, with clusters freezing at the 10% level at a much higher redshift ($z \sim 3$) compared to the field ($z \lesssim 1$).

Two interesting inferences can be drawn from this intriguing cluster-field similarity:

$\star$ **The metallicity of the present universe is $\sim$ 1/3 solar.** The metallicity of the local universe has to be virtually identical to that measured in clusters ($\sim$ 1/3 solar), since star formation, hence the ensuing metal enrichment, have proceeded to the same level of *baryon consumption* ($\sim$ 10%). In an analogy to clusters, a majority share of the metals now reside out of galaxies in a warm/hot intergalactic medium (WHIM) containing the majority of the baryons. Most baryons as well as most metals in the local universe remain unaccounted for, but observational efforts are currently being made to detect them [26, 80].

$\star$ **The thermal energy (temperature) of the local universe is about the same as the pre-heating energy of clusters.** Similar overall star formation activities most likely result not only in similar metal productions but also in similar energy depositions by galactic winds. Hence, the temperature of the local IGM is likely to be $kT \sim 0.1-1$ keV, whatever the physical nature of the cluster pre-heating turns out to be. Again, attempts are currently going on to detect this metal rich WHIM. The detection of OVI absorption clouds, physically located within the Local Group [80] as well as at moderate redshift [101], are important steps in this direction.

At $z \sim 0$ field early-type galaxies (ETG) show very little, yet detectable differences with respect to their cluster analogues [12, 35], implying typical ages only $\sim 1\,\mathrm{Gyr}$ younger than ETGs in clusters. Moreover, bulges appear very similar to ellipticals in their integrated properties, such as the $\mathrm{Mg}_2 - \sigma$ and fundamental plane relations [36, 59]. In the well-studied case of the Milky Way bulge no trace of stars younger than halo-bulge globular clusters could be found [127]. At $z \sim 1$ old ETGs are also found in sizable numbers in the general field, while it appears that star formation may have been a little more extended than in clusters [8, 22, 112].

Therefore, spheroids in the general field appear almost as old as cluster ellipticals, i.e., with the bulk of their stellar populations having formed at $z \gtrsim$ 2–3. Given this, it is estimated that at least 65% of the stellar mass is at least 8 Gyr old, or formed at $z > 1$ [54]. With $\sim 50\%$ of the stellar mass in spheroids that formed $\gtrsim 80\%$ of their mass at $z \gtrsim$ 2–3, one can conclude that $\gtrsim 30\%$ of the stellar mass we see today was already in place by $z \sim 3$ [94]. This *indirect* estimate is ~ 3 times higher than *directly* measured in the HDF-N [32]. However, this latter result may be subject to cosmic variance given the small size of the explored field, and a value as high as $\sim 30\%$ cannot be excluded by current observations [42].

7.1 The Metallicity of the Universe at $z = 3$

With $\sim 30\%$ of all stars having formed by $z = 3$, also $\sim 30\%$ of the metals should have been formed before such an early epoch. I have argued that the global metallicity of the present-day universe is $\sim 1/3$ solar, hence, the metallicity of the $z = 3$ universe should be $\sim 1/10$ solar [94], because by that redshift the universe has experienced only $\sim 1/3$ of the cumulative star formation all the way to the present. This simple argument supports the notion of a *prompt initial enrichment* of the early universe. While the $\sim 10\%$ solar metallicity at $z = 3$ is a very straightforward estimate however, its direct observational test is not so easy.

Figure 18 [82] shows that at $z = 3$ the universe had already developed to become extremely inhomogeneous in chemical composition, with the metallicity ranging from supersolar in the central regions of young/forming spheroids and in QSOs likely hosted by them, down to $\sim 10^{-3}$ solar in the Lyα forest. Making the proper (mass-) average abundance of the heavy elements requires to know the fractional mass of each baryonic component at $z = 3$ – not an easy task. Sometimes the Lyα forest is considered as representative of the global metallicity of the high-z universe, as it may fill most of the volume and perhaps contain most of the baryons. But at best it may provide an estimate of the *volume-averaged* metallicity, which is irrelevant. What matters is in fact the *mass-averaged* metallicity, which I argue can be ~ 100 times higher at $z \sim 3$ than the volume-averaged one. At this early time most metals are likely to be locked into stars, in metal rich winds, and in shocked IGM which has already diluted wind materials, and none of these components qualify as

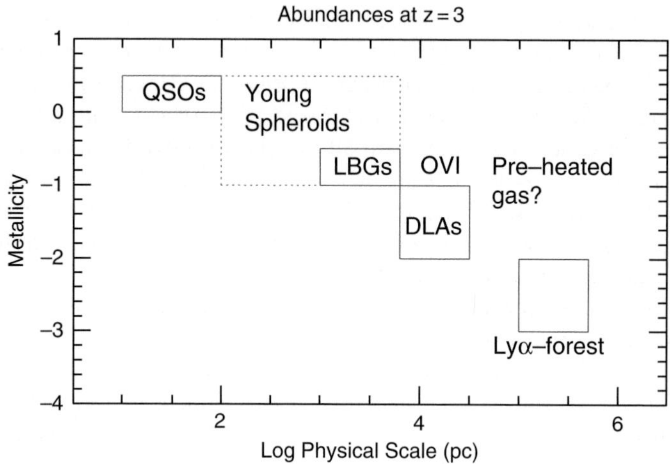

Fig. 18. Summary of current knowledge of metal abundances at $z \sim 3$. On the vertical axis the logarithmic abundance relative to solar is reported. The horizontal axis gives the typical linear dimensions of the structures for which direct abundance measurements are available. This figure has been adapted from [82] by the inclusion of the box for "young spheroidals" for which the estimate is indirect, as based on the present day observed metallicity range and on the estimated redshift of formation. The figure also includes the approximate location of the OVI absorbers [103], and the hypothetical location of the intergalactic medium enriched and pre-heated by early galactic winds

Lyα absorbers. Instead, a metal rich WHIM may have been detected thanks to its OVI absorption [103]. By and large, Lyα absorbers are very poor tracers of cosmic chemical evolution.

8 Clusters vs. the Chemical Evolution of the Milky Way

The Galactic bulge luminosity is $\sim 5.5 \times 10^{10}\ L_{\mathrm{K},\odot}$ [61] and $L_{\mathrm{B}}^{\mathrm{BULGE}} \simeq 6 \times 10^{9} L_{\mathrm{B},\odot}$ respectively in the K and in the B band. If we take the cluster empirical yield of metals ($\sim 0.2 \times L_{\mathrm{B}}\ M_{\odot}$) as universal, it follows that the Galactic bulge has produced $M_{\mathrm{Z}} \simeq 0.2 L_{\mathrm{B}}^{\mathrm{BULGE}} = 0.2 \times 6 \times 10^{9} \simeq 10^{9} M_{\odot}$ of metals. Where are all these metals? One billion solar masses of metals should not be easy to hide: part must be in the stars of the bulge itself, part must have been ejected by winds. The stellar mass of the bulge follows from its K-band mass to light ratio, $M_{*}^{\mathrm{BULGE}}/L_{\mathrm{K}} = 1$ [62, 127], and its luminosity, and hence $M_{*}^{\mathrm{BULGE}} \simeq 10^{10} M_{\odot}$. Its average metallicity is about solar or slightly lower [72, 127], i.e. $Z = 0.02$, and therefore the bulge stars all together contain $\sim 2 \times 10^{8} M_{\odot}$ of metals. Only $\sim 1/5$ of the metals produced when the bulge was actively star forming some 11–13 Gyr ago are still in the bulge! Hence, $\sim 80\%$, or $\sim 10^{9} M_{\odot}$ were ejected into the surrounding space by an early wind.

At the time of bulge formation, such $\sim 10^9 M_\odot$ of metals ran into largely pristine ($Z = 0$) material, experienced R–T instabilities leading to chaotic mixing, and establishing a distribution of metallicities in a largely inhomogeneous IGM surrounding the young Milky Way bulge. For example, this enormous amount of metals was able to bring to a metallicity 1/10 solar (i.e. $Z = 0.002$) about $5 \times 10^{11} M_\odot$ of pristine material, several times the mass of the yet to be formed Galactic disk. Therefore, it is likely that the Galactic disk formed and grew out of such pre-enriched material, which provides a quite natural solution to the classical *G-Dwarf problem* [98].

But there is another very intriguing aspect of chemical evolution that is revealed by the comparison of the Milky Way to clusters of galaxies [88]. Both for the Galactic disk and for clusters one can estimate the *empirical* metal yield, i.e., the ratio of the mass of metals to the mass of stars. Thus, in the MW disk we have:

$$y^{\rm disk} \simeq \frac{Z_\odot(M_* + 0.2M_*)}{M_*} = 1.2Z_\odot \tag{12}$$

assuming that both stars and the ISM are solar metallicity on average and the mass of the disk ISM is $\sim 20\%$ the mass of the stars in the disk. In clusters we have instead:

$$y^{\rm clusters} \simeq \frac{Z_\odot(M_* + 0.3 \times 5 \times M_*)}{M_*} = 2.5Z_\odot\ , \tag{13}$$

where stars in clusters are again assumed to be solar metallicity, the ICM is assumed 0.3 solar, and 5 times more massive than stars in galaxies. Thus, the apparent yield of clusters is about twice that of the Galactic disk. Actually, the difference could be even larger, if one adopts a $M_{\rm ICM}/M_*$ ratio as high as in [88].

Two opposite solutions of the discrepant yields are discussed in [88]:

⋆ Option A: The IMF in galaxy clusters is flatter than in the Galactic disk, hence with a *top-heavy IMF* more massive stars are produced, hence more metals.

⋆ Option B: The IMF is the same in the disk as in clusters, but the discrepancy arises from not having counted metals produced by the MW disk stars which have been ejected by disk winds, i.e., the disk has lost metals (just like the bulge).

Option A is favored by [88] based on two arguments: (1) if B were true then most chemical evolution models of the MW galaxy would be wrong, and (2) there is no evidence for star formation in the disk causing mass loss, but material ejected in galactic *fountains* sooner or later would fall back to the disk. I think that this choice is premature, but cannot be ruled out either. It is certainly true that Option B would cause some problems to chemical evolution models, as they usually rest on three assumptions that may not be valid. Namely, (1) that disks started forming out of pristine ($Z = 0$) material,

(2) that they grow by accumulating pristine ($Z = 0$) material, and (3) that disks don't lose any mass.

In favor of Option A, the argument can be put forward that, even if individual star-formation events follow a universal IMF, the resulting global IMF depends on the distribution function of the mass of individual (star cluster size) formation events [63, 120]. Hence galaxies, where most of star formation originated in powerful starbursts, will closely follow the universal IMF, while quiescent star formation resulting from many small individual events may have a somewhat steeper IMF. Therefore, it is not so inconceivable that elliptical galaxies (which formed in powerful starbursts and which dominate clusters) may have a flatter IMF than the MW disk.

On the other hand, while the present SFR in the MW disk is very low ($\sim 1 - 2M_{\odot}\mathrm{yr}^{-1}$), it may have been much higher in the past. Actually, it has been argued that most of the factor of $\sim$ 10 increase in the global SFR between $z = 0$ and $z \sim 1$ is due to an increase within disks [50, 65]. If so, just a few Gyr ago the MW disk was forming stars much more violently than is currently observed, and substantial ejection of metals from the disk is therefore not at all inconceivable. By the same token, if most of the disk build-up was through stronger bursts than observed today, then the global disk IMF may also be close to that in ellipticals. Thus the choice between Option A and B is still an open debate.

9 Summary

With these lectures I hope to have conveyed the feeling that the chemistry of galaxy clusters is at the crossroads of many interesting astrophysical and cosmological issues, and that we can learn a lot from their study. A number of unexpected inferences are derived, starting from a few empirical facts, namely, the iron and metal content of the ICM and cluster galaxies, the fraction of the baryons locked into stars in clusters and in the field, and the age and baryon-fraction of stellar populations of galactic spheroids. Such inferences include:

- In clusters and in the general field alike there are more metals in the gas that has diffused out of galaxies (ICM and IGM) than there are locked into stars inside galaxies (only $\sim$ 10%). The loss of metals to the surrounding media is therefore a major factor in the chemical evolution of galaxies. Furthermore, at this global level, the outcome of star formation through cosmic time is largely independent of environment, most likely just because a major fraction of all stars formed before cluster formation.

- Various arguments support the notion that the metals now in the ICM/IGM were *ejected* by galactic winds, rather then being *extracted* from galaxies by ram pressure.

• Having processed the same fraction of baryons into stars, the global metallicity of the local universe has to be nearly the same that one can measure in clusters, i.e., $\sim 1/3$ solar.

• For the same reason, one expects the IGM to have experienced nearly the same amount of *pre-heating* as the ICM, and therefore to be at a temperature of $\sim$ 0.1–1 keV, regardless of the amount of pre-heating that is required for clusters.

• Given the predominance and formation redshift of galactic spheroids, both in clusters as well as globally in the universe, it is likely that the universe experienced a prompt metal enrichment, with the global metallicity possibly reaching $\sim 1/10$ solar already by $z \sim 3$. Most metals remain unaccounted for at low-redshift as well as at high-redshift, however, and are likely to reside in a warm/hot IGM (WHIM) whose existence may have been revealed by observational data.

• This same scenario may be valid down to the scale of our own Milky Way galaxy, with early winds from the forming Galactic bulge having pre-enriched to $\sim 1/10$ solar a much greater mass of gas, out of which the Galactic disk started to form and evolve.

• The empirical metal yield of clusters is at least twice that of the MW disk. This signals that either the stellar IMF of the disk is a little steeper than that of ellipticals, or that the disk has lost at least as many metals than it has produced.

Acknowledgements

I would like to thank Manolis Plionis and Omar López-Cruz for having invited me to Tonantzintla to give these lectures, and for their friendly hospitality. I would also like to thank Laura Greggio for having modified her original figures, specifically for these proceedings.

References

1. Aguirre, A., Hernquist, L., Schaye, J., Katz, N., Weinberg, D.H., Gardner, J.: ApJ **561**, 521 (2001)
2. Anders, E., Grevesse, N.: Geochimica et Cosmochimica Acta **53**, 197 (1989)
3. Arimoto, N., Matsushita, K., Ishimaru, Y., Ohashi, T., Renzini, A.: ApJ **477**, 128 (1997)
4. Arnaboldi, M., et al.: AJ **125**, 514 (2003)
5. Arnaud, M., Evrard, A.E.: MNRAS **305**, 631 (1999)
6. Arnaud, M., Rothenflug, R., Boulade,O., Vigroux, R., Vangioni-Flam, E.: A&A **254**, 49 (1992)
7. Baumgartner, W.H., Loewenstein M., Horner D.J., Mushotzsky, R.F.: ApJ **620**, 680 (2005)

8. Bell, E.F., Wolf, C., Meisenheimer, K., Rix, H.-W., Borch, A., Dye, S., Kleinheinrich, M., McIntosh, D.H.: ApJ **608**, 752 (2004)
9. Bennett, C.L., et al.: ApJS **148**, 1 (2003)
10. Benson, A.J., Frenk, C.S., Sharples, R.M.: ApJ **574**, 104 (2002)
11. Bernardi, M., Renzini, A., da Costa, L.N., Wegener, G., Alonso, M.V., Pellegrini, P.S., Rit'e, C., Willmer, C.N.A.: ApJ **508**, L43 (1998)
12. Bernardi, M., et al.: AJ **125**, 1882 (2003)
13. Borgani, S., Governato, F., Wadsley, J., Menci, N., Tozzi, P., Lake, G., Quinn, T., Stadel, J.: ApJ **559**, L71 (2001)
14. Böhringer, H., Matsushita, K., Churazov, E., Finoguenov, A., Ikebe, Y.: A&A **416**, L21 (2004)
15. Böhringer, H., Matsushita, K., Finoguenov, A., Xue, Y., Churazov, E.: AdSpR **36**, 677 (2005)
16. Bower, R.G., Lucey, J.R., Ellis, R.S.: MNRAS **254**, 613 (1992)
17. Buote, D.A.: MNRAS **311**, 176 (2000)
18. Cappellari, M., Bacon R., Bureau, M., Damen, .C., et al.: MNRAS **366**, 1126 (2006)
19. Cappellaro, E., Evans, R., Turatto, M.: A&A **351**, 459 (1999)
20. Cavaliere, A., Colafrancesco, S., Menci, N. ApJ **415**, 50 (1993)
21. Chiosi, C.: A&A **364**, 423 (2000)
22. Cimatti, A., et al.: A&A **381**, L68 (2002)
23. Ciotti, L., D'Ercole, A., Pellegrini, S., Renzini, A.: ApJ **376**, 380 (1991)
24. Ciotti, L., Ostriker, J.P.: ApJ **551**, 131 (2001)
25. Cole, S., et al.: MNRAS **326**, 255 (2001)
26. Danforth, C.W., Shull, J.M.: ApJ **62**, 555 (2005)
27. David, L.P., Jones, C., Forman, W, Daines S.: ApJ **428**, 544 (1994)
28. Davis, D.S., Mulchaey, J.S., Mushotzky, R.F.: ApJ **511**, 34 (1999)
29. De Grandi, S., Ettori, S., Longhetti, M., Molendi, S.: A&A **419**, 7 (2004)
30. De Grandi, S., Molendi, S.: ApJ **551**, 153 (2001)
31. De Grandi, S., Molendi, S.: ASP Conf. Ser. **253**, 3 (2002)
32. Dickinson, M., Papovich, C., Ferguson, H., Budavari, T.: ApJ **587**, 25 (2003)
33. Dressler, A., Gunn, J. E.: ASP. Conf. Ser. **10**, 204 (1990)
34. Dupke, R.A., White, R.E. III: ApJ **537**, 123 (2000)
35. Eisenstein, D.J., Hogg, D.W., Fukugita, M., Nakamura, O., Bernardi, M., et al.: ApJ **585**, 694 (2003)
36. Falc'on-Barroso, J., Peletier, R.F., Balcells, M.: MNRAS **335**, 741 (2002)
37. Ferguson, H.C., Tanvir, N.R., von Hippel, T.: Nature **391**, 461 (1998)
38. Finoguenov, A., Arnaud, M., David, L.P.: ApJ **555**, 191 (2001)
39. Finoguenov, A., Borgani, S., Tornatore, L., B¨ohringer, H.: A&A **398**, L35 (2003)
40. Finoguenov, A., David, L.P., Ponman, T.J.: ApJ **544**, 188 (2000)
41. Finoguenov, A., Jones, C., B¨ohringer, H., Ponman, T.J.: SpJ **578**, 74 (2002)
42. Fontana, A., Pozzetti, L., Donnarumma, I., Renzini, A., Cimatti, A., et al.: A&A **424**, 23 (2004)
43. Fukazawa, Y, Ohashi, T., Fabian, A.C., Canizares, C.R., Ikebe, Y., Makishima, K., Mushotzky, R.F., Yamashita, K.: PASJ **46**, L55 (1994)
44. Fukugita, M., Hogan, C.J., Peebles, P.J.E.: ApJ **503**, 518 (1998)
45. Gal-Yam, A., et al.: AJ **125**, 1087 (2003)
46. Gastaldello, F., Molendi, F.: ApJ **572**, 160 (2002)

47. Gibson, B., Matteucci, F.: MNRAS **291**, L8 (1997)
48. Greggio. L.: ASP Conf. Ser. **342**, 459 (astro-ph/0410187) (2005a)
49. Greggio. L.: A&A **441**, 1055 (2005b)
50. Hammer, F., Flores, H., Elbaz, D., Zheng, X.Z., Liang, Y.C., et al.: A&A **430**, 115 (2005)
51. Hamuy, M.: ApJ **582**, 905 (2003)
52. Heckman, T.M.: Asp. Conf. Ser. **254**, 292 (2002)
53. Heckman, T.M., Lehnert, M.D., Strickland, D.K., Armus, L.: ApJS **129**, 493 (2000)
54. Hogg, D.W., et al.: AJ **124**, 646 (2002)
55. Iben, I. Jr., Renzini, A.: ARA&A **21**, 271 (1983)
56. Iben, I. Jr., Tutukov A.V.: ApJS **54**, 335 (1984)
57. Ishimaru, Y., Arimoto, N.: PASJ **49**, 1 (1997)
58. Iwamoto, K., Brachwitz, F., Nomoto, K., Kishimoto, N. et al.: ApJS **125**, 439 (1999)
59. Jablonka, P., Martin, P., Arimoto, N.: AJ **112**, 1415 (1996)
60. Kaiser, N.: ApJ **383**, 104 (1991)
61. Kent, S.M.: ApJ **387**, 181 (1994)
62. Kent, S.M., Dame, T.M., Fazio, G.: ApJ **378**, 131 (1991)
63. Kroupa, P., Weidner, C.: ApJ **598**, 1076 (2003)
64. Larson, R.B., Dinerstein, H.L.: PASP **87**, 511 (1975)
65. Lilly, S.J., Tresse, L., Hammer, F., Crampton, D., Le F‘evre, O.: ApJ **455**, 108 (1995)
66. Livio, M.: In: Niemeyer, J.C., Truran, J.W. (eds.) The Progenitors of Type Ia Supernovae, p. 33. CUP, Cambridge (2000)
67. Loewenstein, M.: ApJ **557**, 573 (2001)
68. Loewenstein, M., Mushotzky, R.F.: ApJ **466**, 695 (1996)
69. Maraston, C., Greggio, L., Renzini, A., Ortolani, S., Saglia, R. P., Puzia, T.H., Kissler-Patig, M.: A&A **400**, 823 (2003)
70. Matsushita, K., Finoguenov, A., B¨ohringer, H.: A&A **401**, 443 (2003)
71. Matteucci, F., Vettolani, G.: A&A **202**, 21 (1988)
72. McWilliam, A., Rich, R.M.: ApJS **91**, 794 (1994)
73. Mitchell, R.J., Culhane, J.L., Davison, P.J., Ives, J.C.: MNRAS **175**, 29 (1976)
74. Mohr, J.J., Mathiesen, B., Evrard, A.E.: ApJ **627**, 649 (1999)
75. Mulchaey, J.S., Davis, D.S., Mushotsky, R.F., Burnstein, D.: ApJ **404**, L9 (1993)
76. Mulchaey, J.S., Davis, D.S., Mushotsky, R.F., Burnstein, D.: ApJ **456**, 80 (1996)
77. Mushotzky, R.F.: In: Durret, F., Mazure, A., Tran Thanh Van, J. (eds.) Clusters of Galaxies. p. 167. Editions Fronti‘eres, Gyf-sur-Yvette (1994)
78. Mushotsky, R.: Phil. Trans. R. Soc. Lond. A **360**, 2019 (2002)
79. Mushotzky, R.F., et al.: ApJ **466**, 686 (1996)
80. Nicastro, F., Zezas, A., Elvis, M., Mathur, S., Fiore, F., Cecchi-Pestellini, C., Burke, D., Drake, J., Casella, P.: Nature **421**, 719 (2003)
81. Patat, F., Barbon, R., Cappellaro, E., Turatto, M.: A&A **282**, 731 (1994)
82. Pettini, M., In “Cosmochemistry: The melting pot of the elements”, eds. Esteban, C. et al., Cambridge Contemporary Astrophysics, Cambridge. Univ. Press (2004)
83. Pettini, M., Shapley, A.E. Steidel, C.C., Cuby, J.-G., Dickinson, M., Moorwood, A.F.M., Adelberger, K.L., Giavalisco, M.: ApJ **554**, 981 (2001)

84. Pipino, A., Matteucci, F., Borgani, S., Biviano, A.: New Astr. **7**, 227 (2002)
85. Ponman, T.J., et al.: Nature **369**, 462 (1994)
86. Ponman, T.J., Cannon, D.G., Navarro, J.F.: Nature **397**, 135 (1999)
87. Ponman, T.J., Sanderson, A.J.R., Finoguenov, A.: MNRAS **343**, 331 (2003)
88. Portinari, L., et al.: ApJ **604**, 579 (2004)
89. Pratt, G.W., Arnaud, M.: A&A **408**, 1 (2003)
90. Raymond, J.C., Smith, B.W.: ApJS **35**, 419 (1977)
91. Renzini, A., Ciotti, L., D'Ercole, A., Pellegrini, S.: ApJ **419**, 52 (1993)
92. Renzini, A.: In: Durret, F. et al. (eds.) Clusters of Galaxies, p. 221. Edition Fronti'eres, Gyf-sur-Yvette (1994)
93. Renzini, A.: ApJ **488**, 35 (1997)
94. Renzini, A.: ASP Conf. Ser. **146**, 298 (1998a)
95. Renzini, A.: AJ **115**, 2459 (1998b)
96. Renzini, A.: In: Carollo, C.M., et al. (eds.) The Formation of Galactic Bulges, p. 9. CUP, Cambridge (1999)
97. Renzini, A.: In: Plionis, M., Georgantopoulos, I. (eds.) Large Scale Structure in the X-ray Universe, p. 103. Atlantisciences, Paris (2000)
98. Renzini, A.: ASP Conf. Ser. **253**, 331 (2002)
99. Renzini, A.: In: Mulchaey, J.S., Dressler, A., Oemler, A. (eds.) Clusters of Galaxies: Probes of Cosmological Structures and Galaxy Evolution, p. 261. CUP, Cambridge (2004)
100. Renzini, A.: In: Corbelli, E., Palla, F., Zinnecker, H. (eds.) The Initial Mass Function 50 Years Later, p. 221. Springer, Berlin (2005)
101. Richter, P., Savage, B.D., Tripp, T.M., Sembach, K.R.: ApJS **153**, 165 (2004)
102. Schindler, S., et al.: A&A **435**, L25 (2005)
103. Simcoe, R.A., Sargent, W.L.W., Rauch M.: ApJ **578**, 737 (2002)
104. Songaila, A., Cowe, L.L., Lilly, S.J.: ApJ **348**, 371 (1990)
105. Stanford, S.A., Eisenhardt, P.R., Dickinson, M.: ApJ **492**, 461 (1998)
106. Steidel, C.C., et al.: ApJ **604**, 534 (2004)
107. Strolger, L.-G., et al.: ApJ **613**, 200 (2004)
108. Tamura, T, Kaastra, J.S., Makishima, K, Takahashi, I.: A&A **399**, 497 (2004)
109. Thomas, D.: In: Redshift, ed. Walsh, J.R., Rosa, M.R. (eds.) Chemical Evolution from Zero to High, p.197. Springer, Berlin (1999)
110. Tozzi, P., Norman, C.: ApJ **546**, 63 (2001)
111. Tozzi, P., Rosati, P., Ettori, S., Borgani, S., Mainieri, V., Norman, C.: ApJ **503**, 705 (2003)
112. Treu, T., Ellis, R.S., Liao, T.X., van Dokkum, P.G., Tozzi, P., et al.: ApJ **633**, 174 (2005)
113. Tsuru, T.: PhD Thesis, University of Tokyo (1992)
114. Turatto, M., LNP, **598**, p. 21 (Springer)
115. Valageas, P., Silk, J.: A&A **350**, 725 (1999)
116. van der Marel, R.: MNRAS **253**, 710 (1991)
117. van Dokkum, P.G., Stanford, S.S.: ApJ **585**, 78 (2003)
118. Veilleux, S., Cecil, G., Bland-Hawthorn, J.: ARA&A **43**, 769 (2005)
119. Vigroux, L.: A&A **56**, 473 (1977)
120. Weidner, C., Kroupa, P.: MNRAS **365**, 1333 (2006)
121. Whelan, J., Iben, I. Jr.: ApJ **186**, 1007 (1976)
122. White, D.A.: MNRAS **312**, 663 (2000)
123. White, S.D.M., Navarro, J.F., Evrard, A.E., Frenk, C.S.: Nature **366**, 429 (1993)

124. Woosley, S.E., Weaver, T.A.: ApJS **101**, 181 (1995)
125. Wu, K.K.S., Fabian, A.C., Nulsen, P.E.J.: MNRAS **318**, 889 (2000)
126. Yungelson, L.R., Livio, M.: ApJ **528**, 108 (2000)
127. Zoccali, M., Renzini, A., Ortolani, S., Greggio, L., Saviane, I., Cassisi, S., Rejkuba, M., Barbuy, B., Rich, R. M., Bica, E.: A&A **399**, 931 (2003)

Gravitational Lensing by Clusters of Galaxies

J.-P. Kneib

Laboratoire d'Astrophysique de Marseille, Marseille, France
jean-paul.kneib@oamp.fr

1 Introduction

Clusters of galaxies are massive structures found at the intersections of the filaments of the cosmic web, and following General Relativity, they significantly deform locally the Space-Time. Thus light rays of distant objects passing through a cluster are deflected, and the resulting images appear distorted and amplified: hence cluster of galaxies act as powerful gravitational lenses. Cluster lensing comes into two flavors: (1) strong lensing, characterized by effects that can readily be seen by eye: giant arcs, multiple images, and arclets; and (2) weak lensing, which can only be characterized in a statistical way. Science topics using cluster lenses can be divided into three broad categories: (i) study of the lens(es): the understanding of the cluster mass distribution and issues related to cluster formation and evolution, as well as constraining the nature of the (Dark) Matter particles; (ii) study of the lensed objects: the understanding of the lensed galaxy population and issues related to galaxy formation and evolution; and (iii) study of the geometry of the Universe: lens equations depend on angular diameter distances, and thus on the cosmological parameters, offering a possible test of cosmological models.

In these lecture notes, following a historical perspective on lensing, I will introduce the basics of gravitational lensing theory. Then I will discuss measurements of cluster masses using strong and weak lensing and finally, I will discuss the use of cluster lensing as a tool to probe the distant Universe and to constrain cosmology. I will conclude these lectures with a discussion on future prospects.

2 Historical Perspective

When Albert Einstein presented the General theory of Relativity, one of the proposed observational tests was the deflection of light by massive objects. Indeed, the light deflection in General Relativity is twice the value derived

J.-P. Kneib: *Gravitational Lensing by Clusters of Galaxies*, Lect. Notes Phys. **740**, 213–253 (2008)
DOI 10.1007/978-1-4020-6941-3_7

in the Newtonian approximation. In the case of the Sun deflecting light from distant stars, one expects a deflection angle of 1.75 arcsec at the limb of our Sun. The adventurous, Sir Arthur Eddington, led in 1919 an eclipse expedition to Principe Island in West Africa with the aim to verify Einstein's new theory. Eddington's successful experiment, confirming Einstein's theory, was the first astronomical observation of the gravitational lensing phenomenon.

Later on, in the early days of modern cosmology, soon after one realized that the Universe was expanding and that Dark Matter was likely the dominant component in clusters of galaxies [147], Fritz Zwicky [148] suggested that gravitational lensing will be an invaluable tool to: *(i)* trace and measure the amount of Dark Matter (DM), thought to pervade the cosmos; and *(ii)* study magnified distant objects.

Zwicky's courageous predictions were based on a good understanding of the properties of gravitational lensing, but at that time, technology and the lack of understanding of cluster of galaxies and gravitational lensing hampered much progress and discovery.

Although clusters of galaxies are known for about two centuries, first recognized by Messier and Herschel as "remarkable concentrations of nebulae on the sky" (see the review of Biviano [12] and reference therein), their study only really matured in the 1950's. In particular, the publication of the first comprehensive cluster catalogue in the nearby Universe by Abell [1] can be considered as a milestone in the history of cluster of galaxies.

In comparison, gravitational lensing theory only developed latter in the 1960's with a few theoretical studies showing the usefulness of lensing for astronomy. In particular, Sjur Refsdal derived the basic equations of gravitational lens theory [111] and subsequently showed how the gravitational lens effect can be used to determine Hubble's constant by measuring the time delay between two lensed images [112]. Following the discoveries of quasars, Barnothy [7] linked gravitational lensing to the study of quasars. And with the discovery of the first double quasar Q0957+561 by Walsh, Carswell & Weymann [143] gravitational lensing really emerged in astronomy. Interestingly, the large separation (6.1 arcsec) of Q0957+561 ($z = 1.41$) can only be explained with the magnification boost of the cluster in which resides the lensing galaxy.

Despite the fact that clusters of galaxies were starting to be well studied astronomical objects in the late 1970's and early 1980's, in particular thanks to the study of the X-ray emitting intra-cluster with the *Einstein* X-ray telescope and the numerous optical studies of galaxies in clusters, almost nothing was discussed in theoretical papers regarding their lensing effect. The work by Narayan, Blandford & Nityananda [95] is probably one of the earliest account of the possibility that clusters can act as a powerful lens. In particular they proposed that the large separation of the first double quasar Q0957+561 can only be explained if the lensing was "cluster-assisted". The likely explanation of the lack of interest of clusters in lensing research was probably the belief that clusters were sufficiently diffuse and extended systems that they could not act as powerful lenses.

It, thus, came as quite a surprise when, in the mid 80's "giant arcs", strongly elongated images of galaxies in the core of massive clusters were discovered [81, 134]. This new phenomenon was then recognized by Paczynski [103] as the results of gravitational lensing, and soon after confirmed by the redshift measurement of the arc in Abell 370 ([135], Fig. 1). The giant arc discovery was revealing the strong lensing regime, however it only represents the tip of the iceberg! In 1990, Antony Tyson, who was conducting deep CCD imaging of clusters, identified a *systematic alignment* of faint galaxies around cluster cores. He then suggested that this alignment, produced by the cluster lensing distortion, could be used to map dark matter in clusters. These two discoveries opened up a new field in astronomy, the study of "cluster lenses", and stroke the theoretical community who produced in the first half of the 1990's a large number of related papers related.

It is important to underline that these discoveries were made possible by the successful development of CCD imaging that produces deeper and sharper optical images of the sky, as well as by deep spectroscopy – essential to measure the spectrum of faint low-surface brightness galaxies. Another technological revolution was in preparation at that time – the launch of the *Hubble Space Telescope* (*HST*). *HST* impacted dramatically the cluster lensing research (and particularly that related to strong lensing). Although launched in 1991, *HST*

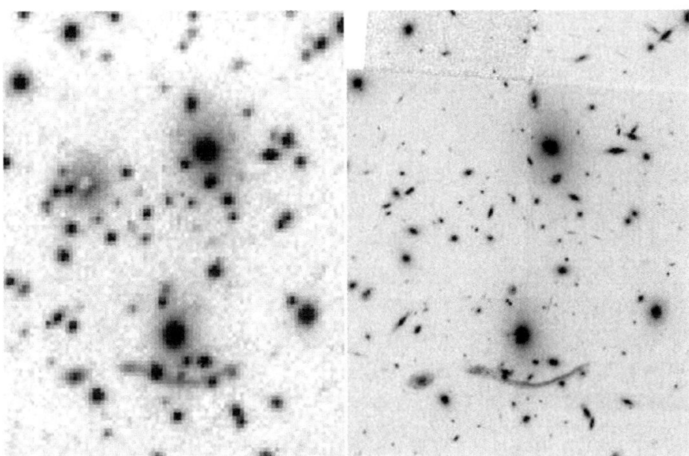

Fig. 1. The galaxy cluster Abell 370 as observed in 1985 (**left**) with one of the first CCD cameras, in which the first gravitationally lensed arc was later identified [81, 134]. The **right** image show the Hubble image of Abell 370. Most of the bright galaxies seen are cluster members at $z = 0.375$, whereas the arc, i.e. the highly elongated feature, is the image of a galaxy at redshift $z = 0.724$ [135]. North is on top, East to the left, field of view is roughly 40×60 arcsec^2

could not immediately provide high-quality images, as its unforeseen "myopia" was blurring the faint images of distant galaxies, making them useless for lensing purposes. With the implementation of COSTAR and the odd-shaped WFPC2 camera, HST recovered all its image sharpness. It is not a surprise that one of the first images to be released was the astonishing view of the Cluster Lens Abell 2218.

During, the second part of the 1990's the wide field imaging camera had been developed (such as: UH8k followed by CFHT12k at CFHT, Suprime at Subaru, and more recently the Megacam camera at CFHT) on large ground based telescopes. These cameras are made from a mosaic of large format CCDs (4k×2k) allowing one to cover large areas on the sky (from a quarter of a square degree up to one square degree). The construction of these instruments was strongly motivated by the detection of the weak lensing distortion of faint galaxies produced by clusters and large scale structures. A number of results on "cosmic" shear measurements have been obtained (e.g., [142]) as well as the direct detection of galaxy clusters via weak lensing [144].

At the turn of the second millennium the new impact of cluster lensing was probably the growing use of clusters, as a natural telescopes, to study the infancy of the Universe, e.g., [36, 38, 104]. This became possible by using deep spectroscopy on 4 m and then 8–10 m class telescopes to probe the high redshift Universe, but also by making observations through these natural telescopes at different wavelengths of the electromagnetic spectrum. In particular, the discoveries and study of the sub-millimeter galaxies using SCUBA at JCMT (see reviews: [11, 139] and references therein, see also [13, 69, 71], the Caltech interferometer at Owens Valley [39, 40], the IRAM interferometer [70, 102], and the Very Large Array (VLA) [138] greatly benefited from the boost of gravitational lensing in cluster fields. Similarly, observations of lensed galaxies in the mid-infrared with the ISO-CAM mid-infrared camera on the ISO satellite, e.g., [4, 89], and now with the *Spitzer* Observatory [34], have allowed us to push further the limits of our knowledge on distant galaxies. Gravitational lensing has now been recognized as a powerful technique to count the faintest galaxies in their different classes (EROs: [130]; Lyman-α emitter at $z \sim 4$–6: [53, 122]; Lyman-break galaxies at $z \sim 6$–10: [115]) as well as to study in details the most magnified sources [34, 68, 105–107].

Recently, the implementation of the new Advanced Camera for Surveys (ACS) on *HST* has provided some new observational advances in the study and use of cluster lenses. These can be envisioned with the very deep ACS images of Abell 1689 [17]. This color image revealed more than 30 faint multiple image systems in the core of the cluster, leading to more than one hundred lensed images. This increase in the number of multiple images and thus of strong lensing constraints in the cluster core, allows us to in principle achieve a better mass modeling and to effectively use strong lensing in clusters to constrain the cosmology [49, 78, 136].

This brief historical account of "cluster lens" research summarizes some of the important scientific results gathered up to now, and demonstrates the importance of cluster lensing in modern cosmology.

When necessary, I will adopt a flat world-model with a Hubble constant $H_0 = 70\,\mathrm{km\,s^{-1}\,Mpc^{-1}}$, a matter density parameter $\Omega_\mathrm{m} = 0.3$ and a cosmological constant $\Omega_\Lambda = 0.7$.

3 Lensing Theory Useful in Cluster Lensing

3.1 General Description

Due to their large mass density, galaxy clusters (as well as galaxies) locally deform the Space-Time (see Fig. 2). Therefore, the wave front of any light coming from a distant source, passing through a galaxy cluster, will be distorted and this happens regardless of its wavelength as the effect is purely geometric. Moreover, for the most massive clusters the mass density in the core is large enough to break the wave front coming from a distant source into pieces, hence producing multiple images, which then usually form these extraordinary gravitational giant arcs (the *strong lensing* domain). Distant galaxies will thus appear distorted, magnified and tangentially aligned toward the cluster center, and we usually call them arclets because of their noticeable elongated shape. Note however that their shape is a combination of the intrinsic shape and the distortion induced by the cluster. If the alignment between the observer, the cluster and the distant galaxies is less perfect then the distortion induced by the cluster will be less important and will not

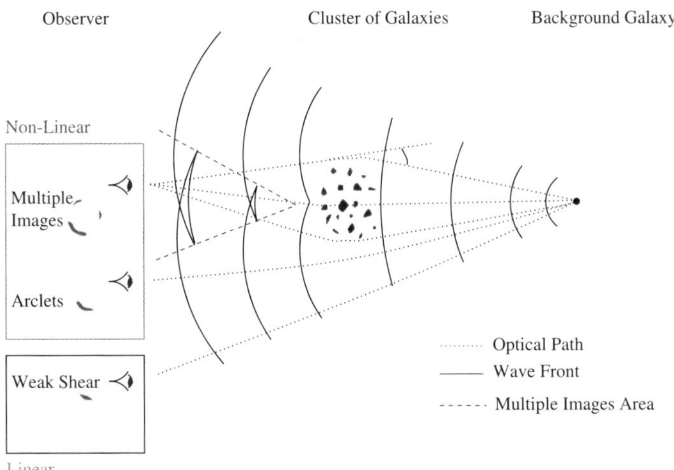

Fig. 2. Gravitational lensing in clusters: A simple representation of how gravitational images are formed (see text for a complete description)

be immediately recognized – statistical methods are required – corresponding to the *weak regime* domain. In this region, the observed shapes of galaxies are dominated by their intrinsic ellipticity and they are also affected by the geometrical distortion and the point spread function (PSF) of the camera and telescope. Thus, only a careful analysis (correcting the observed images for the non-lensing distortion) can reveal the weak lensing signal.

3.2 Gravitational Lens Equation

Before going to the mathematics of lensing, I will first recap the assumptions needed to derive the lens equations. First, it is assumed that the "Cosmological Principle" (the Universe is homogeneous and isotropic) is correct on large scales. The scale to be considered is the one corresponding to the Gravitational force: $L \sim c/\sqrt{G\bar{\rho}} \sim 2\,\mathrm{Gpc}$ where c is the speed of light, G is the gravitational constant and $\bar{\rho}$ is the mean density of the Universe. The large scale distribution of galaxies (as determined by surveys like 2dF and SDSS) and the Cosmic Microwave Background (CMB) (as revealed by COBE and more recently by WMAP) are in good agreement with the "Cosmological Principle".

The metric of the homogeneous Universe is locally perturbed by dense concentrations of mass such as stars, galaxies or clusters of galaxies. The Schwarzschild solution gives the metric near a point mass, and it is easy to generalize it for a stationary weak mass field ($\Phi \ll c^2$) to a continuous mass distribution:

$$\mathrm{d}s^2 = \left(1 + \frac{2\Phi}{c^2}\right) c^2 \mathrm{d}t^2 - \left(1 - \frac{2\Phi}{c^2}\right) \mathrm{d}r^2 \,, \tag{1}$$

where Φ is the 3D gravitational potential of the mass distribution considered.

If we consider a simple configuration of a single thin deflector (Fig. 3), the observer (O) will see the image (I) of the source (S) deflected by the lens (L). The geometrical equation relating the position of the source, $\boldsymbol{\theta}_{\mathrm{S}}$, to the

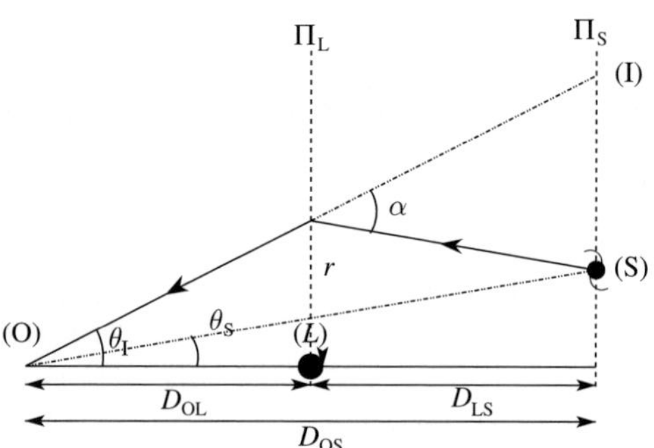

Fig. 3. A single deflector configuration, showing the different angles and distances needed to express the lens equation

position of the image, $\boldsymbol{\theta}_{\rm I}$, depends on the deflection angle, $\boldsymbol{\alpha}$, and the angular diameter distance, D_{ij}, and reads:

$$\boldsymbol{\theta}_{\rm I} = \boldsymbol{\theta}_{\rm S} + \frac{D_{\rm LS}}{D_{\rm OS}}\boldsymbol{\alpha} \ . \tag{2}$$

The value of $\boldsymbol{\alpha}$ depends on the local perturbation of the mass on the space-time. The path of a photon will follow a null geodesic, that is $ds = 0$. Hence from (1), one can determine the travel time t_T for a given path which is a function of the angle $\boldsymbol{\alpha}$. By applying Fermat's principle, which states that the light path is the one with a stationary travel time: $\mathrm{d}t_T/\mathrm{d}\boldsymbol{\theta}_{\rm I} = \mathbf{0}$, we can derive the value of $\boldsymbol{\alpha}$ as a function of the local Newtonian potential:

$$\boldsymbol{\alpha}(\boldsymbol{\theta}_{\rm I}) = \frac{2}{c^2}\frac{D_{\rm LS}}{D_{\rm OS}} \ \boldsymbol{\nabla}_{\boldsymbol{\theta}_{\rm I}}\phi_N^{2D}(\boldsymbol{\theta}_{\rm I}) \tag{3}$$

where ϕ_N^{2D} is the Newtonian projected gravitational potential.

Combining (2) and (3), we thus derive the *thin lens equation approximation* (which holds for stars, galaxies and cluster of galaxies, e.g., Schneider, Ehlers & Falco [118]):

$$\boldsymbol{\theta}_{\rm S} = \boldsymbol{\theta}_{\rm I} - \frac{2\mathcal{E}}{c^2}\boldsymbol{\nabla}\phi_N^{2D}(\boldsymbol{\theta}_{\rm I}) = \boldsymbol{\theta}_{\rm I} - \boldsymbol{\nabla}\varphi(\boldsymbol{\theta}_{\rm I}) \ , \tag{4}$$

where we defined φ as the lensing potential – a lensing normalized version of the Newtonian projected potential, and the distance ratio, $\mathcal{E} = D_{\rm LS}/D_{\rm OS}$, which depends on the redshift of the cluster $z_{\rm L}$ and the background source $z_{\rm S}$, as well as, but only weakly, on the cosmological parameter Ω_m and Ω_Λ. The distance ratio $\mathcal{E}$ is also measuring the efficiency of a given lens at redshift $z_{\rm L}$. Indeed $\mathcal{E}$ is an increasing function of the source redshift $z_{\rm S}$, meaning that the larger the redshift the stronger the deflection and distortion. This can be however slightly more complex in strong lensing regions. Note also that $\mathcal{E}$ is independent of the Hubble constant, *which means that lensing deflection angles are independent of the value of* H_0.

In the real Universe, the mass is not distributed in planes, and we can sometime have multiple deflectors situated at different redshifts. In these cases, a more appropriate formalism should be used, where one puts the mass in different lens planes, and the lensing is then calculated by adding up the different deflections plane by plane. Note that this is generally a non-linear combination, although it can be linearized in the weak lensing regime. In the case of a massive cluster of galaxies, the likelihood of having an important contribution from a second lens plane is very small. So for this reason, I refer the reader to the work of Kochanek & Apostolakis [72] and Moller & Blain [93].

3.3 Gravitational Lens Mapping

The lensing transformation can be seen as a mapping from Source plane to Image plane, and the Hessian of this transformation – also called the magnification matrix – relates (to the first order) a source element of the Image

($\mathrm{d}\boldsymbol{\theta}_\mathrm{I}$) to the Source plane ($\mathrm{d}\boldsymbol{\theta}_\mathrm{S}$) in the following way (written in Cartesian and polar coordinates):

$$\frac{\mathrm{d}\boldsymbol{\theta}_\mathrm{S}}{\mathrm{d}\boldsymbol{\theta}_\mathrm{I}} = \mathcal{A}^{-1} = \begin{pmatrix} 1-\partial_{xx}\varphi & -\partial_{xy}\varphi \\ -\partial_{xy}\varphi & 1-\partial_{yy}\varphi \end{pmatrix} = \begin{pmatrix} 1-\partial_{rr}\varphi & -\partial_r\left(\frac{1}{r}\partial_\theta\varphi\right) \\ -\partial_r\left(\frac{1}{r}\partial_\theta\varphi\right) & 1-\frac{1}{r}\partial_r\varphi-\frac{1}{r^2}\partial_{\theta\theta}\varphi \end{pmatrix} \tag{5}$$

One generally calls this matrix the magnification matrix:

$$\mathcal{A}^{-1} = \begin{pmatrix} 1-\kappa-\gamma_1 & -\gamma_2 \\ -\gamma_2 & 1-\kappa+\gamma_1 \end{pmatrix} , \tag{6}$$

where we have defined the *convergence*, $\kappa = \Delta\varphi/2 = \Sigma/2\Sigma_\mathrm{crit}$, the *shear* vector (also often noted as a complex), $\boldsymbol{\gamma} = (\gamma_1, \gamma_2)$, where:

$$\gamma_1 = (\partial_{yy}\varphi - \partial_{xx}\varphi)/2 \qquad , \gamma_2 = \partial_{xy}\varphi \,. \tag{7}$$

The term Σ_crit is the lensing critical surface density and it is defined by:

$$\Sigma_\mathrm{crit} = \frac{cH_0}{2\pi G}\frac{D_\mathrm{OS}}{D_\mathrm{LS}D_\mathrm{OL}} , \tag{8}$$

and scales like:

$$\Sigma_\mathrm{crit} \simeq 0.14 \left(\frac{H_0}{70\,\mathrm{kms}^{-1}\,\mathrm{Mpc}^{-1}}\right)\left(\frac{D_\mathrm{OS}}{D_\mathrm{LS}D_\mathrm{OL}}\right) \mathrm{gcm}^{-2} . \tag{9}$$

Thus, for $D_\mathrm{OS}/D_\mathrm{LS}D_\mathrm{OL} \simeq 3$, the critical lensing mass density is about 10^{-25} gcm^{-3} which is a few 1000 times larger than the critical density of the Universe, ρ_crit, making massive cluster efficient lenses.

The magnification matrix is real and symmetric and therefore it can be diagonalised and written in its principal axis as:

$$\mathcal{A}^{-1} = \begin{pmatrix} 1-\kappa+\gamma & 0 \\ 0 & 1-\kappa-\gamma \end{pmatrix} = (1-\kappa)\left[\begin{pmatrix} 1 & 0 \\ 0 & 1 \end{pmatrix} + \frac{\gamma}{1-\kappa}\begin{pmatrix} 1 & 0 \\ 0 & -1 \end{pmatrix}\right] . \tag{10}$$

From this equation we see that κ is controlling the isotropic deformation, and that the reduced shear, $g = \frac{\gamma}{1-\kappa}$, is controlling the anisotropic deformation (for a simple example see Fig. 4). The direction of the deformation (or equivalently of the shear) can be written as:

$$\tan 2\theta_\mathrm{shear} = \frac{2\partial_{xy}\varphi}{\partial_{yy}\varphi - \partial_{xx}\varphi} \tag{11}$$

As the direction of the shear is a ratio of the lensing potential, the shear direction θ_shear will be independent (modulus 90 degree) of the distance ratio $\mathcal{E} = D_\mathrm{LS}/D_\mathrm{OS}$, and thus it will be independent of the source redshift z_S. Only the intensity of the shear will change with the source redshift z_S.

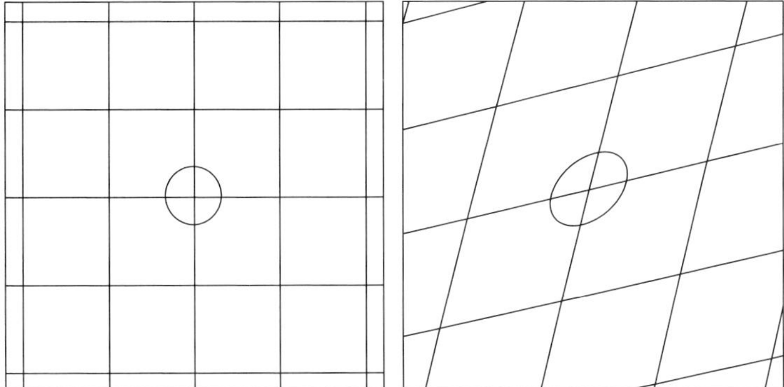

Fig. 4. Local deformation of a regular grid and a circle for a constant value of κ and γ over this region

3.4 Critical and Caustic Lines

The magnification μ is defined as the determinant of the magnification matrix and can be expressed as a function of κ and γ as:

$$\mu^{-1} = \det(\mathcal{A}^{-1}) = (1-\kappa)^2 - \gamma^2 \,. \tag{12}$$

The magnification is infinite if one of the principal values of the magnification matrix is equal to zero, which means that the reduced shear, g, is equal to 1 or -1. Thus, the locus in the image plane of infinite magnification defines two closed lines that do not intersect (i.e., g can not be equal to 1 and -1 at the same location) which are called the "critical lines". Their corresponding lines in the source plane are called "caustic lines", and they are all closed lines, but contrary to the critical lines, they can intersect each other. In general for a simple mass distribution, we can distinguish two critical lines: the external critical line where the deformations are tangential, and the internal critical line where the deformations are radial.

For a circular mass distribution, the equation of the critical lines are simple, as the magnification matrix in polar coordinates reduces to:

$$A^{-1} = \begin{pmatrix} 1 - \partial_{rr}\varphi & 0 \\ 0 & 1 - \frac{1}{r}\partial_r\varphi \end{pmatrix} \,. \tag{13}$$

Thus both the critical and caustic lines (if they exists) are circles (see Fig. 5). In fact, substituting the equation of the tangential critical line, $r = \partial_r\varphi$, into the lensing equation to compute the caustic line, we find that the tangential caustic line is always restricted to a single point (only true for a circular mass distribution).

It is also relatively easy to demonstrate that for a well-behaved mass distribution, the radial critical line is located within the tangential critical line [62].

It is important to notice that for a circular mass distribution, the projected mass within a radius r can be written as

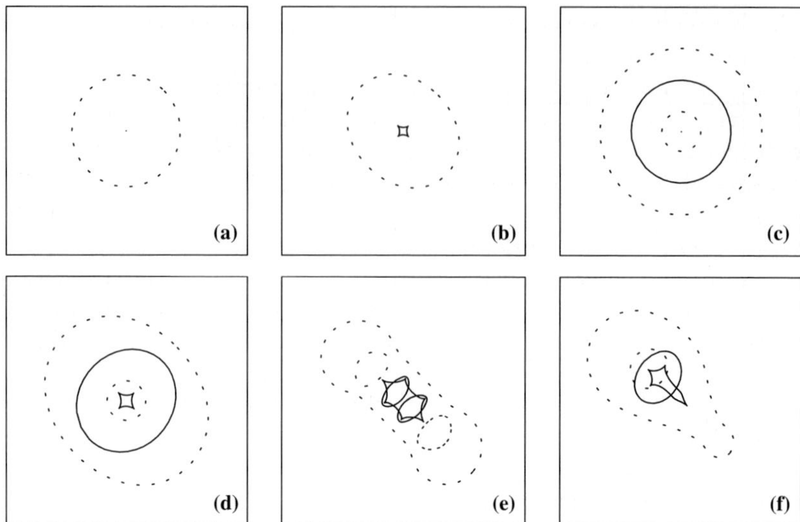

Fig. 5. Critical line (*dashed*) and caustics (*solid*) for different mass model types: (**a**) singular isothermal circular mass distribution, where the radial critical line is the central point, and the corresponding caustic line is at infinity, (**b**) singular isothermal elliptical mass distribution, where the tangential caustic line is then an astroid, (**c**) circular mass distribution with an inner slope shallower than the isothermal mass distribution. In this case a radial critical curve appears, and both caustics are circles. (**d**) same as (**c**) but for an elliptical mass distribution; the relative size of both caustic lines will depend on the mass profile and the ellipticity of the mass distribution, (**e**) bimodal mass distribution of two clumps similar to (**d**) with an equivalent mass, and (**f**) bimodal distribution with different mass clumps

$$M(r) = \frac{c^2}{4G}\frac{D_{\rm OS}D_{\rm OL}}{D_{\rm LS}} r\partial_r\varphi(r) = \pi\Sigma_{\rm crit} r\partial_r\varphi(r) \; . \tag{14}$$

At the tangential critical radius we have $r_{ct} = \partial_r\varphi(r_{ct})$. Thus the mass within the tangential critical radius (also called Einstein radius $r_{\rm E}$) is

$$M(r_{\rm E}) = \pi\Sigma_{\rm crit} r_{\rm E}^2 \; , \tag{15}$$

The critical surface density, $\Sigma_{\rm crit}$, corresponds thus to the mean surface density within the Einstein radius. Thus the more concentrated the cluster the larger the Einstein radius. For a given surface mass-density profile, the size of the Einstein radius will depend on the redshifts of the lens and the source as well as on cosmology. The most effective lens is placed at less than half the source redshift.

It is important to note that the radial critical curve is defined as

$$\partial_{rr}\varphi(r) = \partial_r \frac{M(r)}{\Sigma_{\rm crit}\pi r} = 1 \; , \tag{16}$$

and thus its location depends on the gradient of the mass profile at the location of the radial critical line.

The above considerations are very important properties of lensing, which suggest that one can measure, with the tangential critical curve, the absolute mass distribution within a circular aperture once the redshift of the lens and the redshift of the source are known, regardless of the exact mass profile of the structure. Using the radial critical curve, the slope of the mass profile near the cluster center can be measured.

In a general (non circular) case, the determination of the critical lines can not be addressed analytically (except for certain simple elliptical mass profiles) and thus one has to solve for them numerically. The above property, of linking the total mass within the critical line to the area within the critical line, does not hold exactly for a more general case, but it is still a good approximation if the mass distribution is not too deviant from circular symmetry. Hence identifying the characteristic sizes of the critical lines in an observed cluster is the first step to measuring its central mass and concentration.

3.5 Multiple Images

Critical lines are virtual lines, and thus are not directly visible. What we can identify, however, are multiple images that will straddle critical lines as tangentially or radially distorted images. One often refers to tangential pairs or radial pairs, which are simple configurations easy to recognize, but one can have triplet, quadruplet, quintuplet or even more images coming from the same source depending on the local complexity of the mass distribution.

The number of multiple images is the number of solutions of (4). It can be shown easily, following the catastrophe theory, that each time one crosses the caustic lines, in the source plane, two images are added. Thus for a non-singular mass distribution [19] we expect to always have an odd number of multiple images. However, some images could be much less magnified (or even strongly de-magnified) to the point that they are not observable, making the counting of multiple images not as easy as the simple catastrophe theory predicts.

Multiple images have different symmetries which can be summarized by 2 signs. We have 4 possibilities: $(+,+)$ which correspond to the symmetry of the source; $(+,-)$, $(-,+)$ and $(-,-)$. These symmetries however can only be identified with sufficiently high resolution images, namely those delivered by *HST* (see Fig. 19).

In order to produce multiple images, the cluster surface mass density needs to reach or be larger than the *critical density*. The configuration of multiple images tells us about the structure of the mass distribution. A cluster with one dominant clump of mass will produce (see Fig. 6) *fold*, *cusp* or *radial* arcs (e.g., MS2137.3-2353: [37, 87]; AC114: [98]; A383: [129, 132]). A bimodal cluster can produce straight arcs (e.g., Cl2236-04: [65]), triplets (A370: [10, 63]) or even a triangular image. A very complex structure, with lots of massive haloes in

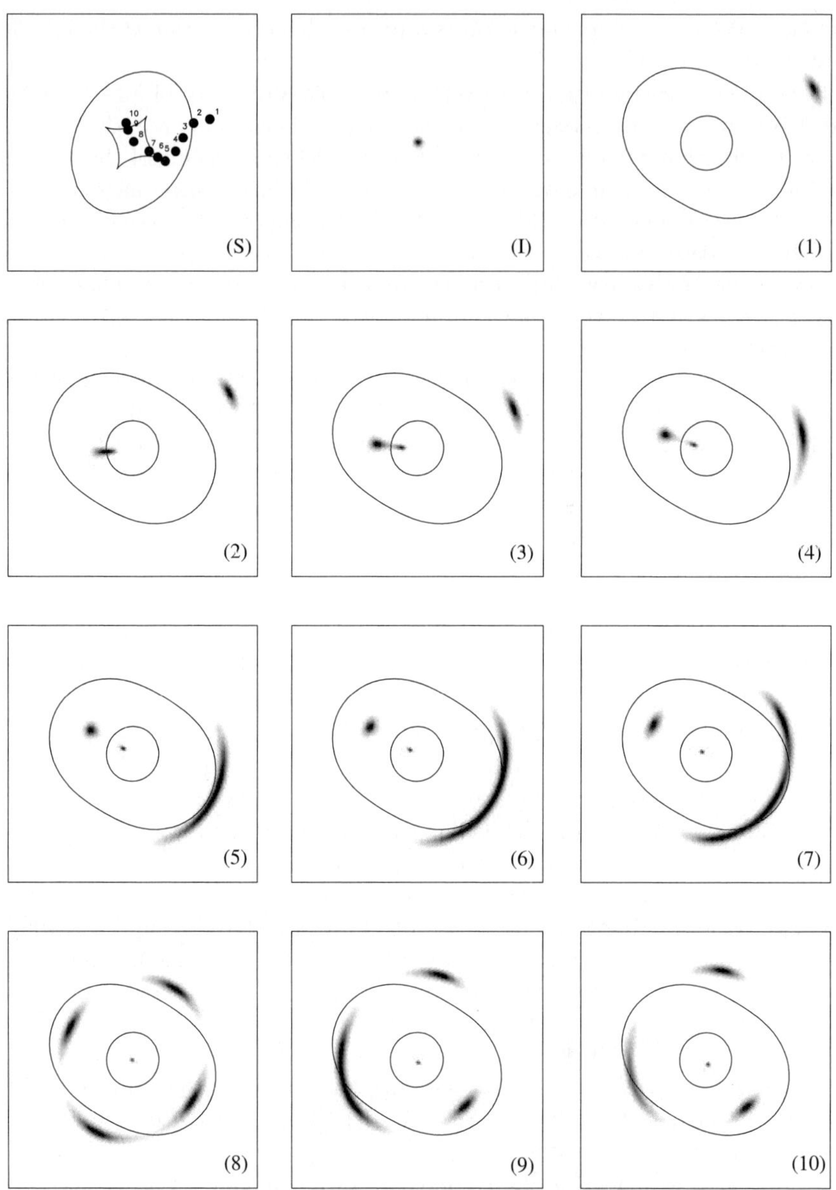

Fig. 6. Multiple-image configuration for a simple elliptical mass distribution. The panel (**S**) shows the caustic lines in the source plane, and panels 1–10 show the source position relative to the caustic lines. The panel (**I**) shows the image of the source without lensing. The panels (1)–(10) show the resulting lensed imaged for the various source positions. Certain configurations are very typical of the lensing effect, and are named as follows: radial arc (3), cusp arc (6), Einstein cross (8), fold arc (10)

the core, can produce a multiple image system with 7 or more images of the same source (e.g., Cl2244-04, A2218 – Fig. 7). The idea is that each massive halo can add 2 extra images to a simple configuration if that halo is well positioned.

Multiple images can be identified by different properties. Originally, multiple images have been recognized as the images forming the giant arcs (3 images in the case of Abell 370, but only 2 images in the case of MS2137.3-2353 or Cl2244-04). However, not every giant arc is made of multiple images, indeed it is most likely that the northern giant arc in Abell 963 is only made of one single image, and its southern arc is made of two or three arclet (single image) at different redshifts.

Multiple images can be recognized in terms of their (mirror) symmetry. In the case of the 'E' multiple image system in AC114, spectroscopy has confirmed a point-like object as a distant quasar at $z = 3.147$ which, after detailed investigation, turned out to be multiply lensed [20].

Moreover, as lensing is achromatic, multiple images can also be recognized as having similar colors, or by being extremely bright at some particular wavelength like in the sub-mm or in the mid-infrared.

Finally, the ultimate way to confirm a multiple image system is through lens modeling. This can allow one, in principle, to test if a set of images, having similar morphology and colors, can effectively be multiple images of the same source. The lens model can then predict the location of counter-images and predict the redshift of the multiply lensed source [63, 66].

Fig. 7. A spectacular case of multiple images in the cluster Abell 2218 seen in a BRI HST image. The distant E/S0 galaxy at z=0.702 is lensed into a 7-image configuration

3.6 First Order Shape Deformation

Distant sources are only multiply-imaged in the central region of the cluster where the surface mass density is sufficiently important. But every observed galaxy image is deformed by the lensing effect whether it is in the strong or weak regime region. To first order, one can approximate a galaxy by an object with elliptical isophotes, and thus its shape and size could be defined in terms of ellipticity, orientation and area enclosed by a boundary isophote.

However, the shapes of galaxies can be quite irregular (specially for late types or irregular galaxies) and they are not well approximated by ellipses. We thus need to express the shape of a galaxy in terms of its pixel surface brightness, as measured by a digital detector. For this purpose, we will use the moments of the light distribution to define the shape parameters. If $I(x,y)$ is the surface brightness distribution of the considered galaxy, we can define the center of the image, $\boldsymbol{\theta_c}{=}(x_c, y_c)$, using the first moment of the $I(x,y)$ distributions:

$$\boldsymbol{\theta}_c = \frac{\int W(I(\boldsymbol{\theta}))\boldsymbol{\theta}\mathrm{d}\boldsymbol{\theta}}{\int W(I(\boldsymbol{\theta}))\mathrm{d}\boldsymbol{\theta}} . \tag{17}$$

Note that $W(I)$ is a weight/window function that allows, in the case of real noisy data, to have finite integrals. The simplest choice of the function $W(I)$ is the Heaviside step function $H(I - I_{\mathrm{detiso}})$ which is equal to 1 for $I(x,y) > I_{\mathrm{detiso}}$, where I_{detiso} is the isophote limiting the detection of the object, and 0 otherwise. The center will then be the center of the detection isophote. Another popular weight function is $W(I) = I \times H(I - I_{\mathrm{detiso}})$, where the center is then weighted by the light distribution within the isophote.

The second order moment matrix of the light distribution, centered on $\boldsymbol{\theta_c}$, is:

$$M_{ij} = \frac{\int W(I(\boldsymbol{\theta}))(\theta_{\mathrm{I}} - \theta_i^C)(\theta_j - \theta_j^C)\mathrm{d}\boldsymbol{\theta}}{\int W(I(\boldsymbol{\theta}))\mathrm{d}\boldsymbol{\theta}} \equiv R_\theta \begin{pmatrix} a^2 & 0 \\ 0 & b^2 \end{pmatrix} R_{-\theta} , \tag{18}$$

allows one to define the size, the axis ratio and the orientation of the corresponding ellipse. Indeed M is definite and positive and can be written in its principal axes, where a and b are the semi major and semi minor axis and θ is the position angle of the equivalent ellipse. Thus M can define the equivalent galaxy ellipse parameters: size, ellipticity and orientation (for an example see Fig. 8).

It is useful to define the equivalent complex ellipticity. Note that various methods exist to define the norm of the complex ellipticity, and different lensing studies have used a multitude of notations:

$$\varepsilon = \chi = \frac{a^2 - b^2}{a^2 + b^2} \quad \tau = \frac{a^2 - b^2}{2ab} \quad \epsilon = \frac{a - b}{a + b} . \tag{19}$$

However, ϵ has now become the standard definition, essentially because it is a direct estimator of the reduced shear g (see below). The ellipticity parameters are of course linked to each other by: $\varepsilon = 2\epsilon/(1 + \epsilon^2)$.

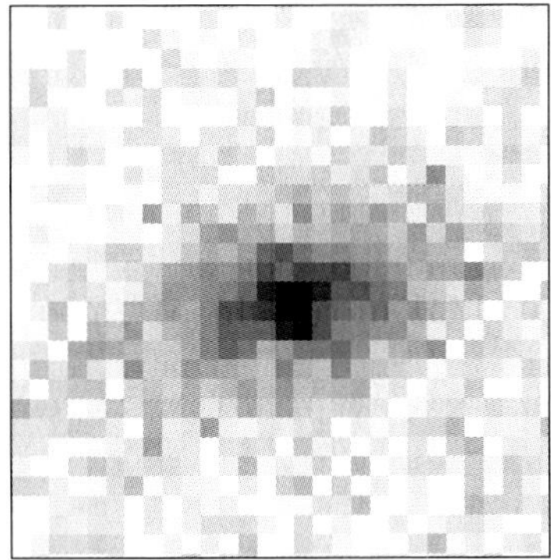
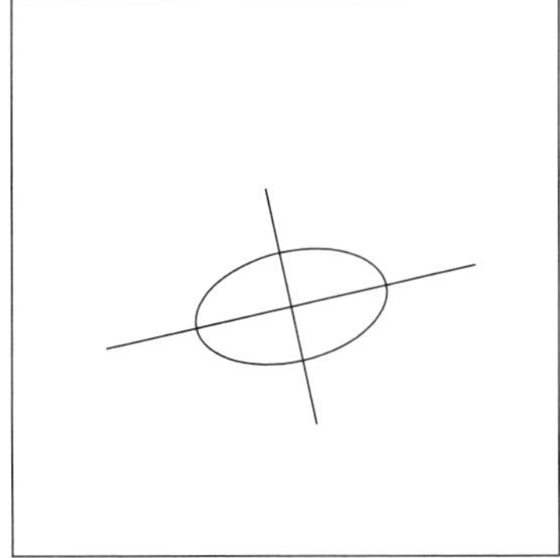

Fig. 8. A typical faint galaxy observed on a CCD image (**left**), and the equivalent ellipse defined from the second order moments (**right**)

Now that we have defined these different parameters, we need to express how gravitational lensing transforms the shape of a galaxy. First it can be showed [73, 90] that the image of the center of the source corresponds to the center of the image in the case where the amplification matrix does not change significantly across the size of the image (which can also be considered as the definition of the weak lensing regime). To demonstrate this, one has to use the fact that the surface brightness is conserved by gravitational lensing, as demonstrated by Etherington in 1933, i.e., $I(\boldsymbol{\theta}_\mathrm{I}) = I(\boldsymbol{\theta}_\mathrm{S})$.

The lens mapping will transform the shape of the galaxy, by amplifying it and stretching it along the shear direction. This transformation can be written in terms of the moment matrix as:

$$M^S = \, A^{-1} \, M^I \; {}^t\!A^{-1} \, . \tag{20}$$

This equation allows one to express how the equivalent ellipse of the source is mapped onto the equivalent ellipse of the image or vice versa.

If we consider the size, $\sigma = a \times b$, of the equivalent ellipse, we can write:

$$\sigma_\mathrm{S}^2 = \det M^S = \det M^I .(\det A^{-1})^2 = \sigma_\mathrm{I}^2 .\mu^{-2} \tag{21}$$

The size σ_S of the source is thus amplified by the magnification factor μ.

For the complex ellipticity ϵ we have:

$$\epsilon_\mathrm{S} = \frac{\epsilon_\mathrm{I} - g}{1 - g\epsilon_\mathrm{I}}, \quad \text{for} \quad |g| < 1 \, , \tag{22}$$

which corresponds to the region external to the critical lines, and

$$\epsilon_\mathrm{S} = \frac{1 - g^* \epsilon_\mathrm{I}^*}{\epsilon_\mathrm{I}^* - g^*}, \quad \text{for} \quad |g| > 1 \, , \tag{23}$$

which corresponds to the region internal to the critical lines (the * denotes the transpose notation of a complex number).

In the weak regime, where the distortions are small ($|g| \ll 1$), the lensing equation simplifies to:

$$\epsilon_{\rm I} = \epsilon_{\rm S} + g^* \ . \tag{24}$$

Thus the ellipticity of the image is just a linear sum of the source ellipticity and the lensing distortion. Averaging the above equation over a number of N sources means that the averaged image ellipticities is a direct measure of the reduced shear, g, provided that the orientation of sources are random and not correlated:

$$\langle \epsilon_{\rm I} \rangle = \langle g^* \rangle \ , \quad \sigma_{\epsilon_{\rm I}} \sim \sigma_g \sim \sigma_{\epsilon_{\rm S}}/\sqrt{N} \ . \tag{25}$$

The error on the measurement is directly proportional to the rms source ellipticity divided by the square root of the number of sources used. Observationally, for typical cluster data, we have $\sigma_{\epsilon_{\rm S}} = 0.3$, which then directly gives the order of sources needed to reach the desired accuracy on $\langle g \rangle$. Note that in practice we need also to take into account the errors on the shape measurements.

3.7 Higher Order Shape Deformation – "Flexion"

The equation of the previous section assumes that κ and γ (and thus the reduced shear g) are constant across an image. This assumption will, however, fail when an image is large and/or when it is close to critical regions, where the lensing distortion is rapidly changing. To simplify the situation, there are two main effects that lensing will produce on an elliptical source: a shift of the flux peak center, compared to the center of the fainter isophotes, and the transformation of the elliptical shape into a "banana" shape.

To determine these transformations numerically one needs to go to higher orders of the lensing transformation using a Taylor's expansion of the image shape. This was first investigated by Goldberg & Natarajan [48], and recently formalized and summarized in two papers [5, 47]. A parallel development of high-order lensing shape deformation, based on the beam-physics formalism, has been recently presented by Irwin & Shmakova [54] which merits to be looked at in detail.

Similarly, the observed flexion will be a combination of the intrinsic flexion (assumed on averaged to be random), the lensing flexion, and the flexion introduced by the telescope/camera. Hence measuring the flexion, correcting for instrumental flexion and averaging it, will lead to a measurement of the lensing flexion. Thus the flexion can usefully complement shear estimators and this is particularly true in the intermediate regime, between the classical strong and weak regimes.

Although not yet applied to cluster lensing, this formalism could in principle be applied to cluster lens modeling by adding further constraints.

4 Constraining the Cluster Mass Distribution

Gravitational lensing, as we have just seen, is a clean way to probe the mass distribution of clusters, regardless of the nature of the matter particles. However comparing lensing estimates to other mass estimates is an excellent way to probe the amount, nature and distribution of dark matter particles, as well as to investigate cluster physics. Strong lensing can only be observed in the core region ($<\sim$ 1 arcmin) of the most massive clusters and much can be learned about the detailed mass distribution in these regions. Weak lensing can only be observed on scales where enough faint galaxies can be averaged to locally measure the reduced shear, typically on scales larger than those of strong lensing ($\gtrsim$ 1 arcmin).

4.1 Strong Lensing

A particular useful and simple mass estimate in the strong lensing regime is the mass within the Einstein radius, $R_{\rm E}$: $M(< R_E) = \pi\Sigma_{\rm crit}\theta_{\rm E}^2$, where $R_{\rm E}$ is the location of the critical line for a circular mass distribution, usually approximated by the arc radius $R_{\rm arc}$. It is a very handy expression, but one should be careful in using it, either because the arc used to derive the mass is at an unknown redshift, or the arc is a single image and thus does not trace the Einstein radius (for a singular isothermal sphere model, a single image can not be closer than twice the Einstein radius or it will have a counter image!), or even because the mass distribution could very complex with a lot of sub-structure. In conclusion, *this estimator generally overestimates the mass.*

The radial critical line can be constrained when a radial arc is observed in the cluster core. This has been the case in a number of cluster lenses [37, 41, 120, 121, 129]. These features are important as they lie very close to the cluster core, and thus provide a *unique* way to probe the central mass surface density, from which we can directly probe the Dark Matter slope in the cluster core.

The only route to accurately constrain the mass in cluster cores is to use multiple images, with spectroscopic redshifts, to absolutely calibrate the mass. As the problem is generally degenerate –*in the sense that there is not a single mass distribution but a family of models that fit the observables*, the best way is to use physically motivated representations of the mass distribution and adjust these in order to best reproduce the different families of multiple images, e.g., [66, 129]. As the position of the images are known to great accuracy, and are usually located in different places of the cluster cores, a simple mass model with one clump usually cannot reproduce the image configuration. The lens model needs to include the cluster galaxies to match-up the image configurations and positions. Since there is not an infinite number of multiple images, the number of constraints is limited. It is therefore important to limit the number of free parameters of the model and keep it physically motivated. An alternative method, using a non-parametric description, has been explored by Abdelsalam et al. [2], and more recently by Diego et al. [31] who described in great details different possible approaches to the non-parametric problem. However, such methods usually lack the high resolution of a parametric form which is needed due to the large dynamical range of the cluster core mass-density.

The strong lensing-mass modeling technique can be seen as an iterative method, in the sense that once a multiple image is securely identified, other multiple image systems can be discovered using morphological/color/redshift-photometric criteria as well as the predictions from the lens model. The lens model can then predict redshifts for these multiple systems [36, 63, 98] as well as for the arclets [64, 66]; on the basis that, on average, a distant galaxy is randomly orientated, and its ellipticity follow a relatively peaked ellipticity distribution. These predictions can then be tested/verified [35] and an improved mass-model can be derived including the new constraints. The LENSTOOL software [62], that performs parametric strong-lensing modeling, is publicly available at: *http://www.oamp.fr/cosmology/lenstool/*.

4.2 Probing the Radial Profile of the Mass in Cluster Cores

One important prediction from *dark matter* only numerical simulations is the value of the slope of the density profile in the central part of relaxed gravitational systems. Although there is still some debate on the exact value of the inner slope [88, 94, 101], the real limitation in such numerical predictions is probably the lack of baryonic matter (both stars and gas). Cosmological simulations that include gas dynamics, radiative cooling and star formation (e.g., [46]) are currently limited by the poorly constrained physical processes acting in the cluster core, and are not able to give a unique prediction. Nevertheless, the radial slope of the total mass-profile is a quantity that lensing observations can constrain. The first attempt of this endeavor was conducted by Smith et al. [129] who characterized the inner slope of the total mass-profile in the Abell 383 cluster by modeling (using both a radial and tangential arcs) the cluster core by a sum of a cD halo and a cluster clump halo.

Once tangential and radial arcs have been identified on the *HST* images, the main observational limitation is to measure the redshift of both multiple arcs to firmly constrain the radial mass-profile (see Fig. 9). Large telescopes (Keck/VLT/Gemini/Subaru) are powerful instruments to measure the redshifts of faint galaxies. Thus these telescopes are now playing a *key role* in cluster lensing, used to measure redshifts of multiple images. Furthermore, working at high spectral resolution with large telescopes allows us to probe the dynamics of cD galaxies [120, 121]. Thus, combining the stellar dynamics constraints and the lensing constraints on the mass distribution of cluster cores, makes it possible to weigh the different mass components. Sand et al. [120] first investigated this technique on the cluster MS2137.3-2353 ($z = 0.313$),

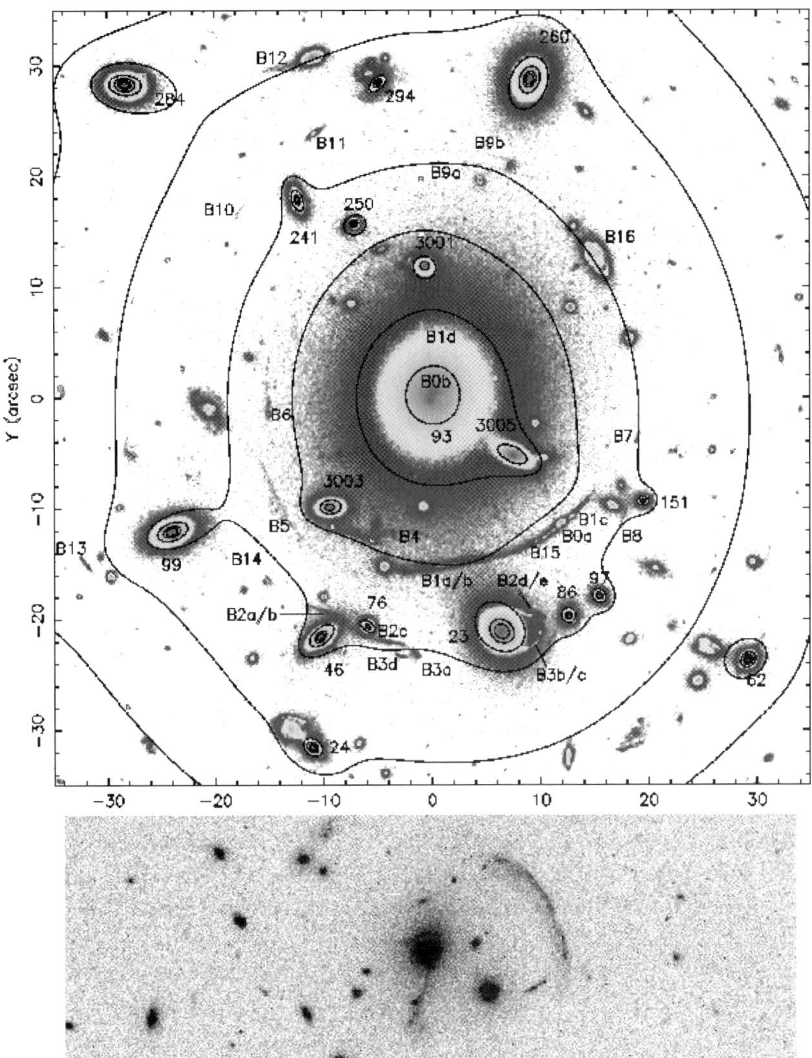

Fig. 9. (**Top**) WFPC2 image of the core of the cluster A383 [129] which shows both a radial and tangential arc. The first constrain of the slope of the total matter in a cluster core was derived from this analysis. (**Bottom**) A recent discovery of a radial arc in the cluster MACSJ1133.2+5008 [121]. In this system, both the radial and tangential arcs are from the same source

where they measured the redshift of both the radial and tangential arc at $z \sim 1.5$, as well as the velocity dispersion of the central cD. A similar analysis was conducted by Gavazzi et al. [41] and on a larger cluster sample, by Sand et al. [121]. The conclusion is that the dark matter slope is shallower than the one predicted by NFW ($\beta = 0.5 \pm 0.3$), but the comparison between numerical simulations and observations is not direct as the stars of the cD galaxies dominate the total cluster mass in the very center (with a possible contribution of the X-ray gas).Thus they do not follow the simple DM physical processes of numerical simulations.

It is also most likely that combined constraints of strong lensing and high resolution images of cluster cores, in X-rays using *Chandra* and/or *XMM-Newton*, will help understand the mass distribution in the inner core and the physical processes present, specially for the most relaxed clusters. Clearly, more work is needed to be done in this direction, as X-ray and lensing analyses are currently conducted separately.

4.3 Weak Lensing

As soon as we look a little bit further away from the highest density region, the lensing distortion (and magnification) gets smaller, and very quickly the shape of faint galaxies are dominated by their intrinsic ellipticities. Thus, in the weak lensing regime the game is very different than in strong lensing where any multiple images will put strong constraints on the mass distribution. Here, we need to measure the *mean* ellipticity or the *mean* number density of faint galaxies (if the idea is to use the magnification effect), in order to relate these statistics to the mean surface mass-density κ of the cluster.

There are two problems in measuring mass from weak lensing:

- for *observers*: How to best determine the "true" ellipticity of a faint galaxy which is smeared by a PSF barely smaller than the object (when using ground-based images) which is not circular (camera distortion, focusing, tracking errors ...) and which is not stable in time? How to best estimate the variation in the number density of faint galaxies due to lensing, taking into account the crowding effect due to the cluster and the intrinsic spatial fluctuations in the distribution of galaxies?
- for *theorists*: What is the best way to reconstruct the mass distribution κ (as a mass map or a radial mass profile) from the "reduced shear field" $\boldsymbol{g}$ and/or the magnification bias?

Various approaches have been proposed to solve these two problems, and two families of methods can be distinguished: **direct** and **inverse** methods.

For the observer, before any data-handling, the first priority is to choose the telescope that will minimize the source of noise in the determination of the ellipticity of faint galaxies. Although *HST* has the best characteristics in terms of the PSF, it has a very limited field of view, and it is not really appropriate to probe the large-scale distribution of a cluster. Wide-field ground-based imagers are much more appropriate in terms of field of view covered, but are lacking the crisp image resolution of *HST* .

Ultimately, what would be really needed is a wide field imager with excellent image resolution and PSF stability: the SNAP satellite concept is matching well the requirements for a (weak) lensing telescope of the future [85, 114].

Once the data have been taken with the best care to minimize the distortion and with, hopefully, the best seeing conditions, the next step is to convert the images of faint galaxies into some valuable lensing constraints. For this, we can use a direct approach, using for example the Kaiser, Squires & Broadhurst [58] method [KSB implemented in the *imcat* software], or any other improvement of it [59, 80, 113] that relates the true ellipticity, to the observed ellipticity correcting it from the smearing of an elliptical PSF (using the second moments of the galaxy and the PSF). Alternatively, one can use the inverse approach, based on the maximum likelihood or the Bayesian method, to find the shape of the source galaxy which, when convolved by the estimated local PSF, reproduces best the observed galaxy (e.g., [76], IM2SHAPE: [16]). These inverse approaches have the advantage of providing directly an uncertainty in the parameter recovery, as illustrated in Fig. 10. Furthermore the extension of this technique, in particular including the SHAPELET techniques [9, 86, 109, 110], may become the standard weak lensing measurement method in the next decade.

To understand more about the different measurement techniques and how they compare, the STEP (Shear TEsting Program) initiative has brought together a number of techniques developed and applied by different people to evaluate, on simulated images, the accuracy of the recovery of weak lensing signals [50]. Note however that the shear measurements are not only the ability of measuring the shapes of galaxies but also: (i) to clean the galaxy catalogue used to measure shear from faint stars, spurious objects or ill-defined shape objects (such as mergers) that will reduce the accuracy of shear measurements; (ii) to select galaxies in redshift-space to minimize foreground and cluster contamination in the shear signal.

From this lensing catalogue (containing information such as position on the sky, shape and redshift information with errors) a mass map can be derived. Again a direct and an inverse approach are possible.

The **direct** approaches are: (i) the Kaiser & Squires [56] method (an integral method, that express κ as the convolution of γ by a kernel) and subsequent refinements, e.g., [123–126]; (ii) the local inversion method [57, 79, 117, 128], that integrates the gradient of γ within the boundary of the observed field to then derive κ.

The **inverse** approach works on either κ or the lensing potential φ, and uses maximum likelihood [8, 61, 119], maximum entropy [15, 82, 127] or atomic inference approaches, coupled with MCMC optimization techniques [84], to determine the most likely mass distribution (as a 2D mass map or a 1D mass profile) that best match the reduced shear signal $\boldsymbol{g}$ in the lensing catalogue, and/or the variation in the faint galaxy number densities. These inverse methods are of great interest as they allow us to quantify the errors in the resultant mass maps or mass estimates [15, 67, 82], and in principle to cope with external constraints (such as strong lensing, X-ray, galaxy dynamics or SZ measurements).

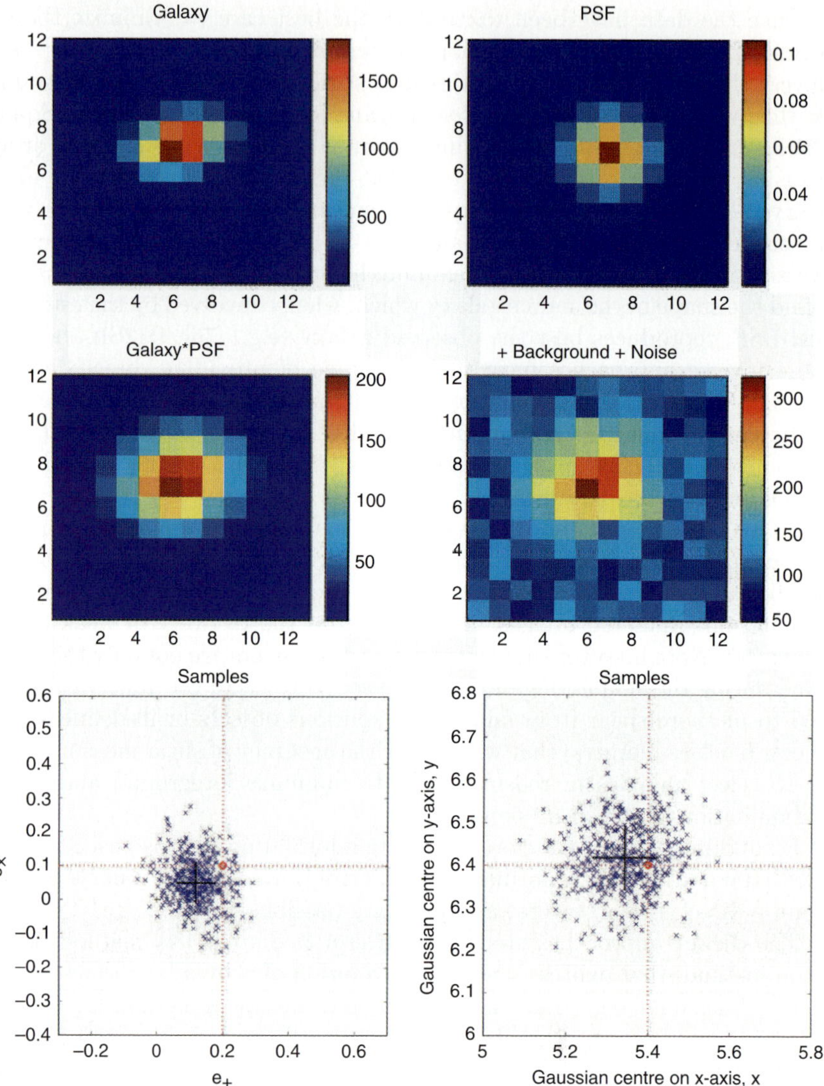

Fig. 10. Simulated images: the galaxy model, the PSF model, the convolution of the two, and the final image when noise is added. These simple simulations allow us to then quickly compare them to the observed faint galaxies, and then find the best galaxy model (taking into account the convolution effect of the measured PSF) that fits the data. *(Bottom panels):* the Markov Chain Monte-Carlo (MCMC) samples fitting the data in the ellipticity and position dimension space of the unknown parameters, from which a best fiducial model with errors can be extracted (black cross)

An important issue is to which resolution a 2D lensing mass map can be reconstructed. Generally, mass maps are reconstructed on a fixed size grid, which then defines the minimum mass resolution that can be obtained. By comparing the likelihood of the different resolution mass maps, we can then calculate to with which resolution the mass map is best estimated (Fig. 11). However, it is most likely that the best scale to which a mass map is to be reconstructed has to adapt itself according to the strength of the lensing signal. As we are limited by the intrinsic ellipticity distribution, it is only by averaging a large number of galaxies that we can reach the 1% shear level. Such low shear levels can thus only be probed on relatively large scales or by

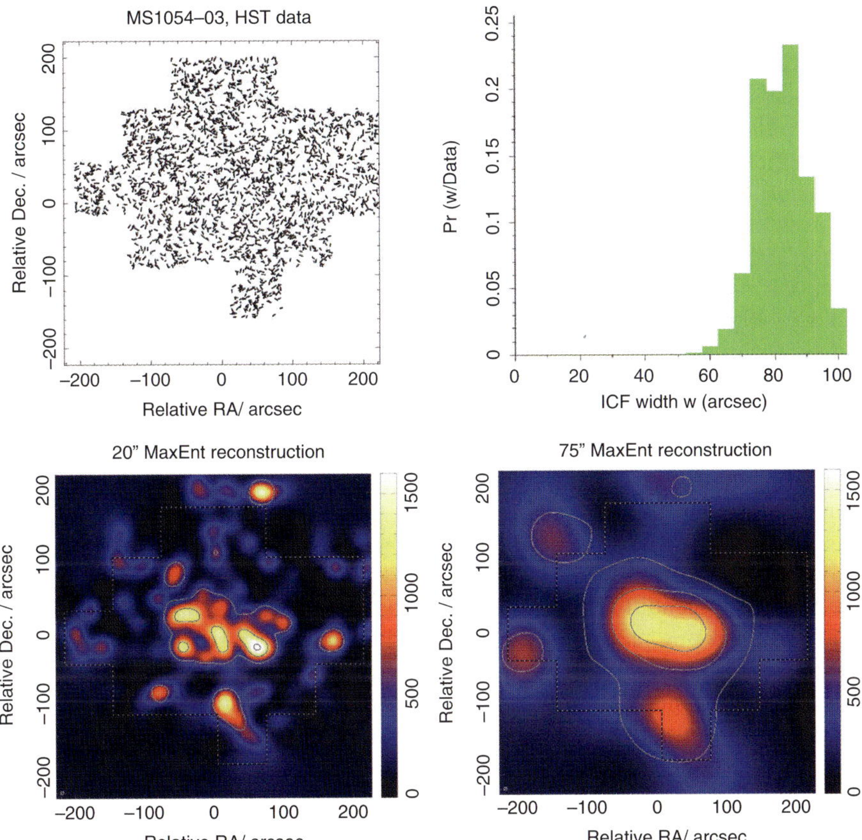

Fig. 11. Maximum entropy mass reconstruction [15] of the X-ray luminous MS1054 at $z = 0.83$ using Hoekstra's HST dataset [51]. (**Top Left**) Distribution of the galaxy positions used in the mass reconstruction. (**Top Right**) Evidence values for different size of the Intrinsic Correlation Function (ICF). (**Bottom**) Two mass reconstructions for 2 different value of the ICF: (**left**) small ICF having a small evidence value, (**right**) large ICF having the largest evidence

radially averaging a large number of galaxies. Furthermore, as the projected cluster mass distribution on large scales falls off relatively slowly ($1/R - 1/R^2$ respectively for an SIS or a NFW profile), it means that cluster masses have a contribution on large scales. It is likely that the best methods to accurately model the cluster mass distribution, on large scales, is probably by using parametric methods (allowing, sufficient freedom in the radial profile and number of substructures to closely map the observed lensing distortions).

The main output of weak lensing measurements in clusters are generally: (i) the detection of mass peaks, (ii) the measurement of the mass profile and total mass.

Although the lensing-mass peaks in clusters are usually centred on the optical and X-ray centers, in rare cases they point to different positions. For example, in the case of the merging cluster 1E0657-558 [25], the two lensing mass clumps are significantly offset compared to the X-ray surface brightness peak, leading to the conclusion that a large amount of dark matter is needed in this system, regardless of the nature of the gravitational force. The merging cluster 1E0657-558, is probably today the *cleanest and unambiguous astrophysical proof for the existence of DM in a cluster.*

In wide-field imaging surveys, cluster weak lensing techniques are applied to find clusters directly, irrespective of the galaxy concentration or X-ray detection. Important current surveys are the Deep Lens Survey [145], the Subaru survey [92], the CFHT Legacy Survey. The ultimate aim is to use cluster lensing counts to probe cosmology and particularly dark matter and dark energy. These will however only be possible with the future dedicated Dark Energy surveys such as DES, and the future novel telescopes, LSST and SNAP.

The weak-shear mass reconstruction techniques have been applied to ground based wide-field camera data (UH8k, CFH12k, ESO-WFI, CTIO-MegaCam) as well as multi-pointing HST data. Impressive results have been published for medium redshift ($z \sim 0.2 - 0.3$) clusters [6, 23, 24, 30] and for low ($z < 0.1$) redshift clusters [55]. For high ($z > 0.5$) redshift clusters, large aperture telescope [22] or *HST* [51, 52, 67] are probably more adequate.

Regarding the determination of the total cluster mass and its profile, weak lensing measurements are plagued by contamination of cluster galaxies in the central part, where the shear profile is generally underestimated. Total masses are less affected, unless the field of view of the camera is limited, in which case the mass-sheet degeneracy can be important. Using color selection to remove cluster contamination, using wide-field cameras or mosaiced images, and combining the weak and strong lensing constraints, can allow one to work around these different issues.

4.4 Haloes of Cluster Galaxies

We know that galaxies are massive and that their stellar content represents only a small part of their total mass. Although the existence of dark haloes around disk galaxies was obvious from very early on, with the study of their flat velocity curve out to large radius (*e.g.* [141]), the existence of elliptical galaxy dark haloes has been accepted only relatively recently, e.g., [74, 116]. These studies have found that the stellar content dominates the central part of the galaxies, but at distances larger than the effective radius the dark halo dominates the total mass. What is less obvious in cluster of galaxies, is how far the galaxy dark haloes extends, as one expects some tidal stripping of the galaxy dark haloes as they pass through the cluster cores. Furthermore, we would ideally like to relate the strong morphological evolution, observed in cluster galaxies, e.g., [75, 140], to their mass properties.

Galaxy lensing effects were first clearly detected in clusters by Kassiola et al. [60] who noted that the lengths of the triple arc in Cl0024+1654 could only be explained if the galaxies near the B image were massive enough. Detailed treatment of the galaxy contribution to the cluster mass became critical with the refurbishment of *HST*, as first shown by Kneib et al. [66] who concluded that galaxies (and their dark haloes) in cluster cores contributes ~10% of the total cluster mass. The theory of what is usually called galaxy-galaxy lensing in clusters was first discussed in detail by Natarajan & Kneib [97], and the first application to clusters followed shortly [42, 43, 98]. A recent analysis of this effect in various cluster-lenses at different redshifts seems to indicate an increase of the dark halo size of cluster ellipticals with increasing local mass density [99, 100]. Clearly more work is to be done in this direction, in particular addressing the variation of the galaxy halo size as a function of distance to the cluster center [77]. Note, that lensing appears here as the best method to probe such an effect, and the next step in attempting to improve such measurement is to analyze an even larger number of cluster galaxies, and combine those with velocity dispersion measurements of cluster galaxies involved in the lensing distortion.

It is important to realize, that the standard direct *weak shear* methods miss the small scale fluctuations (typically of galaxy halo scales) because of the *averaging* of the galaxy ellipticities. Thus, only dedicated methods can probe this effect in the weak shear method. The only practical route are the inverse approaches, using parametric mass models for cluster galaxies and the unbinned lensing catalogue (e.g., [98]).

4.5 Strong Plus Weak Lensing Cluster Modeling

A number of recent developments have shown the importance of combining strong and weak lensing to constrain the mass distribution in clusters [14, 67]. Indeed, strong and weak constraints can be combined together in a very complementary fashion, since strong lensing addresses the detailed and absolute mass distribution in cluster cores, while weak lensing probes more the large scale distribution. Putting both constraints together, lead to strong constraints on the mass profile on large scale.

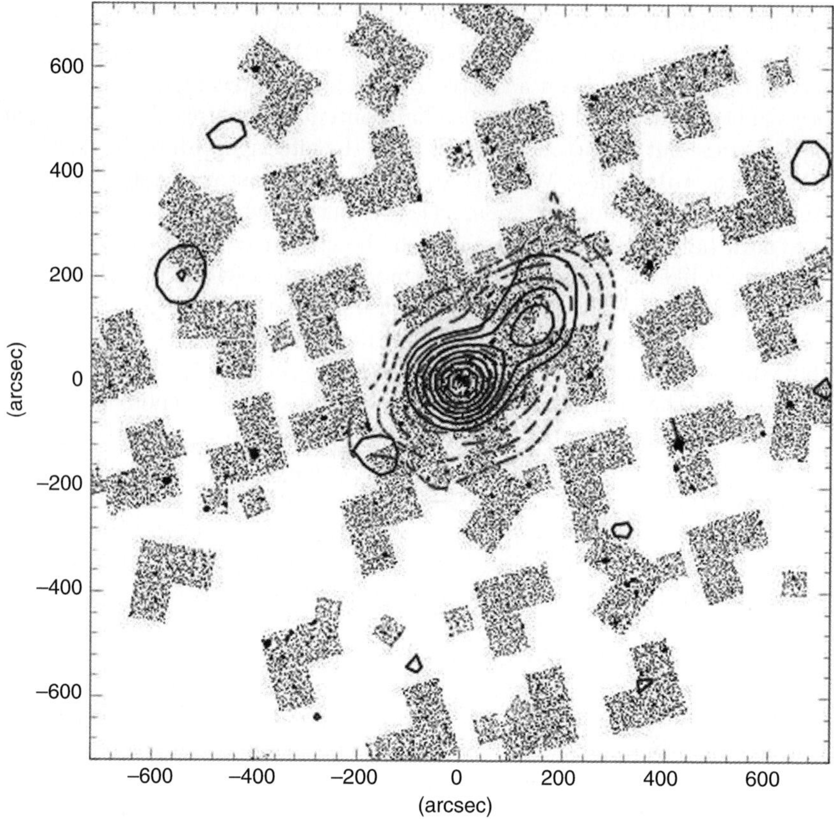

Fig. 12. The 39 WFPC2/F814W, and the 38 STIS/50CCD pointings, sparsely covering the Cl0024+1654 cluster. The (red) dashed contours represent the number density of cluster members as derived by Czoske et al. [28] . The (blue) solid contour is the mass map built from the joint WFPC2/STIS analysis derived using the LensEnt software [15, 82]

For example, important results have been obtained on the high redshift cluster Cl0024+1654, using 39 WFPC2/F814W pointings sparsely distributed around the cluster center, as shown in Fig. 12. By measuring the weak lensing distortion out to $\sim$ 5 Mpc and by taking into account the strong lensing constraints (a 5-image multiple image system at redshift 1.675). Kneib et al. [67] found a close correlation between the lensing mass distribution and the light distribution (Fig. 13). Furthermore, a clear substructure is detected in the 2D-mass reconstruction, of which the M/L ratio is similar to the main cluster clump. Finally, the fit of the radial mass profile (using both weak and strong lensing constraints) rejects, with high confidence, a SIS model but favors a cored-power-law model or a NFW model.

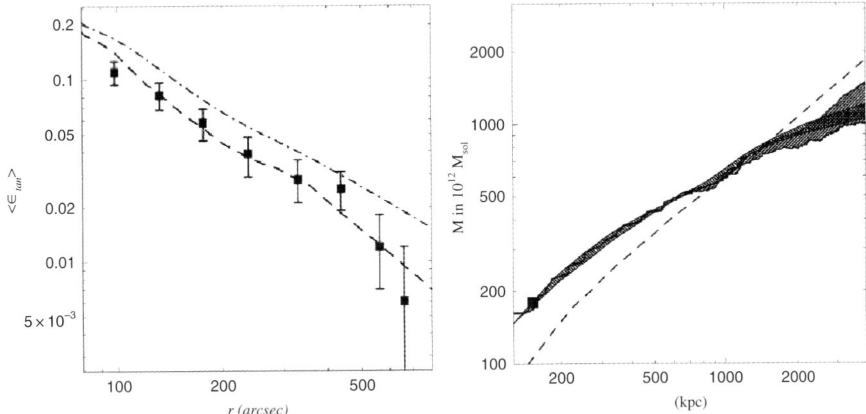

Fig. 13. (**Left**) The cluster radial shear profile as fitted by a NFW model. The shear expectation for a SIS model, fitting the strong lensing constraints, is shown. (**Right**) The enclosed mass as a function of radius: dark hashed region represents the lensing mass while the red hashed regions represent the light profile multiplied by a fix constant to match the mass profile. The blue dashed line show the mass profile of an SIS model fitting the weak lensing data only. The conclusion is that there is a good match between the light and mass distribution [67]

Similar results have been published for the clusters: MS2137.3-2353 [41] and Abell 1689 [18]. All three analyses discussed above find a much higher concentration for the NFW parameterization than the one found in (dark matter only) numerically simulated clusters. The origin of this high concentration may be explained by:

- projection effects along the line of sight: if the cluster under study is elongated along the line of sight, probably linked to an observational bias in the cluster selection. However, this is not likely to be the case for MS2137.3-2353 which is X-ray selected.
- an earlier cluster formation than the one assumed in the numerical simulation, as the concentration parameter increases with the age of the structure.
- the lack of baryons in simulations, missing some important dynamical effects (ram pressure, baryon/dark-matter interaction), that may impact the distribution of matter in cluster cores.
- DM particles being self-interacting thus changing the mass profile particularly in the highest density regions.

These different possibilities clearly need to be checked with larger cluster samples in order to better appreciate each of them.

4.6 Mass Distribution of Cluster Samples

Although the individual modeling of cluster cores is of great importance in order to characterize the best fitting mass distribution model (see previous sections), a lot can be learned by conducting statistical analyses of well defined cluster samples. The aim of such studies is to characterize the cluster lensing mass distribution and compare these results to other measured physical properties, like the X-ray temperature, the X-ray luminosity, the velocity dispersion of cluster galaxies as well as the measured SZ decrement. Understanding the different scaling relations between the various parameters should allow us to better understand clusters physics and dynamics. What is the repartition of mass between the different components. Are clusters relaxed? How important is substructure? How triaxial are the clusters? How many mergers occurred in a cluster? How important are projection effects? What was the evolutionary history of clusters? When did they first form?

To answer these questions, only statistical approaches are possible, and they should be based on a comprehensive multi-wavelength dataset, while ideally they should cover the various cluster scales. Collecting such a dataset is a **huge** challenge as it needs very broad skills and thus should involve a large number of people. First steps in this direction are now starting to produce interesting results, as shown by Dahle et al. [30] and Smith et al. [132, 133]. Many more results will follow based on data already collected, and on future data.

As an example of such developments, we started a thorough analysis of a sample of 12 $z \sim 0.2$ X-ray luminous clusters of galaxies selected from the XBACS catalog (see Fig. 14). These clusters have been imaged with the WFPC2 camera [132, 133]. Most of these clusters have been observed with the wide field CFHT12k camera in 3 colors (BRI) in order to probe the wide field mass distribution of these clusters.

This cluster sample has also been observed with the two X-ray satellites *Chandra* and *XMM-Newton*. A simple comparison of the derived lensing mass estimates of the cluster core and the *Chandra* X-ray temperature shows a somewhat loose correlation between mass and X-ray temperature [133]. However, by splitting the clusters into two samples, relaxed and un-relaxed clusters, a better picture of the M–T relation seems to arise and favors a low value for σ_8 (Fig. 15). But considering the various possible biases and uncertainties, it is clear that this first analysis needs to be improved by looking at a larger cluster set, with possibly a view on the 3D distribution. As an example, this can be acquired from the measurement of a large number of velocities of cluster galaxies (see [28, 29]) and by taking into account the weak lensing measurements that probe the mass on larger scales.

Cypriano et al. [27] have also conducted a weak lensing analysis on a much larger cluster sample using VLT/FORS and Gemini/GMOS. The idea is to extensively image the XBACS/BCS cluster sample searching for lensing effects in these clusters (see the Abell 2029 analysis – Fig. 16). At this time only

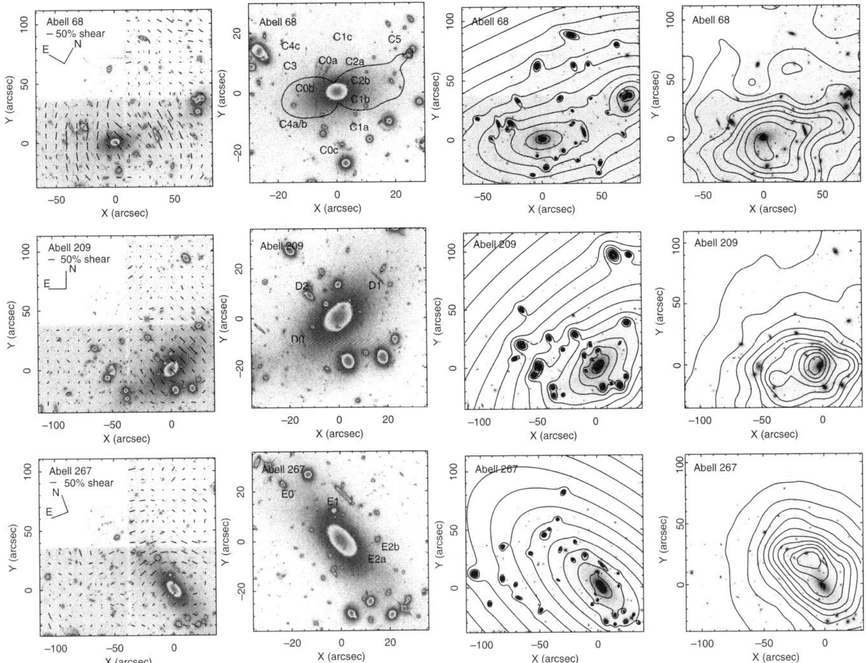

Fig. 14. Three of the 12 $z \sim 0.2$ X-ray luminous clusters of galaxies selected from the XBACS catalog [32] observed with the HST/WFPC2 camera. Top row is Abell 68, second row is Abell 209 and last row Abell 267. First column is the weak shear field, second column is a zoom on the cluster core, third column is the lensing mass reconstruction and last column is the overlay of the *Chandra* X-ray map [132]

half of the targeted clusters have been published (24 out of the 50 at VLT). Nevertheless, the interesting results [27] show a clear correlation between lensing mass and X-ray temperature, except for the most massive/X-ray luminous clusters, for which the clean correlation breaks (likely due to merger activity that temporally boost or decrease the X-ray signal, as well as affecting the lensing measurements depending on the projection angle).

4.7 New and Larger Cluster Samples – Looking for New Lenses

Understanding lensing clusters will be possible only if more observations are accumulated, but also if more clusters can be detected and studied. The number of published massive clusters, for which lensing analysis can be conducted on, is relatively limited (probably much less than one hundred). This, however, is beginning to change slowly – but the most massive clusters are in any case very rare – hence only sensitive enough surveys, covering a large fraction of sky will allow us to probe such clusters.

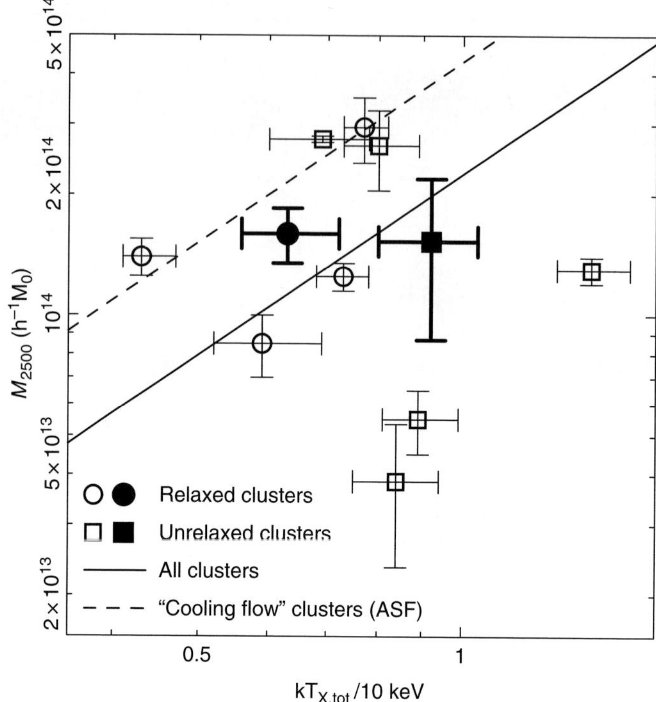

Fig. 15. Mass–Temperature relation of a sample of 10 X-ray luminous clusters at $z \sim 0.2$ [132]. Circles correspond to relaxed systems, squares to un-relaxed clusters, and the filled symbols to the mean of each category

There are different strategies to find massive clusters, and each technique has its advantages and weaknesses. One could list four techniques that are being, or will be, used to search for clusters in the near future:

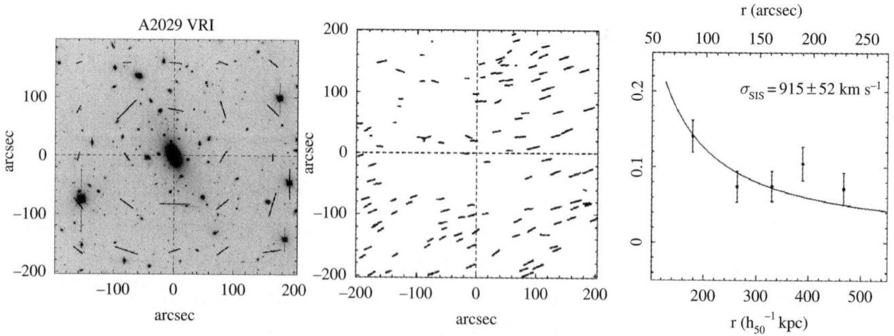

Fig. 16. (**Left**) The strong shear field around the cluster core of Abell 2029. (**Middle**) The star ellipticities as measured from the data. (**Right**) The SIS model fitting of the radial shear profile

- Photometric searches in wide field surveys, like the SDSS, the Red Cluster Sequence survey (Fig. 17), the CFHT Legacy Survey (CFHTLS), the DPOS survey.
- Weak lensing searches: the CFHTLS wide survey, the Deep Lens Survey [144, 145].
- SZ searches: based on the AMI and AMIBA projects, the bolometer array to be installed on the APEX antenna and in a more distant future on the *Planck* mission.
- X-ray selected cluster searches: in particular the MAssive Cluster Survey (MACS) [33] which is to-date the only survey capable to detect a large number of the most massive clusters at $z > 0.3$ (Figs. 17, 18), and other surveys like: REFLEX, WARPS, SHARC and XMM-LSS.

These surveys will produce a larger catalog of clusters, particularly at redshifts $z \gtrsim 0.3$. Those dedicated lensing studies will enable us to more closely investigate their mass distribution, as well as reveal, from time to time, very efficient lenses that can be used for other purposes, like studying the distant Universe thanks to the gravitational magnification effect.

4.8 Lensing and Other Mass Estimators

Gravitational lensing allow us to measure the *total* mass distribution of clusters without making any assumption on the cluster physical state. Other estimators always require some assumption when trying to relate the observables to the *total* mass. Generally these assumptions look reasonable but may suffer strong biases due to the unknown physical state of the clusters. By providing the *total* mass, lensing does constitute a **key** tool to understand cluster

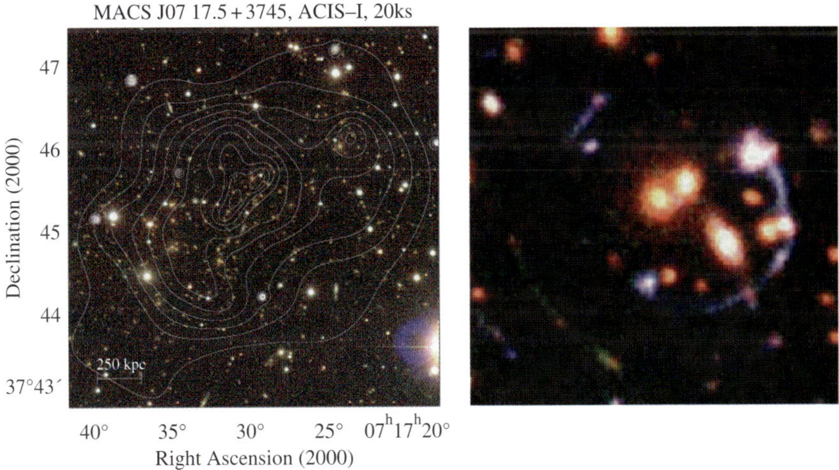

Fig. 17. Examples of newly found high redshift clusters. (**Left**) The cluster MACS0717.5+3745 as observed by Chandra, showing a very extended X-ray emission following the very extended distribution of red galaxies. (**Right**) A very compact cluster detected in the RCS survey [44] showing prominent arcs, one of them at a redshift of 4.8

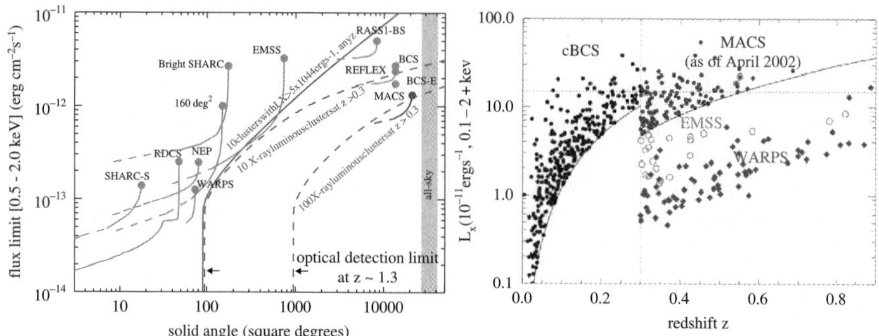

Fig. 18. (**Left**) A comparison between the different X-ray surveys showing the flux limit versus the area on the sky covered by the survey. Also indicated is the iso-line showing the number of X-ray luminous clusters expected, depending on the redshift range probed. This demonstrates that the MACS survey is the most sensitive one to discover the most massive clusters in the Universe [33]. (**Right**) The luminosity versus redshift plot, comparing a number of X-ray surveys: EMSS, BCS and MACS. It is particularly evident that MACS is very efficient in detecting the most X-ray luminous clusters at $z > 0.3$

physics. It is probably best then to first derive the total lensing mass using lensing, and then from other observations derive various physical cluster properties, like the cluster dynamical state using galaxy velocities [96], the X-ray gas temperature profile [108] and the baryon fraction or the equilibrium status of the cluster. Lensing mass estimates, however, have also their limitations (in particular due to line of sight projection effects).

The ultimate way is to compare the different mass estimators in a joint analysis (e.g., a joint analysis of lensing and SZ observations of clusters [83]). As an example, X-ray mass estimates generally differ sensibly from the strong/weak lensing estimates – although not always. The differences could be due to different reasons, depending on the cluster studied (e.g., [91]): (i) projection effects: 2 clusters can be aligned along the line of sight and boost the lensing mass; (ii) simple X-ray modeling: for example a multi-phase gas distribution is necessary in cooling flow clusters (e.g., [3]); (iii) non-thermal effects can modify the central mass estimates; (iv) the clusters may have just suffered a major merger event and the dynamical state of the gas can not be considered as being in thermal equilibrium. Another important issue that can affect the X-ray mass estimate is the exact form of the total mass profile (generally assumed to be an isothermal sphere or to follow a NFW form).

The canonical lensing clusters Cl0024+1654 is one example where the X-ray mass and lensing mass do not agree. In this cluster, the X-ray emission is weak compared to the large Einstein ring observed. The recent redshift survey of ~300 cluster galaxies [28, 29] and the recent lensing analysis [67] reveals that this system is very complex, showing different substructures, indicating that this cluster is certainly not yet fully relaxed. Simple X-ray estimates give a factor of 3 between the observed lensing mass and the one derived from X-ray data alone. A proper 3D model of this cluster and its X-ray gas may solve the discrepancy and allow a better understanding of its dynamical status.

The Sunyaev-Zel'dovich (SZ) effect is now routinely measured towards the most X-ray luminous clusters [21]. As the SZ is probing the intra-cluster gas in a different way than X-ray observations, it is important to use SZ as a complementary approach to the lensing, X-ray and galaxy velocities estimators, since a detailed comparison could teach us a lot about cluster physics.

5 Clusters Lenses as Natural Telescopes

Cluster lenses magnify and distort the galaxies behind them. For efficient lenses (massive clusters with intermediate redshift, $z \sim 0.2 - 0.4$), the magnification factor for the faint galaxy population is typically ~ 2 for a few square arc-minutes. This gain would correspond to a factor ~ 1.5 in the diameter of a telescope or an increase of a factor ~ 4 in exposure time. Clearly, looking through cluster lenses can yield great rewards when studying the faint (and thus distant) galaxy population, as it allows us to observe intrinsically fainter objects than would otherwise be possible.
Cluster lenses magnify, but also distort, the shape of distant galaxies; the further the sources, the stronger the distortions. Hence the shape of a lensed galaxy, and whether it is multiple or not, is generally a good distance indicator.

Of course, the most interesting regions are those having the largest magnification (also called the critical line regions). As the magnification is independent of wavelength, the benefit of using cluster lenses as natural telescopes has been used from X-ray to radio wavelengths, e.g., [4, 89, 137]. It was first investigated in the optical/NIR domain, where a number of the most distant galaxies (at their time of discovery) were found in the cluster magnified regions, e.g., [36, 38, 53, 68, 105, 146]. There are two interests here: to discover the most distant objects, and to study the morphology which otherwise would not have been possible (Fig. 19).

Lensing has been beneficially used in the searches of EROs in a sample of 10 X-ray luminous galaxy clusters [130, 131], where about 60 EROs were identified allowing to compute the ERO counts down to a very faint limit. It has also permitted the study, with more accuracy, of the morphology of these peculiar galaxies, revealing in some cases, a spectacular disky component (Fig. 19).

One of the exiting current focuses, is to map the critical region searching for Lyman-alpha emitters at very large redshift ($z > 7$), in order to compute their number density and luminosity function. One hope is to discover a population III galaxy, which will allow us to put strong constraints on galaxy formation in the early Universe, as well as determine the epoch of re-ionization. First results of this systematic search found a lensed pair at $z = 5.58$ [36], and a

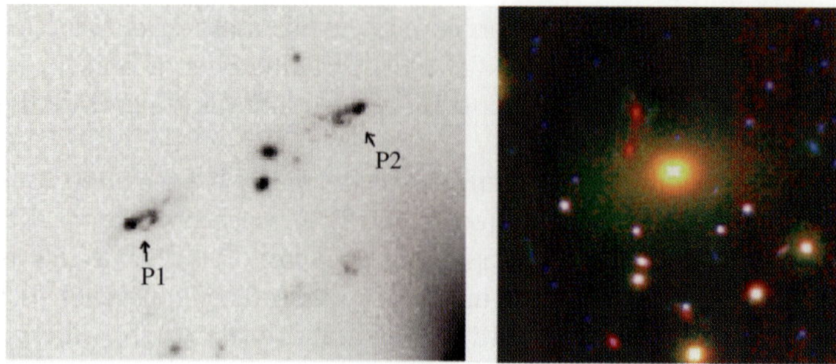

Fig. 19. (**Left**) The lensed pair P1–P2 in AC114. This galaxy at redshift 1.67 display a surprising morphology, similar to a hook, in this HST/WFPC2 R-band observation. (**Right**) The triple ERO galaxy in the core of Abell 68 at redshift $z \sim 1.6$. The large magnification allows to detect the blue light coming from the disk of that galaxy [131]

detailed description of a sample of lensed Lyman-alpha emitters were then obtained allowing to determine their luminosity function for $4 < z < 6$ [122]. With the development of 3D – integral field unit spectrograph – such like the VIMOS/IFU instrument, it is now becoming possible to study these high magnification region more systematically [26].

Other important searches are to conduct deep ACS z-band and ground based JHK observations of these massive clusters to have a better knowledge of the SED of those high redshift galaxies. Certainly, we will discover a (small) number of I-band, z-band or J-band dropouts that should put to even larger redshift the detection of the most distant object. We may however have to wait for JWST and the ELTs in order to collect a large number of these very high redshift systems.

6 Cosmological Constraints

The ultimate step of strong lensing modeling is to constrain the cosmological parameters that enters into the lensing equations through the $D_{\rm LS}/D_{\rm OS}$ term. This can be undertaken, when a sufficient number of multiple images (> 3) are identified in a cluster core, and for which spectroscopic information can be obtained [49].

As shown by Golse et al. [49] a sample of three multiply imaged pairs in a cluster is in principle sufficient to decrease the effect of uncertainties in the cluster mass modeling to sufficiently low level and to provide interesting constraints on Ω_m. This has been recently attempted on the famous lensing cluster Abell 2218 (see Fig. 7), but clearly this can be improved a lot by adding more constrains to the mass model and by measuring the redshift of the other multiple images identified in the core of Abell 2218, thanks to exquisite deep multicolor HST data (the number of multiple images systems is of the order of seven). Thus, it is now reasonable to attempt to use cluster-lenses to constrain not only the mass distribution but also the cosmological parameters Ω_m and Ω_Λ, as numerical simulations suggest [45] (see Fig. 20).

To reach this goal, we need to improve as much as possible the constraints on the mass distribution of the cluster. This could be done by 3 different approaches: (**1**) determine the redshift of the multiple images not yet known (**2**) measure the radial velocity dispersion profile of key elliptical galaxies, in particular the cD galaxy (**3**) enlarge the number of cluster members and their

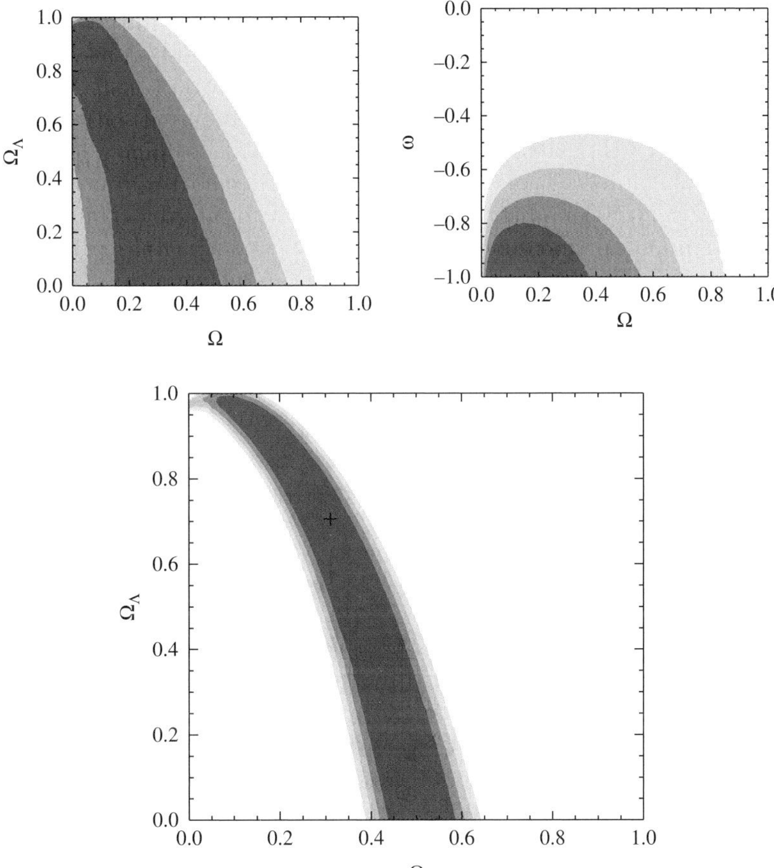

Fig. 20. (**Top Left**) The current cosmological constraints in the $(\Omega_m, \Omega_\Lambda)$ based on the currently known four multiple images with redshifts (at $z = 0.702$, $z = 1.04$, $z = 2.55$ and $z = 5.56$) in A2218 [136]. (**Top Right**) Respective constraints but in the (Ω_m, w) plane assuming a flat Universe. (**Bottom**) Confidence contours of the cosmological constrains expected for a cluster lens similar to Abell 2218 assuming that five multiple images with (different) redshift are known. All contours are 68,3%, 95,5%, 99,7% et 99,99% confidence level

velocity dispersion by extending current redshift surveys in this field (this will allow to check/confirm any substructure in redshift space).

7 Future and Prospects

Since the discovery of giant arcs in the late 1980's, gravitational lensing in cluster of galaxies has become a powerful *cosmological* tool.

- We are now able to reconstruct the mass distribution in clusters in great detail from the galaxy scale to the virial radius. The lensing mass estimates are usefully compared to other mass estimators to provide critical information on the cluster physics (from the largest cluster scales to galaxy scales) on well defined cluster samples. This is allowing us to give a direct proof of the existence of Dark Matter and hopefully will allow us to put constraints on the nature of the dark matter in clusters.
- Wide field surveys of *mass selected* clusters, using lensing techniques, will allow us to make a direct comparison to the analytic/numerical models of the Universe and thus better understand the growth of structure and the large scale distribution of mass. It will also confirm or not the existence of dark lumps of mass, as well as determine how massive are the filaments between galaxy clusters. Ultimately it could provide a complementary probe of Dark Energy.
- Multiple images in cluster cores are about to measure directly the cosmological parameters through an optical geometrical test of the curvature of the Universe [136]. Although more spectroscopic and mass modeling are needed, it is a very clean method to tackle this problem.
- Likewise, time dependent phenomena, like supernovae or AGNs fluctuations, if observed behind well-known lensing clusters, may prove to be a very accurate way to probe the Hubble constant on cosmological scales, as it has been initiated using multiple quasars. However, the likelihood of having multiple images of such transient phenomena is weak. The recently found SDSS large separation quadruple quasar will certainly be one such case to study in detail.
- Finally, massive clusters will always be the *unique place* to look at in order to boost telescope and instrument sensitivities at *all wavelength*, to push ahead the discoveries, to reach the faintest detection levels and explore in detail the morphology of distant galaxies.

The understanding of cluster lenses has greatly improved in the last 20 years, and will continue to progress with the current and future planned instruments.

Acknowledgments

This work is heavily based on results obtained since I started my PhD on cluster lenses. A large number of people have been involved on different aspects of my research on cluster lenses, and I would like to thank them all for the fruitful work and discussions that have promoted a better understanding of our Universe, using cluster lenses as a powerful cosmological tool. I acknowledge support from CNRS.

References

1. Abell, G.O.: ApJS **3**, 211 (1958)
2. Abdelsalam, H.M., Saha, P., Williams, L.L.R.: MNRAS **294**, 734 (1998)
3. Allen, S.W., Fabian, A.C., Kneib, J.P.: MNRAS **279**, 615 (1996)
4. Altieri, B., et al.: A&A **343**, L65 (1999)
5. Bacon, D.J., Goldberg, D.M., Rowe, B.T.P., Taylor, A.N. Mon. Not. R. Astron. Soc. **365**, 414–428 (2006)
6. Bardeau, S., Kneib, J.-P., Czoske, O., Soucail, G., Smail, I., Ebeling, H., Smith, G.P.: A&A **434**, 433 (2005)
7. Barnothy, J.M.: AJ **70**, 666 (1965)
8. Bartelmann, M., Narayan, R., Seitz, S., Schneider, P.: ApJ **464**, L115 (1996)
9. Bernstein, G.M., Jarvis, M.: AJ **123**, 583 (2002)
10. Bézecourt, J., Soucail, G., Ellis, R.S., Kneib, J.-P.: A&A **351**, 433 (1999)
11. Blain, A.W., Smail, I., Ivison, R.J., Kneib, J.-P., Frayer, D.T.: Phys. Rep. **369**, 111 (2002)
12. Biviano, A.: From Messier to Abell: 200 Years of Science with Galaxy Clusters. Constructing the Universe with Clusters of Galaxies (2000)
13. Borys, C., et al.: MNRAS **352**, 759 (2004)
14. Brada, M., et al.: A&A **437**, 49 (2005)
15. Bridle, S.L., Hobson, M.P., Lasenby, A.N., Saunders, R.: MNRAS **299**, 895 (1998)
16. Bridle, S., Kneib, J.-P., Bardeau, S., Gull, S.: The shapes of galaxies and their dark haloes. Proceedings of the Yale Cosmology Workshop "The Shapes of Galaxies and Their Dark Matter Haloes", New Haven, Connecticut, USA, 28–30 May 2001. Edited by Priyamvada Natarajan. Singapore: World Scientific, 2002, ISBN 9810248482, p.38, 38 (2002)
17. Broadhurst, T., et al.: ApJ **621**, 53 (2005)
18. Broadhurst, T., Takada, M., Umetsu, K., Kong, X., Arimoto, N., Chiba, M., Futamase, T.: ApJ **619**, L143 (2005)
19. Burke, W.L.: ApJ **244**, L1 (1981)
20. Campusano, L.E., Pelló, R., Kneib, J.-P., Le Borgne, J.-F., Fort, B., Ellis, R., Mellier, Y., Smail, I.: A&A **378**, 394 (2001)
21. Carlstrom, J.E., Holder, G.P., Reese, E.D.: ARAA **40**, 643 (2002)
22. Clowe, D., Luppino, G.A., Kaiser, N., Gioia, I.M.: ApJ **539**, 540 (2000)
23. Clowe, D., Schneider, P.: A&A **379**, 384 (2001)
24. Clowe, D., Schneider, P.: A&A **395**, 385 (2002)
25. Clowe, D., Gonzalez, A., Markevitch, M.: ApJ **604**, 596 (2004)
26. Covone, G., Kneib, J.-P., Soucail, G., Richard, J., Jullo, E., Ebeling, H.: A&A, astro-ph/0511332 (2006)
27. Cypriano, E.S., Sodré, L.J., Kneib, J.-P., Campusano, L.E.: ApJ **613**, 95 (2004)

28. Czoske, O., Kneib, J.-P., Soucail, G., Bridges, T.J., Mellier, Y., Cuillandre, J.-C.: A&A **372**, 391 (2001)
29. Czoske, O., Moore, B., Kneib, J.-P., Soucail, G.: A&A **386**, 31 (2002)
30. Dahle, H., Kaiser, N., Irgens, R.J., Lilje, P.B., Maddox, S.J.: ApJS **139**, 313 (2002)
31. Diego, J.M., Sandvik, H.B., Protopapas, P., Tegmark, M., Ben'ıtez, N., Broadhurst, T.: MNRAS **362**, 1247 (2005)
32. Ebeling, H., Voges, W., Bohringer, H., Edge, A.C., Huchra, J.P., Briel, U.G.: MNRAS **281**, 799 (1996)
33. Ebeling, H., Edge, A.C., Henry, J.P.: ApJ **553**, 668 (2001)
34. Egami, E., et al.: ApJ **618**, L5 (2005)
35. Ebbels, T., Ellis, R., Kneib, J.-P., Leborgne, J.-F., Pello, R., Smail, I., Sanahuja, B.: MNRAS **295**, 75 (1998)
36. Ellis, R., Santos, M.R., Kneib, J.-P., Kuijken, K.: ApJ **560**, L119–L122 (2001)
37. Fort, B., Le Fevre, O., Hammer, F., Cailloux, M.: ApJ **399**, L125 (1992)
38. Franx, M., Illingworth, G.D., Kelson, D.D., van Dokkum, P.G., Tran, K.-V.: ApJ **486**, L75 (1997)
39. Frayer, D.T., Ivison, R.J., Scoville, N.Z., Yun, M., Evans, A.S., Smail, I., Blain, A.W., Kneib, J.-P.: ApJ **506**, L7 (1998)
40. Frayer, D.T., et al.: ApJ **514**, L13 (1999)
41. Gavazzi, R., Fort, B., Mellier, Y., Pelló, R., Dantel-Fort, M.: A&A **403**, 11 (2003)
42. Geiger, B., Schneider, P.: MNRAS **295**, 497 (1998)
43. Geiger, B., Schneider, P.: MNRAS **302**, 118 (1999)
44. Gladders, M.D., Yee, H.K.C., Ellingson, E.: AJ **123**, 1 (2002)
45. Gilmore, J., Natarajan, P.: ApJ, astro-ph/0605245 (2006)
46. Gnedin, O.Y., Kravtsov, A.V., Klypin, A.A., Nagai, D.: ApJ **616**, 16 (2004)
47. Goldberg, D.M., Bacon, D.J.: ApJ **619**, 741–748 (2005)
48. Goldberg, D.M., Natarajan, P.: ApJ **564**, 65–72 (2002)
49. Golse, G., Kneib, J.-P., Soucail, G.: A&A **387**, 788 (2002)
50. Heymans, C., et al.: MNRAS **368**, 1323 (2006)
51. Hoekstra, H., Franx, M., Kuijken, K.: ApJ **532**, 88 (2000)
52. Hoekstra, H., Yee, H.K.C., Gladders, M.D., Barrientos, L.F., Hall, P.B., Infante, L.: ApJ **572**, 55 (2002)
53. Hu, E.M., Cowie, L.L., McMahon, R.G., Capak, P., Iwamuro, F., Kneib, J.-P., Maihara, T., Motohara, K.: ApJ **568**, L75 (2002)
54. Irwin, J., Shmakova, M.: ApJ **645**, 17–43 (2006)
55. Joffre, M., et al.: ApJ **534**, L131 (2000)
56. Kaiser, N., Squires, G.: ApJ **404**, 441 (1993)
57. Kaiser, N.: ApJ **439**, L1 (1995)
58. Kaiser, N., Squires, G., Broadhurst, T.: ApJ **449**, 460 (1995)
59. Kaiser, N.: ApJ **537**, 555 (2000)
60. Kassiola, A., Kovner, I., Fort, B.: ApJ **400**, 41 (1992)
61. King, L.J., Schneider, P.: A&A **369**, 1 (2001)
62. Kneib, J.-P.: Ph.D. Thesis (1993)
63. Kneib, J.P., Mellier, Y., Fort, B., Mathez, G.: A&A **273**, 367 (1993)
64. Kneib, J.-P., Mathez, G., Fort, B., Mellier, Y., Soucail, G., Longaretti, P.-Y.: A&A **286**, 701–717 (1994)
65. Kneib, J.P., Melnick, J., Gopal-Krishna.: A&A **290**, L25–L28 (1994)

66. Kneib, J.-P., Ellis, R.S., Smail, I., Couch, W.J., Sharples, R.M.: ApJ **471**, 643 (1996)
67. Kneib, J.-P., Hudelot, P., Ellis, R.S., Treu, T., Smith, G.P., Marshall, P., Czoske, O., Smail, I., Natarajan, P.: ApJ **598**, 804–817 (2003)
68. Kneib, J.-P., Ellis, R.S., Santos, M.R., Richard, J.: ApJ **607**, 697–703 (2004)
69. Kneib, J.-P., van der Werf, P.P., Kraiberg Knudsen, K., Smail, I., Blain, A., Frayer, D., Barnard, V., Ivison, R.: Mon. Not. R. Astron. Soc. **349**, 1211–1217 (2004)
70. Kneib, J.-P., Neri, R., Smail, I., Blain, A., Sheth, K., van derWerf, P., Knudsen, K.K.: A&A **434**, 819–825 (2005)
71. Knudsen, K.K., et al.: MNRAS **368**, 487 (2006)
72. Kochanek, C.S., Apostolakis, J.: MNRAS **235**, 1073 (1988)
73. Kochanek, C.S.: MNRAS **247**, 135 (1990)
74. Kochanek, C.S.: ApJ **445**, 559 (1995)
75. Kodama, T., Smail, I., Nakata, F., Okamura, S., Bower, R.G.: ApJ **562**, L9 (2001)
76. Kuijken, K.: A&A **352**, 355 (1999)
77. Limousin, M., Kneib, J.-P., Natarajan, P., MNRAS, **356**, 309 (2005)
78. Link, R., Pierce, M.J.: ApJ **502**, 63 (1998)
79. Lombardi, M., Bertin, G.: A&A **330**, 791 (1998)
80. Luppino, G.A., Kaiser, N.: ApJ **475**, 20 (1997)
81. Lynds, R., Petrosian, V.: Bull. Am. Astron. Soc. **18**, 1014 (1986)
82. Marshall, P.J., Hobson, M.P., Gull, S.F., Bridle, S.L.: MNRAS **335**, 1037 (2002)
83. Marshall, P.J., Hobson, M.P., Slosar, A.: MNRAS **346**, 489 (2003)
84. Marshall, P.J.: astro-ph/0511287 (2006)
85. Massey, R., et al.: AJ **127**, 3089 (2004)
86. Massey, R., Refregier, A.: MNRAS **363**, 197 (2005)
87. Mellier, Y., Fort, B., Kneib, J.-P.: ApJ **407**, 33 (1993)
88. Merritt, D., Navarro, J.F., Ludlow, A., Jenkins, A.: ApJ **624**, L85 (2005)
89. Metcalfe, L., et al.: A&A **407**, 791 (2003)
90. Miralda-Escude, J.: ApJ **380**, 1 (1991)
91. Miralda-Escude, J., Babul, A.: ApJ **449**, 18 (1995)
92. Miyazaki, S., et al.: ApJ **580**, L97 (2002)
93. Moller, O., Blain, A.W.: MNRAS **299**, 845 (1998)
94. Moore, B., Governato, F., Quinn, T., Stadel, J., Lake, G.: ApJ **499**, L5 (1998)
95. Narayan, R., Blandford, R., Nityananda, R.: Nature **310**, 112 (1984)
96. Natarajan, P., Kneib, J.-P.: Mon. Not. R. Astron. Soc. **283**, 1031–1046 (1996)
97. Natarajan, P., Kneib, J.-P.: Mon. Not. R. Astron. Soc. **287**, 833–847 (1997)
98. Natarajan, P., Kneib, J.-P., Smail, I., Ellis, R.S.: ApJ **499**, 600 (1998)
99. Natarajan, P., Kneib, J.-P., Smail, I.: ApJ **580**, L11–L15 (2002)
100. Natarajan, P., Loeb, A., Kneib, J.-P., Smail, I.: ApJ **580**, L17–L20 (2002)
101. Navarro, J.F., Frenk, C.S., White, S.D.M.: ApJ **462**, 563 (1996)
102. Neri, R., et al.: ApJ **597**, L113 (2003)
103. Paczynski, B.: Nature **325**, 572 (1987)
104. Pelló, R., et al.: A&A **346**, 359 (1999)
105. Pelló, R., Schaerer, D., Richard, J., Le Borgne, J.-F., Kneib, J.-P.: A&A **416**, L35 (2004)
106. Pettini, M., Steidel, C.C., Adelberger, K.L., Dickinson, M., Giavalisco, M.: ApJ **528**, 96 (2000)

107. Pettini, M., Rix, S.A., Steidel, C.C., Adelberger, K.L., Hunt, M.P., Shapley, A.E.: ApJ **569**, 742 (2002)
108. Pierre, M., Le Borgne, J.F., Soucail, G., Kneib, J.P.: A&A **311**, 413 (1996)
109. Refregier, A.: MNRAS **338**, 35 (2003)
110. Refregier, A., Bacon, D.: MNRAS **338**, 48 (2003)
111. Refsdal, S.: MNRAS **128**, 295 (1964)
112. Refsdal, S.: MNRAS **128**, 307 (1964)
113. Rhodes, J., Refregier, A., Groth, E.J.: ApJ **536**, 79 (2000)
114. Rhodes, J., et al.: Astropart. Phys. **20**, 377 (2004)
115. Richard, J., et al.: A&A, astro-ph/0606134 (2006)
116. Rix, H.-W., de Zeeuw, P.T., Cretton, N., van der Marel, R.P., Carollo, C.M.: ApJ **488**, 702 (1997)
117. Schneider, P.: MNRAS **283**, 837 (1996)
118. Schneider, P., Ehlers, J., Falco, E.E., "Gravitational Lenses", **560**, p. 112, Springer-Verlag (1992)
119. Schneider, P., King, L., Erben, T.: A&A **353**, 41 (2000)
120. Sand, D.J., Treu, T., Ellis, R.S.: ApJ **574**, L129 (2002)
121. Sand, D.J., Treu, T., Smith, G.P., Ellis, R.S.: ApJ **604**, 88 (2004)
122. Santos, M.R., Ellis, R.S., Kneib, J.-P., Richard, J., Kuijken, K.: ApJ **606**, 683–701 (2004)
123. Seitz, C., Schneider, P.: A&A **297**, 287 (1995)
124. Seitz, S., Schneider, P.: A&A **305**, 383 (1996)
125. Seitz, C., Kneib, J.-P., Schneider, P., Seitz, S.: A&A **314**, 707 (1996)
126. Seitz, C., Schneider, P.: A&A **318**, 687 (1997)
127. Seitz, S., Schneider, P., Bartelmann, M.: A&A **337**, 325 (1998)
128. Seitz, S., Schneider, P.: A&A **374**, 740 (2001)
129. Smith, G.P., Kneib, J.-P., Ebeling, H., Czoske, O., Smail, I.: ApJ **552**, 493 (2001)
130. Smith, G.P., et al.: MNRAS **330**, 1 (2002)
131. Smith, G.P., Smail, I., Kneib, J.-P., Davis, C.J., Takamiya, M., Ebeling, H., Czoske, O.: MNRAS **333**, L16 (2002)
132. Smith, G.P., Edge, A.C., Eke, V.R., Nichol, R.C., Smail, I., Kneib, J.-P.: ApJ **590**, L79 (2003)
133. Smith, G.P., Kneib, J.-P., Smail, I., Mazzotta, P., Ebeling, H., Czoske, O.: MNRAS **359**, 417 (2005)
134. Soucail, G., Fort, B., Mellier, Y., Picat, J.P.: A&A **172**, L14 (1987)
135. Soucail, G., Mellier, Y., Fort, B., Mathez, G., Cailloux, M.: A&A **191**, L19 (1988)
136. Soucail, G., Kneib, J.-P., Golse, G.: A&A **417**, L33 (2004)
137. Smail, I., Ivison, R.J., Blain, A.W., Kneib, J.-P.: ApJ **507**, L21–L24 (1998)
138. Smail, I., Ivison, R.J., Owen, F.N., Blain, A.W., Kneib, J.-P.: ApJ **528**, 612–616 (2000)
139. Smail, I., Ivison, R.J., Blain, A.W., Kneib, J.-P.: Mon. Not. R. Astron. Soc. **331**, 495–520 (2002)
140. Treu, T., Ellis, R.S., Kneib, J.-P., Dressler, A., Smail, I., Czoske, O., Oemler, A., Natarajan, P.: ApJ **591**, 53 (2003)
141. van Albada, T.S., Bahcall, J.N., Begeman, K., Sancisi, R.: ApJ **295**, 305 (1985)
142. Van Waerbeke, L., et al.: A&A **374**, 757 (2001)
143. Walsh, D., Carswell, R.F., Weymann, R.J.: Nature **279**, 381 (1979)

144. Wittman, D., Tyson, J.A., Margoniner, V.E., Cohen, J.G., Dell'Antonio, I.P.: ApJ **557**, L89–L92 (2001)
145. Wittman, D., Dell'Antonio, I.P., Hughes, J.P., Margoniner, V.E., Tyson, J.A., Cohen, J.G., Norman, D.: ApJ **643**, 128 (2006)
146. Yee, H.K.C., Ellingson, E., Bechtold, J., Carlberg, R.G., Cuillandre, J.-C.: AJ **111**, 1783 (1996)
147. Zwicky, F.: Helv. Phys. Acta **6**, 110–127 (1933)
148. Zwicky, F.: ApJ **86**, 217 (1937)

The Sunyaev–Zel'dovich Effect in Cosmology and Cluster Physics

M. Birkinshaw and K. Lancaster

Department of Physics, University of Bristol, Tyndall Avenue, Bristol BS8 1TL, UK
Mark.Birkinshaw@Bristol.ac.uk and Katy.Lancaster@Bristol.ac.uk

1 Introduction

Rich clusters of galaxies contain extensive atmospheres of hot gas. Emission from this gas in X-rays was one of the major discoveries of the first generation of X-ray telescopes. High-quality images and spectra of the X-rays from clusters are now available from the *Chandra* and *XMM-Newton* satellites.

A second method of imaging these atmospheres is provided by the Sunyaev-Zel'dovich (SZ) effect [35]. As the cosmic microwave background (CMB) radiation propagates through a cluster of galaxies towards us, photons have a small probability of being inverse-Compton scattered by electrons in the cluster gas. Since the microwave background radiation has a temperature of about 2.7 K, while the gas in a cluster of galaxies may have a temperature as high as 10^8 K, scatterings tend to increase the photon energies, so causing a change in the microwave background radiation intensity and spectrum towards the cluster.

The power of the SZ effect comes about because the effect is caused by scattering, rather than emission, and so scales with the density of the scattering electrons. A cluster of galaxies may therefore appear quite different in its X-ray and SZ effect structures, and a comparison of those structures can provide interesting information on the physics of clusters and their atmospheres.

This introduction to the use of the SZ effect describes the underlying physics of the thermal, kinematic, and other aspects of the effect (Sect. 2), the techniques used for observing the small signal produced by the effect (Sect. 3), the information on clusters of galaxies that can be obtained from the data (Sect. 4), and the cosmological information available by such work (Sect. 5), and then describes some of the coming generation of instruments designed for detailed work on the SZ effect, to take us beyond the exploratory phase of SZ effect science. More details about the effect can be found in recent reviews by Rephaeli [33], Birkinshaw [4], and Carlstrom et al. [9], but

M. Birkinshaw and K. Lancaster: *The Sunyaev–Zel'dovich Effect in Cosmology and Cluster Physics*, Lect. Notes Phys. **740**, 255–285 (2008)
DOI 10.1007/978-1-4020-6941-3_8

progress on detecting and using SZ effects is currently rapid, and even the present introduction will be out of date soon.

2 The Physics of the Sunyaev-Zel'dovich Effect

2.1 Inverse-Compton Scatterings

Inverse-Compton scattering is a process in which energetic electrons give up energy to photons. In the electron rest frame it is a simple exercise to show that an electron/photon scattering in which a photon is deflected by angle θ_{12} (in the rest frame of the electron before the encounter) changes the photon energy from ϵ to ϵ', where

$$\frac{\epsilon'}{\epsilon} = \left(1 + \frac{\epsilon}{m_{\rm e}c^2}\,(1 - \cos\theta_{12})\right)^{-1} . \tag{1}$$

If the electrons scatter photons of the microwave background radiation, which have a characteristic energy of less than an MeV, clearly $\epsilon \ll m_{\rm e}c^2$, and the scattering is almost elastic in the electron rest frame.

In this (Thomson) limit, the photon energy as seen by the observer changes only because of the change of photon direction in the scattering. If the direction cosines of the motion of the photon before and after collision, in the rest frame of the electron, are μ and μ', then the photon energy changes from ϵ to ϵ'' where

$$\frac{\epsilon''}{\epsilon} = e^s = \frac{1 + \beta\mu'}{1 - \beta\mu} \tag{2}$$

in the observer's frame. Here $\beta = v_{\rm e}/c$ is the dimensionless speed of the electron, which is typically ~ 0.1 for the electrons in a cluster of galaxies with gas temperature ~ 5 keV. Equation (2) defines the logarithmic energy shift factor, s, which is often useful in performing integrations.

The probability that an inverse-Compton scattering causes a shift s is

$$P(s;\beta) = \int p(\mu)\,{\rm d}\mu\,\phi(\mu';\mu)\left(\frac{{\rm d}\mu'}{{\rm d}s}\right) , \tag{3}$$

where $\phi(\mu';\mu)$ is the function

$$\phi(\mu';\mu) = \frac{3}{8}\left(1 + \mu^2\mu'^2 + \frac{1}{2}(1-\mu^2)(1-\mu'^2)\right) \tag{4}$$

and $p(\mu)$ is the probability of a scattering occurring at direction cosine μ,

$$p(\mu)\,{\rm d}\mu = \left(2\gamma^4(1-\beta\mu)^3\right)^{-1}\,{\rm d}\mu . \tag{5}$$

γ is the Lorentz factor of the electron.

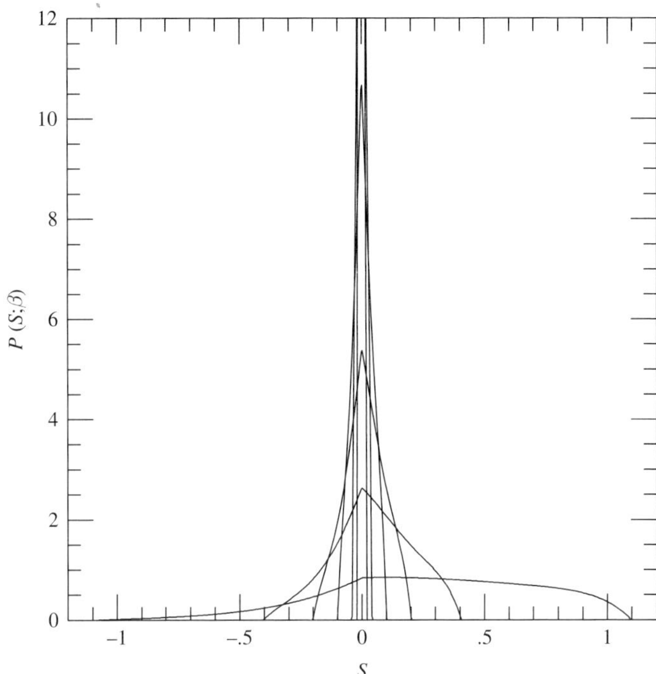

Fig. 1. The function $P(s;\beta)$ for $\beta = 0.01$, 0.02, 0.05, 0.10, 0.20, and 0.50. $P(s;\beta)$ becomes increasingly asymmetric and broader as β increases, with $P(s;\beta) > 0$ for $|s| < \ln((1+\beta)/(1-\beta))$

The function $P(s;\beta)$ is shown in Fig. 1. It is narrow and symmetrical in β at small β, but as β increases it broadens and develops a significant asymmetry, with a preference to positive values of s, corresponding to energy gains by the scattered photons. For 5 keV electrons the asymmetry is already significant.

2.2 Thermal Sunyaev–Zel'dovich Effect

The thermal SZ effect results from the scattering of the microwave background radiation by the thermal gas in a cluster of galaxies. To calculate this effect we need to obtain the probability distribution for s from single scatterings by electrons in gas at temperature T_{e}, $P_1(s)$. We calculate P_1 by integrating the function $P(s;\beta)$ over the distribution function of electron speeds.

$P_1(s)$ is shown for a population of electrons with temperature 5 keV in Fig. 2. At higher temperatures the function becomes more asymmetric, with a stronger tail at positive values of s, as relativistic effects become more important. By a temperature of 15 keV $P_1(s)$ is clearly asymmetric.

The specific intensity of the microwave background, $I_0(\nu)$ at frequency ν, is changed to

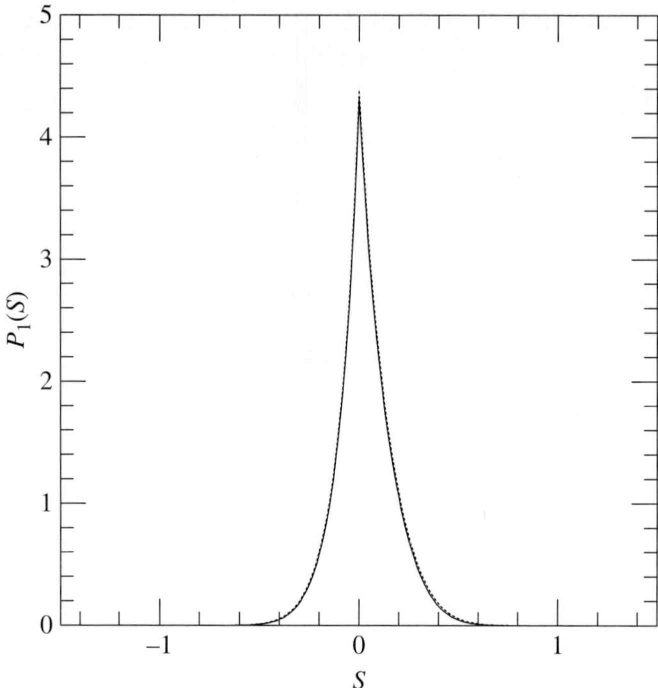

Fig. 2. The function $P_1(s)$ for a population of electrons with temperature 5 keV. The function is not symmetrical about $s = 0$, with a higher positive wing indicating that it is more likely that scatterings will cause the photon energy to increase

$$I(\nu'') = \int_{-\infty}^{\infty} P_1(s)\, I_0(\nu)\, \mathrm{d}s \quad . \tag{6}$$

at frequency $\nu'' = \nu e^s$ if every photon is scattered only once as the radiation propagates towards us through the population of electrons. In most cases we expect a low scattering probability, as expressed by the smallness of the electron scattering optical depth

$$\tau_\mathrm{e} = \int n_\mathrm{e} \sigma_\mathrm{T} \mathrm{d}l \ll 1 \, , \tag{7}$$

where the integral is along the line of sight, n_e is the number of electrons per unit volume along that line of sight, and σ_T is the Thomson cross-section. If $\tau_\mathrm{e} \ll 1$, then the change in the specific intensity of the radiation field, $\Delta I_\mathrm{T}(\nu) = I(\nu) - I_0(\nu)$ is

$$\Delta I_\mathrm{T}(\nu) = \tau_\mathrm{e} \int \mathrm{d}s\, P_1(s)\, \left(I(\nu_0) - I(\nu)\right) \, . \tag{8}$$

At low temperature this integral can be performed analytically, since the scattering kernel, $P_1(s)$ can be approximated as symmetrical and Gaussian. This Kompaneets form,

$$P_{\mathrm{K}}(s) = \frac{1}{\sqrt{4\pi y_{\mathrm{e}}}} \exp\left(-\frac{(s+3y_{\mathrm{e}})^2}{4y_{\mathrm{e}}}\right) , \tag{9}$$

where

$$y_{\mathrm{e}} = \int n_{\mathrm{e}}\, \sigma_{\mathrm{T}}\, \mathrm{d}l\, \frac{k_{\mathrm{B}}\, T_{\mathrm{e}}}{m_{\mathrm{e}}\, c^2} \tag{10}$$

is the Comptonization parameter, is obtained as the solution of a diffusion problem for the photon occupation number as a result of inverse-Compton scatterings, and can be thought of as encoding only the mean and width of the $P_1(s)$ distribution.

For such a thermal electron population and a Planckian microwave background spectrum, with radiation temperature T_{rad}, the single-scattering change in the specific intensity is given by the Kompaneets spectrum for the SZ effect,

$$\Delta I_{\mathrm{T}}(\nu) = \Delta\, I_{\mathrm{T0}} \frac{x^4\, e^x}{(e^x - 1)^2} \left(x \coth\left(\frac{x}{2}\right) - 4\right) , \tag{11}$$

where x is the scaled frequency

$$x = \frac{h\nu}{k_{\mathrm{B}} T_{\mathrm{rad}}} = 0.0176\ (\nu/\mathrm{GHz}) \tag{12}$$

and ΔI_{T0} is the specific intensity scale

$$\Delta I_{\mathrm{T0}} = I_0\, y_{\mathrm{e}} , \tag{13}$$

with I_0 being the specific intensity scale of the CMB itself

$$I_0 = \frac{2h}{c^2} \left(\frac{k_{\mathrm{B}} T_{\mathrm{rad}}}{h}\right)^3 = 2.7 \times 10^{-18}\ \mathrm{W\,Hz^{-1}\,m^{-2}\,sr^{-1}} . \tag{14}$$

At low frequencies, or for moderate cluster temperatures, this spectrum is a close approximation to the relativistically-correct form that can be calculated by numerical integration. However, precise measurements of the SZ effect, and measurements at high frequency, should use the full form (and include the effects of multiple scatterings), since the change in the spectral shape with temperature is surprisingly fast. Figure 3 compares the spectrum of ΔI_{T} for $T_{\mathrm{e}} = 15$ keV as calculated from (11) and as calculated using the full numerical integral. There is a significant difference between the high-frequency shapes of the two spectra at $x \gtrsim 1$ ($\nu \gtrsim 50$ GHz). Approximations for the relativistic corrections to the thermal SZ effect have been given by a number of authors, for example [17], and provide convenient ways of avoiding the full calculation.

For a rich cluster of galaxies, the electron scattering optical depth along a line of sight through the centre of the cluster is typically of order 10^{-2},

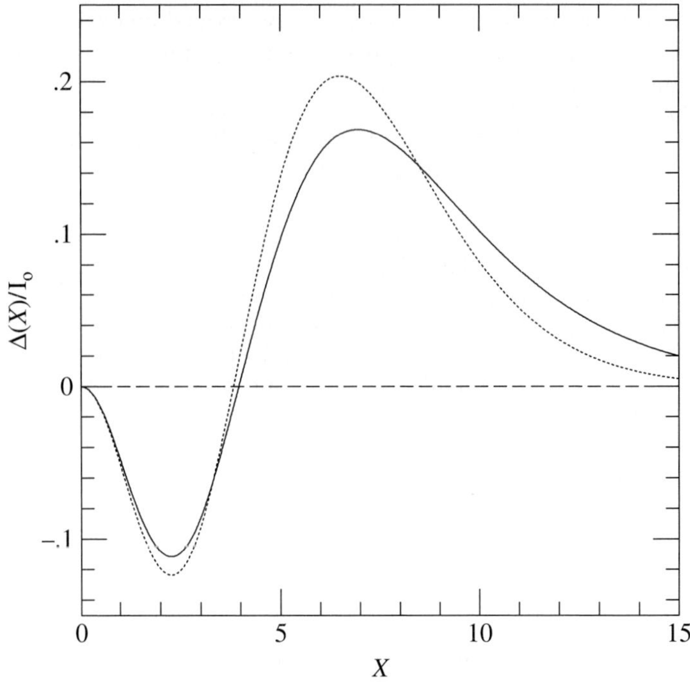

Fig. 3. The spectral distortion of the microwave background spectrum, $\Delta I_{\rm T}/\Delta I_{\rm T0}$, as a function of $x = 0.0176(\nu/{\rm GHz})$ for a population of electrons with temperature 15 keV. The dotted line shows the Kompaneets spectrum, while the solid line shows the spectrum calculated by numerical integration

implying that only about 1% of CMB photons are scattered, and the single-scattering approximation should be good for most purposes. Since the typical electron temperature of a rich cluster is about 5 keV, the corresponding value of the Comptonization parameter, $y_{\rm e} \sim 10^{-4}$, indicates that after passage through the cluster the CMB photons are far from being in equilibrium with the cluster gas.

The shape of $y_{\rm e}$ on the sky gives the angular structure of the SZ effect. For an isothermal cluster, the shape of $y_{\rm e}$ as a function of angle from the cluster centre, θ, is the same as the shape of the $\tau_{\rm e}(\theta)$ function, and is given by the projected electron density. The isothermal β model [10] provides a simple description of the electron density in a cluster atmosphere, and often gives a reasonable model for the shape of the X-ray surface brightness from a cluster. In this model the gas density follows

$$\rho_{\rm g}(r) = \rho_{\rm g0} \left(1 + \frac{r^2}{r_{\rm c}^2}\right)^{-\frac{3}{2}\beta} \tag{15}$$

and the resulting SZ effect has shape

$$\frac{y_{\rm e}}{y_{\rm e0}} = \frac{\tau_{\rm e}}{\tau_{\rm e0}} = \left(1 + \frac{\theta^2}{\theta_{\rm c}^2}\right)^{\frac{1}{2} - \frac{3}{2}\beta} \quad . \tag{16}$$

The value of β for a cluster of galaxies is typically $0.65 - 0.75$, so that the SZ effect of a cluster falls off relatively slowly, approximately as θ^{-1}, at large angles from the cluster centre. This is a much slower decrease with angle than the X-ray profile of a cluster (which falls off approximately as θ^{-3}), and so we expect the SZ effects of clusters of galaxies have larger angular sizes than their X-ray structures.

At large angles from the cluster centre (16) must cease to be valid, or the total (negative) flux density of a cluster would become infinite. We expect the temperature profile and density profile to change so that the simple isothermal beta-model ceases to be valid.

2.3 Kinematic Sunyaev–Zel'dovich Effect

The thermal SZ effect of a cluster of galaxies will be confused by its kinematic SZ effect. The kinematic effect is produced by clusters which are in motion relative to the frame in which the CMB has zero dipole. In the cluster frame, the CMB develops a dipole term proportional to the cluster's speed, and the scattering of this anisotropic radiation field causes it to become slightly more isotropic (by an amount $\propto \tau_{\rm e}\beta$, and hence generates a kinematic SZ effect). Transforming back to the observer's frame, there will be a change in the brightness of the radiation towards the cluster centre. The spectrum of this effect is

$$\Delta I_{\rm K}(\nu) = -\Delta\, I_{\rm K0} \frac{x^4\, e^x}{\left(e^x - 1\right)^2} \tag{17}$$

where x is again the dimensionless frequency and the scale of the effect is

$$\Delta I_{\rm K0} = I_0\, \beta_{\rm z}\, \tau_{\rm e} = \frac{2h}{c^2} \left(\frac{k_{\rm B} T_{\rm rad}}{h}\right)^3 \beta_{\rm z}\, \tau_{\rm e}\ , \tag{18}$$

where $\beta_{\rm z}$ is the component of the cluster velocity along the line of sight. This spectral shape is valid in the same limits as the Kompaneets spectrum: corrections for higher cluster temperatures are given in [30].

At low frequencies the ratio of the kinematic and thermal SZ effects is

$$\frac{\Delta I_{\rm K}(\nu = 0)}{\Delta I_{\rm T}(\nu = 0)} = \frac{1}{2} \frac{\Delta I_{\rm K0}}{\Delta I_{\rm T0}} \approx 0.085 \left(v_{\rm z}/1000\,{\rm km\,s^{-1}}\right) \left(k_{\rm B} T_{\rm e}/10\,{\rm keV}\right)^{-1} \tag{19}$$

which suggests that the kinematic effect will be hard to detect against the more intense thermal effect. Separation of the effects relies on their different spectra (Fig. 4). If the total SZ effect of a cluster can be measured over a wide enough range of frequencies, then spectral decomposition can be used to separate the effects. The thermal effect is zero at roughly (exactly, in the

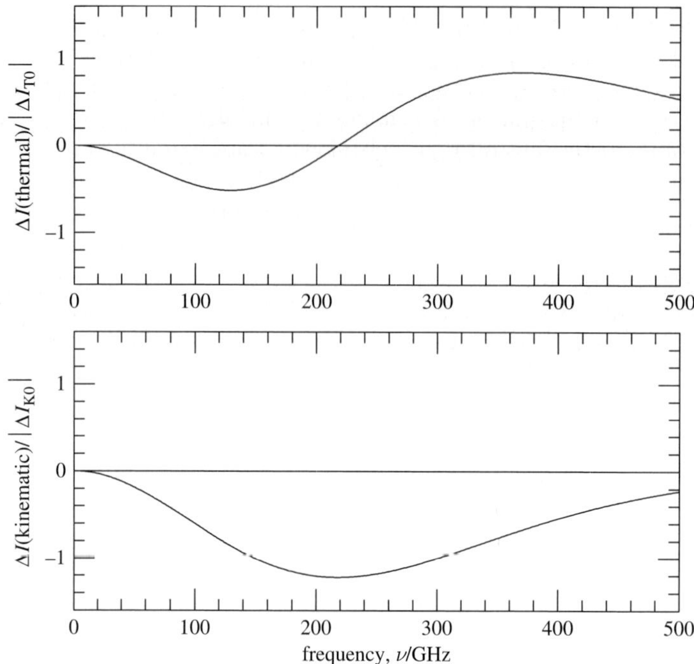

Fig. 4. The spectra of the thermal and kinematic Sunyaev–Zel'dovich effects. The two effects are separable if the spectrum is observed over a wide enough range of frequencies

Kompaneets spectrum) the frequency at which the kinematic effect has maximum amplitude, ~ 220 GHz, so precise measurements near 220 GHz have the potential of measuring the radial components of the velocities of clusters of galaxies. In fact any attempt to measure the velocity of a cluster in this way will suffer from the difficulty of extracting the signal of the kinematic effect in the presence of primordial structure in the CMB, which has the same spectrum and is likely to be of larger amplitude on angular scales of a few arcmin on which cluster SZ effects are significant.

2.4 Non-Thermal Sunyaev–Zel'dovich Effect

Just as scattering from thermal electrons gives an SZ effect, so does scattering from non-thermal electrons. The spectrum of the scattered radiation in this case will differ from that shown in Fig. 4, and will depend on the shape of the electron spectrum. In the limiting case where the electrons are highly relativistic, we might expect inverse-Compton scatterings to shift photons completely out of the radio band and into the X-ray (or beyond). In that case the SZ effect will be proportional to $\tau_{\rm e}$ rather than $y_{\rm e}$. Details of the calculation of the spectrum for a non-thermal electron population are given by [4, 12].

While the non-thermal SZ effect is observable in principle, it is expected to be difficult to detect in practice because the population of non-thermal electrons is likely to be associated with direct non-thermal radiation, such as synchrotron radiation. Nevertheless, measurements of the non-thermal SZ effect may be of interest in studying the energetics of radio sources [27].

2.5 Polarization

Along with the intensity signals, the SZ effects also contain polarization signals. The simplest to understand is the polarization caused by multiple scatterings within the cluster, where one expects a radial pattern of polarization with intensity $\propto \tau_e \Delta I$. Other polarization signals are associated with motion of the cluster across the line of sight (with intensity $\propto \Delta I v_\perp / c$), and from scattering the CMB quadrupole. All these effects are extremely small, inaccessible to the current generation of instruments, and generally badly confused by background structures in the CMB. However, this channel of the SZ effect may become amenable to study with the next generation of instruments. Details of the amplitude and spectrum of these effects may be found in [11].

3 Observing the Sunyaev–Zel'dovich Effect

SZ observations can be undertaken using radiometric detectors, interferometers or bolometers. Each requires specific techniques and has associated with it a particular set of systematic errors.

3.1 Fundamental Considerations in Observation Design

Before embarking upon an SZ observing program, various factors must be considered. Estimation of the angular size and amplitude of SZ signals is imperative. For a beam-switching experiment, the observer must account for any SZ signal present in the "off" beam in order to correctly measure the central decrement. High-resolution interferometers are only sensitive to the extended cluster signal on their shortest baselines: reconfiguration may be an option. Sensitivity requirements apply to all instruments.

An isothermal β-model of the form (15), although not strictly physically correct, can be applied to estimate the extent of Comptonization for a particular cluster. In a rich cluster with a high X-ray luminosity a typical core radius would be ~ 250 kpc, corresponding to an angular core radius θ_c of a few arcmin for a cluster at redshift ~ 0.1. The β parameter for a rich cluster is typically $\approx 2/3$. The central Comptonization parameter $y_{e0} \approx 10^{-4}$. Note that the SZ effect is larger in angular size than the X-ray signal, perhaps by a factor of four in full-width to half-maximum (FWHM).

The fundamental quantity of interest is the flux density, S_ν. This is the specific intensity integrated over the solid angle observed by the telescope.

In the Kompaneets limit the flux density of the thermal effect produced by a cluster is

$$S_{\mathrm{T},\nu} = \int \Delta I_{\mathrm{T}}(\nu)\, \mathrm{d}\Omega = I_0 \frac{x^4 e^{\mathrm{x}}}{(e^{\mathrm{x}}-1)^2} \left(x \coth\frac{x}{2} - 4 \right) \int y_{\mathrm{e}} \mathrm{d}\Omega \,, \tag{20}$$

where x is the scale frequency (12) and I_0 is the specific intensity scale of the CMB (14).

As SZ effects are usually of larger angular size than the FWHM of the telescope beam, a measure of the surface brightness is useful. Interferometer observers commonly adopt the flux density per unit solid angle, Σ_Ω, and single dish observers tend to prefer the brightness temperature, T_{RJ}. The fractional change in the radiation temperature along the line of sight through the cluster centre from the thermal SZ effect, and the corresponding beam-averaged brightness temperature change, are

$$\left(\frac{\Delta T}{T} \right)_{\mathrm{T}} (\nu) = y_{\mathrm{e0}} \left(x \coth\frac{x}{2} - 4 \right) \tag{21}$$

$$\Delta T_{\mathrm{RJ,T}}(\nu) = y_{\mathrm{e0}} T_{\mathrm{rad}} \frac{x^2 e^x}{(e^x-1)^2} \left(x \coth\frac{x}{2} - 4 \right) \,. \tag{22}$$

Both these temperature quantities are redshift independent, but the flux density $S_{\mathrm{T},\nu}$, and practical measures of the temperature quantities are not, because of the introduction of a distance scale associated with the ratio of the linear size of the cluster ($r_{\mathrm{c}} = D_{\mathrm{A}}\theta_{\mathrm{c}}$, where D_{A} is the angular diameter distance of the cluster) and the angular size of the telescope beam.

The flux density per unit solid angle and the brightness temperature are related by

$$\Sigma_\Omega = \frac{2k_{\mathrm{B}}}{\lambda^2} T_{\mathrm{RJ},\nu} \,, \tag{23}$$

where Σ_Ω and T_{RJ} are functions of both frequency and position.

Figure 3 shows the specific intensity spectra for the thermal and kinematic SZ effects. The thermal effect is zero around 220 GHz, which leaves a useful window for observing the kinematic effect alone, but both the thermal and kinematic effects may be confused by primordial structure in the CMB.

Sensitivity Requirements and Limits

In order to determine the feasibility of observing a particular cluster, some assessment of the expected SZ signal is required. The richest clusters of galaxies typically have $y_{\mathrm{e0}} \approx 10^{-4}$. So for exploration of the thermal SZ effect, sensitivity in $\Delta T/T$ units of around 10^{-5} is useful. Thus the minimum required sensitivity must be a fraction of

$$\Delta I_{\mathrm{T0}} = I_0\, y_{\mathrm{e0}} \approx 2\ \mathrm{mJy\,arcmin^{-2}} \,, \tag{24}$$

which is a rough estimate of the amplitude of the thermal effect at the negative and positive peaks of the spectrum (Fig. 3). A telescope with a 1 arcmin beam (i.e. solid angle ≈ 0.5 arcmin^2) will therefore need a sensitivity of 0.2 mJy or better if it is to detect a rich cluster's thermal SZ effect with high significance. This is more than feasible at cm and mm wavelengths, although still challenging in the sub-mm. If work further down the cluster luminosity function is of interest, then a sensitivity of 20 μJy might be regarded as a typical requirement.

In terms of brightness temperature, in the Rayleigh–Jeans region, a sensitivity $y_{e0} \approx 10^{-5}$ corresponds to

$$\Delta T_{\mathrm{RJ,T0}} = -2 y_{e0} T_{\mathrm{rad}} = -55\ \mu\mathrm{K}\ . \tag{25}$$

The brightness temperature signal is smaller at higher frequencies, although work in the secondary peak at about 350 GHz (Fig. 3) may be feasible.

In practice, the sensitivities in (24) and (25) are barely adequate for observing the thermal SZ effect. Extra sensitivity is always useful in the light of systematic problems with data, and also for probing quantities such as the cluster luminosity function. To detect the kinematic effect a factor of ten further improvement in sensitivity is required. Since spectral techniques must be used to separate the thermal and kinematic effects, this sensitivity must be available in several bands that cover a wide frequency range.

3.2 Basic Observation Types

Radiometers

Single-dish systems may have a receiver mounted at either primary or secondary focus, depending on the size of the array. The simplest case would be to observe with a single beam, however the data obtained are then likely to be severely contaminated by atmospheric variations, ground spillover, and other parasitic signals.

The sensitivity of a radiometer system is

$$\Delta T_{\mathrm{A}} = \frac{T_{\mathrm{sys}}}{\sqrt{2\,\Delta\nu\,\tau}}\ , \tag{26}$$

where T_{sys} is the system noise temperature, $\Delta\nu$ is the bandwidth of the receiver, and τ is the integration time used. For systems operating at $\sim$30 GHz, values of $T_{\mathrm{sys}} \sim 30$ K and $\Delta\nu \sim 1$ GHz are readily obtained, so that the antenna temperature noise after about 1 hour of integration should be $\Delta T_{\mathrm{A}} \sim 11\ \mu$K, and the corresponding sky noise (for an antenna with efficiency 0.6) would be $\Delta T_{\mathrm{sky}} \sim 19\ \mu$K. In reality, the noise does not decrease as $\tau^{-1/2}$ because of the varying ground and atmosphere contributions, and because the receiver noise cannot be made "white" over such a long period.

In order to overcome these problems, it has been found useful to combine techniques of *position switching* and *beam switching*. We discuss here only a simple switching strategy based on a two-beam system, but the concept can be extended to more complicated switching patterns and arrays of beams, as described in [7], for example. Position switching involves physically moving the telescope between the target and a reference background region every few seconds. The main limitation of using this technique alone is that the switching occurs at a rate far slower than that at which observing conditions may be changing. Beam switching involves using two beams, provided by two separate feeds, one to observe the target (beam A) and one to observe the reference background (beam B). The resulting measurement of the differential power, $\Delta P_{\mathrm{AB}} = P_{\mathrm{A}} - P_{\mathrm{B}}$ is averaged over the desired integration time. The receiver system can switch between the two beams on millisecond time-scales, which is sufficiently fast to freeze out atmosphere, ground, and receiver fluctuations. The remaining problem is that the two regions are being observed with different feeds, which may have systematically different responses. Combining the two switching strategies improves things further. Now an observing cycle of duration t_{cy} is broken into three segments

1. beam A is off target, and beam B is pointed at the target with the difference signal ΔP_{AB} integrated over time $1/4 t_{\mathrm{int}}$ ($s_1 = \int \Delta P_{\mathrm{AB}}\, \mathrm{d}t$);
2. beam A is pointed at the target, and beam B is off target with the difference signal integrated over $1/2 t_{\mathrm{int}}$ (s_2); and
3. beam A is off target, and beam B is pointed at the target with the difference signal integrated over $1/4 t_{\mathrm{int}}$ (s_3)

and then the best estimate of the sky brightness difference between the target and the average brightness of two reference regions offset to either side of the target is proportional to

$$s = s_2 - s_1 - s_3 \ . \tag{27}$$

This symmetrical switching pattern is relatively efficient at reducing noise from parasitic signals and changing receiver characteristics, since it takes out linear drifts in the behaviour of the system. Typically the integration time, t_{int}, is (80–90)% of the time taken for the complete observing cycle, t_{cy}, with the lost time being taken up by moving the telescope. It is, however, still necessary to design the equipment to reduce non-ideal behaviour as far as possible.

Since the observation of an SZ effect may take a number of hours, spread over a number of days, the positions of the reference beams rotate on the sky about the target to populate *reference arcs* which may extend into a full circle for a circumpolar source. As with all astronomical measurements, SZ observations are subject to contamination by foreground and background sources. The problem is exaggerated for switching strategies, but fortunately switching in azimuth provides helpful modulation according to the parallactic angle, p. Data contaminated by sources located in the reference arcs can thus be filtered and removed, but this is not possible for clusters where significant radio sources lie near the cluster centre — in such cases the source flux densities must be subtracted from the measurements, and the observation should be designed to avoid the effects of the sources as much as possible, by pointing away from them, or by choosing a frequency at which their effect is reduced.

By adopting the such strategies, single-dish observations can potentially provide rapid detections of SZ effects, particularly if the telescope is equipped with an array receiver, but the technique faces three generic problems.

1. *Cluster selection.* Switching techniques introduce a selection effect because they limit the range of angular sizes that can be observed efficiently. Small SZ effects may only fill a fraction of the beam, and so produce a relatively small signal. Larger objects may fill both the target and reference beams giving a small detectable signal after differencing. Between these limits there is an optimal range of angular sizes (and hence redshifts) for a particular system.
2. *Calibration.* The brightness scale must be well-calibrated if the absolute value of the SZ effect is of interest. This is difficult, as few bright radio sources have well-known flux densities and many of the contenders are variable or have polarisation issues. In addition, the bandpass of the receiver must be well-known for spectral studies, the gain of the telescope may change with elevation, and the beam-shape must be well-known (including any variations) for correct interpretation of the results. Finally, real-time calibration is necessary to account for variations in the opacity of the atmosphere.
3. *Confusion.* As with all astronomical observations, SZ effects are liable to be confused with other foreground or background structures. Primordial anisotropies in the CMB may be problematic. If sensitive data at two or more frequencies are available, this contamination can be removed as it has a different spectrum from thermal SZ effects. The kinematic effect has the same spectrum as the CMB, so will also be removed by this method. Foreground radio sources make an important contribution to the confusion level. A large fraction of these objects will be steep-spectrum and thus less important at high frequencies, although a significant fraction of the remaining sources will then be variable. Some improvement can be made by detecting potentially confusing sources using a high-resolution interferometer map, and then subtracting their contribution from the SZ data.

Many reliable SZ measurements have been obtained using single-dish systems, with random measurement errors $<100\ \mu$K, and only low-level residual systematic errors (for example from radio source confusion). Some example results (from [4]) are shown in Fig. 5. It can be seen that good measurements of the amplitudes and angular sizes of the SZ effects of X-ray bright clusters can be made using the beam- and position-switching.

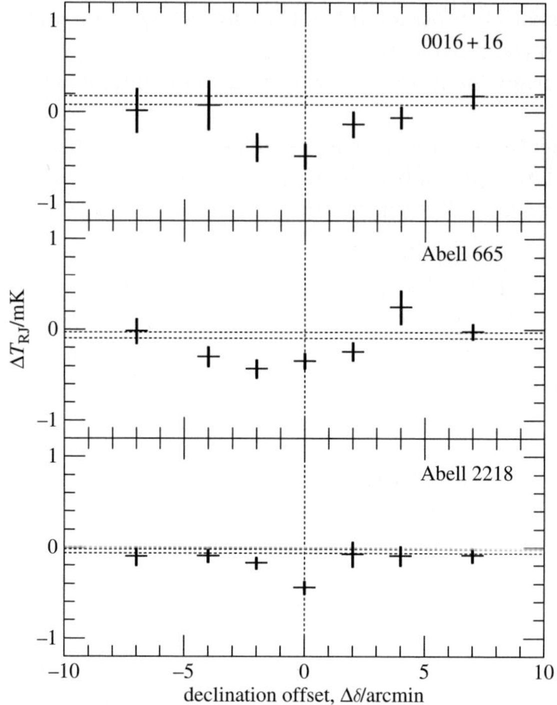

Fig. 5. Measurements of changes in the apparent brightness temperature of the CMB as a function of declination in the three X-ray bright clusters CL 0016+16, Abell 665, and Abell 2218. The largest SZ effects are seen near the centres of the clusters, and the angular sizes of the effects are consistent with predictions based on the X-ray images. The horizontal lines mark the range of possible systematic errors in the zero levels on the data, and the error bars contain both random and systematic components. The brightness temperature scale is subject to a 5% systematic error

Interferometers

Interferometers offer a natural improvement to single-dish radiometers by virtue of their ability to control various sources of systematic error. Decreasing sensitivity away from the pointing centre means that contaminating signals such as ground spillover and terrestrial interference will be attenuated. Signals from the Sun, Moon and planets can be filtered due to different modulating fringe patterns. Also, instruments with a wide range of baselines allow *simultaneous* observations of foreground radio sources whose contribution can subsequently be separated from the SZ signal, automatically taking account of possible source variability.

An interferometer measures the product of voltages between pairs of antennas. A simple case is presented in Fig. 6. Two antennas of area a are separated by a distance b and observing a source at an angle θ. The energy

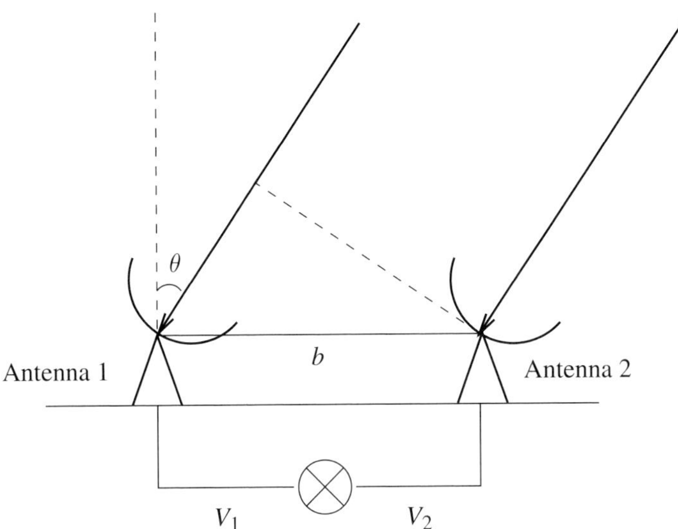

Fig. 6. A simple one-dimensional interferometer. Radiation from the source must travel an extra distance $b \sin\theta$ to reach antenna 1

received per unit area from the sources is S, giving an output

$$\mathcal{A} \propto aS \cos\left(2\pi \frac{b}{\lambda} \sin\theta\right) \quad . \tag{28}$$

Phases are usually not measured absolutely, but relative to some reference direction, θ_0. For a source offset by a small angle $\Delta\theta$ from θ_0, we have $\theta = \theta_0 + \Delta\theta$ and (28) becomes

$$\mathcal{A} \propto aS \cos\left(2\pi \frac{b}{\lambda} \Delta\theta \cos\theta_0\right) . \tag{29}$$

The correlated output differs at different antenna separations, so that the angular resolution of this simple interferometer is proportional to λ/b. A more complicated multi-baseline instrument is sensitive to a range of scales determined by the set of baseline lengths defined by the antenna locations. The shortest baseline defines the maximum scale which can be sampled. Sky structures on larger angular scales will not modulate $\mathcal{A}$ with θ_0 (and hence with time), and so will not produce a detected signal.

The interferometer response can be expressed more generally — see [37] for a full treatment. We can write the baseline as a vector (u, v, w), where w is towards the source and u and v point East and North respectively as seen from the source position. The position of the source on the sky is usually described in terms of co-ordinates (l, m, n). The response becomes

$$\mathcal{A} \propto \int \mathrm{d}l \int \mathrm{d}m \, a(l,m) \, I(l,m) \, \frac{e^{-2\pi i(ul+vm+w(n-1))}}{n} \, , \tag{30}$$

where $a(l, m)$ is the effective total area of the antennas in the direction (l, m) and $I(l, m)$ is the brightness distribution on the sky. $n = (1 - l^2 - m^2)^{1/2} \approx 1$ for small angles, simplifying the Fourier inversion required in (30) to produce a sky map of $I(l, m)$. A map made from interferometer data contains structures which are modulated by the *synthesized beam*. This is given by the Fourier transform of the telescope aperture, which is (30) above with the sky brightness replaced by a two-dimensional δ function.

Of course, interferometers also come with their own set of problems.

1. *Bandwidth smearing.* In practice an interferometer observes over a range $\Delta\nu$ about some central value ν_0, rather than at a single frequency. The values of (u, v, w) change across the passband, limiting the field size and sensitivity. To avoid this, the band $\Delta\nu$ may be split and each channel correlated separately.
2. *Time constant.* Integrating over a few seconds per measurement causes off-axis sources to be smeared in arcs in the image plane, reducing the peak signal. Loss of precision can be minimised by reducing integration times appropriately.
3. *Temperature sensitivity.* The temperature sensitivity of an interferometer is given by

$$\Delta T_{\rm A} \propto \frac{T_{\rm sys}}{\sqrt{\Delta\nu\, t_{\rm int}\, N_{\rm corr}}} \frac{1}{\Omega_{\rm synth}} \tag{31}$$

(compare 26) where $N_{\rm corr}$ is the number of antenna–antenna correlations used in making the synthesized beam of solid angle $\Omega_{\rm synth}$. However, for a source of size θ, baselines longer than λ/θ "resolve out" the signal, and thus only the shorter baselines contribute to the sensitivity. Most interferometers are designed to have high resolution, and may not be efficient when observing the extended SZ effect.
4. *Cross-talk.* For compact configurations, microwave signals can leak into adjacent antennas, giving a cross-talk signal which can easily dominate the signals expected from SZ effects. This can sometimes be filtered out using the differing modulation rates, although again this may increase the noise.

The first interferometric map of the SZ effect, shown in Fig. 7, was made by [18] using a Ryle telescope observation of the cluster Abell 2218 which lies at $z \approx 0.17$. When making a map of such data, it is normal to include only the short baselines where the SZ signal is strongest: longer baselines contribute extra thermal noise, and have already been used to locate and remove a number of confusing radio sources. The result of censoring the baselines is that the maps have limited angular dynamic range. The agreement of the SZ brightness recorded for Abell 2218 with previous single-dish measurements established the credibility of SZ effect research. Many other interferometric SZ detections have been made since, by instruments including the highly successful OVRO and BIMA arrays [9].

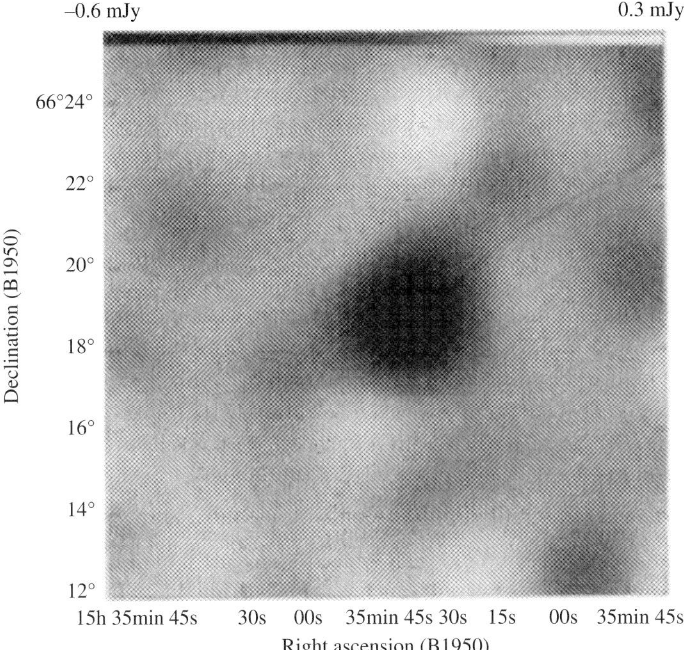

Fig. 7. This Ryle telescope observation of Abell 2218 was the first interferometric detection of the SZ effect

The quality of SZ observations can obviously be improved by designing an instrument to overcome the difficulties of using "normal" interferometers. Baselines can be tuned for optimum SZ effect detection over some redshift range (i.e. range of angular sizes), and longer baselines can be added to facilitate the removal of radio sources. An example of such a system now in operation is the Very Small Array (VSA [19]; Fig. 8).

Bolometers

Bolometric observations are fairly similar to those made using radiometers in a number of ways. They should be useful for SZ surveying as it is possible to build large array detectors (with corresponding improvements in efficiency). It is possible to cover a wide frequency range on a single telescope thus facilitating the subtraction of primordial CMB contamination, and also the search for kinematic SZ effects. Bolometers have their own set of problems, including atmospheric effects and confusion (mostly from star-forming galaxies). A number of projects have measured the SZ effect, notably SCUBA, MITO and VIPER+ACBAR [13, 34, 41].

The ACBAR instrument makes use of the significant advantage of bolometers in that it takes simultaneous multi-wavelength data in several millimetre

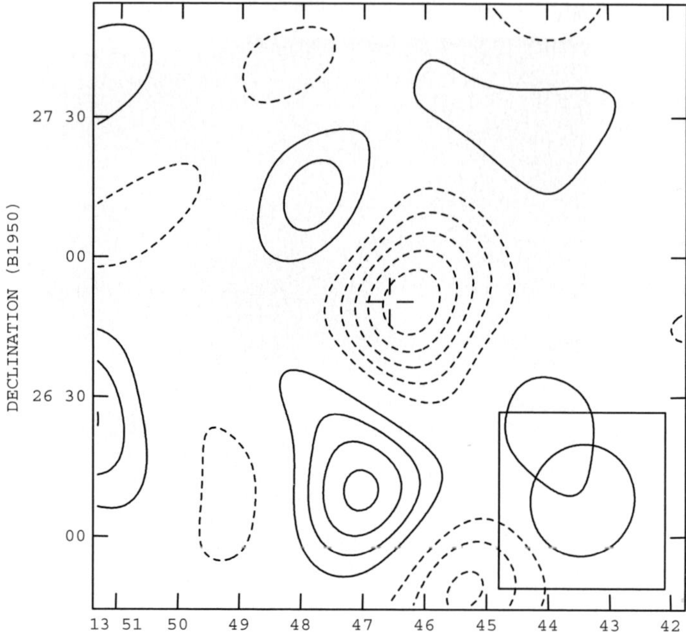

Fig. 8. The VSA was designed specifically to observe both galaxy clusters in SZ, and primordial anisotropies in the CMB. This observation of Abell 1795 clearly shows as SZ detection, but also illustrates the problem of contamination by the primordial features. Multi-wavelength observations are required in order to remove this confusion. The problem is particularly bad on these large angular scales–higher resolution instruments do not suffer so badly

bands. For this to work effectively excellent conditions are required. The VIPER telescope, for which the ACBAR detector was designed, is located in Antarctica. The bolometric arrays have 16 pixels and work in three frequency channels (with an additional channel in the 2001 season). The conditions are in Antarctica are often excellent, with exceptionally dry air. The telescope has FWHM 4–5 arcmin, chops over 3°, has a large ground shield, and is well-suited to for SZ effect work on relatively low-redshift clusters. The three frequency channels are positioned above, on, and below the null in the thermal SZ effect, allowing a spectral separation of the thermal SZ effect from primordial anisotropies in the CMB. A recent result, for the cluster 1E 0657-56 at redshift $z \approx 0.3$, is shown in Fig. 9, where it can be seen that a signal with the unusual positive, zero, negative signature of the thermal SZ effect is located at the position of the X-ray cluster.

Summary of Observational Issues

In summary, there are some generic and some specific issues that must be considered when an SZ observing programme is begun.

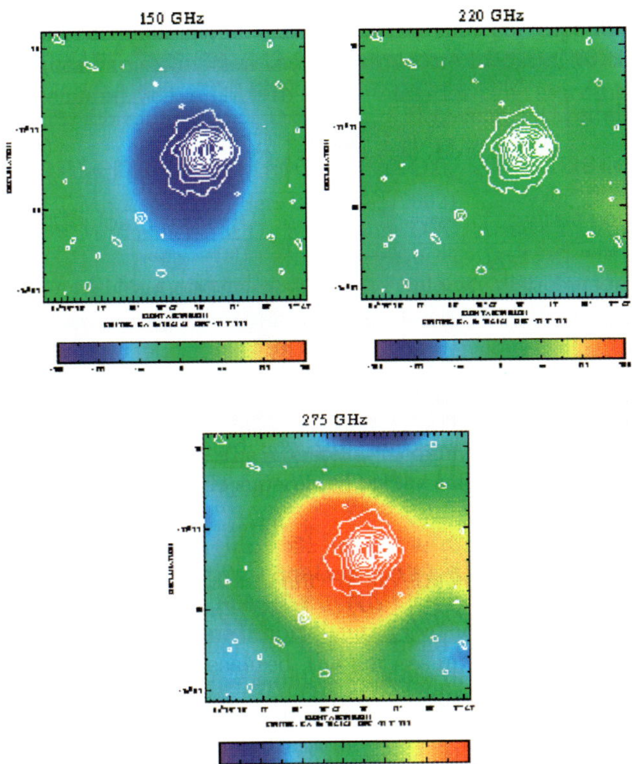

Fig. 9. Images of the cluster 1E 0657-56 from the ACBAR channels at 150, 220 and 275 GHz. The three frequencies clearly show a decrement, no detection, and an increment respectively, as expected. ROSAT X-ray contours are overlaid

1. Observation design must take account of angular dynamic range and sensitivity
2. Radiometric observations are efficient, but usually limited by systematic errors. Their strength lies in simple detections.
3. Interferometric observations come with different systematic errors, should be good for mapping resolved structures, but are currently limited in terms of angular dynamic range.
4. Bolometers are excellent for spectral work and have high sensitivity. Their major limitation comes from atmospheric noise.

4 Cluster Science from the Sunyaev–Zel'dovich Effect

Applications of the thermal SZ effect for studies of cluster properties are often associated with its proportionality to the line-of-sight integrated electron pressure, and hence to the thermal energy content of a cluster, while the kinematic SZ effect has the potential of providing information on cluster velocities. A number of reviews of such applications of the SZ effect have appeared (e.g., [4, 5]), and can be consulted for details beyond the outline here.

4.1 Energy Content

As is evident from Sect. 3, measurements of the thermal SZ effect measure the integrated y_{e} over the solid angle of some telescope beam. From (10), the integrated SZ effect flux density over an entire cluster is

$$S_{\mathrm{T}} \propto \int \mathrm{d}\Omega \int n_{\mathrm{e}} T_{\mathrm{e}} \mathrm{d}l \propto \int n_{\mathrm{e}} T_{\mathrm{e}} \mathrm{d}V \tag{32}$$

and hence is proportional to the total thermal energy content in the cluster atmosphere, U_{gas}. Since the atmosphere should respond quickly (in a sound crossing time or two) to changes in its shape or mass, a measurement of the total SZ effect from a cluster should provide a good measure of the gravitational potential energy, and hence of the degree to which this cluster has assembled. Such a measurement should be possible to high redshift, so that SZ effect data should provide a rather direct measurement of the rate of cluster formation.

In this respect the SZ effects of clusters are easier to use than the X-ray surface brightnesses or temperatures, which are based on the n_{e}^2 emissivity of cluster gas, and hence which require careful analysis to recover linear measures of gas properties, such as the thermal energy content. However, the SZ effects are generally rather small except for the most massive clusters, and so it will be difficult to use this test with the current generation of SZ effect instruments. Furthermore, the measurement requires an integration over the entire cluster, and this would be likely to lead to significant confusion at low redshifts. This will make it difficult to assemble a low-redshift SZ effect sample for comparison with clusters selected at higher redshift.

4.2 Baryon Content

If an X-ray spectral observation has been made of a cluster of galaxies for which an integrated SZ effect measurement has been made, and if the cluster is close to isothermal, then the integrated thermal SZ effect flux density (32) can be written

$$S_{\mathrm{T}} \propto N_{\mathrm{e}} T_{\mathrm{e}} \ , \tag{33}$$

where the electron temperature is measured by the X-ray spectrum, and N_{e} is the total number of electrons in the cluster atmosphere. Since the cluster gas is almost completely ionized, it is straightforward to relate N_{e} to the total number of baryons in the gas (with a slight dependence on the metallicity of the gas). The bulk of the baryonic material in a cluster is in the atmosphere, rather than in galaxies, so that $N_{\rm e}$ is a good indicator of the total baryon content of a cluster. Since the total mass of a cluster can generally be derived from the X-ray image and spectrum, it is possible to combine the value of $N_{\rm e}$ with the X-ray derived mass to obtain a good estimate of $f_{\rm b}$, the fraction of the mass in the cluster which is baryonic. We expect to see gradients in $f_{\rm b}$ with radius, but the integrated value of $f_{\rm b}$ over the entire cluster (or at least over a large part of the cluster) should be close to the universal baryon fraction, of about 12%, if clusters are representative samples of the mass content of the Universe, and the processes of cluster formation have not separated baryonic and dark matter. A detailed study of CL 0016+16 has found good agreement with the universal baryon mass fraction [40], and studies of populations of clusters of different masses over a range of redshifts are in progress [9].

4.3 Gravitational Lensing and Cluster SZ Effects

An alternative measure of the total mass content of a cluster of galaxies is provided by gravitational lensing, where the surface mass density is derived from the shear field by an integral of the form

$$\Sigma = -\frac{1}{\pi}\,\Sigma_{\rm crit} \int d^2\theta'\,\kappa_i(\theta',\theta)\,e_i(\theta')\;, \tag{34}$$

where κ_i is some kernel and e_i is the measured shear field (with summation over two components of κ and $\mathbf{e}$ implied). In principle an SZ effect map of a cluster of galaxies could be divided by the map of Σ to provide an image of the line-of-sight integrated $f_{\rm b}$ as a function of position, since the relationship between $y_{\rm e}/\Sigma$ and $f_{\rm b}$ depends only on quantities which are constant over the cluster, provided that the cluster gas is isothermal. A map of $f_{\rm b}$ might be expected to reveal some evidence of non-gravitational processes that occur in cluster formation. For clusters at redshifts of a few tenths, where gravitational lensing is most effective, maps with a resolution of about half an arcminute would provide a few tens of resolution elements across a cluster. So far, only integrated comparisons of the baryonic mass fraction derived in this way have been possible (e.g., [40]), but comparisons of this type are likely to become useful in the future. While similar tests are possible with X-ray data, the $n_{\rm e}^2$ dependence of the X-ray emissivity makes the extraction of a bias-free baryonic density more prone to systematic errors, but with smaller random errors, than with SZ data.

4.4 Cluster Gas Structure

Just as X-ray data can be used to study the structure of cluster atmospheres, so potentially can the SZ effect data. However, as described in Sect. 3.2, most current SZ data for clusters have poor angular dynamic range, so detailed structural studies have not been possible. For example, Lancaster et al. [19]

were able only to constrain a simple angular size measure for the low-redshift clusters observed by the VSA, rather than to separate the effects of β and θ_c in (15).

Although the n_e^2 dependence of X-ray data makes X-ray images far more sensitive to the core than to the outer parts of clusters, the overall sensitivity of X-ray satellites is higher than of SZ effect measurements at moderate redshift, so that X-ray data provide far superior information on cluster gas structures than SZ data. This is not necessarily the case at high redshift, where, because of the redshift-independence of the SZ effect surface brightness, one might expect the SZ effect to provide the best information on cluster formation, if sensitivities in the μK range on angular scales of order 10 arcsec can be achieved, because of the redshift-independence of the SZ effect surface brightness and the scaling of the SZ effect with n_e.

4.5 Velocity Structure and the Growth of Clusters

The kinematic SZ effect (Sect. 2.3) can be separated from the thermal SZ effect by spectral studies, and so could be used to measure the velocities of clusters of galaxies. Such velocities might be expected to be small, except where the clusters are in particularly massive superclusters or are about to merge, and so the kinematic SZ effects would be expected to be of order 50 μK or less. Detection of SZ effects of this amplitude will be difficult in the presence of the thermal effect and confusion with primordial CMB structures. While the thermal effect can, in principle, be removed to high accuracy if precise spectral measurements are made (but see [7]), primordial CMB structures impose a limit of about 150 $\mathrm{km\,s^{-1}}$ on the accuracy with which cluster velocities might be measured (for clusters with arcminute-scale SZ effects, at redshifts $\gtrsim 0.2$ — the confusion is higher at lower redshift). While this is the limit for individual clusters, it is possible that a statistical measure of the random velocities of clusters might be measured by comparing the scatter of apparent SZ effects from cluster and non-cluster fields. Measurements to date have not approached this accuracy (e.g., [20]).

A further element of confusion will arise from differential motions of gas within the clusters. While the total radial velocity of a cluster will be measured with a larger error in this case, a higher angular resolution measurement of the kinematic SZ effect could show the motions of gas within the cluster atmospheres. The largest such signals will come from motions near the cluster cores, associated with cooling flows or with the injection of kinetic energy that seems to prevent runaway cooling flows, since the density of the moving gas is likely to be highest in the core.

A further source of kinematic SZ effect signals will arise from material falling into the clusters, and measurements of these velocities would provide useful information on the physics of cluster growth. Here the confusion from CMB structures could be small, because the angular sizes of the infalling clumps will be small, but the amplitudes of the kinematic SZ effects will be only at the μK level unless the infalling lumps of material are both fast-moving and dense. An example of a case where such a measurement might be possible is 1E 0657-56 [22], where a fast-moving bullet of material is traversing an X-ray atmosphere.

4.6 Supercluster Gas

On the largest angular scales we might expect to see thermal SZ effects from the diffuse gas in superclusters of galaxies, where the X-ray surface brightness is too low to detect, because of the n_e^2 emissivity of the gas, but the line-of-sight integrated electron pressure is high. Measurements at a single frequency are unlikely to be able to detect the SZ effect, since the angular sizes of superclusters are large, and so they will be badly confused by primordial CMB structures. However, the spectral signature of the thermal SZ effect is distinctive, and should allow measurements of the SZ effects of supercluster gas, provided that other sources of confusion are sufficiently small. Such supercluster SZ effects might be detectable by the Planck satellite, and would provide measurements of the residual baryonic material outside the dense parts of cluster atmospheres. The integrated amount of such material cannot be too large, or the overall Comptonization of the CMB would have been detected as a spectral distortion by the COBE satellite (e.g., [4]).

5 Cosmology from the SZ Effect

Cosmology with the SZ effects is based upon two major attributes of the effects—their redshift-independence, which should allow the study of cluster atmospheres at high redshift, and the contrast between X-ray and SZ effect images of clusters, which allow absolute measurements of distance. Reviews of some of these applications can be found in [4, 5, 7, 9].

5.1 The Cluster Hubble Diagram

The best-known use of the thermal SZ effect is to measure the distances of clusters of galaxies. In its simplest form, the method compares the X-ray surface brightness, b_{X0}, of a cluster on some fiducial line of sight,

$$b_{\mathrm{X0}} \propto n_{\mathrm{e0}}^2 \, \Lambda(T_{\mathrm{e0}}) \, L \; , \tag{35}$$

where $\Lambda(T_{\mathrm{e0}})$ is the emissivity at electron temperature T_{e0}, with the thermal SZ effect on the same line of sight,

$$\Delta T_{\mathrm{RJ,T0}} \propto n_{\mathrm{e0}} \, T_{\mathrm{e0}} \, L \; , \tag{36}$$

and eliminates the scale (here written as the central) electron density from (35) and (36) to obtain a relation for the path length along this line of sight

$$L \propto \frac{\Delta T_{\mathrm{RJ},0}^2}{b_{\mathrm{X}0}} \cdot \frac{\Lambda(T_{\mathrm{e}0})}{T_{\mathrm{e}0}^2} \; . \tag{37}$$

The path length, L, is a linear measure of the size of the cluster, and can be related to the corresponding angular measure, θ_{L}, that can be obtained from a cluster image (under some assumption about the shape of the cluster along the line of sight). The angular diameter distance of the cluster is then obtained as $D_{\mathrm{A}} = L/\theta_{\mathrm{L}}$. The most difficult step in this process is that of determining the constant of proportionality in (37). This constant depends on the structure of the gas, and must be based on a detailed model of the cluster, derived from and consistent with all the information available. It is necessary in using this method to treat each cluster as an individual, and derive an individual constant of proportionality.

This technique has been used for many clusters, for example [15, 25, 32]. A recent calculation of the distance to CL 0016+16 (Fig. 10) using this technique gave $D_{\mathrm{A}} = 1.16 \pm 0.15$ Gpc [40], implying a Hubble constant of $68 \pm 8 \pm 18 \ \mathrm{km\,s^{-1}\,Mpc^{-1}}$ if it is assumed that the density parameters $\Omega_m = 0.3$ and $\Omega_\Lambda = 0.7$.

Because this distance-measuring technique relies on comparing a line-of-sight depth of the cluster, L, with the apparent angular size, θ, Hubble constant estimates based on single clusters are likely to be prone to large systematic errors because of their unknown three-dimensional shapes (this is responsible for much of the systematic error quoted for CL 0016+16).

Fig. 10. A vignetting-corrected 0.3 – 5.0 keV image of CL 0016+16, from an *XMM-Newton* observation [40]. A quasar 3 arcmin north of the cluster [23], and a second cluster, 9 arcmin to the south–west [16], are at similar redshifts to CL 0016+16

Clumping of the intracluster medium, poorly-determined thermal substructure, and a number of other problems can also cause systematic errors. Exhaustive discussions of such problems may be found in papers on such distance measurements, such as [6].

Perhaps the most serious issue in using this technique is the orientation bias in the distances that is introduced if the sample of clusters is selected according to their central surface brightnesses, since such clusters will preferentially be elongated along the line of sight, with L being too large for the measured $\theta_{\rm L}$. SZ effect surveys in total SZ effect flux density (Sect. 5.2) will provide an excellent sample of clusters without such a bias over a wide redshift range. Molnar et al. [29] have shown that a set of about 70 clusters could provide useful measure of the equation-of-state parameter, w, as well as the Hubble constant. A recent review of the state of measurement of $D_{\rm A}(z)$ using this technique is given by Carlstrom et al. [9].

If distances are to be measured by this technique, then it is critical that the absolute calibrations of the X-ray and SZ data are excellent, and that cluster substructure is well modeled. Since clusters are relatively young structures, and likely to be changing significantly with redshift, variations in the amount of substructure with redshift might be a significant source of systematic error—this has not yet been sufficiently studied, because of the lack of a well-defined sample of clusters spanning a sufficient redshift range with high-quality X-ray imaging and spectroscopy.

5.2 Surveys for Clusters at High Redshift

It is difficult to detect high-redshift clusters by their X-ray emission, since the X-ray flux of a cluster of given X-ray luminosity decreases steeply with increasing redshift. Thus only the highest-luminosity clusters are seen at the highest redshifts, and studies of the evolution of the luminosity function are correspondingly restricted. By contrast, the SZ effect flux density of a cluster, as observed by most telescopes, is a weak function of redshift over a wide range of redshifts (e.g., Fig. 11), so that the sample of clusters obtained from a blind survey for SZ effects should be almost mass limited, provided that the cluster structures don't change too much with redshift.

Since the SZ effect survey sensitivity is such a flat function of redshift, SZ effect techniques should prove more effective at finding high-redshift clusters, and at locating clusters over a wider range of gas mass, than X-ray surveys: for surveys such as the XMM-LSS [31], plausible SZ effect surveys are more effective at $z \gtrsim 0.7$, if the clusters at such redshifts resemble those at low redshift. However, since we know that clusters assembled relatively recently, we might expect that SZ-selected samples of clusters will be limited by the changing cluster sizes, gas contents, and coherence, and that there will be some maximum detectable redshift at which the gas in clusters first gains a high enough pressure to become the source of a significant SZ effect.

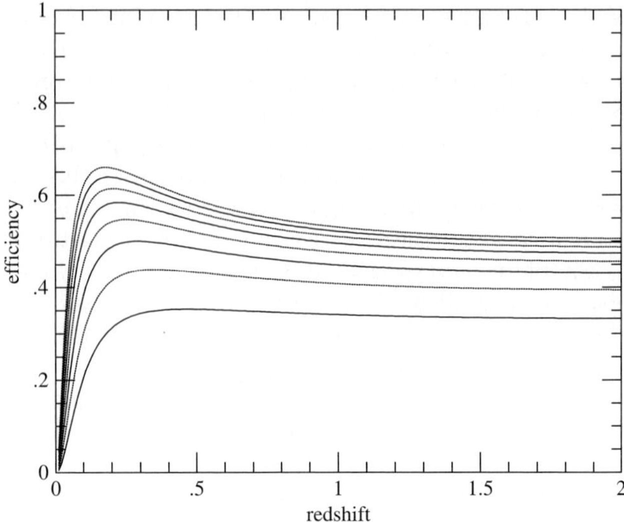

Fig. 11. The observable signal as a fraction of the central SZ effect, for measurements with a 1 arcmin beam and 1.5–5 arcmin beam-switching angle. Note the long region of almost constant efficiency at redshift $\gtrsim 0.3$, which causes surveys using such configurations to generate almost mass-limited samples of clusters

The distribution of clusters by redshift can provide useful measures of cosmological parameters and tests of our models of cluster heating and evolution (e.g., [2, 14, 29]). Follow-up studies of a subsample of the clusters, with long X-ray observations and high-sensitivity, high angular-resolution, SZ-effect mapping, should provide excellent information on the physics of cluster formation – for example on the evolution of the cluster baryon fraction (Sect. 4.2, [9]), or the changing distribution of cluster velocities (Sect. 4.5). It would also provide an ideal set of clusters for the measurement of cluster distances and the determination of cosmological parameters, provided that enough structure information is available to allow good models of the gas distribution to be deduced for each cluster. It has even been shown that a comparison of a map of the SZ effect with a gravitational lensing map can provide a rich set of information on cluster properties without conducting a redshift survey [39].

5.3 SZ Effect Confusion and the Primordial Background

Just as primordial structures in the CMB confuse measurements of the SZ effect, and limit the detectability of the effect on some angular scales unless careful spectral decomposition is undertaken, so the presence of a foreground of SZ effects limits the quality of the information that can be obtained on the power spectrum of the CMB at high multipole numbers. The thermal SZ effect can be removed using spectral techniques, but the kinematic SZ effect provides a source of confusion that can be reduced only by excluding parts of a CMB map showing significant thermal SZ effects under the assumption that there are no fast, low-temperature, gas clouds that could produce large kinematic effects without a corresponding thermal effects: see (19). Studies of SZ effect confusion suggest that it can be important only at multipole numbers, $l \gtrsim 2000$ (e.g., [28]), but the excess power recently seen in high-l measurements of the power spectrum of the CMB may indicate that there is a larger contribution from early hot gas atmospheres than had been thought. Information on this will come with the next generation of multi-frequency CMB surveys, which are also effectively SZ effect surveys.

5.4 Sampling the Cosmic Microwave Background

Finally, a variety of techniques have been proposed for using SZ effects to study the intrinsic properties of the CMB. Thus, for example, the ratio of the thermal SZ effects from a single cluster at two different frequencies provides a measure of the temperature of the CMB, $T_{\rm rad}$ [3]. Measurements of the SZ effects of a sample of clusters at different redshifts allow a test of the cosmological change of CMB temperature with redshift,

$$T_{\rm rad} = T_{\rm rad,0} \, (1 + z) \; . \tag{38}$$

No deviation from this relation was found by Battistelli et al. [3].

Another application of SZ effects to the study of the CMB could be to check the universality of the low quadrupolar term that has been measured in the CMB power spectrum. This is possible by examining the SZ effect polarization of a distant cluster, since one of the polarization terms arises from the conversion of the local CMB quadrupole to linear polarization by scattering in the cluster atmosphere. Multiple samples of the CMB quadrupole at distant places in the Universe could then be used to reduce cosmic variance in our local measurement of the quadrupole. Unfortunately the polarization signal is weak, and is expected to be confused by other polarization signals, including the gravitational lensing of the CMB polarization by the mass of the cluster. No sufficiently sensitive work on the SZ effect polarization channel has yet tested the feasibility of this technique.

6 The Next Generation of Instruments

SZ studies to date have largely been performed on an ad-hoc set of clusters, with few studies of representative samples of clusters selected without orientation bias (such as, e.g., [24]). SZ effects are linear probes of cluster properties so should provide well-defined samples for both cosmological studies and investigating cluster physics. A major focus of current SZ work is to construct instruments capable of performing blind surveys, and then to use those instruments to develop SZ-selected samples of clusters. The full range of observing techniques are suitable to some extent, but it is likely that array receivers (both radiometric and bolometric) mounted on single dishes will provide the fastest means of surveying many square degrees with arcminute resolution.

Most SZ work to date has focussed on simple detections. In the new era, with purpose-built instruments and SZ-selected samples, high angular resolution follow-up will be required. Such studies will return detailed information about the cluster gas. It may be possible to detect kinematic effects from rapidly-infalling filaments, or to see substructure from the accretion of subclusters. Polarisation measurements of the brightest SZ clusters will yield yet further information. Measurements of the radial and transverse velocities would provide the dynamical information required to test models of cluster formation.

Distinct types of telescope are required for blind surveys and detailed studies. Survey work requires a reasonable field of view and high sensitivity. This may be simplest using bolometer or radiometer arrays, though tailored interferometers should also be suitable. Detailed studies will require high angular resolution (synthesized beams of 10 arcsec or less) which may best be achieved via bolometer arrays on large single dishes, or interferometers.

6.1 Proposed Instruments

Most SZ work to date has been undertaken using non-ideal instruments such as the Ryle interferometer [18] and the OVRO 40 m [15]. Some more suitable instruments also exist, such at the VSA [19], ACBAR [34], CBI [38] and SuZIE [26]. There is, however, still room for improvement.

New interferometers currently in the construction phase include AMiBA [21], AMI [1] and the SZA [36]. We here discuss AMiBA as an example. AMiBA is an project of ASIAA and the National Taiwan University, and the instrument is situated on Mauna Loa, Hawaii. AMiBA is a dedicated CMB interferometer, designed for rapid surveys of CMB structures including SZ effects. The operational array is nearing completion, and replaces the prototype which previously occupied the site. The AMiBA design involves 19 dishes of two diameters (0.3 and 1.2 m), resulting in baselines over the range 1.2−6 m ($380\lambda < b < 1875\lambda$ at 95 GHz). With its 20 GHz bandwidth and dual polarisation capabilities, AMiBA should achieve a sensitivity of ~ 1.3 mJy in 1 hour. The problem of radio source confusion is not expected to be significant at such a high observing frequency, with an estimated 0.3 sources above ~ 1 mJy per survey field. Neither is CMB primordial anisotropy expected to be a significant contaminant on these angular scales, and most can be filtered in mosaic mode. For redshifts > 0.7, AMiBA will be more sensitive than the XMM Large Scale Structure Survey [31], and an AMiBA-survey selected sample of clusters will be close to mass-limited. AMiBA also has the potential of being able to find a new class of objects, which are X-ray dark but SZ bright, if such objects exist.

Planned radiometers for survey work are arrays mounted on large single dishes. An example is OCRA, which is expected to be the fastest survey instrument of its kind [8]. The two-beam prototype receiver, OCRA-p, is mounted on the 32-m Torun telescope, and a 100-beam array is planned. OCRA-p has observed four clusters in SZ during its check-out phase, achieving detections with $\sim$ 800 s integration times. Operating at 30 GHz, OCRA-p has FWHM 1 arcmin and achieves 5 mJy sensitivity in 5 minutes. The full OCRA array would be sufficiently sensitive to map 100 deg^2 to the confusion limit in a matter of months, and should generate a catalogue of clusters with significant gas contents over the entire redshift range by virtue of the redshift independence of the SZ effect. In addition, the full array will be ideal for detailed mapping of clusters at lower redshifts.

Large bolometer arrays are also an excellent way to make SZ effect surveys. Good examples are provided by the the planned SPT and APEX surveys, and bolometer arrays on large single dishes may also provide high-quality SZ effect images of clusters detected in finding surveys made using smaller antennas — the AzTEC array on the LMT may be of great use in this respect.

6.2 The Future: A Hypothetical Ideal Instrument

The next level in terms of SZ surveys will be to achieve sky coverage of 100 deg^2 or more to a sensitivity of 30 μK or better. For this to be really effective, surveys of this scale should be made on timescales of a year or less. With the expected thousands of cluster detections from the Planck satellite, it will also be desirable to have the capability of providing detailed follow-up observations of SZ effect clusters. In addition, the next generation of X-ray telescopes (*Con-X* and *XEUS*) will have much improved sensitivity and spectral capabilities, thus producing far superior images of clusters and their thermal substructures. In order to push SZ science forwards, and to further exploit the power of combining SZ and X-ray data, it is necessary to begin considering an SZ telescope capable of matching these instruments.

The desired improvements in SZ surveying will require multi-wavelength observations with channels well matched in resolution and astrometric accuracy to overcome the problem of confusion imposed by CMB features, and to separate the thermal and SZ effects. The ideal facility may be a bolometer array mounted on a large telescope located at an excellent site, with simultaneous imaging in several passbands in the mm and sub-mm ranges. For substructure studies, sub-arcmin resolution will be required over fields several arcmin in size. This may also be achievable via bolometric techniques, and multi-wavelength capabilities would facilitate separation of SZ signals from foreground contamination. The addition of some polarization capability to these instruments would also open new avenues of exploration.

Scaled interferometric arrays may also provide a route towards matching standards set by the next X-ray missions. The general design would involve a configuration such that comparable sky areas could be synthesised in each frequency channel (chosen to coincide with atmospheric windows), each with the same angular resolution. For each array, the ideal antenna size would be around 300λ in order to retain sensitivity to scales ~ 10 arcmin by means of close packing the antennas, while also achieving maximum resolution up to 20 arcsec on the longest baselines $\sim 10^4\lambda$. Alternatively, multi-frequency bolometric arrays composed of about 1000 elements could achieve similar results if mounted on a telescope of 50-m class, such as the LMT. Either type of system could potentially provide the necessary follow-up to Planck SZ detections.

All proposed advances involve large-scale projects and would require a collaborative effort, perhaps in the form of an "International SZ Observatory". Of course this concept requires considerable work to reach the level of a costed proposal, but it is clear that such an observatory will be required within the next 10 years in order to take SZ science to the next level, and provide SZ effect data complementary to the X-ray studies that will be possible with the next generation of satellite observatories.

References

1. The AMI Collaboration Preprint astro-ph/0509215 (2005)
2. Angulo, R., et al.: MNRAS **362**, L55 (2005)
3. Battistelli, E.S., et al.: ApJ **580**, L101 (2002)
4. Birkinshaw, M.: PhR **310**, 97 (1999)
5. Birkinshaw, M.: Carnegie observatories astrophysics series. In Mulchaey, J.S., Dressler, A., Oemler, A., Clusters of Galaxies: Probes of Cosmological Structure and Galaxy Evolution, vol. 3, p. 161. Cambridge University Press, Cambridge (2004)
6. Birkinshaw, M., Hughes, J.P., Arnaud, K.A.: ApJ **379**, 466 (1991)
7. Birkinshaw, M., Lancaster, K.: Preprint astro-ph/0410336 (2004)
8. Browne, I.W.A., et al.: Proceedings of SPIE, vol. 4015, p. 299. SPIE, Washington (2000)
9. Carlstrom, J.E., Holder, G.P., Reese, E.D.: ARAA **40**, 643 (2002)
10. Cavaliere, A., Fusco-Femiano, R.: A&A **70**, 677 (1978)
11. Challinor, A.D., Lasenby, A.N.: ApJ **510**, 930 (1999)
12. Colafrancesco, S., Marchegiani, P., Palladino, E.: A&A **397**, 27 (2003)
13. De Petris, M., et al.: ApJ **574**, L119 (2002)
14. Fan, Z., Chiueh, T.: ApJ **550**, 547 (2001)
15. Hughes, J.P., Birkinshaw, M.: ApJ **501**, 1 (1998)
16. Hughes, J.P., Birkinshaw, M., Huchra, J.P.: ApJ **448**, L93 (1995)
17. Itoh, N., Kohyama, Y., Nozawa, S.: ApJ **502**, 7 (1998)
18. Jones, M., et al.: Nature **365**, 320 (1993)
19. Lancaster, K., et al.: MNRAS **359**, 16 (2005)
20. LaRoque, S.J., et al.: Preprint (2004)
21. Lo, K.Y., et al.: IAU Symp. **201**, 31 (2001)
22. Markevitch, M., et al.: ApJ **567**, L27 (2002)
23. Margon, B., Downes, R.A., Spinrad, H.: Nature **301**, 221 (1983)
24. Mason, B.S., Myers, S.T.: ApJ **540**, 614 (2000)

25. Mason, B.S., Myers, S.T., Readhead, A.C.S.: ApJ **555**, L11 (2001)
26. Mauskopf, P.D., et al.: ApJ **538**, 505 (2000)
27. McKinnon, M.M., Owen, F.N., Eilek, J.A.: AJ **101**, 2026 (1990)
28. Molnar, S.M., Birkinshaw, M.: ApJ **537**, 542 (2000)
29. Molnar, S.M., Birkinshaw, M., Mushotzky, R.F.: ApJ **570**, 1 (2002)
30. Nozawa, S., Itoh, N., Kohyama, Y.: ApJ **508**, 17 (1998)
31. Pierre, M., et al.: JCAP **9**, 011 (2005)
32. Reese, E.D., et al.: ApJ **581**, 53 (2002)
33. Rephaeli, Y.: ARAA **33**, 541 (1995)
34. Runyan, M.C., et al.: NewAR **47**, 915 (2002)
35. Sunyaev, R.A., Zel'dovich Ya.B.: CoASP **4**, 173 (1972)
36. The SZA Collaboration: Proc. ASPC **257**, 43 (2002)
37. Thompson, A.R., Moran, J.M., Swenson, G.W.: Interferometry and Synthesis in Radio Astronomy. John Wiley and Sons, Inc., New York (1986)
38. Udomprasert, P.S., et al.: ApJ **615**, 63 (2004)
39. Umetsu, K. et al., JKAS, **38**, 191 (2005)
40. Worrall, D.M., Birkinshaw, M.: MNRAS **340**, 1261 (2003)
41. Zemcov, M., et al.: MNRAS **346**, 1179 (2003)

Cosmology with Clusters of Galaxies

S. Borgani

Department of Astronomy, University of Trieste, via Tiepolo 11, I-34131
Trieste, Italy
borgani@ts.astro.it

1 Introduction

Clusters of galaxies occupy a special place in the hierarchy of cosmic structures. They arise from the collapse of initial perturbations having a typical comoving scale of about $10\,h^{-1}$Mpc[1]. According to the standard model of cosmic structure formation, the Universe is dominated by gravitational dynamics in the linear or weakly non–linear regime and on scales larger than this. In this case, the description of cosmic structure formation is relatively simple since gas dynamical effects are thought to play a minor role, while the dominating gravitational dynamics still preserves memory of initial conditions. On smaller scales, instead, the complex astrophysical processes, related to galaxy formation and evolution, become relevant. Gas cooling, star formation, feedback from supernovae (SN) and active galactic nuclei (AGN) significantly change the evolution of cosmic baryons and, therefore, the observational properties of the structures. Since clusters of galaxies mark the transition between these two regimes, they have been studied for decades both as cosmological tools and as astrophysical laboratories.

In this Chapter I concentrate on the role that clusters play in cosmology. I will highlight that, in order for them to be calibrated as cosmological tools, one needs to understand in detail the astrophysical processes which determine their observational characteristics, i.e. the properties of the cluster galaxy population and those of the diffuse intra–cluster medium (ICM).

Constraints of cosmological parameters using galaxy clusters have been placed so far by applying a variety of methods. For example:

1. The mass function of nearby galaxy clusters provides constraints on the amplitude of the power spectrum at the cluster scale (e.g., [138, 164] and references therein). At the same time, its evolution provides constraints on the linear growth rate of density perturbations, which translate into

[1] Here h is the Hubble constant in units of $100\,\mathrm{km\,s^{-1}\,Mpc^{-1}}$

S. Borgani: *Cosmology with Clusters of Galaxies*, Lect. Notes Phys. **740**, 287–334 (2008)
DOI 10.1007/978-1-4020-6941-3_9

dynamical constraints on the matter and Dark Energy (DE) density parameters.

2. The clustering properties (i.e., correlation function and power spectrum) of the large–scale distribution of galaxy clusters provide direct information on the shape and amplitude of the underlying DM distribution power spectrum. Furthermore, the evolution of these clustering properties is again sensitive to the value of the density parameters through the linear growth rate of perturbations (e.g., [26, 114] and references therein).
3. The mass-to-light ratio in the optical band can be used to estimate the matter density parameter, Ω_m, once the mean luminosity density of the Universe is known and under the assumption that mass traces light with the same efficiency both inside and outside clusters (see [9, 35, 70], as examples of the application of this method).
4. The baryon fraction in nearby clusters provides constraints on the matter density parameter, once the cosmic baryon density parameter is known, under the assumption that clusters are fair containers of baryons (e.g., [60, 168]). Furthermore, the baryon fraction of distant clusters provide a geometrical constraint on the DE content and equation of state, under the additional assumption that the baryon fraction within clusters does not evolve (e.g., [6, 58]).

An extensive presentation of all these methods would probably require a dedicated book. For this reason, in this contribution I will mostly concentrate on the method based on the evolution of the cluster mass function. A substantial part of my Lecture will concentrate on the different methods that have been applied so far to weight galaxy clusters. Since all the above cosmological applications rely on precise measurements of cluster masses, this part of my contribution will be of general relevance for cluster cosmology.

Also, since most of the cosmological applications of galaxy clusters have been based so far on X–ray surveys, my discussion will be definitely X–ray biased, although I will discuss in some detail methods based on optical observations and what present and future optical surveys are expected to provide. I refer to the lecture by Roy Gal in this volume for more details regarding the properties of galaxy clusters in the optical band. Also, I will refer to the Lectures by M. Birkinshaw for cosmological studies of clusters based on the Sunyaev–Zel'dovich (SZ) effect, to the Lectures by J.–P. Kneib for cluster studies and mass measurement through gravitational lensing, and to the Lectures by C. Jones and by C. Sarazin for more details about the cosmological application of the baryon fraction method.

The structure of this Chapter will be as follows. I provide in Sect. 2 a short introduction to the basics of cosmic structure formation. I will shortly review the linear theory for the evolution of density perturbations and the spherical collapse model. In Sect. 3 I will describe the Press–Schechter (PS) formalism to derive the cosmological mass function. I will then introduce extensions of the PS approach and present the most recent calibrations of the mass function from N–body simulations. In Sect. 4 I will review the methods to build samples of galaxy clusters, based on optical and X–ray observations, while I will only briefly discuss the SZ methodology for cluster surveys. Section 5 is devoted to the discussion of different methods to derive cluster masses and to review the results of the application of these methods. In Sect. 6 I will describe the cosmological constraints, which have been obtained so far by tracing the cluster mass function with a variety of methods: distribution of velocity dispersions, X–ray temperature and luminosity functions, and gas mass function. In this Section I will also critically discuss the reasons for the different, sometimes discrepant, results that have been obtained in the literature and I will highlight the relevance of properly including the analysis of the cluster mass function all the statistical and systematic uncertainties in the relation between mass and observables. Finally, I will describe in Sect. 7 the future perspectives for cosmology with galaxy clusters and which are the challenges for clusters to keep playing an important role in the era of precision cosmology.

2 A Concise Handbook of Cosmic Structure Formation

In this section I will briefly review the basic concepts of cosmic structure formation, which are relevant for the study of galaxy clusters as tools for precision cosmology through the evolution of their mass function. A complete treatment of models of structure formation can be found in classical cosmology textbooks (e.g., [42, 123, 124]).

2.1 The Statistics of Cosmic Density Fields

Let $\rho(\mathbf{x})$ be the matter density field, which is a continuous function of the position vector $\mathbf{x}$, $\bar{\rho} = \langle \rho \rangle$ its average value computed over a sufficiently large (representative) volume of the Universe and

$$\delta(\mathbf{x}) = \frac{\rho(\mathbf{x}) - \bar{\rho}}{\bar{\rho}} \tag{1}$$

the corresponding relative density contrast. By definition, it is $\bar{\delta} = 0$ and $\delta(\mathbf{x}) \geq -1$. If the density field is traced by a discrete distribution of points (i.e., galaxies or galaxy clusters) all having the same weight (mass), then $\rho(\mathbf{x}) = \sum_{\mathrm{i}} \delta_{\mathrm{D}}(\mathbf{x} - \mathbf{x}_{\mathrm{i}})$, where $\delta_{\mathrm{D}}(\mathbf{x})$ is the Dirac delta–function. The Fourier representation of the density contrast is given by

$$\tilde{\delta}(\mathbf{k}) = \frac{1}{(2\pi)^{3/2}} \int \mathrm{d}\mathbf{x}\, \delta(\mathbf{x}) \mathrm{e}^{\mathrm{i}\mathbf{k}\cdot\mathbf{x}}\,, \tag{2}$$

with the corresponding dual relation for the inverse Fourier transform.

The 2–point correlation function for the density contrast is defined as

$$\xi(r) = \langle \delta(\mathbf{x}_1)\delta(\mathbf{x}_2) \rangle \,, \tag{3}$$

which only depends on the modulus of the separation vector, $r = |\mathbf{x}_1 - \mathbf{x}_2|$, under the assumption of statistical isotropy of the density field. Therefore, it can be shown that the power spectrum of the density fluctuations is the Fourier transform of the correlation function, so that

$$P(k) = \langle |\tilde{\delta}(\mathbf{k})|^2 \rangle = \frac{1}{2\pi^2} \int \mathrm{dr\, r^2} \xi(\mathrm{r}) \frac{\sin \mathrm{kr}}{\mathrm{kr}} \,, \tag{4}$$

which, again, depends only on the modulus of the wave-vector **k**.

In case we are interested in the study of a class of observable structures of mass M, which arise from the collapse of initial perturbations having size $R \propto (M/\bar{\rho})^{1/3}$, then it is common to introduce the smoothed density field, which is defined as

$$\delta_R(\mathbf{x}) = \delta_{\mathrm{M}}(\mathbf{x}) = \int \delta(\mathbf{y}) \mathrm{W_R}(|\mathbf{x} - \mathbf{y}|)\, \mathrm{d}\mathbf{y} \,. \tag{5}$$

As such, it is given by the convolution of the density fluctuation field with a window function, which filters out the fluctuation modes having wavelength $\lesssim R$. Equation (5) allows us to introduce the variance of the fluctuation field computed at the scale R, defined as

$$\sigma_R^2 = \sigma_M^2 = \langle \delta_R^2 \rangle = \frac{1}{2\pi^2} \int \mathrm{d}k\, k^2 P(k)\, \tilde{W}_R^2(k) \,, \tag{6}$$

where $\tilde{W}_R(k)$ is the Fourier transform of the window function.

The shape of the window function defines the exact relation between mass and smoothing scale. For instance, for the top–hat window it is

$$\tilde{W}_R(k) = \frac{3[\sin(kR) - kR\cos(kR)]}{(kR)^3} \tag{7}$$

while the Gaussian window gives

$$\tilde{W}_R(k) = \exp\left(-\frac{(kR)^2}{2}\right) ,. \tag{8}$$

The corresponding relations between mass scale and smoothing scale are $M = (4\pi/3)R^3\bar{\rho}$ and $M = (2\pi R^2)^{3/2}\bar{\rho}$ for the top–hat and Gaussian filters, respectively.

The shape of the power spectrum is (essentially) fixed once the matter density parameter, Ω_m, that associated to the baryonic component, Ω_{bar}, and the Hubble parameter, H_0, are specified (e.g., [53]). However, its normalization can only be fixed through a comparison with observational data of the large–scale structure of the Universe or of the anisotropies of the Cosmic Microwave Background (CMB). A common way of parametrizing this normalization is through the quantity σ_8, which is defined as the variance, computed for a top–hat window having comoving radius $R = 8\,\mathrm{h}^{-1}\mathrm{Mpc}$ (given in (6)). The historical reason for this choice of the normalization scale is that the variance of the galaxy number counts, within the first redshift surveys, was observed to be about unity inside spheres of that radius (e.g., [46]). In this way, the value of σ_8 for a given cosmology directly provides a measure of the biasing parameter relating the galaxy and mass distribution, expected for that model. Furthermore, a top–hat sphere of $8\,\mathrm{h}^{-1}\mathrm{Mpc}$ radius contains a mass $M \simeq 5.9\times10^{14}\Omega_m M_\odot$, which is the typical mass of a moderately rich galaxy cluster. Therefore, as we shall see in Sect. 3, the mass function of galaxy clusters provides a direct measure of σ_8.

2.2 The Linear Evolution of Density Perturbations

Let us assume that the matter content of the Universe is dominated by a pressurless and self–gravitating fluid. This approximation holds if we are dealing with the evolution of the perturbations in the dark matter (DM) component or in case we are dealing with structures whose size is much larger than the typical Jeans scale–length of baryons. Let us also define $\mathbf{x}$ to be the comoving coordinate and $\mathbf{r} = \mathrm{a(t)}\mathbf{x}$ the proper coordinate, $a(t)$ being the cosmic expansion factor. Furthermore, if $\mathbf{v} = \dot{\mathbf{r}}$ is the physical velocity, then $\mathbf{v} = \dot{\mathrm{a}}\mathbf{x} + \mathbf{u}$, where the first term describes the Hubble flow, while the second term, $\mathbf{u} = \mathrm{a(t)}\dot{\mathbf{x}}$, gives the peculiar velocity of a fluid element which moves in an expanding background.

In this case the equations that regulate the Newtonian description of the evolution of density perturbations are the continuity equation:

$$\frac{\partial \delta}{\partial t} + \nabla \cdot [(1+\delta)\mathbf{u}] = 0\,, \tag{9}$$

which gives the mass conservation, the Euler equation

$$\frac{\partial \mathbf{u}}{\partial \mathrm{t}} + 2H(t)\mathbf{u} + (\mathbf{u}\cdot\nabla)\mathbf{u} = -\frac{\nabla\phi}{\mathrm{a}^2}\,, \tag{10}$$

which gives the relation between the acceleration of the fluid element and the gravitational force, and the Poisson equation

$$\nabla^2\phi = 4\pi G\bar{\rho}a^2\delta \tag{11}$$

which specifies the Newtonian nature of the gravitational force. In the above equations, ∇ is the gradient computed with respect to the comoving coordinate $\mathbf{x}$, $\phi(\mathbf{x})$ describes the fluctuations of the gravitational potential and

$H(t) = \dot{a}/a$ is the Hubble parameter at the time t. Its time–dependence is given by $H(t) = E(t)H_0$, where

$$E(z) = [(1+z)^3\Omega_m + (1+z)^2(1-\Omega_m-\Omega_{\rm DE}) + (1+z)^{3(1+w)}\Omega_{\rm DE}]^{1/2} \quad (12)$$

is related to the density parameter contributed by non–relativistic matter, Ω_m, and by Dark Energy (DE), $\Omega_{\rm DE}$, with equation of state $p = w\rho c^2$ (if the DE term is provided by cosmological constant then $w = -1$).

In the case of small perturbations, these equations can be linearized by neglecting all the terms which are of second order in the fields δ and $\mathbf{u}$. In this case, after further differentiating (9) with respect to time, using the Euler equation to eliminate the term $\partial\mathbf{u}/\partial$t, and using the Poisson equation to eliminate $\nabla^2\phi$, one ends up with:

$$\frac{\partial^2\delta}{\partial t^2} + 2H(t)\frac{\partial\delta}{\partial t} - 4\pi G\bar{\rho}\delta = 0\,. \quad (13)$$

This equation describes the Jeans instability of a pressurless fluid, with the additional "Hubble drag" term $2H(t)\partial\delta/\partial t$, which describes the counter–action of the expanding background on the perturbation growth. Its effect is to prevent the exponential growth of the gravitational instability taking place in a non–expanding background [14]. The solution of the above equation can be casted in the form:

$$\delta(\mathbf{x},\mathrm{t}) = \delta_+(\mathbf{x},\mathrm{t_i})\mathrm{D}_+(\mathrm{t}) + \delta_-(\mathbf{x},\mathrm{t_i})\mathrm{D}_-(\mathrm{t})\,, \quad (14)$$

where D_+ and D_- describes the growing and decaying modes of the density perturbation, respectively. In the case of an Einstein–de-Sitter (EdS) Universe ($\Omega_m = 1$, $\Omega_{\rm DE} = 0$), it is $H(t) = 2/(3t)$, so that $D_+(t) = (t/t_i)^{2/3}$ and $D_-(t) = (t/t_i)^{-1}$. The fact that $D_+(t) \propto a(t)$ for an EdS Universe should not be surprising. Indeed, the dynamical time–scale for the collapse of a perturbation of uniform density ρ is $t_{\rm dyn} \propto (G\rho)^{-1//2}$, while the expansion time scale for the EdS model is $t_{\rm exp} \propto (G\bar{\rho})^{-1//2}$, where $\bar{\rho}$ is the mean cosmic density. Since for a linear (small) perturbation it is $\rho \simeq \bar{\rho}$, then $t_{\rm dyn} \sim t_{\rm exp}$, thus showing that the cosmic expansion and the perturbation evolution take place at the same pace. This argument also leads to understanding the behaviour for a $\Omega_m < 1$ model. In this case, the expansion time scale becomes shorter than the above one at the redshift at which the Universe recognizes that $\Omega_m < 1$. This happens at $1+z \simeq \Omega_m^{-1/3}$ or at $1+z \simeq \Omega_m^{-1}$ in the presence or absence of a cosmological constant term, respectively. Therefore, after this redshift, cosmic expansion takes place at a quicker pace than gravitational instability, with the result of freezing the perturbation growth.

The exact expression for the growing model of perturbations is given by

$$D_+(z) = \frac{5}{2}\Omega_m E(z)\int_z^\infty \frac{1+z'}{E(z')^3}\,\mathrm{d}z' \quad (15)$$

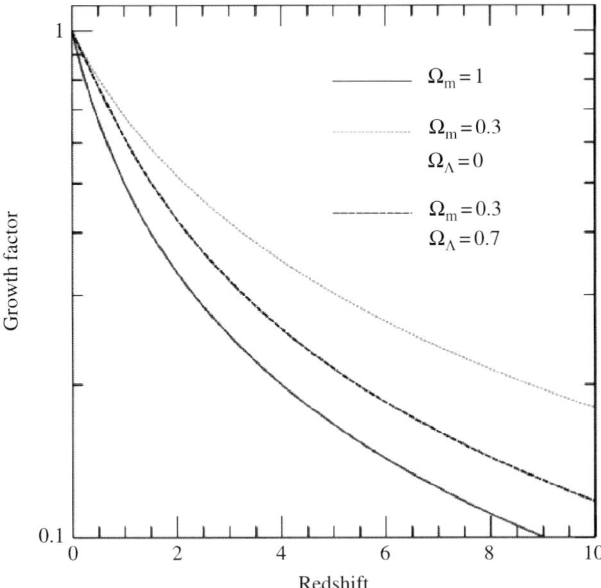

Fig. 1. The redshift dependence of the linear growth mode of perturbations for a flat model with $\Omega_m = 1$ (solid curve), for a flat $\Omega_m = 0.3$ model with a cosmological constant (dashed curve) and for an $\Omega_m = 0.3$ open model with vanishing cosmological constant (dotted curve)

(e.g., [124]). I show in Fig. 1 the redshift dependence of the linear growth factor for an Eds model and for two models with $\Omega_m = 0.3$ both with and without a cosmological constant term to restore spatial flatness. Quite apparently, the EdS has the faster evolution, while the slowing down of the perturbation growth is more apparent for the open low–density model, the presence of cosmological constant providing an intermediate degree of evolution. A more pictorial view is provided in Fig. 2, where we show the dark matter density fields for two different cosmologies and at different epochs, as obtained from N–body simulations. The two models, an EdS one and a flat low–density one with $\Omega_m = 0.3$, have been tuned so as to have a similar appearance at $z = 0$. This figure clearly shows that any observational probe of the degree of evolution of density perturbations would correspond to a sensitive probe of cosmological parameters. Such a cosmological test is conceptually different to those provided by the standard geometrical tests based on luminosity and angular–size distances.

As we shall discuss in the following, clusters of galaxies provide such a probe, since the evolution of their number density is directly related to the growth rate of perturbations.

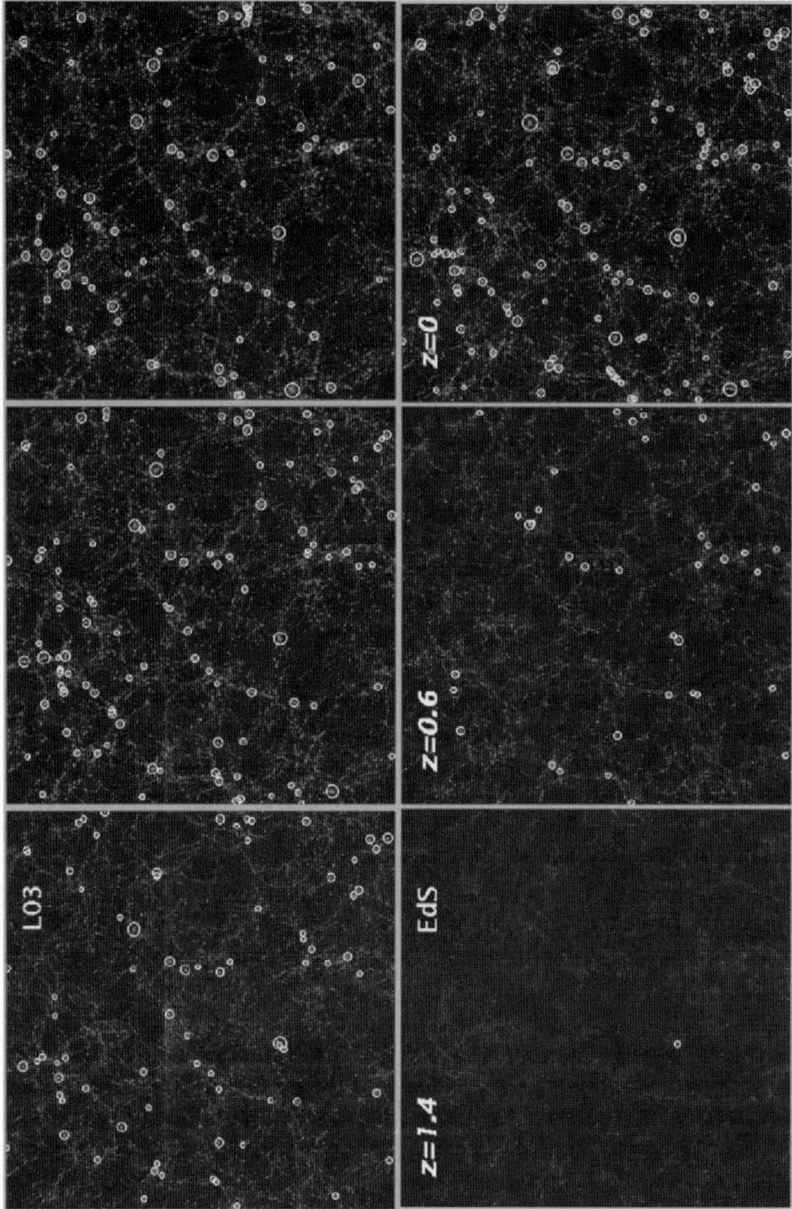

Fig. 2. The evolution of the cluster population from N–body simulations in two different cosmologies [26]. Left panels describe a flat, low–density model with $\Omega_m = 0.3$ and $\Omega_\Lambda = 0.7$ (L03); right panels are for an Einstein–de-Sitter model (EdS) with $\Omega_m = 1$. Superimposed on the dark matter distribution, the yellow circles mark the positions of galaxy clusters with virial temperature $T > 3$ keV, the size of the circles is proportional to temperature. Model parameters have been chosen to yield a comparable space density of clusters at the present time. Each snapshot is 250 h^{-1} Mpc across and 75 h^{-1} Mpc thick (comoving with the cosmic expansion)

2.3 The Spherical Top-Hat Collapse

A spherical perturbation at constant density represents the only case in which the evolution can be exactly computed. Although the assumptions on which this model is based are quite restrictive, nevertheless it serves as a very useful guideline to characterize the process of evolution and formation of virialized DM halos. This approach is based on treating the perturbation as a separate Friedmann–Lemaitre–Robertson–Walker (FLRW) universe, with the constraint of null velocity at the boundary of the perturbation. Here we will sketch the derivation in the case of $\Omega_m = 1$ (e.g., [42], while an extension of this derivation to more general cosmologies can be found in [54] and [93], with useful fitting functions provided in [31].

Assuming null velocities at an initial time t_i provides the relation $D_+(t_i) = (3/5)\delta(t_i)$, between the linear growth mode of the perturbation and the initial overdensity. The initial density parameter, which characterizes this separate Universe, is then $\Omega_p(t_i) = \Omega(t_i)(1 + \delta_i)$. Therefore, the condition for the perturbation to re-collapse will be $\Omega_p(t_i) > 1$. If this condition is satisfied, then we can derive the density within the perturbation at the time t_m of its maximum expansion (turn–around) as

$$\rho_p(t_m) = \rho_c(t_i)\Omega_p(t_i)\left[\frac{\Omega_p(t_i) - 1}{\Omega_p(t_i)}\right]^3 . \tag{16}$$

The time t_m is given by the solution of the Friedmann equations for a closed Universe:

$$t_m = \frac{\pi}{2H_i}\frac{\Omega_p(t_i)}{[\Omega_p(t_i) - 1]^{3/2}} = \left[\frac{3\pi}{32G\rho_p(t_m)}\right]^{1/2} , \tag{17}$$

where H_i is the Hubble parameter within the perturbation. At the same epoch t_m, the density of the general cosmic background is $\rho(t_m) = (6\pi G t_m^2)^{-1}$. Therefore, the exact value for the perturbation overdensity at the turn–around is

$$\delta_+(t_m) = \frac{\rho_p(t_m)}{\rho(t_m)} - 1 = \left(\frac{3\pi}{4}\right)^2 - 1 \simeq 4.6 . \tag{18}$$

On the other hand, the linear–theory extrapolation to t_m would give

$$\delta_+(t_m) = \delta_+(t_i)\left(\frac{t_m}{t_i}\right)^{2/3} = \frac{3}{5}\left(\frac{3\pi}{4}\right)^{2/3} \simeq 1.07 . \tag{19}$$

This demonstrates that the linear–theory extrapolation significantly underestimates overdensities at the turn-around.

After reaching the maximum expansion, the perturbation then evolves by detaching from the general Hubble expansion and then re-collapses, reaching virial equilibrium supported by the velocity dispersion of DM particles. This happens at the virialization time $t_{\rm vir}$, at which the perturbation meets by definition the virial condition $E = K + U = -K$, being E, K and U the total, the kinetic and the potential energy, respectively.

At the turn–around point, the perturbation has no kinetic energy, so that the total energy is

$$E_m = U = -\frac{3}{5}\frac{GM^2}{R_m}\,, \tag{20}$$

where we have used the expression for the potential energy of a uniform spherical density field of radius R_m and total mass M. In a similar manner, the total energy at the virialization is

$$E_{\rm vir} = \frac{U}{2} = -\frac{1}{2}\frac{3}{5}\frac{GM^2}{R_{\rm vir}}\,. \tag{21}$$

Therefore, the condition of energy conservation in a dissipationless collapse gives $R_m = 2R_{\rm vir}$ for the relation between the radii at turn-around and at virial equilibrium. This allows us to compute the overdensity at $t_{\rm vir}$ as

$$\frac{\rho_p(t_{\rm vir})}{\rho(t_{\rm vir})} = \left(\frac{t_{\rm vir}}{t_m}\right)^2 \left(\frac{R_m}{R_{\rm vir}}\right)^3 \frac{\rho_p(t_m)}{\rho(t_m)} = 2^2 2^3 \left(\frac{3\pi}{4}\right)^2 = 18\pi^2 \simeq 178\,, \tag{22}$$

where we have accounted for both the compression of the perturbation density, due to its shrinking, and of the dilution of the background density as the Universe expands from t_m to $t_{\rm vir}$. Equation (22) shows why an overdensity of about 200 is usually considered as typical for a DM halo which has reached the condition of virial equilibrium. As for the extrapolation of linear–theory prediction, it would have given

$$\delta_+(t_{\rm vir}) = \left(\frac{t_{\rm vir}}{t_m}\right)^{2/3} \delta_+(t_m) \simeq 1.69\,. \tag{23}$$

The above equation shows the derivation of another fundamental number that will be used in what follows in order to characterize the mass function of virialized halos. It gives the overdensity that a perturbation in the initial density field must have for it to end up in a virialized structure. While the above derivation holds for an EdS Universe, it can be generalized to any generic cosmology. For $\Omega_m < 1$ the increased expansion rate of the Universe causes a faster dilution of the cosmic density from t_m to $t_{\rm vir}$ and, as a consequence, a larger value of the overdensity at virialization.

In the following, we will indicate with $\Delta_{\rm vir}$ the overdensity at virial equilibrium, computed with respect to the background density, and with Δ_c the same quantity expressed in units of the critical density ρ_c. As a reference, a flat low–density model with $\Omega_m = 0.3$ has $\Delta_c \simeq 100$ and $\Delta_{\rm vir} \simeq 330$. Also, we will use in the following the notation R_N to indicate the radius of a halo encompassing an average overdensity equal to $N\rho_c$, so that M_N will denote the halo mass contained within that radius. As we shall see in the following, values often used in the literature are $N = 200$, 500 and 2500.

3 The Mass Function

The mass function (MF) at redshift z, $n(M,z)$, is defined as the number density of virialized halos found at that redshift with mass in the range $[M, M+\mathrm{d}M]$. In this section I will derive the MF expression following the approach originally devised by Press and Schechter [132] (PS hereafter). After commenting on the limitations of this approach, I will discuss the accuracy with which improved derivations of the MF reproduce the "exact" predictions from N-body simulations.

3.1 The Press-Schechter Mass Function

The PS derivation of the MF is based on the assumption that the fraction of matter ending up in objects of a given mass M can be found by looking at the portion of the initial (Lagrangian) density field, smoothed on the mass–scale M, lying at an overdensity exceeding a given critical threshold value, δ_c. Under the assumption of Gaussian perturbations, the probability for the linearly-evolved smoothed field δ_M to exceed at redshift z the critical density contrast δ_c reads

$$p_{>\delta_c}(M,z) = \frac{1}{\sqrt{2\pi}\sigma_M(z)} \int_{\delta_c}^{\infty} \exp\left(-\frac{\delta_M^2}{2\sigma_M(z)^2}\right) \mathrm{d}\delta_M = \frac{1}{2}\mathrm{erfc}\left(\frac{\delta_c}{\sqrt{2}\sigma_M(z)}\right), \tag{24}$$

where $\mathrm{erfc}(x)$ is the complement error function and $\sigma_M(z) = \delta_+(z)\sigma_M$ is the variance at the mass scale M linearly extrapolated at redshift z. Under the assumption of spherical collapse, the critical overdensity δ_c is given by the linear extrapolation of the overdensity at virial equilibrium, as derived in the previous section. In this case, it will be $\delta_c = \delta_c(z)$ with a weak dependence upon redshift and cosmological parameters, with $\delta_c \simeq 1.69$ independent of z only in the case of an EdS cosmology. By definition, the above equation provides the fraction of unity volume, which ends up by redshift z in objects with mass above M. Therefore, the fraction of Lagrangian volume in objects with mass in the range $[M, M+\mathrm{d}M]$ is

$$\mathrm{d}p_{>\delta_c}(M,z) = \left|\frac{\partial p_{>\delta_c}(M,z)}{\partial M}\right| \mathrm{d}M\,. \tag{25}$$

Since the probability of (24) is a decreasing function of mass, the absolute value is required in order to have a positive–defined differential probability. Equation (25) shows a fundamental limitation of the PS derivation of the MF. Indeed, we expect that, as we take the limit of arbitrarily small limiting mass, we should recover the whole mass content of the Universe. This is to say that, in the hierarchical clustering picture, all the mass is contained within halos of arbitrarily small mass. However, integrating (25) over the whole mass range gives $\int_0^\infty \mathrm{d}p_{>\delta_c}(M,z) = 1/2$. This implies that the PS derivation of the mass function only accounts for half of the total mass at disposition. The basic reason for this is that, in this derivation, we give zero probability for a point with $\delta_M < \delta_c$, for a given filtering mass scale M, to have $\delta_{M'} > \delta_c$ for some larger filtering scale $M' > M$. This means that the PS approach neglects the possibility for that point to end up in a collapsed halo of larger mass. A more rigorous derivation of the mass function, which is based on the excursion–set formalism [23], correctly accounts for the missing factor 2, at least for the particular choice of a sharp–k filter (i.e., a top–hat window function in Fourier space).

Since (25) provides the fraction of volume in objects of a given mass, the number density of such objects will be obtained after dividing it by the volume, $V_M = M/\bar{\rho}$, occupied by each object. Therefore, after accounting for the missing factor 2, the expression for the mass function reads

$$\begin{aligned}\frac{\mathrm{d}n(M,z)}{\mathrm{d}M} &= \frac{2}{V_M}\frac{\partial p_{>\delta_c}(M,z)}{\partial M} \\ &= \sqrt{\frac{2}{\pi}}\frac{\bar{\rho}}{M^2}\frac{\delta_c}{\sigma_M(z)}\left|\frac{\mathrm{d}\log\sigma_M(z)}{\mathrm{d}\log M}\right|\exp\left(-\frac{\delta_c^2}{2\sigma_M(z)^2}\right) . \end{aligned} \quad (26)$$

This is the expression for the PS mass function. Although we will present below a more accurate expressions for the MF, this equation already demonstrates the reason for which the mass function of galaxy clusters is a powerful probe of cosmological models. Cosmological parameters enter in (26) through the mass variance σ_M, which depends on the power spectrum and on the cosmological density parameters, through the linear perturbation growth factor, and, to a lesser degree, through the critical density contrast δ_c. Taking this expression in the limit of massive objects (i.e., rich galaxy clusters), the MF shape is dominated by the exponential tail. This implies that the MF becomes exponentially sensitive to the choice of the cosmological parameters. In other words, a reliable observational determination of the MF of rich clusters would allow us to place tight constraints on cosmological parameters.

3.2 Extensions of the PS Approach and N-body Tests

Following [89], an alternative way of recasting the mass function is

$$f(\sigma_M, z) = \frac{M}{\bar{\rho}}\frac{\mathrm{d}n(M,z)}{\mathrm{d}\ln\sigma_M^{-1}} . \quad (27)$$

In this way, the PS expression is recovered by setting

$$f(\sigma_M, z) = \sqrt{\frac{2}{\pi}}\frac{\delta_c}{\sigma_M}\exp\left(-\frac{\delta_c^2}{2\sigma_M^2}\right) \quad (28)$$

Despite its subtle simplicity (e.g., [112]), the PS MF has served for more than a decade as a guide to constrain cosmological parameters from the mass distribution of galaxy clusters. Only with the advent of a new generation of N–body simulations, which are able to cover a very large dynamical range, have significant deviations of the PS expression from the exact numerical description been noticed (e.g., [59, 76, 77, 89, 150, 165]). Such deviations have been usually interpreted in terms of corrections to the PS approach.

Incorporating the effect of non–spherical collapse, the PS expression has been generalized [146] to

$$f(\sigma_M, z) = \sqrt{\frac{2a}{\pi}} C \left[1 + \left(\frac{\sigma_M^2}{a\delta_c^2}\right)^q\right] \frac{\delta_c}{\sigma_M} \exp\left(-\frac{a\delta_c^2}{2\sigma_M^2}\right) . \qquad (29)$$

These authors also compared this expression with results from N–body simulations, in which the mass of the clusters were estimated with a spherical overdensity (SO) algorithm, by computing the mass within the radius encompassing a mean overdensity equal to the virial one. As a result, they found the best-fitting values $a = 0.707$, $q = 0.3$, with the normalization constant $C = 0.3222$ obtained from the normalization requirement $\int_0^\infty f(\sigma_M)\mathrm{d}\nu = 1$ (note that the PS expression is recovered for $a = 1$, $q = 0$ and $C = 1/2$; see also [147]).

Jenkins et al. [89] proposed an alternative expression for the mass function:

$$f(\sigma_M, z) = 0.315 \exp(-|\ln \sigma_M^{-1} + 0.61|^{3.8}) , \qquad (30)$$

which has been obtained as the best fit to the results of a combination of different simulations, covering a wide dynamical range. More recently, Springel et al. [150] used the largest available single N–body simulation to verify in detail the accuracy of (30). The result of this comparison, which is reported in Fig. 3, demonstrates that this mass function reproduces remarkably well numerical results over a wide range of sampled halo masses and redshifts, thereby representing a substantial improvement with respect to the PS mass function. The accuracy of (30) in reproducing results of numerical experiments has been also discussed in [59], where it is also pointed out the role of different algorithms to identify clusters and to estimate their mass in simulations, in [166], where the universality of this expression for a generic cosmology is discussed, and in [165], where the widest dynamical range to date has been samples by combining a series of N-body simulations.

In practical applications, the observational mass function of clusters is usually determined over about one decade in mass. Therefore, it probes the power spectrum over a relatively narrow dynamical range, and does not provide strong constraints on the shape of the power spectrum. Using only the number density of nearby clusters of a given mass M, one can constrain the amplitude of the density perturbation at the physical scale $R \propto (M/\Omega_m \rho_{\rm crit})^{1/3}$ which contains this mass. Since such a scale depends both on M and on Ω_m, the mass function of nearby ($z \lesssim 0.1$) clusters is only able to constrain a relation between σ_8 and Ω_m. In the left panel of Fig. 4 we show that, for a fixed value

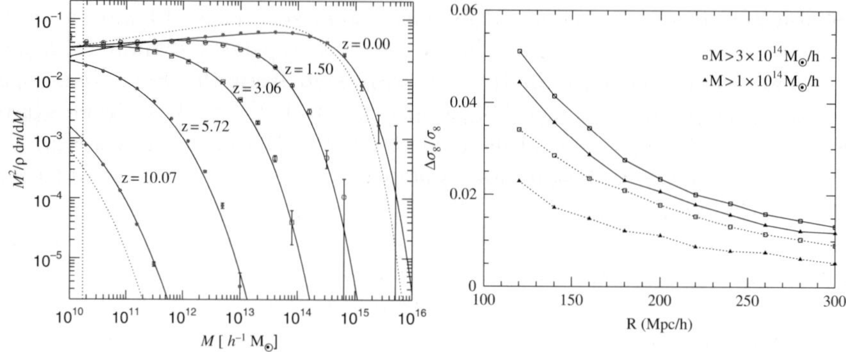

Fig. 3. Left panel: the mass function of DM halos (*dots with errorbars*) identified at different redshifts in the Millenniun Run [150], compared to the predictions of the mass function by [89] and by [132]. The two model mass functions are plotted with solid and dotted curves, respectively. **Right panel**: the relative standard deviation of σ_8 as a function of the sample size [166] from the cumulative mass function above two different mass limits. The solid lines indicate the width of the distribution when including the clustering of clusters

of the observed cluster mass function, the implied value of σ_8 from (29) increases as the density parameter decreases. Determinations of the cluster mass function in the local Universe using a variety of samples and methods indicate that $\sigma_8\Omega_m^\alpha = 0.4-0.6$, where $\alpha \simeq 0.4-0.6$, almost independent of the presence of a cosmological constant term providing spatial flatness. As for the

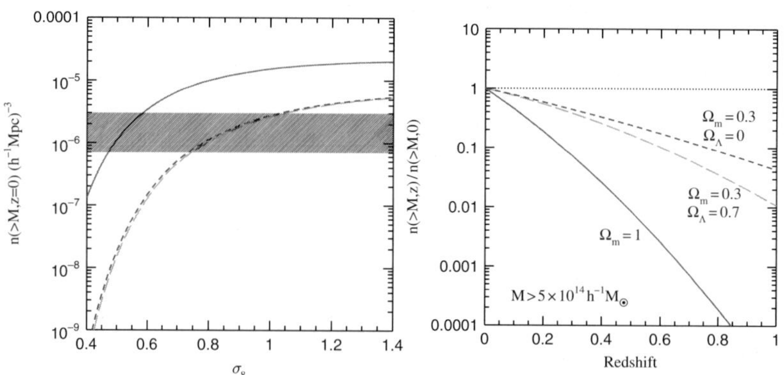

Fig. 4. The sensitivity of the cluster mass function to cosmological models [138]. **Left panel**: The cumulative mass function at $z=0$ for $M > 5\times10^{14}h^{-1}M_\odot$ for three cosmologies, as a function of σ_8, with shape parameter $\Gamma = 0.2$; *solid line*: $\Omega_m = 1$; short–dashed line: $\Omega_m = 0.3$, $\Omega_\Lambda = 0.7$; long–dashed line: $\Omega_m = 0.3$, $\Omega_\Lambda = 0$. The shaded area indicates the observational uncertainty in the determination of the local cluster space density. **Right panel**: Evolution of $n(>M, z)$ for the same cosmologies and the same mass–limit, with $\sigma_8 = 0.5$ for the $\Omega_m = 1$ case and $\sigma_8 = 0.8$ for the low–density models

evolution with redshift, the growth rate of the density perturbations depends primarily on Ω_m and, to a lesser extent, on Ω_Λ, at least out to $z \sim 1$, where the evolution of the cluster population is currently studied. Therefore, following the evolution of the cluster space density over a large redshift baseline, one can break the degeneracy between σ_8 and Ω_m. This is shown in a pictorial way in Fig. 2 and quantified in the right panel of Fig. 4: models with different values of Ω_m, which are normalized to yield a comparable number density of nearby clusters, predict cumulative mass functions that progressively differ by up to orders of magnitude at increasing redshifts.

Although (30) provides a very accurate and flexible tool to constrain the parameter space of cosmological models using the mass function of collapsed halos, nevertheless a further source of uncertainty may arise from the effect of cosmic variance. Fluctuations modes, with wavelength exceeding the size of the volumes sampled by observations, induces appreciable changes in the number counts of halos of a given mass. This effect has been thoroughly discussed in [87, 166]. In the right panel Fig. 3 (from [166]) I report the relative variation of the power spectrum normalization, σ_8, induced by cosmic variance, as a function of the sample size, for halos having two different mass limits. As expected, the variance decreases with the sample size (fluctuations on larger scales have a smaller effect), while it increases with the halo mass (the distribution of rarer objects suffer for a more pronounced large–scale modulation). This result demonstrates that a precision calibration of cosmological parameters requires properly accounting for the effect of cosmic variance.

4 Building a Cluster Sample

4.1 Identification in the Optical / Near IR Band

Abell [1] provided the first extensive, statistically complete sample of galaxy clusters, later extended to the Southern hemisphere [2]. Based on purely visual inspection, clusters were identified as enhancements in the galaxy surface density and were characterized by their *richness* and estimated distance. The Abell catalog has been for decades the prime source for detailed studies of individual clusters and for characterizing the large scale distribution of matter in the nearby Universe. Several variations of the Abell criteria defining clusters were used in an automated and objective fashion when digitized optical plates became available (e.g., [45, 102]). Deep optical plates were used successfully to search for more distant clusters, out to $z \simeq 0.9$, with purely visual techniques (e.g., [43, 78]). These searches for distant clusters became much more effective with the advent of CCD imaging. Postman et al. [131] were the first to carry out a V&I-band survey over 5 deg^2 (the Palomar Distant Cluster Survey, PDCS). This technique enhances the contrast of galaxy overdensity at a given position, utilizing prior knowledge of the luminosity profile typical of galaxy clusters. Dalcanton [44] proposed another method of optical selection of clusters, in which drift scan imaging data from relatively small telescopes is used to detect clusters as positive surface brightness fluctuations in the background sky. Gonzalez et al. [75] applied a technique based on surface brightness fluctuations from drift scan imaging data to build a sample of $\sim$1000 cluster candidates over 130 deg^2.

A common feature of all these methods of cluster identification is that they classify clusters according to definitions of richness, which generally have a loose relation with the actual cluster mass. This represents a serious limitation for any cosmological application, which requires the observable, on which the cluster selection is based, to be a reliable proxy of the cluster mass.

An improved definition of richness, based on the amplitude of the galaxy–cluster cross–correlation function, has been applied [74] to clusters identified in a large area survey in R and z bands (the Red Sequence Cluster Survey). This survey, whose optical and X-ray follow-up, is currently underway, promises to unveil a fairly large number of clusters out to $z \sim 1.5$.

By increasing the number of observed passbands and using red colors one can increase the contrast with which clusters are seen in color space. In this way, one can increase the efficiency of cluster selection also at high redshift (e.g., [74, 151, 152]) and the accuracy of their estimated redshifts through spectro-photometric techniques. In this way, Miller et al. [111] designed a cluster-finding algorithm which makes full use of information of both position and color space to detect clusters of galaxies from the SDSS. They were able to identify about 750 clusters out to $z \lesssim 0.2$, and assessed the degree of completeness by resorting to a comparison with mock SDSS surveys extracted from large N–body simulations. Once completed, the search of clusters over the entire SDSS sample will provide about 2500 nearby and medium–distant objects. At the same time the next generation of wide field (> 100 deg^2) deep multicolor surveys in the optical and especially the near-infrared will powerfully enhance the search for distant clusters, out to $z \gtrsim 1$.

4.2 Identification in the X-ray Band

Already from the first pioneering attempts to map the X-ray sky ([66], see [138] for a historical review), clusters were associated with extended sources, whose dominant emission mechanism was recognized to be thermal bremsstrahlung from optically thin plasma at a temperature of several keV [40, 61]. The all-sky survey conducted by the the HEAO-1 X-ray Observatory was the first to provide a flux–limited sample of X-ray identified clusters, for which both the flux number counts and the X-ray luminosity function have been computed for the first time [126]. However, it is only thanks to the much improved sensitivity of the *Einstein Observatory* [65] that X-ray surveys were recognized as an efficient means of constructing samples of galaxy clusters out to cosmologically interesting redshifts.

First, the X-ray selection has the advantage of revealing physically-bound systems, because diffuse emission from a hot ICM is the direct manifestation of the existence of a potential-well within which the gas is in dynamical equilibrium with the cool baryonic matter (galaxies) and the dark matter. Second, the X-ray luminosity is well correlated with the cluster mass (see Fig. 11). Third, the X-ray emissivity is proportional to the square of the gas density, hence cluster emission is more concentrated than the optical bidimensional galaxy distribution. In combination with the relatively low surface density of X-ray sources, this property makes clusters high contrast objects in the X-ray sky, and alleviates problems due to projection effects that affect optical selection. Finally, an inherent fundamental advantage of X-ray selection is the ability to define flux-limited samples with well-understood selection functions. This leads to a simple evaluation of the survey volume and therefore to a straightforward computation of space densities. Nonetheless, there are some important caveats described below. Pioneering work in this field [67, 84] was based on the *Einstein Observatory* Extended Medium Sensitivity Survey (EMSS). The EMSS survey covered over 700 square degrees and lead to the construction of a flux-limited sample of 93 clusters out to $z = 0.58$, allowing the cosmological evolution of clusters to be investigated.

The *ROSAT* satellite, launched in 1990, allowed a significant step forward in X-ray surveys of clusters. The *ROSAT* All-Sky Survey (RASS, [155]) was the first X-ray imaging mission to cover the entire sky, thus paving the way to large contiguous-area surveys of X-ray selected nearby clusters. In the northern hemisphere, the largest compilations with virtually complete optical identification include, the Bright Cluster Sample (BCS, [51]), and the Northern *ROSAT* All Sky Survey (NORAS, [22]). In the southern hemisphere, the *ROSAT*-ESO flux limited X-ray (REFLEX) cluster survey [21] has completed the identification of 452 clusters, the largest, homogeneous compilation to date. The Massive Cluster Survey (MACS, [52]) is aimed at targeting the most luminous systems at $z > 0.3$ which can be identified in the RASS at the faintest flux levels. The deepest area in the RASS, the North Ecliptic Pole (NEP, [85]) which *ROSAT* scanned repeatedly during its All-Sky survey, was used to carry out a complete optical identification of X-ray sources over a 81 deg^2 region. This study yielded 64 clusters out to redshift $z = 0.81$.

In total, surveys covering more than 10^4 deg^2 have yielded over 1000 clusters, out to redshift $z \simeq 0.5$. A large fraction of these are new discoveries, whereas approximately one third are identified as clusters in the Abell or Zwicky catalogs. For the homogeneity of their selection and the high degree of completeness of their spectroscopic identifications, these samples are now the basis for a large number of follow-up investigations and cosmological studies.

Besides the all-sky surveys, the *ROSAT-PSPC* archival pointed observations were intensively used for serendipitous searches of distant clusters. These projects, which are now completed, include: the RIXOS survey [38], the *ROSAT* Deep Cluster Survey (RDCS, [138, 139]), the Serendipitous High-Redshift Archival *ROSAT* Cluster survey (SHARC, [32], the Wide Angle *ROSAT* Pointed X-ray Survey of clusters (WARPS, [125]), the 160 deg^2 large area survey [117], the *ROSAT* Optical X-ray Survey (ROXS, [49]).

ROSAT-HRI pointed observations have also been used to search for distant clusters in the Brera Multi-scale Wavelet catalog (BMW, [113]).

In Fig. 5, we give an overview of the flux limits and surveyed areas of all major cluster surveys carried out over the last two decades. RASS-based surveys have the advantage of covering contiguous regions of the sky so that the clustering properties of clusters (e.g., [143]) can be investigated. They also have the ability to unveil rare, massive systems albeit over a limited redshift and X-ray luminosity range. Serendipitous surveys which are at least a factor of ten deeper but cover only a few hundreds square degrees, provide complementary information on lower luminosities, more common systems and are well suited for studying cluster evolution on a larger redshift baseline.

A number of systematic studies have been carried out to compare the nature of clusters identified with the optical and the X–ray technique (e.g., [13, 48, 129]). The general conclusion of these studies is that optically selected

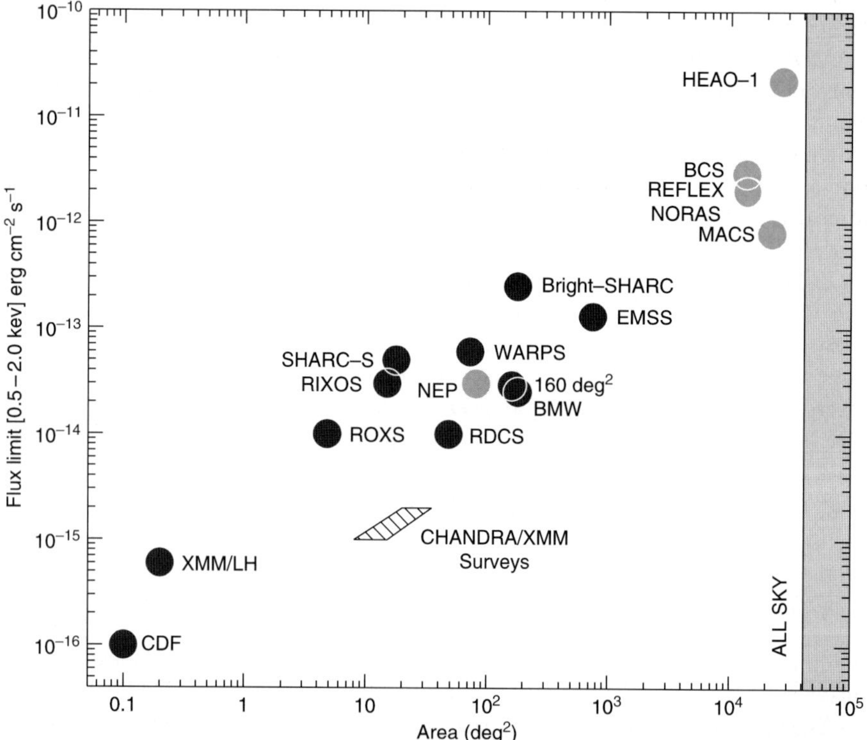

Fig. 5. Solid angles and flux limits of X-ray cluster surveys carried out over the last two decades. Dark filled circles represent serendipitous surveys constructed from a collection of pointed observations. Light shaded circles represent surveys covering contiguous areas. The hatched region is a predicted locus of current serendipitous surveys with *Chandra* and *Newton-XMM*. From [138]

clusters are on average underluminous in the X-ray band. This suggests that optical selection tends to pick up objects which have not yet reached a high enough density to make the ICM lighting up in X–rays.

In order for a survey to be used for cosmological applications, one needs to know not only how many clusters it contains, but also the volume within which each of them is found. In other words, one needs to define the selection function of the survey, which depends on the survey strategy and on the details of the adopted cluster finding algorithm (see [138], for a review). An essential ingredient for the evaluation of the selection function of X-ray surveys is the computation of the sky coverage: the effective area covered by the survey as a function of flux. In general, the exposure time, as well as the background and the PSF are not uniform across the field of view of X-ray telescopes, which introduces vignetting and a degradation of the PSF at increasing off-axis angles. As a result, the sensitivity to source detection varies significantly across the survey area so that only bright sources can be detected over the entire solid angle of the survey, whereas at faint fluxes the effective area decreases. An example of survey sky coverage is given in the left panel of Fig. 6. By covering different solid angles at varying fluxes, these surveys probe different volumes at increasing redshift and therefore different ranges in X-ray luminosities at varying redshifts.

Once the survey flux-limit and the sky coverage are defined one can compute the maximum search volume, $V_{\rm max}$, within which a cluster of a given luminosity is found in that survey:

$$V_{\rm max} = \int_0^{z_{\rm max}} S[f(L,z)] \left(\frac{d_L(z)}{1+z}\right)^2 \frac{c\,{\rm d}z}{H(z)} \, . \tag{31}$$

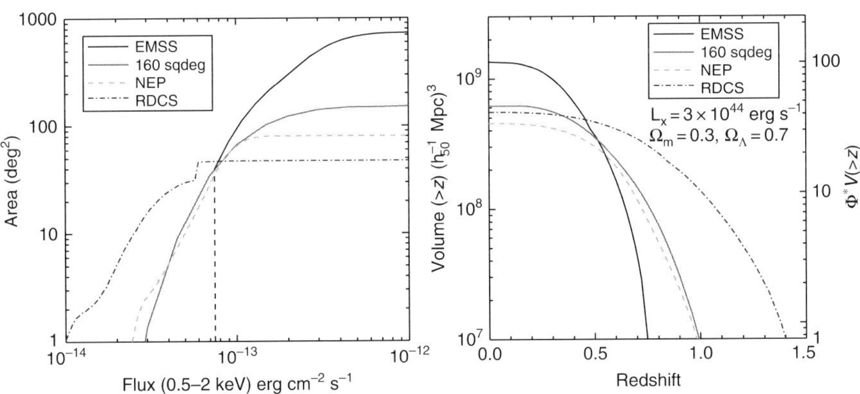

Fig. 6. Left panel: sky coverage as a function of X-ray flux of several serendipitous surveys. **Right panel**: corresponding search volumes, $V(>z)$, for a cluster of given X-ray luminosity ($L_X = 3 \times 10^{44}$ [0.5–2 keV] $\simeq L^*_X$). From [138]

Here $S(f)$ is the survey sky coverage, which depends on the flux $f = L/(4\pi d_L^2)$, $d_L(z)$ is the luminosity distance, and $H(z)$ is the Hubble constant at z. We define $z_{\max}$ as the maximum redshift out to which the flux of an object of luminosity L lies above the flux limit. The corresponding survey volumes are shown in the right panel of Fig. 6.

Once again, I emphasize that one of the main advantages of the X–ray selection lies in the fact that the survey selection function can be precisely computed, thus allowing reliable comparisons between the observed and the predicted evolution of the cluster population.

4.3 Identification Through the SZ Effect

The Sunyaev–Zel'dovich (SZ) effect [154] allows to observe galaxy clusters by measuring the distortion of the CMB spectrum owing to the hot ICM. This method does not depend on redshift and provides in principle a reliable estimate of cluster masses. For these reasons, it is now considered as one of the most powerful means to find distant clusters in the years to come. For a detailed discussion of the SZ technique for cluster identification and for the ongoing and future surveys, I refer to the lectures by Mark Birkinshaw and to the reviews in [15, 37]. For the purpose of the present discussion, I show in the left panel of Fig. 7 a comparison between the limiting mass as a function of redshift, expected for a X–ray and for a SZ cluster survey (from [80]). While the standard flux dimming with the luminosity distance, $f_X \propto d_L^2(z)$, causes

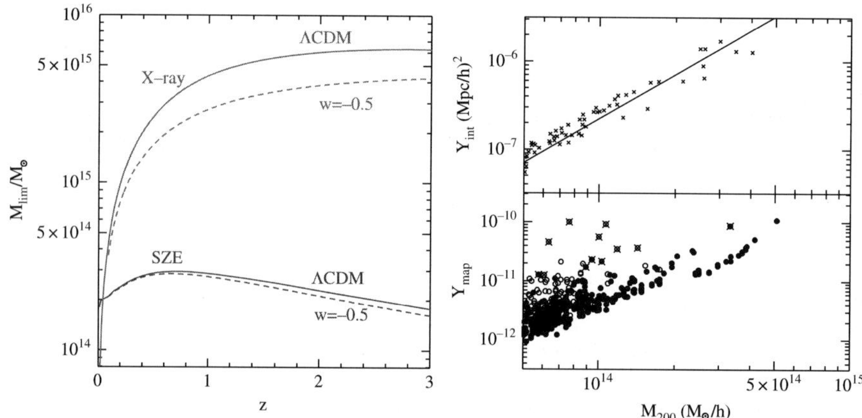

Fig. 7. Left panel: limiting cluster virial mass for detection in an X–ray and in a SZ survey (from [80]). Each pair of curves show the results for two $\Omega_m = 0.3$ cosmologies, having $w = -1$ and $w = -0.5$ for the DE equation of state. **Right panel**: the relation between the Comptonization parameter and M_{200}, from [167]. The upper panel shows the decrement contributed from the gas within $0.5R_{200}$. The lower panel indicates the signal from noise–free maps projected on the light cone

the limiting mass to quickly increase with distance for the X–ray selection, this limiting mass has a much less sensitive dependence on redshift for the SZ selection. This is the reason why SZ surveys are generally considered as essentially providing mass–limited cluster samples.

It has been recently pointed out [115] that the integrated SZ flux–decrement has a very tight correlation with the total cluster mass (see also [47]). This fact, joined with the redshift–independence of the SZ selection, makes the SZ identification a promising route toward precision cosmology with galaxy clusters.

A potential problem with the SZ identification of clusters resides in the possible contamination of the signal from foreground/background structures. Diffuse gas, residing in large–scale filaments, are likely to provide a negligible contamination, as a consequence of the comparatively low density and temperature which characterize such structures. However, small halos, which are expected to be present in large number, contain gas at the virial overdensity. Since they are not resolved in current SZ observations, their integrated contribution may provide a significant contamination. Using cosmological hydrodynamical simulations, White et al. [167] have created SZ sky maps with the aim of correlating the SZ signal seen in projection with the actual mass of clusters.The result of this test is shown in the right panel of Fig. 7. The upper panel shows the relation between the integrated SZ signal contributed only from the gas within $0.5R_{200}$ and M_{200}, while the lower panel is when using the actual Compton-y parameter measured from the projected maps. Quite apparently, the scatter in the relation is significantly increased in projection. Part of the scatter is due to the different redshifts at which clusters seen in projection are placed. This contribution to the scatter can be removed once redshifts of clusters are known from follow–up optical observations. However, a significant contribution to the overall scatter is contributed by cluster asphericity and by contamination from fore/background structures. This highlights the relevance of keeping this scatter under control for a full exploitation of the SZ signal as a tracer of the cluster mass.

5 Methods to Estimate Cluster Masses

5.1 The Hydrostatic Equilibrium

The condition of hydrostatic equilibrium determines the balance between the pressure force and the gravitational force: $\nabla P_{\rm gas} = -\rho_{\rm gas}\nabla\phi$, where $P_{\rm gas}$ and $\rho_{\rm gas}$ are the gas pressure and density, respectively, while ϕ is the underlying gravitational potential. Under the assumption of a spherically symmetric gas distribution, the above equations read:

$$\frac{{\rm d}P_{\rm gas}}{{\rm d}r} = -\rho_{\rm gas}\frac{{\rm d}\phi}{{\rm d}r} = -\rho_{\rm gas}\frac{GM(<r)}{r^2}\,, \tag{32}$$

where r is the radial coordinate (cluster-centric distance) and $M(<r)$ is the total mass contained within r. Using the equation of state of ideal gas to relate pressure to gas density and temperature, the mass is then given by

$$M(<r) = -\frac{r}{G}\frac{k_B T}{\mu m_p}\left(\frac{\mathrm{d}\ln\rho_{\mathrm{gas}}}{\mathrm{d}\ln r} + \frac{\mathrm{d}\ln T}{\mathrm{d}\ln r}\right), \tag{33}$$

where μ is the mean molecular weight of the gas ($\mu \simeq 0.59$ for primordial composition) and m_p is the proton mass. An often used mass estimator is based on assuming the β–model for the gas density profile,

$$\rho_{\mathrm{gas}}(r) = \frac{\rho_0}{[1+(r/r_c)^2]^{3\beta/2}} \tag{34}$$

[39]. In the above equation, r_c is the core radius, while β is the ratio between the kinetic energy of any tracer of the gravitational potential (e.g. galaxies) and the thermal energy of the gas, $\beta = \mu m_p \sigma_v^2/(k_B T)$ (σ_v: one–dimensional velocity dispersion). By further assuming a polytropic equation of state, $\rho_{\mathrm{gas}} \propto P_{\mathrm{gas}}^{\gamma}$ (γ: polytropic index), (33) becomes

$$M(<r) \simeq 1.11\times 10^{14}\beta\gamma\frac{T(r)}{\mathrm{keV}}\frac{r}{h^{-1}\mathrm{Mpc}}\frac{(r/r_c)^2}{1+(r/r_c)^2}M_\odot\,, \tag{35}$$

where $T(r)$ is the temperature at the radius r. In its original derivation, the β–model was aimed at representing the distribution of isothermal gas sitting in hydrostatic equilibrium within a King–like potential. The corresponding mass estimator is recovered from (35) by setting $\gamma = 1$ and replacing $T(r)$ with the global ICM temperature, T_0. In the absence of accurately resolved temperature profiles from X–ray observations, (35) has been used to estimate cluster masses both in its isothermal (e.g., [136]) and in its polytropic form (e.g., [56, 62, 120]).

Thanks to the much improved sensitivity of the Chandra and XMM–Newton X–ray observatories, temperature profiles are now resolved with high enough accuracy to allow the application of more general methods of mass estimation (providing tight M–L_x relations; see for example Fig. 8), not necessarily bound to the assumptions of β–model and of an overall polytropic form for the equation of state (e.g., [5, 8, 56, 160]).

An alternative way of recasting the isothermal version of (35) between temperature and mass is based on expressing the mass according to the virial theorem as $M_{\mathrm{vir}} = \sigma_v^2 R_{\mathrm{vir}}/G$, so that

$$k_B T = \frac{1.38}{\beta}\left(\frac{M_{\mathrm{vir}}}{10^{15}M_\odot}\right)^{3/2}[\Omega_m\Delta_{\mathrm{vir}}(z)]^{1/3}(1+z)\,\mathrm{keV}\,. \tag{36}$$

This expression, originally introduced in [54], has been sometimes used to express the M–T relation as obtained from hydrodynamical simulations of galaxy clusters (e.g., [25, 31]).

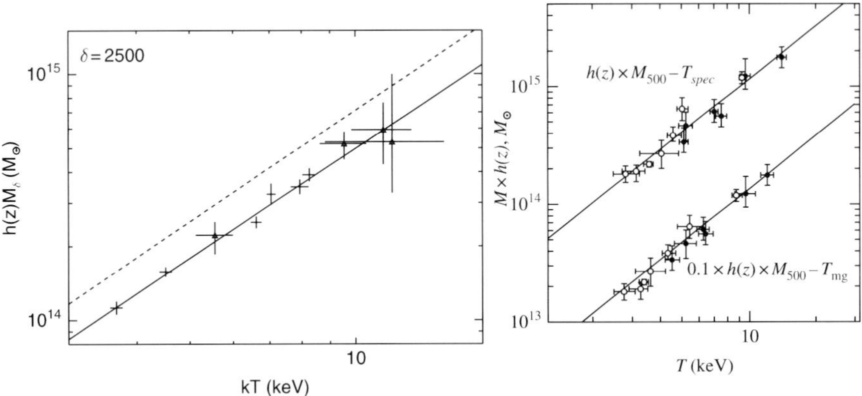

Fig. 8. The mass-temperature relation for nearby clusters (from [8]) and for distant clusters (from [95]), based on a combination of Chandra and XMM–Newton data

It is clear that the two crucial assumptions underlying any mass measurements based on the ICM temperature concerns the existence of hydrostatic equilibrium and of spherical symmetry. While effects of non–spherical geometry can be averaged out by performing the analysis over a large enough number of clusters, the former can lead to systematic biases in the mass estimates (e.g., [133]) and references therein). So far, ICM temperature measurements have been based on fits of the observed X–ray spectra of clusters to plasma models, which are dominated at high temperatures by thermal bremsstrahlung. However, local deviations from isothermality, e.g. due to the presence of merging cold gas clumps, can bias the spectroscopic temperature with respect to the actual electron temperature (e.g., [108, 110, 159]). This bias directly translates into a comparable bias in the mass estimate through hydrostatic equilibrium (see Sect. 7, below).

5.2 The Dynamics of Member Galaxies

From a historical point of view, the dynamics traced by member galaxies, has been the first method applied to measure masses of galaxy clusters [148, 171]. Under the assumption of virial equilibrium, the mass of the cluster can be estimated by knowing position and redshift for a high enough number of member galaxies:

$$M = \frac{\pi}{2}\frac{3\sigma_v^2 R_V}{G} \tag{37}$$

(e.g., [99]), where the first factor accounts for the geometry of projection, σ_r is the line-of-sight velocity dispersion and R_V is the virialization radius, which depends on the positions of the galaxies with measured redshifts and recognized as true cluster members:

$$R_V = N^2 \left(\sum_{i>j} r_{ij}^{-1} \right)^{-1} , \tag{38}$$

where N is the total number of galaxies, and r_{ij} the projected separation between the i-th and j-th galaxies. This method has been extensively applied to measure masses for statistical samples of both nearby (e.g., [16, 17, 71, 130, 137]) and distant (e.g., [36, 73]) clusters.

Besides the assumption of virial equilibrium, which may be fulfilled to different degrees by different populations of galaxies (e.g., late vs. early type), a crucial aspect in the application of the dynamical mass estimator concerns the rejection of interlopers, i.e. of back/foreground galaxies which lie along the line-of-sight of the cluster without belonging to it. A spurious inclusion of non–member galaxies in the analysis leads in general to an overestimate of the velocity dispersion and, therefore, of the resulting mass. A number of algorithms have been developed for interlopers rejection, whose reliability must be judged on a case-by-case basis (e.g., [68, 156]). A further potential problem of this analysis concerns the possibility of realizing a uniform sampling of the cluster potential using galaxies with measured redshifts. For instance, the technical difficulty of packing slits or fibers in optical spectroscopic observations may lead to an undersampling of the cluster central regions. In turn, this leads to an overestimate of R_V and, again, of the collapsed mass.

Tests of the accuracy of mess estimates based on the dynamical virial method have been performed by using hydrodynamical simulations of galaxy clusters, in which galaxies are identified from gas cooling and star formation [18, 63]. For instance, [18] have shown that galaxies identified in the simulations are fair tracers of the underlying dynamics, with no systematic bias in the estimate of cluster masses, although a rather large scatter between true and recovered masses is induced mostly by projection effects.

Quite reassuringly, despite all the assumptions and possible systematics affecting both dynamical optical and X-ray mass estimates, these two methods provide in general fairly consistent results for both nearby (e.g., [71, 130]) and distant (e.g., [96]) clusters. Two examples of such comparisons are shown in Fig. 9. In the left panel, we report the comparison between X–ray and optical dynamical masses [71]. This plot shows a reasonable agreement among the two mass estimates, although with some scatter. The right panel reports the comparison presented in [130]. In this plot, the triangles indicates the cluster with clear evidences of complex dynamics. Quite interestingly, the agreement between the two mass estimates is acceptable, with a few outliers which are generally identified with non–relaxed clusters.

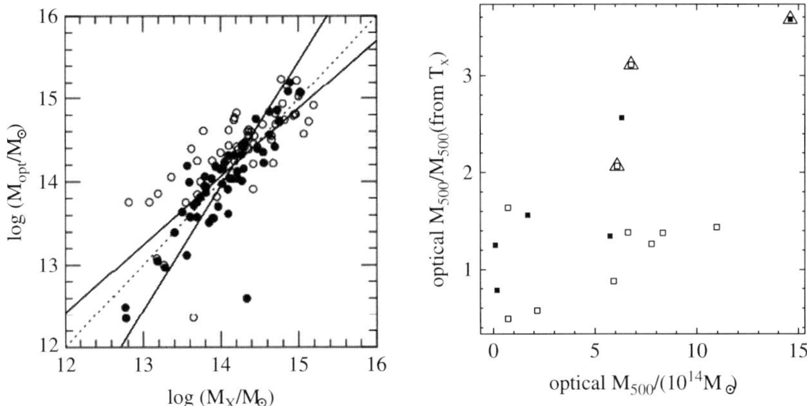

Fig. 9. The relation between dynamical optical masses and masses derived from the X-ray temperature by assuming hydrostatic equilibrium (from [71], left panel, and from [130], right panel, based on SDSS spectroscopic data)

5.3 The Self–Similar Scaling

The simplest model to explain the physics of the ICM is based on the assumption that gravity only determines the thermodynamical properties of the hot diffuse gas [90]. Since gravity does not have a preferred scale, we expect clusters of different sizes to be the scaled version of each other as long as gravity only determines the ICM evolution and there are no preferred scales in the underlying cosmological model. This is the reason why the ICM model based on the effect of gravity only is said to be self-similar.

If we define M_{Δ_c} as the mass contained within the radius R_{Δ_c}, encompassing a mean density Δ_c times the critical density, then $M_{\Delta_c} \propto \rho_c(z)\Delta_c R^3_{\Delta_c}$. Here $\rho_c(z)$ is the critical density of the universe which scales with redshift as $\rho_c(z) = \rho_{c,0}E^2(z)$, where $E(z)$ is given by (12). On the other hand, the cluster size R scales with z and M_{Δ_c} as $R \propto M^{1/3}E^{-2/3}(z)$. Therefore, assuming hydrostatic equilibrium, the cluster mass scales with the temperature T as

$$M_{\Delta_c} \propto T^{3/2}E^{-1}(z)\,. \tag{39}$$

If $\rho_{\rm gas}$ is the gas density, the corresponding X-ray luminosity for pure thermal bremsstrahlung emission is

$$L_X = \int_V \left(\frac{\rho_{\rm gas}}{\mu m_p}\right)^2 \Lambda(T)\,{\rm d}V\,, \tag{40}$$

where $\Lambda(T) \propto T^{1/2}$. Further assuming that the gas distribution traces the dark matter distribution, $\rho_{\rm gas}(r) \propto \rho_{DM}(r)$, then

$$L_X \propto M_{\Delta_c}\rho_c T^{1/2} \propto T^2 E(z)\,. \tag{41}$$

As for the CMB intensity decrement due to the thermal SZ effect we have

$$\Delta S \propto \int y(\theta) \mathrm{d}\Omega \propto d_A^{-2} \int T n_e \mathrm{d}^3 r \propto d_A^{-2} T^{5/2} E^{-1}(z) , \qquad (42)$$

where y is the Comptonization parameter, d_A is the angular size distance and n_e is the electron number density. We can also write ΔS in a different way to get the explicit dependence on y_0:

$$\Delta S \propto y_0 d_A^{-2} \int \mathrm{d}\Omega \propto y_0 d_A^{-2} M^{2/3} E^{-4/3}(z) \propto y_0 d_A^{-2} T E^{-2}(z) . \qquad (43)$$

In this way, we obtain the following scalings for the central value of the Comptonization parameter:

$$y_0 \propto T^{3/2} E(z) \propto L_X^{3/4} E^{1/4}(z) . \qquad (44)$$

Equations (39), (41) and (44) are unique predictions for the scaling relations among ICM physical quantities and, in principle, they provide a way to relate the cluster masses to observables at different redshifts. As we shall discuss in the following, deviations with respect to these relations witness the presence of more complex physical processes, beyond gravitational dynamics only, which affect the thermodynamical properties of the diffuse baryons and, therefore, the relation between observables and cluster masses.

5.4 Phenomenological Scaling Relations

Using the X-ray Luminosity

The relation between X-ray luminosity and temperature of nearby clusters is considered as one of the most robust observational facts against the self–similar model of the ICM. A number of observational determinations now exist, pointing toward a relation $L_X \propto T^\alpha$, with $\alpha \simeq 2.5$–3 (e.g., [170]), possibly flattening towards the self–similar scaling only for the very hot systems with $T \gtrsim 10$ keV [3]. While in general the scatter around the best–fitting relation is non negligible, it has been shown to be significantly reduced after excising the contribution to the luminosity from the cluster cooling regions [106] or by removing from the sample clusters with evidence of cooling flows [7]. As for the behaviour of this relation at the scale of groups, $T \lesssim 1$ keV, the emerging picture now is that it lies on the extension of the L_X–T relation of clusters, with no evidence for a steepening [116], although with a significant increase of the scatter [121], possibly caused by a larger diversity of the groups population when compared to the cluster population. This result is reported in the left panel of Fig. 10 (from [121]), which shows the L_X–T relation for a set of clusters with measured ASCA temperatures and for a set of groups.

As for the evolution of the L_X–T relation, a number of analyses have been performed, using Chandra [57, 86, 109, 161] and XMM–Newton [95, 101] data.

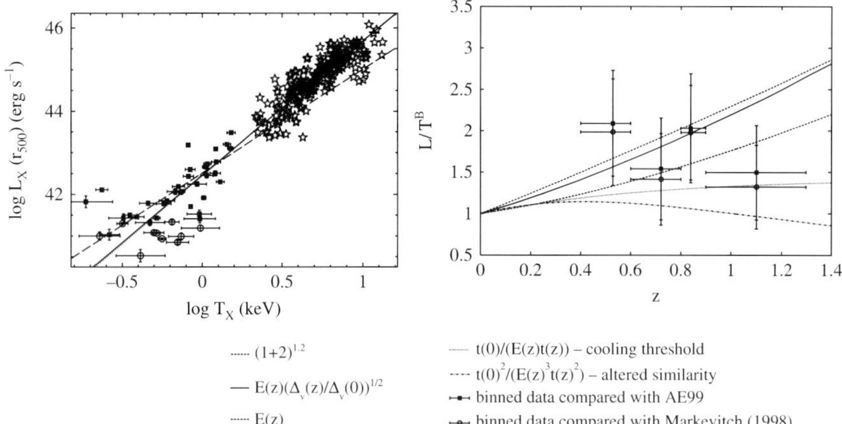

Fig. 10. Left panel: the L_X–T relation for nearby clusters and groups (from [121]). The star symbols are the for the sample of clusters, with temperature measured from ASCA, while the filled squares and open circles are for a sample of groups, also with ASCA temperatures. **Right panel**: the evolution of the L_X–T relation, normalized to the local relation (from [109]), using Chandra temperatures of clusters at $z > 0.4$

Although some differences exist between the results obtained from different authors, such differences are most likely due to the convention adopted for the radii within which luminosity and temperature are estimated. In general, the emerging picture is that clusters at high redshift are relatively brighter, at fixed temperature. The resulting evolution for a cosmology with $\Omega_m = 0.3$ and $\Omega_\Lambda = 0.7$ is consistent with the predictions of the self–similar scaling, although the slope of the high–z L_X–T relation is steeper than predicted by self–similar scaling, in keeping with results for nearby clusters. The left panel of Fig. 10 shows the evolution of the L_X–T relation from [109], where Chandra and XMM–Newton observations of 11 clusters with redshift $0.6 < z < 1.0$ were analyzed. The vertical axis reports the quantity L_X/T^B, where B is the slope of the local relation. Quite apparently, distant clusters are systematically brighter relatively to the local ones. However, the uncertainties are still large enough not to allow the determination of a precise redshift dependence of the L_X–T normalization.

As for the relation between X–ray luminosity and mass, its first calibration has been presented in [136], for a sample of bright clusters extracted from the ROSAT All Sky Survey (RASS). In their analysis, these authors derived masses by using temperatures derived from ASCA observations and applying the equation of hydrostatic equilibrium, (33), for an isothermal β–model. The resulting M–L_X relation is shown in Fig. 11 (see also Fig. 8). From the one hand, this relation demonstrates that a well defined relation between X–ray luminosity and mass indeed exist, although with some scatter, thus confirming that L_X can indeed be used as a proxy of the cluster mass. From the other

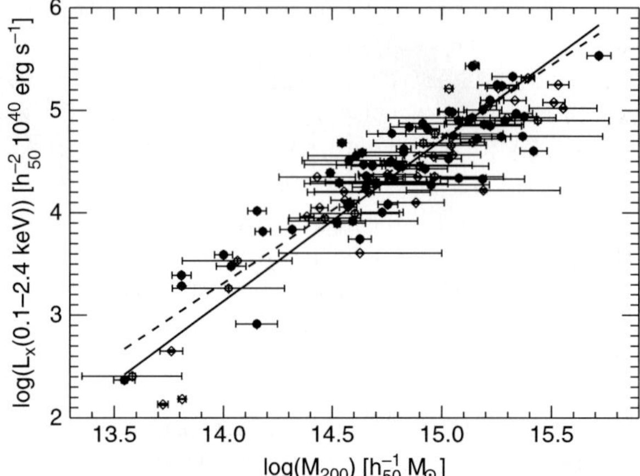

Fig. 11. The L_X–M relation for nearby clusters (from [136]). X–ray luminosities are from the RASS, while masses are estimated using ASCA temperatures and assuming hydrostatic equilibrium for isothermal gas

hand, the slope of the relation is found to be steeper than the self–similar scaling, thus consistent with the observed L_X–T relation.

Using the Optical Luminosity

The classical definition of optical richness of clusters is known to be a poor tracer of the cluster mass (e.g., [26]). However, the increasing quality of photometric data for the cluster galaxy population and the ever improving capability of removing fore/background galaxies thanks to larger spectroscopic galaxy samples have recently allowed different authors to demonstrate the optical/near-IR luminosities to be as reliable tracers of the cluster mass as the X-ray luminosity.

Two examples of recent calibrations between optical/near–IR luminosity and mass are shown in Fig. 12. In the left panel we report the result presented in [100], based on K–band luminosites from the 2MASS and masses obtained by applying the M–T relation by [62]. Although the data points are rather scattered, they define a clear correlation. Quite interestingly, the best–fitting relation has a slope shallower than unity, thus indicating the K–band mass-to-light ratio is a (slightly) increasing function of the cluster mass. This result is in line with previous results using optical luminosities (e.g., [72]) who found an increasing M/L when passing from galaxy groups to clusters of increasing richness.

Popesso et al. [130] analysed SDSS data for a set of clusters which have been identified in the RASS. Their mass estimates come from both X–ray temperature [136] and from the velocity dispersions as estimated from the SDSS spectroscopic data. The results of their analysis for the i band are shown

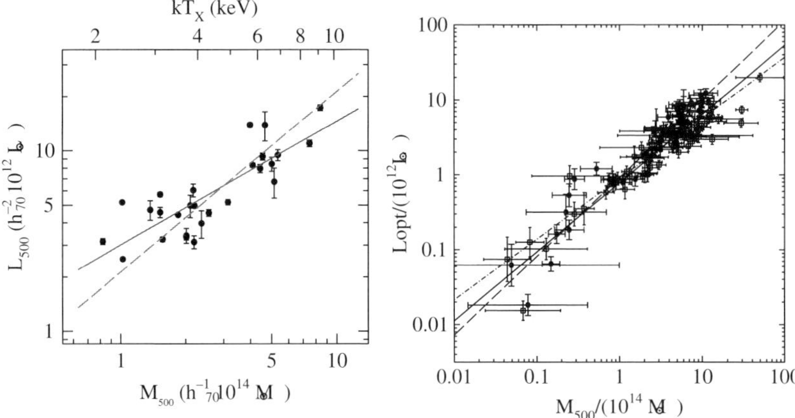

Fig. 12. The relation between cluster masses and optical/near-IR luminosities. **Left panel**: M_{500}–L_{500} relation from [100], using K–band data from the 2MASS survey and masses from X–ray data. The solid line is the best–fit power law, while the dashed line marks the unity slope. **Right panel**: the same relation, but in the iSloan band. Open and filled points corresponds to mass estimates based on the SDSS spectroscopic data and on the M–T_X relation, respectively

in the right panel of Fig. 12. Again, the optical luminosity correlates quite tightly with the cluster mass, with an intrinsic scatter which is comparable to, or even smaller than that of the correlation between X–ray luminosity and mass.

These results highlight how cluster samples with precisely measured optical luminosities can in principle be usefully employed to constrain cosmological parameters. However, while X-ray luminosity provides at the same time a tracer of cluster mass and a criterion to precisely determine the sample selection function, the latter quantity can be extracted from an optically selected sample only in a rather indirect way.

6 Constraints on Cosmological Parameters

In this section we will review critically results on cosmological constraints derived from different ways of tracing the cosmological mass function of galaxy clusters.

6.1 The Distribution of Velocity Dispersions

A first determination of the mass function from velocity dispersions, σ_v, of member galaxies has been attempted in [17]. Girardi et al. [69] used a much larger sample of nearby clusters with measured velocity dispersions to compare the resulting mass function with predictions from cosmological models. The resulting relation between σ_8 and Ω_m was such that $\sigma_8 \simeq 1$ for a fiducial value of the density parameter $\Omega_m = 0.3$. More recently, data for nearby clusters, identified in the SDSS, have been used to calibrate a relation between richness and velocity dispersion [10]. They compared the resulting σ_v–distribution to the prediction of cosmological models and found a significantly lower normalization of the power spectrum, $\sigma_8 \simeq 0.7$ for $\Omega_m = 0.3$. Such differences from different analyses highlight the presence of systematic uncertainties in the relation between mass and observables (i.e., velocity dispersion and richness).

The application of this method to distant clusters has been applied so far only to the CNOC sample [34], which comprises 17 clusters selected from the EMSS out to $z \simeq 0.6$. Still to date, this is the only sample of distant clusters, with calibrated selection function, for which velocity dispersions have been reliably measured. Bahcall et al. [11] pointed out that the resulting evolution of the mass function is consistent with a low–density Universe. Borgani et al. [24] reanalysed this same sample and emphasised that the uncertainties in the local normalization of the mass function are large enough to make any constraints on Ω_m not significant.

6.2 The Temperature Function

The X-ray Temperature Function (XTF) is defined as the number density of clusters with given temperature, $n(T)$. As long as a one-to-one relation exist between temperature and mass, the XTF can be related to the mass function, $n(M)$, by the relation

$$n(T) = n[M(T)] \frac{\mathrm{d}M}{\mathrm{d}T} . \tag{45}$$

In this equation, the ratio $\mathrm{d}M/\mathrm{d}T$ is provided by the relation between ICM temperature and cluster mass.

Measurements of cluster temperatures for flux-limited samples of nearby clusters were first presented in [83]. These results have been subsequently refined and extended to larger samples with the advent of *ROSAT*, *Beppo–SAX* and, especially, *ASCA*. XTFs have been computed for both nearby (e.g., [88, 106, 127, 128]) and distant (e.g., [50, 55, 81, 82]) clusters, and used to constrain cosmological models. The starting point in the computation of the XTF is inevitably a flux-limited sample for which the searching volume of each cluster can be computed. Then the $L_X - T_X$ relation and its scatter is used to derive a temperature limit from the sample flux limit.

Once the XTF is measured from observations, (45) is used to infer the mass function and, therefore, to constrain cosmological models. A slightly different but conceptually identical approach, has been followed in [136], where masses for a flux–limited sample of nearby bright RASS clusters have been computed by applying the assumption of hydrostatic equilibrium, thereby expressing their results directly in terms of mass function, rather than of XTF.

Oukbir and Blanchard [122] first suggested to use the evolution of the XTF as a way to constrain the value of Ω_m. Several independent analyses converge now towards a mild evolution of the XTF, which is interpreted as a case for a low–density Universe, with $0.2 \lesssim \Omega_m \lesssim 0.6$. An example is reported in the right panel of Fig. 13 (from [82]), which shows the comparison between the XTFs of the sample of nearby clusters [83] and a sample of EMSS clusters with ASCA temperatures (see also Fig. 14).

A limitation of the XTFs presented so far is the limited sample size (with only a few $z \gtrsim 0.5$ measurements), as well as the lack of a homogeneous sample selection for local and distant clusters. By combining samples with different selection criteria one runs the risk of altering the inferred evolutionary pattern of the cluster population. This can even give results consistent with a critical–density Universe [19, 41, 158].

Besides the determination of the matter density parameter, the observational determination of the XTF also allows one to measure the normalization of the power spectrum, σ_8. Assuming a fiducial value of $\Omega_m = 0.3$, different (sometimes discrepant) determinations of σ_8 have been reported by different authors, ranging from $\sigma_8 \simeq 0.7$–0.8 (e.g., [55, 82, 136]) to $\sigma_8 \simeq 1$ (e.g., [106, 128]). Ikebe et al. [88] compared different observational determinations of the XTF for nearby clusters (see left panel of Fig. 13) and established that they all agree with each other reasonably well. Although quite comfortable, this result highlights that the discrepant results on the normalization of the σ_8–Ω_m relation comes from the cosmological interpretation of the observed XTF, and not from observational uncertainties in its calibration. While the different model mass functions (i.e., whether Press–Schechter, Jenkins et al. or Sheth–Tormen) can in some cases account for part of the difference, more in general the different results are interpreted in terms of the different

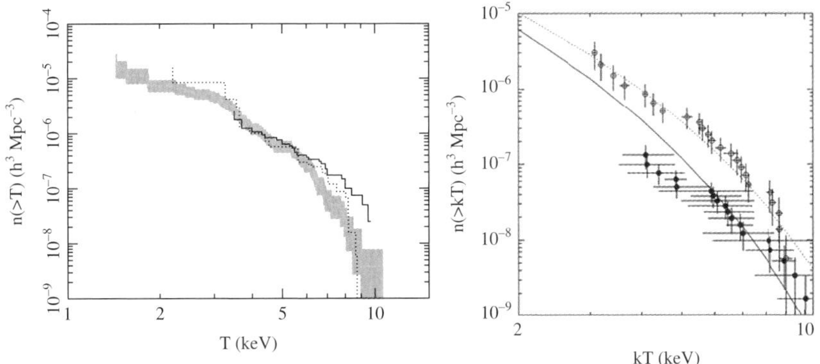

Fig. 13. Left panel: a comparison between the XTF for nearby clusters from [88] (shaded area), [106] (*solid line*)and [81] (*dotted line*). **Right panel**: the evolution of the XTF from [82]. Open and filled circles are for the local and the distant cluster sample, respectively

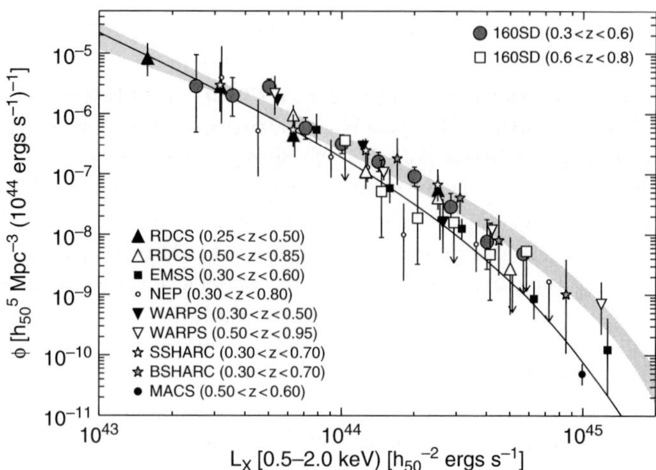

Fig. 14. A compilation of XLF within different redshift intervals for independent X-ray flux-limited surveys (from [119]). The shaded area shows the range of determinations of the local XLF, while the solid curve is the best–fitting evolving XLF from [119]

normalization of the M–T relation to be used in (45) or to the way in which the intrinsic scatter and the statistical uncertainties in this relation are included in the analysis. We shall critically discuss these issues in Sect. 6.5 below.

Substantially improved observational determinations of the XTF, and correspondingly tighter cosmological constraints, are expected to emerge with the accumulations of data on the ICM temperature from the Chandra and XMM–Newton satellites. Thanks to the much improved sensitivity of these X–ray telescopes with respect to ASCA, temperature gradients can be measured for fairly large sets of nearby and medium–distant ($z \lesssim 0.4$) clusters, thus allowing more precise determinations of cluster masses. At the same time, reliable measurements of global temperatures are now emerging for clusters out to the highest redshifts where they have been secured (e.g., [140]). At the time of writing, several years after the advent of the new generation of X–ray telescopes, no determinations of the XTF from Chandra and XMM–Newton data have been presented, a situation that is expected to change quite soon.

6.3 The Luminosity Function

Another method to trace the evolution of the cluster number density is based on the X–ray luminosity function (XLF), $\phi(L_X)$, which is defined as the number density of galaxy clusters having a given X–ray luminosity. Similarly to (45), the XLF can be related to the cosmological mass function of collapsed halos as

$$\phi(L_X) = n[M(L_X)] \frac{\mathrm{d}M}{\mathrm{d}L_X}, \tag{46}$$

where $M(L_X)$ provides the relation between the observable L_X and the cluster mass. The above relation needs to be suitably modified in case an intrinsic scatter exists in the relation between mass and temperature (see Sect. 6.5, here below).

A useful observational quantity, that is related to the XLF, is given by the flux number–counts, $n(S)$, which is defined as the number of clusters per steradian, having measured flux S:

$$n(S) = \left(\frac{c}{H_0}\right)^3 \int_0^\infty \mathrm{d}z\, \frac{r^2(z)}{E(z)}\, n[M(S,z);z]\, \frac{\mathrm{d}M}{\mathrm{d}S} \tag{47}$$

(e.g., [94]) where $r(z)$ is the radial coordinate appearing in the Friedmann–Robertson–Walker metric:

$$r(z) = \int_0^z \mathrm{d}z\, E^{-1}(z) \quad ; \quad \Omega_\Lambda = 1 - \Omega_m$$

$$r(z) = \frac{2\left[\Omega_m z + (2-\Omega_m)\left(1 - \sqrt{1+\Omega_m z}\right)\right]}{\Omega_m^2 (1+z)} \quad ; \quad \Omega_\Lambda = 0\,. \tag{48}$$

The flux S is related to the luminosity according to

$$S = \frac{L_X}{4\pi d_L^2(z)}, \tag{49}$$

where $d_L(z) = r(z)(1+z)$ is the luminosity distance at redshift z.

This quantity can be measured for a flux–limited samples without having information on cluster redshift and provides useful cosmological information in the absence of any spectroscopic optical follow–up. A comparison between different observational determinations of the flux number counts for both nearby and distant cluster samples (e.g., [138]) show indeed a quite good agreement.

Another quantity, which has been used to derive cosmological constraints from flux–limited surveys, is the redshift distribution, $n(z)$, which is defined as the number of clusters found in a survey at a given redshift z:

$$n(z) = \left(\frac{c}{H_0}\right)^3 \frac{r^2(z)}{E(z)} \int_{S_{\rm lim}}^\infty \mathrm{d}S\, f_{\rm sky}(S)\, n[M(S,z);z]\, \frac{\mathrm{d}M}{\mathrm{d}S}\,. \tag{50}$$

In the above expression $S_{\rm lim}$ is the limiting completeness flux of the survey, while $f_{\rm sky}(S)$ is the effective flux–dependent sky–coverage appropriate for the considered survey. Convolving the mass function with the sky coverage inside the integral in the above equation is essential to properly account for the different effective area covered at different fluxes, an aspect which is apparently overlooked in some analyses (e.g., [157]).

In general, the advantage of using X-ray luminosity as a tracer of the mass is that L_X is measured for a much larger number of clusters in samples with well-defined selection properties. As discussed in Sect. 4.2, the most recent flux–limited cluster samples contain now a fairly large (~ 100) number of objects, which are homogeneously identified over a broad redshift baseline, out to $z \simeq 1.3$. This allows nearby and distant clusters to be compared within the same sample, i.e. with a single selection function. However, since the X-ray emissivity depends on the square of the gas density, the relation between L_X and $M_{\rm vir}$, which is based on additional physical assumptions, is more uncertain than the $M_{\rm vir}$–σ_v or the $M_{\rm vir}$–T relations.

A useful parametrization for the relation between temperature and bolometric luminosity can be casted in the form

$$L_{\rm bol} = L_6 \left(\frac{T_X}{6{\rm keV}}\right)^{\alpha} (1+z)^A \left(\frac{d_L(z)}{d_{L,{\rm EdS}}(z)}\right)^2 10^{44}{\rm h}^{-2}\,{\rm erg\,s^{-1}}\,, \qquad (51)$$

with L_6 defining the normalization of the relation and $d_L(z)$ the luminosity–distance at redshift z for a given cosmology.

Analyses of the number counts from different X-ray flux–limited cluster surveys showed that the resulting constraints on Ω_m are rather sensitive to the evolution of the mass–luminosity relation [24, 94, 107]. On the other hand, other authors [135, 141, 157] analysed different flux–limited surveys and found results consistent with $\Omega_m = 1$. Quite intriguingly, this conclusion is common to analyses which combine a normalization of the local mass function, using nearby clusters, and to the evolution of the mass function using deep surveys. Clearly, any uncertainty in the calibration of the selection functions when combining different surveys may induce a spurious signal of evolution of the cluster population, possibly misinterpreted as an indication for high Ω_m.

In order to overcome this potential problem, Borgani et al. [29] analyzed the RDCS sample to trace the cluster evolution over the entire redshift range, $0.05 \lesssim z \lesssim 1.3$, probed by this survey, without resorting to any external normalization from a different survey of nearby clusters [28]. They found $0.1 \lesssim \Omega_m \lesssim 0.6$ at the 3σ confidence level, by allowing the M–L_X relation to change within both the observational and the theoretical uncertainties. In Fig. 15 we show the resulting constraints on the σ_8–Ω_m plane (from [138]) and how they vary by changing the parameters defining the M–L_X relation: the slope α and the evolution A of the L_X–T relation (see Equation 51), the normalization β of the M–T relation (see (36)), and the overall scatter Δ_{M-L_X}. Flat geometry is assumed here, i.e. $\Omega_m + \Omega_\Lambda = 1$.

Similar results have been obtained by combining information on clustering properties and the redshift distribution from the the REFLEX cluster survey [144], thus providing $\sigma_8 \simeq 0.7$ and $\Omega_m \simeq 0.35$. One should however notice that, since these constraints are derived from nearby clusters, the corresponding estimate of Ω_m comes from the shape of the CDM power spectrum, rather than from the growth rate of perturbations. It is rather reassuring that dynamical and geometrical constraints on Ω_m are in fact consistent with each other.

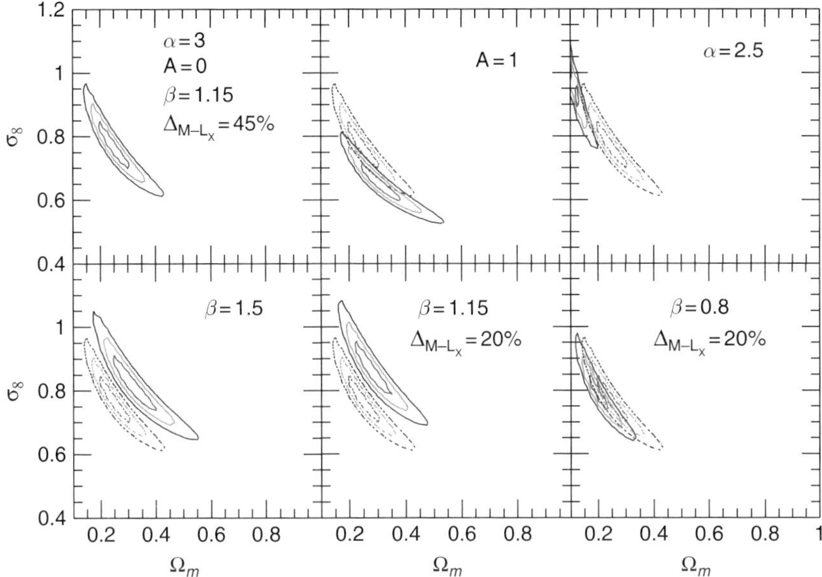

Fig. 15. Probability contours in the σ_8–Ω_m plane from the evolution of the X-ray luminosity distribution of RDCS clusters. The shape of the power spectrum is fixed to $\Gamma = 0.2$ [138]. Different panels refer to different ways of changing the relation between cluster virial mass, M, and X-ray luminosity, L_X, within theoretical and observational uncertainties (see also [29]). The upper left panel shows the analysis corresponding to the choice of a reference parameter set. In each panel, we indicate the parameters which are varied, with the dotted contours always showing the reference analysis

Constraints on Ω_m from the cluster X–ray luminosity and temperature distribution are thus in line with the completely independent constraints derived from the baryon fraction in clusters, f_b (e.g., [4, 58, 168]).

6.4 The Gas Mass Function

An alternative way of tracing the mass function of galaxy clusters is based on using as its proxy the mass function of the cluster gas content [162]. This method is based on the assumption that galaxy clusters are fair containers of cosmic baryons. Similarly to the method based on the baryon fraction, it relies on the knowledge of the cosmic baryon fraction, either provided by data on the deuterium abundance in high–redshift absorption systems (e.g., [92]) combined with predictions of primordial nucleosynthesis, or from the spectrum of CMB anisotropies (e.g., [103, 149]). This method has the potential advantage that cluster gas mass is an easier quantity to measure than the total collapsed mass, since it is essentially related to the total cluster emissivity.

If we define $n_b(M_b)$ to be the baryonic mass function and $n(M)$ the total mass function, by definition we have $n_b(M_b) = n(\Omega_m M_b/\Omega_b)$. Therefore, once $n_b(M_b)$ and Ω_b are known from observations, the total mass function can be computed as a function of Ω_m, thereby treated as a fitting parameter.

While this method has the remarkable advantage of avoiding the uncertainties related to direct estimates of the total collapsed mass, it is affected by possible violations of the assumption of universality of the baryon content of clusters. Indeed, while this assumption should be valid for suitably relaxed and massive clusters, it may be less so when considering all objects belonging to a flux–limited sample, thus including also relatively small clusters and structures with a complex non-relaxed dynamics.

This method was applied [163] to a set of bright clusters selected from the RASS [136] and found $\sigma_8 = 0.72 \pm 0.04$ with $\Omega_m h = 0.13 \pm 0.07$. Chandra observations were included in the analysis for a set of clusters extracted from the 160 deg^2 survey [162]. They found an evolution of the gas mass function, which is consistent with a flat cosmological model with $\Omega_m = 0.3$.

We emphasize here that the above different methods used to reconstruct the mass function of galaxy clusters consistently prefer relatively low values of σ_8, in the range 0.7–0.8. Quite remarkably, such values have been shown now to be required by the 3–years WMAP data release [149].

6.5 Including Uncertainties in the Analysis

As we have discussed in the previous sections, most of the analyses of the cluster populations converge toward a low-density model, with $\Omega_m \sim 0.3$. However, significant differences exist between different determinations of the normalization of the power spectrum, σ_8, which amount to up to ~ 20 per cent. These differences are much larger that the statistical uncertainties associated to the finite number of clusters included in the samples, thus indicating that they arise from unaccounted sources of error, which affects the analyses.

For instance, the role of the uncertain normalization of the mass–temperature relation in the determination of σ_8 (at fixed Ω_m) from the XTF analysis has been emphasized by different authors (e.g., [88, 127, 145]). Increasing the normalization of the M–T relation implies that a larger mass corresponds to a fixed temperature value. As a consequence, an observed XTF translates into a larger mass function, therefore implying a larger σ_8 (at fixed Ω_m). Since results from hydrodynamical simulations generally imply a larger M–T normalization, a larger σ_8 is expected when using in the analysis the simulation predictions. The left panel of Fig. 16 (from [88]) shows how the best fitting values of σ_8 and Ω_m from the local XTF change as one uses different mass–temperature relations, taken from both observations and simulations. The right panel of Fig. 16 (from [127]) show the dependence of σ_8 on the normalization of the mass–temperature relation. Note that this normalization

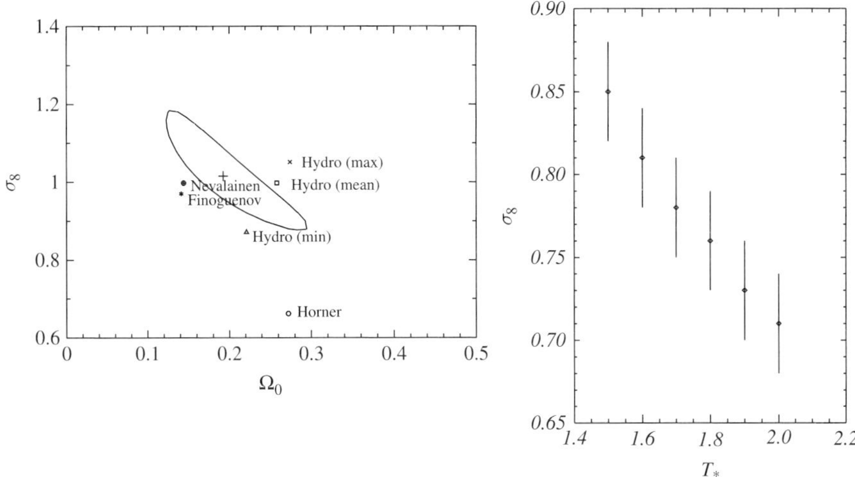

Fig. 16. Left panel: the dependence of the best-fitting values of Ω_m and σ_8, from the XTF of [88], upon different determinations of the mass–temperature relation, from both observational data and from hydrodynamical simulations of galaxy clusters. **Right panel**: the dependence of σ_8 on the normalization of the mass–temperature relation, in the XTF analysis by [127]

is allowed here to vary over the range encompassed by observational and simulation results. The range of variation of σ_8 induced by the uncertainty in the M–T relation is at least comparable to the purely statistical uncertainty, as indicated by the errorbars.

One may wonder why not relying only on the observational determination of the M–T relation, instead of considering also simulation results. As we shall discuss in Sect. 7, observational results can not necessarily provide the most reliable determination of the M–T scaling.

The effect of changing the normalization of the M–T relation on the analysis of flux–limited samples through the XLF evolution is shown in the lower left panel of Fig. 15. If this normalization is reduced by $\sim 30\%$, the resulting σ_8 decreases by $\sim 20\%$.

It is clear that, any uncertainty, both statistical and systematic, in the fitting parameters describing the scaling relations between mass and observable, must be included in the analysis by marginalizing over the probability distribution function of these parameters. Let $\mathbf{\Omega}$ be the set of cosmological parameters that we want to constrain, and $\mathbf{W}$ the set of parameters which define a scaling relation between mass M and an observable X (i.e., σ_v, L_X or T). Let us also call $P(\mathbf{W})$ the prior distribution for the uncertainties in the M–X relation. If $\chi^2(\mathbf{\Omega}, \mathbf{W})$ gives the goodness of fit provided by the choice $\mathbf{\Omega}$ of the cosmological parameters, for a given M–X relation, then the goodness of fit after marginalizing over the uncertainties in M–X reads

$$\chi^2(\boldsymbol{\Omega}) = \frac{\int \chi^2(\boldsymbol{\Omega}, \mathbf{W})\, P(\mathbf{W})\, \mathrm{d}\mathbf{W}}{\int P(\mathbf{W})\, \mathrm{d}\mathbf{W}} . \tag{52}$$

Of course, the marginalization generally induces an increase of the uncertainties in the cosmological parameters. Furthermore, one needs to have a reliable modeling of both size and distribution of the errors (i.e. whether they have a uniform, a Gaussian, or a more peculiar distribution).

Besides the errors in the parameters defining the scaling relations, a different source of uncertainty is provided by the intrinsic scatter in these relations. Intrinsic scatter has the effect of widening the range of possible masses which correspond to a given value of the observable quantity. This effect can be included in the analysis by convolving the theoretical mass function with the distribution of the scatter itself (e.g., [29, 98]). Let Ψ be an observable quantity and $\phi(\Psi)$ its distribution (i.e., XLF or XTF), to be compared with observations. Also, let $P(M_\Psi|M;z)$ be the probability of assigning a mass M_Ψ to a cluster of true mass M, at redshift z, from the observable Ψ, for a given M–Ψ relation. Therefore, the model prediction for the distribution $\phi(\Psi)$, to be compared with its observational determination, is given by the convolution of the cosmological mass function with the distribution of the intrinsic scatter:

$$\phi(\Psi)\mathrm{d}\Psi = \int \mathrm{d}M_\Psi n(M_\Psi, z)\, P(M_\Psi|M;z)\, \frac{\mathrm{d}M}{\mathrm{d}\Psi}\, \mathrm{d}\Psi \tag{53}$$

where $n(M, z)$ is the cosmological mass function at redshift z. If one makes the standard assumption of Gaussian scatter in the log–log plane, then

$$P(M_\Psi|M) = \left(2\pi\sigma^2_{\ln M}\right)^{-1/2} \exp\left[-x^2(M_\Psi)\right] , \tag{54}$$

where $x(M_\Psi) = (\ln M_\Psi - \ln M)/(\sqrt{2}\sigma_{\ln M})$ and $\sigma_{\ln M}$ is the r.m.s. intrinsic scatter. The effect of this convolution is that of increasing $\phi(\Psi)$, for a fixed $n(M)$, as the scatter increases. Therefore, assuming a progressively larger scatter in the M–Ψ relation implies a progressively lower σ_8 (at fixed Ω_m). An illustrative example of the effect of intrinsic scatter on the determination of σ_8 is reported in Fig. 17. In the left panel we show the REFLEX XLF [20] along with the prediction of the best fitting cosmological model for a given choice of the M–L_X relation, after assuming vanishing intrinsic scatter in this relation. In the left panel, we show the same comparison, but assuming an Gaussian-distributed intrinsic scatter of 40 per cent in the M–L_X scaling. As expected, adding the scatter has the effect of increasing the predicted luminosity function, so that σ_8 has to be lowered from 0.8 to 0.65 to recover the agreement with observations. This example highlights that a good calibration of the intrinsic scatter in the scaling relations can be as important as determining the best-fitting amplitude and slope of these relations.

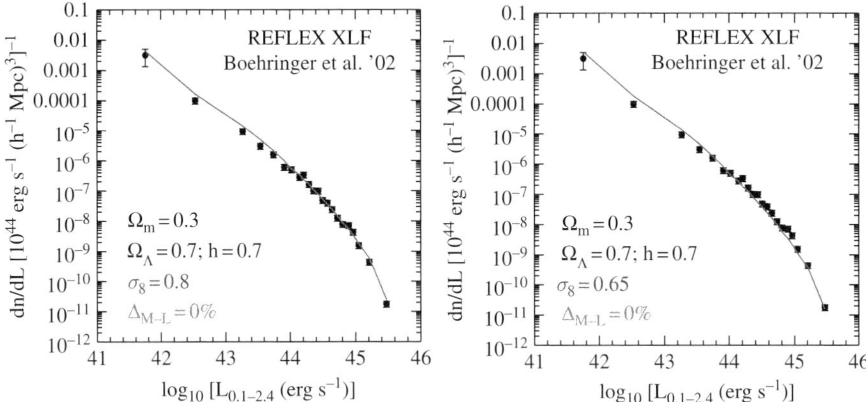

Fig. 17. The dependence of σ_8 on the intrinsic scatter assumed in the relation between X–ray luminosity and mass. The two panels show the comparison between model predictions and the observed XLF from the REFLEX sample [20]. The left panel assumes vanishing scatter while the right panel assumes a 40 per cent intrinsic scatter. The best fitting values of σ_8 are reported in both cases

7 The Future

A new era for cosmology with galaxy clusters is now starting. High sensitivity surveys for blind SZ identification over fairly large contiguous area, $\sim$ 100 deg^2, have have already started or are planned in the coming years (see the lectures by M. Birkinshaw in this volume). Also, the Planck satellite will survey the whole sky, although at a much lower sensitivity, and provide a large set of clusters identified through the SZ effect. These surveys promise to identify several thousands clusters, with a fair number of objects expected to be found at $z > 1$. In the optical/near-IR bands, imaging with dedicated telescopes with large field of view will also allow to secure a large number of distant clusters. At the same time, X–ray observations over contiguous area (e.g., [169] and "serendipitous" searches from XMM–Newton (e.g., [118]) and Chandra (e.g., [30]) archives will ultimately cover several hundreds deg^2 down to flux limits fainter than those reached by the deepest *ROSAT* pointings. Preliminary results suggest that identification of $z > 1$ clusters may eventually become routine [118]. Ultimately, they will lead to the identification of several thousands clusters.

Optimized optics for wide-field X-ray imaging have been originally described in a far-reaching paper by Burrows et al. [33] and proposed for the first time to be implemented in a dedicated satellite mission in the mid 90's. There is no doubt that this would be the right time to plan a dedicated wide-field X-ray telescope, which should survey the sky over an area of several thousands deg^2, with a relatively good, XMM–like or better, point spread function and a low background. This instrument would be invaluable for studies of galaxy clusters, thanks to its ability of both identifying extended sources with low surface brightness. Several missions with a similar profile have been proposed, although none has been approved so far. Still, the community working on galaxy clusters is regularly proposing this idea of satellite to different Space Agencies (e.g., [79]).

The samples of galaxy clusters obtainable from large-area SZ and X-ray surveys contain in principle so much information to allow one to constrain not only Ω_m and σ_8, but also the Dark Energy (DE) content of the Universe (see, e.g., [142, 153] for introductory reviews on Dark Energy). The equation of state of DE is written in the form $p = w\rho$, where p and ρ are the pressure and density terms, respectively. The parameter w must take values in the range $-1/3 > w \geq -1$ for the DE to provide an accelerated cosmic expansion. Constraining the value of w and its redshift evolution is currently considered one of the most ambitious targets of modern cosmology. Ones hope is to unveil the nature of the energy term which dominates the overall dynamics of the Universe at the present time.

As we have mentioned, the limited statistics prevents current cluster surveys to place significant constraints on Ω_Λ. Although this limitation will be overcome with future cluster surveys, the question remains as to whether the systematic effects, discussed in Sect. 6.5, can be sufficiently understood. Different lines of attack have been proposed in the literature, which should not considered as alternative to each other.

Majumdar and Mohr [105] have proposed the approach based on the so–called self-calibration (see also [97, 98, 104]). The idea underlying this approach is that of parametrizing in a sensible way the scaling relations between cluster observables and mass, including the corresponding intrinsic scatter and its distribution. In this way, the parameters describing these relations can be considered as fitting parameter to be added to the cosmological parameters. As long as the cluster samples are large enough, one should be able to fit at the same time both cosmological parameters and those parameters related to the physical properties of clusters. Figure 18 (from [105]) shows the constraints that one can place on the w–Ω_m (left panel) and on the σ_8–Ω_m planes, after marginalizing over the other fitting parameters, from different SZ and X–ray surveys. In each panel different contours indicate the constraints that one can place by progressively adding information in the analysis. The main message here is that combining information on the evolution of the cluster population and on its clustering can place precision constraints on cosmological parameters. These constraints can be further tightened if follow up observations are available to precisely measure masses for 100 clusters.

These forecasts nicely illustrates the potentiality of the self–calibration approach for precision cosmology with future surveys of galaxy clusters. Clearly, the robustness of these predictions is inextricably linked to the possibility of accurately modeling the relations between mass and observables.

A line of attack to this problem is based on using detailed hydrodynamical simulations of galaxy clusters. The great advantages of using simulations is

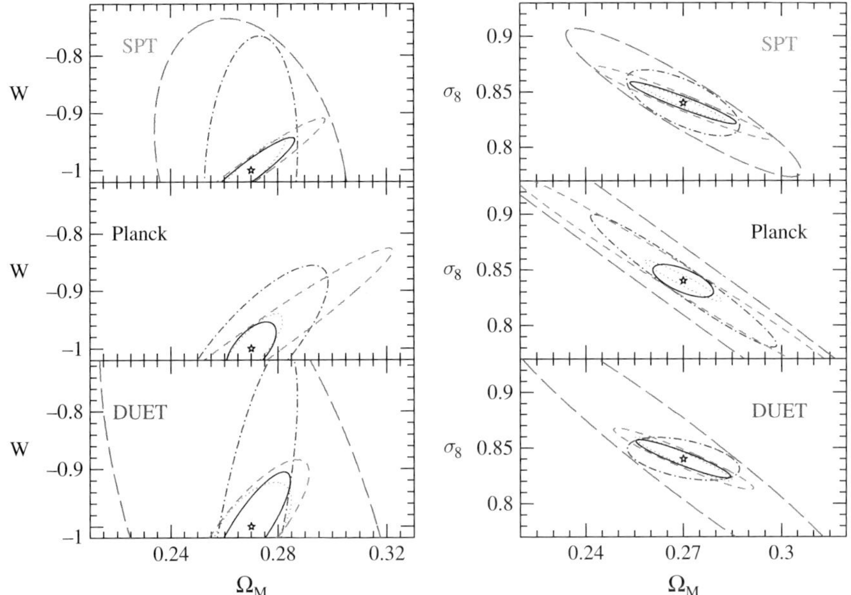

Fig. 18. Constraints on cosmological parameters from the SZ SPT survey, from the SZ Planck survey and from a wide-area deep X-ray survey (from top to bottom panels; from [105]). Dotted lines: no uncertainty in the relation between cluster mass and observables, using only the dn/dz information; long-dashed line: using self-calibration on the dn/dz; dot–dashed line: as before, but also including the information on the cluster power spectrum; short–dashed line: using dn/dz and a calibration of the mass–observable relation for 100 clusters; solid line: all the information combined together

that both cluster mass and observable quantities can be exactly computed. Furthermore, the effect of observational set–ups (e.g., response functions of detectors, etc.) can be included in the analysis and their effect on the scaling relations quantified. This approach has been applied by different groups in the case of X–ray observations [64] to understand the relation between the ICM temperature, as measured from the fitting of the observed spectrum, and the "true" mass–weighted temperature (e.g., [108, 110, 159]). Furthermore, simulations can also be used to verify in detail the validity of assumptions on which the mass estimators, applied to observations, are based. The typical example is represented by the assumption of hydrostatic equilibrium, discussed in Sect. 5.1, for which violations in simulations at the 10–20 per cent level have been found (e.g., [12, 27, 91, 134]).

An example of calibration of observational biases in the mass–temperature relation, using hydrodynamical simulations, is shown in Fig. 19 (from [133]), which provides a comparison between the observed and the simulated M–T relation. Simulations here include radiative cooling, star formation and the

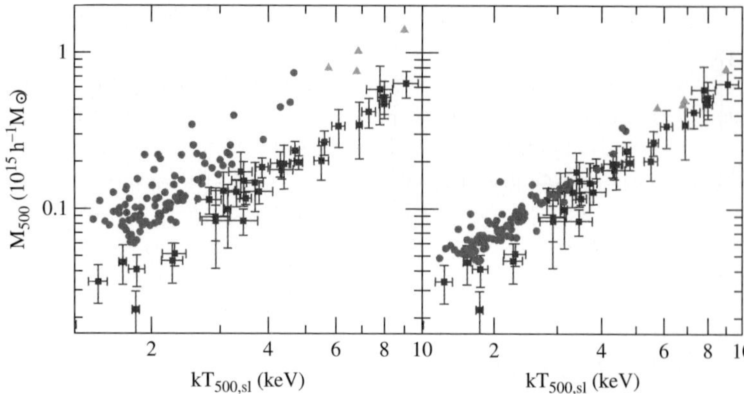

Fig. 19. The mass–temperature relation at $\bar{\rho}/\rho_{\rm cr} = 500$, in simulations (filled circles and triangles) and for the observational data (squares with errorbars, [62]). The left panel is for the true masses of simulated clusters; the right panel is for masses of simulated clusters estimated by adopting the same procedure applied by Finoguenov et al. to observational data (from [133])

effect of galactic winds powered by supernovae, and, as such, provide a realistic description of the relevant physical processes. The observational results, which are taken from [62], corresponds to mass estimates based on the hydrostatic equilibrium for a polytropic β–model of the gas distribution [(33)]. In both panels, the temperature has been computed by using a proxy to the actual spectroscopic temperature, the so–called spectroscopic–like temperature [110]. The left panel shows the results when exact masses of simulated clusters are used for the comparison. Based on this result only, the conclusion would be that simulations do indeed produce too high a M–T relation, even in the presence of a realistic description of gas physics. In the left panel, masses of simulated clusters are computed instead by using the same procedure as for observed clusters, i.e. by applying (34) for the hydrostatic equilibrium of a polytropic β–model. Quite remarkably, the effect of applying the observational mass estimator has two effects. First, the overall normalization of the M–T relation is decreased by the amount required to attain a reasonable agreement with observations. Second, the scatter in the simulated M–T relation is substantially suppressed. This is the consequence of the fact that (34) provides a one-to-one correspondence between mass and temperature, while only the cluster-by-cluster variations of β and γ account for the intrinsic diversity of the cluster thermal structure.

This example illustrates how simulations can be usefully employed as guidelines to study possible biases on observational mass estimates. However, it is worth reminding that the reliability of simulation results depends on our capability to correctly provide a numerical description of all the relevant physical process. In this sense, understanding in detail the (astro)physics of clusters is mandatory in order to calibrate them as tools in the era of precision cosmology.

As a concluding remark, we emphasize once more that a number of independent analyses of the cluster mass function, which have been realized so far, favor a relatively low normalization of the power spectrum, with $\sigma_8 \simeq 0.7$–0.8 for $\Omega_m \simeq 0.3$, thus in agreement with the most recent WMAP results [149]. This agreement must be considered as a success for cluster cosmology and a strong encourgement for future applications to large cluster surveys of the next generation.

Acknowledgments

I would like to thank the organizers of the GH2005 School, David Hughes, Omar López–Cruz and Manolis Plionis for having provided an enjoyable and stimulating environment. I also warmly thank Manolis Plionis and Piero Rosati for a careful reading of the manuscript and useful suggestions to improve it.

References

1. Abell, G.O.: ApJS **3**, 211 (1958)
2. Abell, G.O., Corwin, H.G., Olowin, R.P.: ApJS **70**, 1 (1989)
3. Allen, S.W., Fabian, A.C.: MNRAS **297**, L57 (1998)
4. Allen, S.W., Schmidt, R.W., Ebeling, H., Fabian, A.C., van Speybroeck, L.:MNRAS **353**, 457 (2004)
5. Allen, S.W., Schmidt, R.W., Fabian, A.C.: MNRAS **328**, L37 (2001)
6. Allen, S.W., Schmidt, R.W., Fabian, A.C.: MNRAS **334**, L11 (2002)
7. Arnaud, M., Evrard, A.E.: MNRAS **305**, 631 (1999)
8. Arnaud, M., Pointecouteau, E., Pratt, G.W.: A&A **441**, 893 (2005)
9. Bahcall, N.A., Cen, R., Davé, R., Ostriker, J.P., Yu, Q.: ApJ **541**, 1 (2000)
10. Bahcall, N.A., Dong, F., Bode, P., Kim, R., Annis, J., McKay, T.A., Hansen, S., Schroeder, J., Gunn, J., Ostriker, J.P., Postman, M., Nichol, R.C., Miller, C., Goto, T., Brinkmann, J., Knapp, G.R., Lamb, D.O., Schneider, D.P., Vogeley, M.S., York, D.G.: ApJ **585**, 182 (2003)
11. Bahcall, N.A., Fan, X., Cen, R.: ApJ **485**, L53+ (1997)
12. Bartelmann, M., Steinmetz, M.: MNRAS **283**, 431 (1996)
13. Basilakos, S., Plionis, M., Georgakakis, A., Georgantopoulos, I., Gaga, T., Kolokotronis, V., Stewart, G.: MNRAS **351**, 989 (2004)
14. Binney, J., Tremaine, S.: Galactic Dynamics, p. 747. Princeton University Press, Princeton, NJ (1987)
15. Birkinshaw, M.: Phys. Rep. **310**, 97 (1999)
16. Biviano, A., Girardi, M.: ApJ **585**, 205 (2003)
17. Biviano, A., Girardi, M., Giuricin, G., Mardirossian, F., Mezzetti, M.: ApJ **411**, L13 (1993)

18. Biviano, A., Murante, G., Borgani, S., Diaferio, A., Dolag, K., Girardi, M.: ArXiv Astrophysics e-prints (2006)
19. Blanchard, A., Sadat, R., Bartlett, J.G., Le Dour, M.: A&A **362**, 809 (2000)
20. Böhringer, H., Collins, C.A., Guzzo, L., Schuecker, P., Voges, W., Neumann, D.M., Schindler, S., Chincarini, G., De Grandi, S., Cruddace, R.G., Edge, A.C., Reiprich, T.H., Shaver, P.: ApJ **566**, 93 (2002)
21. Böhringer, H., Schuecker, P., Guzzo, L., Collins, C.A., Voges, W., Cruddace, R.G., Ortiz-Gil, A., Chincarini, G., De Grandi, S., Edge, A.C., MacGillivray, H.T., Neumann, D.M., Schindler, S., Shaver, P.:A&A **425**, 367 (2004)
22. Böhringer, H., Voges, W., Huchra, J.P., McLean, B., Giacconi, R., Rosati, P., Burg, R., Mader, J., Schuecker, P., Simiç D., Komossa, S., Reiprich, T.H., Retzlaff, J.,Trümper, J.: ApJS **129**, 435 (2000)
23. Bond, J.R., Cole, S., Efstathiou, G., Kaiser, N.: ApJ **379**, 440 (1991)
24. Borgani, S., Girardi, M., Carlberg, R.G., Yee, H.K.C., Ellingson, E.: ApJ **527**, 561 (1999)
25. Borgani, S., Governato, F., Wadsley, J., Menci, N., Tozzi, P., Quinn, T., Stadel, J., Lake, G.: MNRAS **336**, 409 (2002)
26. Borgani, S., Guzzo, L.: Nature **409**, 39 (2001)
27. Borgani, S., Murante, G., Springel, V., Diaferio, A., Dolag, K., Moscardini, L., Tormen, G., Tornatore, L., Tozzi, P.: MNRAS **348**, 1078 (2004)
28. Borgani, S., Rosati, P., Tozzi, P., Norman, C.: ApJ **517**, 40 (1999)
29. Borgani, S., Rosati, P., Tozzi, P., Stanford, S.A., Eisenhardt, P.R., Lidman, C., Holden, B., Della Ceca, R., Norman, C., Squires, G.: ApJ **561**, 13 (2001)
30. Boschin, W.: A&A **396**, 397 (2002)
31. Bryan, G.L., Norman, M.L.: ApJ **495**, 80 (1998)
32. Burke, D.J., Collins, C.A., Sharples, R.M., Romer, A.K., Holden, B.P., Nichol, R.C.: ApJ **488**, L83+ (1997)
33. Burrows, C.J., Burg, R., Giacconi, R.: ApJ **392**, 760 (1992)
34. Carlberg, R.G., Morris, S.L., Yee, H.K.C., Ellingson, E.: ApJ **479**, L19+ (1997)
35. Carlberg, R.G., Yee, H.K.C., Ellingson, E., Abraham, R., Gravel, P., Morris, S., Pritchet, C.J.: ApJ **462**, 32 (1996)
36. Carlberg, R.G., Yee, H.K.C., Ellingson, E., Morris, S.L., Abraham, R., Gravel, P., Pritchet, C.J., Smecker-Hane, T., Hartwick, F.D.A., Hesser, J.E., Hutchings, J.B., Oke, J.B.: ApJ **476**, L7+ (1997)
37. Carlstrom, J.E., Holder, G.P., Reese, E.D.: ARAA **40**, 643 (2002)
38. Castander, F.J., Bower, R.G., Ellis, R.S., Aragon-Salamanca, A., Mason, K.O., Hasinger, G., McMahon, R.G., Carrera, F.J., Mittaz, J.P.D., Perez-Fournon, I., Lehto, H.J.: Nature **377**, 39 (1995)
39. Cavaliere, A., Fusco-Femiano, R.: A&A **49**, 137 (1976)
40. Cavaliere, A., Gursky, H., Tucker, W.: Nature **231**, 437 (1971)
41. Colafrancesco, S., Mazzotta, P., Vittorio, N.: ApJ **488**, 566 (1997)
42. Coles, P., Lucchin, F.: Cosmology: The Origin and Evolution of Cosmic Structure, Second Edition. Cosmology: The Origin and Evolution of Cosmic Structure, Second Edition, by Peter Coles, Francesco Lucchin, pp. 512. ISBN 0-471-48909-3. Wiley-VCH, July 2002.
43. Couch, W.J., Ellis, R.S., MacLaren, I., Malin, D.F.: MNRAS **249**, 606 (1991)
44. Dalcanton, J.J.: ApJ **466**, 92 (1996)
45. Dalton, G.B., Maddox, S.J., Sutherland, W.J., Efstathiou, G.: MNRAS **289**, 263 (1997)

46. Davis, M., Peebles, P.J.E.: ApJ **267**, 465 (1983)
47. Diaferio, A., Borgani, S., Moscardini, L., Murante, G., Dolag, K., Springel, V., Tormen, G., Tornatore, L., Tozzi, P.: MNRAS **356**, 1477 (2005)
48. Donahue, Megan, Scharf, Caleb, A., Mack, Jennifer, Lee, Y. Paul, Postman, Marc, Rosati, Piero, Dickinson, Mark, Voit, G. Mark, Stocke, John, T.: ApJ **569**, 689 (2002)
49. Donahue, M., Mack, J., Scharf, C., Lee, P., Postman, M., Rosati, P., Dickinson, M., Voit, G.M., Stocke, J.T.: ApJ **552**, L93 (2001)
50. Donahue, M., Voit, G.M.: ApJ **523**, L137 (1999)
51. Ebeling, H., Edge, A.C., Allen, S.W., Crawford, C.S., Fabian, A.C., Huchra J.P.: MNRAS **318**, 333 (2000)
52. Ebeling, H., Edge, A.C., Henry, J.P.: ApJ **553**, 668 (2001)
53. Eisenstein, D.J., Hu, W.: ApJ **511**, 5 (1999)
54. Eke, V.R., Cole, S., Frenk, C.S.: MNRAS **282**, 263 (1996)
55. Eke, V.R., Cole, S., Frenk, C.S., Patrick Henry, J.: MNRAS **298**, 1145 (1998)
56. Ettori, S., De Grandi, S., Molendi, S.: A&A **391**, 841 (2002)
57. Ettori, S., Tozzi, P., Borgani, S., Rosati, P.: A&A **417**, 13 (2004)
58. Ettori, S., Tozzi, P., Rosati, P.: A&A **398**, 879 (2003)
59. Evrard, A.E., MacFarland, T.J., Couchman, H.M.P., Colberg, J.M., Yoshida, N., White, S.D.M., Jenkins, A., Frenk, C.S., Pearce, F.R., Peacock, J.A., Thomas, P.A.: ApJ **573**, 7 (2002)
60. Fabian, A.C.: MNRAS **253**, 29P (1991)
61. Felten, J.E., Gould, R.J., Stein, W.A., Woolf, N.J.: ApJ **146**, 955 (1966)
62. Finoguenov, A., Reiprich, T.H., Böhringer, H.: A&A **368**, 749 (2001)
63. Frenk, C.S., Evrard, A.E., White, S.D.M., Summers, F.J.: ApJ **472**, 460 (1996)
64. Gardini, A., Rasia, E., Mazzotta, P., Tormen, G., De Grandi, S., Moscardini, L.: MNRAS **351**, 505 (2004)
65. Giacconi, R., Branduardi, G., Briel, U., et al.: ApJ **230**, 540 (1979)
66. Giacconi, R., Murray, S., Gursky, H., Kellogg, E., Schreier, E., Tananbaum, H.: ApJ **178**, 281 (1972)
67. Gioia, I.M., Henry, J.P., Maccacaro, T., Morris, S.L., Stocke, J.T., Wolter, A.: ApJ **356**, L35 (1990)
68. Girardi, M., Biviano, A., Giuricin, G., Mardirossian, F., Mezzetti, M.: ApJ **404**, 38 (1993)
69. Girardi, M., Borgani, S., Giuricin, G., Mardirossian, F., Mezzetti, M.: ApJ **506**, 45 (1998)
70. Girardi, M., Borgani, S., Giuricin, G., Mardirossian, F., Mezzetti, M.: ApJ **530**, 62 (2000)
71. Girardi, M., Giuricin, G., Mardirossian, F., Mezzetti, M., Boschin, W.: ApJ **505**, 74 (1998)
72. Girardi, M., Manzato, P., Mezzetti, M., Giuricin, G., Limboz, F.: ApJ **569**, 720 (2002)
73. Girardi, M., Mezzetti, M.: ApJ **548**, 79 (2001)
74. Gladders, M.D., Yee, H.K.C.: ApJS **157**, 1 (2005)
75. Gonzalez, A.H., Zaritsky, D., Dalcanton, J.J., Nelson, A.: ApJS **137**, 117 (2001)
76. Governato, F., Babul, A., Quinn, T., Tozzi, P., Baugh, C.M., Katz, N., Lake, G.: MNRAS **307**, 949 (1999)
77. Gross, M.A.K., Somerville, R.S., Primack, J.R., Holtzman, J., Klypin, A.: MNRAS **301**, 81 (1998)

78. Gunn, J.E., Hoessel, J.G., Oke, J.B.: ApJ **306**, 30 (1986)
79. Haiman, Z., Allen, S., Bahcall, N., et al.: ArXiv Astrophysics e-prints (2005)
80. Haiman, Z., Mohr, J.J., Holder, G.P.: ApJ **553**, 545 (2001)
81. Henry, J.P.: ApJ **534**, 565 (2000)
82. Henry, J.P.: ApJ **609**, 603 (2004)
83. Henry, J.P., Arnaud, K.A.: ApJ **372**, 410 (1991)
84. Henry, J.P., Gioia, I.M., Maccacaro, T., Morris, S.L., Stocke, J.T., Wolter, A.: ApJ **386**, 408 (1992)
85. Henry, J.P., Gioia, I.M., Mullis, C.R., Voges, W., Briel, U.G., Böhringer, H., Huchra, J.P.: ApJ **553**, L109 (2001)
86. Holden, B.P., Stanford, S.A., Squires, G.K., Rosati, P., Tozzi, P., Eisenhardt, P., Spinrad, H.: AJ **124**, 33 (2002)
87. Hu, W., Kravtsov, A.V.: ApJ **584**, 702 (2003)
88. Ikebe, Y., Reiprich, T.H., Böhringer, H., Tanaka, Y., Kitayama, T.: A&A **383**, 773 (2002)
89. Jenkins, A., Frenk, C., White, S., Colberg, J., Cole, S., Evrard, A., Couchman, H., Yoshida, N.: MNRAS **321**, 372 (2001)
90. Kaiser, N.: MNRAS **222**, 323 (1986)
91. Kay, S.T., Thomas, P.A., Jenkins, A., Pearce, F.R.: MNRAS **355**, 1091 (2004)
92. Kirkman, D., Tytler, D., Suzuki, N., O'Meara, J.M., Lubin, D.: ApJS **149**, 1 (2003)
93. Kitayama, T., Suto, Y.: ApJ **469**, 480 (1996)
94. Kitayama, T., Suto, Y.: ApJ **490**, 557 (1997)
95. Kotov, O., Vikhlinin, A.: ApJ **633**, 781 (2005)
96. Lewis, A.D., Ellingson, E., Morris, S.L., Carlberg, R.G.: ApJ **517**, 587 (1999)
97. Lima, M., Hu, W.: Phys. Rev. D **70**, 043504 (2004)
98. Lima, M., Hu, W.: Phys. Rev. D **72**, 043006 (2005)
99. Limber, D.N., Mathews, W.G.: ApJ **132**, 286 (1960)
100. Lin, Y.-T., Mohr, J.J., Stanford, S.A.: ApJ **591**, 749 (2003)
101. Lumb, D.H., Bartlett, J.G., Romer, A.K., Blanchard, A., Burke, D.J., Collins, C.A., Nichol, R.C., Giard, M., Marty, P.B., Nevalainen, J., Sadat, R., Vauclair, S.C.: A&A **420**, 853 (2004)
102. Lumsden, S.L., Nichol, R.C., Collins, C.A., Guzzo, L.: MNRAS, 258, 1 (1992)
103. MacTavish, C., Ade, P.A.R., Bock, J.J., et al.: ArXiv Astrophysics e-prints (2005)
104. Majumdar, S., Mohr, J.J.: ApJ **585**, 603 (2003)
105. Majumdar, S., Mohr, J.J.: ApJ **613**, 41 (2004)
106. Markevitch, M.: ApJ **504**, 27 (1998)
107. Mathiesen, B., Evrard, A.E.: MNRAS **295**, 769 (1998)
108. Mathiesen, B.F., Evrard, A.E.: ApJ **546**, 100 (2001)
109. Maughan, B.J., Jones, L.R., Ebeling, H., Scharf, C.: MNRAS **365**, 509 (2006)
110. Mazzotta, P., Rasia, E., Moscardini, L., Tormen, G.: MNRAS **354**, 10 (2004)
111. Miller, C.J., Nichol, R.C., Reichart, D., et al.: AJ **130**, 968 (2005)
112. Monaco, P.: FCPh **19**, 157 (1998)
113. Moretti, A., Guzzo, L., Campana, S., Lazzati, D., Panzera, M.R., Tagliaferri, G., Arena, S., Braglia, F., Dell'Antonio, I., Longhetti, M.: A&A **428**, 21 (2004)
114. Moscardini, L., Matarrese, S., Mo, H.J.: MNRAS **327**, 422 (2001)
115. Motl, P.M., Hallman, E.J., Burns, J.O., Norman, M.L.: ApJ **623**, L63 (2005)
116. Mulchaey, J.S., Zabludoff, A.I.: ApJ **496**, 73 (1998)

117. Mullis, C.R., McNamara, B.R., Quintana, H., Vikhlinin, A., Henry, J.P., Gioia, I.M., Hornstrup, A., Forman, W., Jones, C.: ApJ **594**, 154 (2003)
118. Mullis, C.R., Rosati, P., Lamer, G., Böhringer, H., Schwope, A., Schuecker, P., Fassbender, R.: ApJ **623**, L85 (2005)
119. Mullis, C.R., Vikhlinin, A., Henry, J.P., Forman, W., Gioia, I.M., Hornstrup, A., Jones, C., McNamara, B.R., Quintana, H.: ApJ **607**, 175 (2004)
120. Nevalainen, J., Markevitch, M., Forman, W.: ApJ **532**, 694 (2000)
121. Osmond, J.P.F., Ponman, T.J.: MNRAS **350**, 1511 (2004)
122. Oukbir, J., Blanchard, A.: A&A **262**, L21 (1992)
123. Peacock, J.A.: Cosmological Physics. Cosmological Physics, by John A. Peacock, pp. 704. ISBN 052141072X. Cambridge, UK: Cambridge University Press, January 1999.
124. Peebles, P.J.E.: 1993, Principles of physical cosmology. Princeton Series in Physics, Princeton, NJ: Princeton University Press, —c1993
125. Perlman, E.S., Horner, D.J., Jones, L.R., Scharf, C.A., Ebeling, H., Wegner, G., Malkan, M.: ApJS **140**, 265 (2002)
126. Piccinotti, G., Mushotzky, R.F., Boldt, E.A., Holt, S.S., Marshall, F.E., Serlemitsos, P.J., Shafer, R.A.: ApJ **253**, 485 (1982)
127. Pierpaoli, E., Borgani, S., Scott, D., White, M.: MNRAS **342**, 163 (2003)
128. Pierpaoli, E., Scott, D., White, M.: MNRAS **325**, 77 (2001)
129. Plionis, M., Basilakos, S., Georgantopoulos, I., Georgakakis, A.: ApJ **622**, L17 (2005)
130. Popesso, P., Biviano, A., Böhringer, H., Romaniello, M., Voges, W.: A&A **433**, 431 (2005)
131. Postman, M., Lubin, L.M., Gunn, J.E., Oke, J.B., Hoessel, J.G., Schneider, D.P., Christensen, J.A.: AJ **111**, 615 (1996)
132. Press, W., Schechter, P.: ApJ **187**, 425 (1974)
133. Rasia, E., Mazzotta, P., Borgani, S., Moscardini, L., Dolag, K., Tormen, G., Diaferio, A., Murante, G.: ApJ **618**, L1 (2005)
134. Rasia, E., Tormen, G., Moscardini, L.: MNRAS **351**, 237 (2004)
135. Reichart, D.E., Nichol, R.C., Castander, F.J., Burke, D.J., Romer, A.K., Holden, B.P., Collins, C.A., Ulmer, M.P.: ApJ **518**, 521 (1999)
136. Reiprich, T., Böhringer, H.: ApJ **567**, 716 (2002)
137. Rines, K., Geller, M.J., Kurtz, M.J., Diaferio, A.: AJ **126**, 2152 (2003)
138. Rosati, P., Borgani, S., Norman, C.: ARAA **40**, 539 (2002)
139. Rosati, P., della Ceca, R., Norman, C., Giacconi, R.: ApJ **492**, L21+ (1998)
140. Rosati, P., Tozzi, P., Ettori, S., Mainieri, V., Demarco, R., Stanford, S.A., Lidman, C., Nonino, M., Borgani, S., Della Ceca, R., Eisenhardt, P., Holden, B.P., Norman, C.: AJ **127**, 230 (2004)
141. Sadat, R., Blanchard, A., Oukbir, J.: A&A **329**, 21 (1998)
142. Sahni, V.: LNP Vol. 653: The Physics of the Early Universe **653**, 141 (2005)
143. Schuecker, P., Böhringer, H., Guzzo, L., Collins, C.A., Neumann, D.M., Schindler, S., Voges, W., De Grandi, S., Chincarini, G., Cruddace, R., M¨uller, V., Reiprich, T.H., Retzlaff, J., Shaver, P.: A&A **368**, 86 (2001)
144. Schuecker, P., Guzzo, L., Collins, C.A., Böhringer, H.: MNRAS **335**, 807 (2002)
145. Seljak, U.: MNRAS **337**, 769 (2002)
146. Sheth, R., Tormen, G.: MNRAS **308**, 119 (1999)
147. Sheth, R.K., Tormen, G.: MNRAS **329**, 61 (2002)
148. Smith, S.: ApJ **83**, 23 (1936)

149. Spergel, D.N., Bean, R., Doré, O., Nolta, M.R., Bennett, C.L., Hinshaw, G., Jarosik, N., Komatsu, E., Page, L., Peiris, H.V., Verde, L., Barnes, C., Halpern, M., Hill, R.S., Kogut, A., Limon, M., Meyer, S.S., Odegard, N., Tucker, G.S., Weiland, J.L., Wollack, E., Wright, E.L.: ArXiv Astrophysics e-prints (2006)
150. Springel, V., White, S.D.M., Jenkins, A., Frenk, C.S., Yoshida, N., Gao, L., Navarro, J., Thacker, R., Croton, D., Helly, J., Peacock, J.A., Cole, S., Thomas, P., Couchman, H., Evrard, A., Colberg, J., Pearce, F.: Nature **435**, 629 (2005)
151. Stanford, S.A., Holden, B., Rosati, P., Eisenhardt, P.R., Stern, D., Squires, G., Spinrad, H.: AJ **123**, 619 (2002)
152. Stanford, S.A., et al.: ApJ **634**, L129 (2005)
153. Steinhardt, P.J.: RSPTA **361**, 2497 (2003)
154. Sunyaev, R.A., Zel'dovich, Y.B.: Comments on Astrophysics and Space Physics **4**, 173 (1972)
155. Truemper, J.: Science **260**, 1769 (1993)
156. van Haarlem, M.P., Frenk, C.S., White, S.D.M.: MNRAS **287**, 817 (1997)
157. Vauclair, S.C., Blanchard, A., Sadat, R., Bartlett, J.G., Bernard, J.-P., Boer, M., Giard, M., Lumb, D.H., Marty, P., Nevalainen, J.: A&A **412**, L37 (2003)
158. Viana, P.T.P., Liddle, A.R.: MNRAS **303**, 535 (1999)
159. Vikhlinin, A.: ApJ **640**, 710 (2006)
160. Vikhlinin, A., Kravtsov, A., Forman, W., Jones, C., Markevitch, M., Murray, S.S., Van Speybroeck, L.: ApJ **640**, 691 (2005)
161. Vikhlinin, A., VanSpeybroeck, L., Markevitch, M., Forman, W.R., Grego, L.: ApJ **578**, L107 (2002)
162. Vikhlinin, A., Voevodkin, A., Mullis, C.R., VanSpeybroeck, L., Quintana, H., McNamara, B.R., Gioia, I., Hornstrup, A., Henry, J.P., Forman, W.R., Jones, C.: ApJ **590**, 15 (2003)
163. Voevodkin, A., Vikhlinin, A.: ApJ **601**, 610 (2004)
164. Voit, G.M.: Reviews of Modern Physics **77**, 207 (2005)
165. Warren, M.S., Abazajian, K., Holz, D.E., Teodoro, L.: ArXiv Astrophysics e-prints (2005)
166. White, M.: ApJS **143**, 241 (2002)
167. White, M., Hernquist, L., Springel, V.: ApJ **579**, 16 (2002)
168. White, S.D.M., Navarro, J.F., Evrard, A.E., Frenk, C.S.: Nature **366**, 429 (1993)
169. Willis, J.P., Pacaud, F., Valtchanov, I., Pierre, M., Ponman, T., Read, A., Andreon, S., Altieri, B., Quintana, H., Dos Santos, S., Birkinshaw, M., Bremer, M., Duc, P.-A., Galaz, G., Gosset, E., Jones, L., Surdej, J.: MNRAS **363**, 675 (2005)
170. Xue, Y.-J., Wu, X.-P.: ApJ **538**, 65 (2000)
171. Zwicky, F.: ApJ **86**, 217 (1937)

Clusters and the Theory of the Cosmic Web

R. van de Weygaert[1] and J. R. Bond[2]

[1] Kapteyn Astronomical Institute, University of Groningen, P.O. Box 800, 9700 AV Groningen, The Netherlands
weygaert@astro.rug.nl
[2] Canadian Institute for Theoretical Astrophysics, University of Toronto, Toronto, ON M5S 3H8, Canada
bond@cita.utoronto.ca

1 Outline: The Cosmic Web

The spatial cosmic matter distribution on scales from a few to more than a hundred Megaparsecs has emerged over the past 30 years through ever more ambitious redshift survey campaigns. From the first hints of superclustering in the seventies to the progressively larger and more detailed three-dimensional maps of interconnected large scale structure that emerged in the eighties, nineties and especially post-2000, we now have a clear paradigm: galaxies and mass exist in a wispy weblike spatial arrangement consisting of dense compact clusters, elongated filaments, and sheetlike walls, amidst large near-empty void regions, with similar patterns existing at higher redshift, albeit over smaller scales. The hierarchical nature of this mass distribution, marked by substructure over a wide range of scales and densities, has been clearly demonstrated. The large scale structure morphology is indeed that of a *Cosmic Web* Bond et al. [18].

Complex macroscopic patterns in nature arise from the action of basic, often even simple, physical forces and processes. In many physical systems, the spatial organization of matter is one of the most readily observable manifestations of the nonlinear collective actions forming and moulding them. The richly structured morphologies are a rich source of information on the physical forces at work and the conditions under which the systems evolved. In many branches of science the study of geometric patterns has therefore developed into a major industry for exploring and uncovering the underlying physics (see e.g., Balbus & Hawley [5]).

The vast Megaparsec cosmic web is one of the most striking examples of complex geometric patterns found in nature, and certainly the largest in terms of sheer size. Computer simulations show the observed cellular patterns can arise naturally through gravitational instability e.g., [62], with the emergent structure growing from tiny density perturbations and the accompanying tiny

R. van de Weygaert and J. R. Bond: *Clusters and the Theory of the Cosmic Web*, Lect. Notes Phys. **740**, 335–407 (2008)
DOI 10.1007/978-1-4020-6941-3_10

velocity perturbations generated in the primordial Universe. Supported by an impressive body of evidence, primarily those of temperature fluctuations in the cosmic microwave background e.g., [9, 45, 78, 79], the character of the primordial random density and velocity perturbation field is that of a *homogeneous and isotropic spatial Gaussian process*. Such fields of primordial Gaussian perturbations in the gravitational potential are a natural product of an early inflationary phase of our Universe.

The early linear phase of pure Gaussian density and velocity perturbations has been understood in great depth. This knowledge has been exploited extensively to extract from CMB data probing the linear regime half a dozen cosmological parameters. Notwithstanding these successes, the more advanced phases of cosmic structure formation are still in need of substantially better understanding. Observables of the mildly nonlinear regime also offer a wealth of information, probing a stage when individually distinct features start to emerge. The anisotropic filamentary and planar structures, the characteristic large underdense void regions and the hierarchical clustering of matter marking the weblike spatial geometry of the Megaparsec matter distribution are typical manifestations of weak to moderate nonlinearity. The existence of the intriguing foamlike network representative of this early nonlinear phase of evolution was revealed by major campaigns to map the galaxy distribution on Megaparsec scales while ever larger computer N-body simulations demonstrated that such matter distributions are indeed typical manifestations of gravitational instability.

The theoretical understanding of the nature of the emergent web is now reasonably well developed, but the development of quantitatively accurate analytic approximations is impeded by the lack of symmetries, strong nonlocal influences, and the hierarchical nature of the gravitational clustering process, with many spatial scales simultaneously relevant. Computer simulations are relied upon to provide the quantitative basis. However, analytic descriptions provide the physical insight into the complex interplay of emerging structures. An area that is still developing is the morphological analysis of the observed and simulated patterns that develop.

This first lecture notes develops the theoretical framework for our understanding of the Cosmic Web. We outline the various formalisms that have been developed to describe the hierarchical nature, the anisotropic geometry of its elements, the intrinsic and intimate relationship with clusters of galaxies, and the predominance of filaments consisting of galaxies, largely in groups, connecting the clusters. Even though we concentrate on the analytical framework, we also describe and illustrate the related generic situations on the basis of computer simulations of cosmic structure formation.

In the accompanying second set of lecture notes (Van de Weygaert & Bond, 2008), we give an overview of Cosmic Web observations. We focus on the morphology of the Cosmic Web and the role of voids within establishing this fundamental aspect of the Megaparsec Universe.

2 Cosmic Structure Formation: From Primordial Quantum Noise to the Cosmic Web

The weakly nonlinear *Cosmic Web* comprises features on scales of tens of Megaparsecs, in which large structures have not lost memory of the nearly homogeneous primordial state from which they formed, and provide a direct link to early Universe physics.

In our exploration of the cosmic web and the development of appropriate tools towards the analysis of its structure, morphology and dynamics we start from the the assumption that the cosmic web is traced by a population of discrete objects, either galaxies in the real observational world or particles in that of computer simulations. Even though individual dynamically relaxed galaxies were the most notable features historically, followed by collapsed clusters, the deepest large potential wells in the universe, we will pursue the view that filaments are basic elements of the cosmic web. Most matter assembles along the filaments, providing channels along which mass is transported towards the highest density knots within the network, the clusters of galaxies. Likewise we will emphasize the crucial role of the voids – the large underdense and expanding regions occupying most of space – in the spatial organization of the various structural elements in the cosmic web. A goal is the construction of the continuous density and velocity fields from the initial condtions, or the reconstruction of these from data, retaining the geometry and morphology of the weblike structures in all its detail.

2.1 Gravitational Instability

In the gravitational instability scenario, e.g., [62], cosmic structure grows from primordial density and velocity perturbations. It has long been assumed that the initial fluctuations were those of a *homogeneous and isotropic spatial Gaussian process.* There is good evidence for this, most notably from the cosmic microwave background. Zero point quantum noise is ubiquitous, and in particular will exist in any fields present in the early universe. In an early period of cosmic acceleration, these fluctuations and the accompanying perturbations in geometrical curvature freeze out as the universe inflates, providing the Gaussian proto-web for growth after matter is created and cosmic deceleration begins. Here we establish the nomenclature and notation for the initial gravitational potential and density fields. For the study of the developing cosmic web at late times, we can ignore relativistic photons and neutrinos, and focus on gas, dark matter and dark energy.

The formation and molding of structure is fully described by three equations, the *continuity equation*, expressing mass conservation, the *Euler equation* for accelerations driven by the gravitational force for dark matter and gas, and pressure forces for the gas, and the *Poisson–Newton equation* relating the gravitational potential to the density.

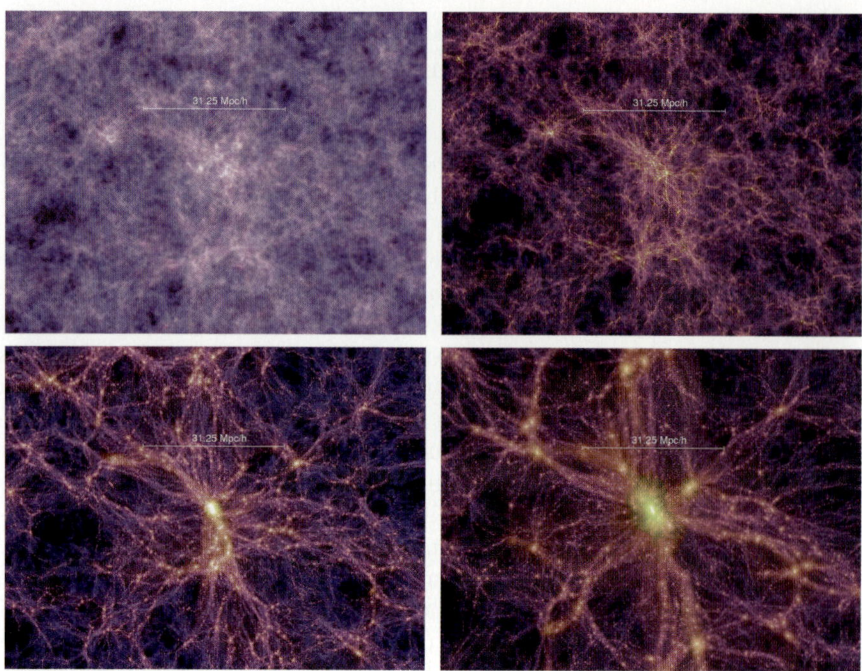

Fig. 1. The Cosmic Web in a box: a set of each four time slices from the Millennium simulation of the ΛCDM model. The frames show the projected (dark) matter distribution in slices of thickness $15\,\mathrm{h}^{-1}$ Mpc, extracted at $z = 8.55$, $z = 5.72$, $z = 1.39$ and $z = 0$. These redshifts correspond to cosmic times of 600 Myr, 1 Gyr, 4.7 Gyr and 13.6 Gyr after the Big Bang. The set of four frames have a size $31.25\,\mathrm{h}^{-1}$ Mpc zooms in on the central cluster. The evolving mass distribution reveals the major characteristics of gravitational clustering: the formation of an intricate filamentary web, the hierarchical buildup of ever more massive mass concentrations and the evacuation of large underdense voids. Image courtesy of Springel & Virgo consortium, also see Springel et al. [80]

A general density fluctuation field for a component of the universe with respect to its cosmic background mass density ρ_{u} is defined by

$$\delta(\mathbf{r},t) = \frac{\rho(\mathbf{r}) - \rho_{\mathrm{u}}}{\rho_{\mathrm{u}}}\,. \tag{1}$$

Here $\mathbf{r}$ is comoving position, with the average expansion factor $a(t)$ of the universe taken out. Although there are fluctuations in photons, neutrinos, dark energy, etc., we focus here on only those contributions to the mass which can cluster once the relativistic particle contribution has become small, valid for redshifts below 100 or so. A non-zero $\delta(\mathbf{r},t)$ generates a corresponding total peculiar gravitational acceleration $\mathbf{g}(\mathbf{r})$ which at any cosmic position $\mathbf{r}$ can be written as the integrated effect of the peculiar gravitational attraction exerted by all matter fluctuations throughout the Universe,

$$\mathbf{g}(\mathbf{r},t) = -4\pi G\bar{\rho}_m(t)a(t)\int \mathrm{d}\mathbf{r}'\,\delta(\mathbf{r}',t)\,\frac{(\mathbf{r}-\mathbf{r}')}{|\mathbf{r}-\mathbf{r}'|^3}\,. \tag{2}$$

Here $\bar{\rho}_m(t)$ is the mean density of the mass in the universe which can cluster (dark matter and baryons). The cosmological density parameter $\Omega_m(t)$ is defined by ρ_u, via the relation $\Omega_m H^2 = (8\pi G/3)\bar{\rho_m}$ in terms of the Hubble parameter H.[1] The relation between the density field and gravitational potential Φ is established through the Poisson–Newton equation,

$$\nabla^2\Phi = 4\pi G\bar{\rho}_m(t)a(t)^2\,\delta(\mathbf{r},t). \tag{3}$$

The peculiar gravitational acceleration is related to $\Phi(\mathbf{r},t)$ through $\mathbf{g} = -\nabla\Phi/a$.

The gravitational perturbations $\mathbf{g}$ induce corresponding perturbations to the matter flows, best expressed in terms peculiar velocities $\mathbf{v}$ rather than total velocities $\mathbf{u}$ which include the average Hubble expansion:

$$\mathbf{u}(\mathbf{r},t) = \frac{\mathrm{d}a(t)\mathbf{r}}{\mathrm{d}t} = H(t)\,a(t)\mathbf{r} + \mathbf{v}(\mathbf{r},t)\,. \tag{4}$$

The equation of motion for these velocity perturbations from the Hubble expansion is a recasting of the *Euler equation*:

$$\frac{\partial\mathbf{v}}{\partial t} + \frac{\dot{a}}{a}\mathbf{v} + \frac{1}{a}\,(\mathbf{v}\cdot\nabla)\mathbf{v} = -\frac{1}{a}\,\nabla\Phi\,. \tag{5}$$

This is appropriate for a pressureless medium. For gas, an additional $-\frac{1}{\rho a}\,\nabla p$ appears, along with possible viscosity and other gasdynamical forces. The mass conservation is expressed by the *Continuity equation*:

$$\frac{\partial\delta}{\partial t} + \frac{1}{a}\,\nabla\cdot(1+\delta)\mathbf{v} = 0\,. \tag{6}$$

In slightly overdense regions around density peaks, the excess gravitational attraction slows down the expansion relative to the mean, while underdense regions expand more rapidly. When a positive density fluctuation becomes sufficiently overdense it can come to a halt, turn around and start to contract. As long as pressure forces do not counteract the infall, the overdensity will grow without bound, assembling more and more matter by accretion from the surroundings, ultimately fully collapsing in a gravitationally bound and

[1] There are other contributions to the density, such as relativistic particles and dark energy which either have negligible energy density or do not effectively cluster and so do not contribute to the local peculiar gravitational acceleration, but of course do contribute to the mean acceleration value, $-(4\pi G/3)(\bar{\rho}+3\bar{p})a\mathbf{x}$, where p is the total pressure. It is conventional to parameterize the mean dark energy pressure by $p_\mathrm{de} = w\rho_\mathrm{de}$. For the cosmological constant, $w = -1$. Any $w > -1/3$ will give an accelerating term, whereas zero or positive pressure terms appropriate for dark matter and baryons give a deceleration contribution.

virialized object. By contrast the underdense regions around density minima expand relative to the background, forming deep voids. Of course, negative δ's cannot become too negative, constrained to be $\delta > -1$, so the void structure is fundamentally different than the cluster structure.

In this way the primordial overdensity field evolves into the collapsed-peak/void structure we observe, with their precise nature of the collapsed objects, dwarf galaxies, galaxies, groups, clusters, and determined by the scale, mass and surroundings of the initial fluctuation.

2.2 Primordial Origins: Gaussian Noise

There are both physical and statistical arguments in favour of the assumption that the primordial density field in the Universe was a stochastic Gaussian random field. These were applied before the observational evidence emerged for this hypothesis.

For over 25 years, the leading paradigm for explaining the large scale smoothness of the universe has been the inflation hypothesis, in which the very early Universe went through an accelerated expansion driven by an effective scalar field dominating the mass-energy. During an extremely rapid nearly exponential (nearly de Sitter) phase the Universe could have expanded by at least $\sim e^{60}$ within a time measured in Planck time units of 10^{-43} s, the details depending upon the specific particle physics realization of the inflation phenomenon. The inflation ends when preheating occurs, namely when the coherent inflaton field begins to decelerate and can then decay into particles. The density and velocity perturbations that finally evolved into the macroscopic cosmic structures in the observable Universe were generated during this phase as quantum zero point fluctuations in the inflaton, with associated small-amplitude curvature fluctuations since the inflaton carries the dominant source of mass-energy. Most inflation models, even radically different ones, predict similar properties for the fluctuations: adiabatic or curvature, Gaussian and nearly scale-invariant (see Sect. 2.2). The Gaussian nature of the perturbations is a simple consequence of the ground state harmonic oscillator wave function for the fluctuations (the zero point oscillations). Field interactions do generate calculable small deviations from Gaussianity, but except in quite contrived cases these are too tiny to effectively nullify the Gaussian hypothesis. Similarly radical deviations can exist from the simple near-scale-invariance in rather baroque models, but now these are quite constrained by the observation of near-scale-invariance in the cosmic microwave background data.

But even if inflation is not invoked, there was an argument from the *Central Limit Theorem* that Gaussian could still arise if the density field $\delta(\mathbf{x})$ is a superposition of independent stochastic processes, each with their own (non-Gaussian) probability distribution. The Fourier components $\hat{\delta}(\mathbf{k})$ are defined by

$$\delta(\mathbf{x}) = \int \frac{\mathrm{d}\mathbf{k}}{(2\pi)^3}\, \hat{\delta}(\mathbf{k})\, e^{-i\mathbf{k}\cdot\mathbf{x}}, \tag{7}$$

where $\mathbf{x}$ is comoving position and $\mathbf{k}$ is comoving wavenumber, will be independent, with random phases. There have been models in which Gaussianity does not follow, in situations where the primordial structure is created in phase transitions, e.g. associated with topological entities such as cosmic strings and domain walls.

Gaussian Random Fields

The statistical nature of a random field $f(\mathbf{x})$ is defined by its set of N-point joint probabilities. For a Gaussian random field, this takes the simple form:

$$\mathcal{P}_N = \frac{\exp\left[-\frac{1}{2}\sum_{i=1}^{N}\sum_{j=1}^{N} f_i\,(\mathsf{M}^{-1})_{ij}\,f_j\right]}{\left[(2\pi)^N\,(\det\mathsf{M})\right]^{1/2}} \prod_{i=1}^{N} \mathrm{d}f_i \tag{8}$$

where $\mathcal{P}_N$ is the probability that the field f has values in the range $f(\mathbf{x}_j)$ to $f(\mathbf{x}_j) + \mathrm{d}f(\mathbf{x}_j)$ for each of the $j = 1, \ldots, N$ (with N an arbitrary integer and $\mathbf{x}_1, \mathbf{x}_2, \ldots, \mathbf{x}_N$ arbitrary locations in the field). (We have assumed zero mean in this expression, as would be the case for $\delta, \mathbf{g}$ and $\mathbf{v}$.)

The matrix M^{-1} is the inverse of the $N \times N$ covariance matrix M,

$$M_{ij} \equiv \langle f(\mathbf{x}_i) f(\mathbf{x}_j)\rangle = \xi(\mathbf{x}_i - \mathbf{x}_j)\ , \tag{9}$$

in which the brackets $\langle \ldots \rangle$ denote an ensemble average over the probability distribution. In effect, M is the generalization of the variance σ^2 in a one-dimensional normal distribution. The equation above shows that a Gaussian distribution is fully specified by the matrix M, whose elements consist of specific values of the autocorrelation function $\xi(r)$, the Fourier transform of the power spectrum $P_f(k)$ of the fluctuations $f(\mathbf{r})$,

$$\xi(\mathbf{r}) = \xi(|\mathbf{r}|) = \int \frac{\mathrm{d}\mathbf{k}}{(2\pi)^3}\, P_f(k) \mathrm{e}^{-\mathrm{i}\mathbf{k}\cdot\mathbf{r}}\,. \tag{10}$$

Notice that the identity of $\xi(\mathbf{r})$ and $\xi(|\mathbf{r})|)$ is assumed, not required. A homogeneous and isotropic Gaussian random field f is statistically fully characterized by the power spectrum $P_f(k)$.

Power Spectrum of Density Fluctuations

The main agent for formation of structure in the Universe is a gravitationally dominant dark matter constituent of the Universe. Within the currently most viable cosmology, often called *Concordance Cosmology*, the dark matter is taken to be Cold Dark Matter: a species of non-baryonic, dissipationless and collisionless matter whose thermal properties are marked by their non-relativistic peculiar velocity (cold) at the time of radiation-matter equality. The popular candidate for this, for which a number of ambitious experiments in deep mines are in place to directly detect it through its very weak non-gravitational interactions, is the lightest supersymmetric partner of ordinary fermions, e.g. the neutralino, a scalar field partner of the neutrino, the photino, the fermionic partner of the photon, or some linear combination of other partners.

The primordial spectrum $P_p(k)$ of density perturbations in the CDM spectrum directly follows from the post-inflation form of the graviational potential fluctuations through the Poisson–Newton relation, $\delta_\rho(k) = -k^2\Phi(k)/(4\pi G\bar{\rho}a^2)$. Scale-invariant means that there is equal power per decade in the gravitational potential fluctuations, $\langle|\Phi(k)|^2\rangle d^3k/(2\pi)^3 \sim k^{n-1} d\ln k$ is $\propto d\ln k$, where n is a power law index measuring deviation from the scale-invariant unity. The corresponding form for the initial density power spectrum is $P^i_\rho(k) \equiv \langle|\delta(k)|^2\rangle \propto k^n$. Current CMB data supports an index n close to the scale-invariant unity, but slightly deviating from it, $n \approx 0.96 \pm 0.02$ Kuo et al. [45] and Spergel et al. [79]. This nearly scale invariant nature is a natural outcome of large classes of inflationary models. The expectation is that there are at least logarithmic deviations from s constant n, and it possible to get more radical deviations, as expressed by the running of the index, $dn/d\ln k \neq 0$. (There are hints of running from CMB observations, -0.06 ± 0.03 without gravity wave perturbations Kuo et al. [45], -0.04 ± 0.03 with them included Bond et al. [20].) Even before inflation theory or the data focussed attention on n nearly one, the scale-invariance was considered a natural property to assume to avoid a power spectrum with large rises either at large wavenumbers ($n > 1$) or small wavenumbers ($n > 1$), since δ could otherwise exceed unity and nonlinear collapsed structures (e.g. primordial black holes) could form in the ultra-early universe. Thus $n = 1$ was recognized as a possibility from the early seventies, defining the Harrison–Zel'dovich–Peebles spectrum.

During acceleration Ha increases and what has often been called the instantaneous horizon over which signals can propagate in a Hubble time, $(Ha)^{-1}$ decreases, and wave structure with $k/Ha < 1$ can no longer communicate, the fluctuations freeze out at their inflationary values. Once preheating occurs and radiation and matter some to dominate the energy density, the universe decelerates, Ha decreases and frozen-in perturbation patterns can respond to forces associated with their gradients once k goes above Ha. The combination of gravity and the opposing radiation pressure cause these sub-horizon fluctuations in radiation and baryon density to respond as sound waves. Meanwhile, positive fluctuations in the cold dark matter have no pressure forces and can grow, however they must do so in an expanding environment dominated by radiation which impedes the rate of growth of δ (called Hubble drag). It is only after the dynamics of cosmic expansion becomes dominated by matter following the *matter-radiation equality*, at $z_{\rm eq} \approx 3450$, when CDM density perturbations can grow rapidly, impeded only by its own Hubble drag to grow at a power law rate $\propto t^{2/3}$ rather than exponentially. The evolutionary history of fluctuations of a wavenumber k then depends upon whether k exceeds Ha in the radiation or matter dominated phase. This is encapsulated in the power spectrum transfer function $T(k)$, defined by the deviation from the primordial power spectrum shape, $P_{\rm CDM}(k) \propto k^n T^2(k)$. Corresponding to the redshift $z_{\rm eq}$ is a characteristic wavenumber scale, $k_{\rm Heq} = Ha(z_{\rm eq})$. For a CDM model with vary small baryon content, the transfer function is a unique function of $k/k_{\rm Heq}$.

From the early 1980s, much effort has gone into computing the transfer functions in terms of the material content of the universe, varying amounts of dark matter, massive neutrinos, baryons, relativistic matter, dark energy, etc. An example of a much-used numerical fitting formula for the CDM class of models which is accurate for low baryon density parameters Ω_b is that given by Bardeen et al. [6],

$$P_{\rm CDM}(k) \propto \frac{k^n}{[1+3.89q+(16.1q)^2+(5.46q)^3+(6.71q)^4]^{1/2}} \times \frac{[\ln(1+2.34q)]^2}{(2.34q)^2},$$

$$q = k/\Gamma\,, \quad \Gamma = \Omega_{m,0}h \exp\left\{-\Omega_b - \frac{\Omega_b}{\Omega_{m,0}}\right\}$$

where $k = 2\pi/\lambda$ is the wavenumber in units of $\mathrm{h\,Mpc}^{-1}$ and Γ the shape parameter. It is indeed a function of $k/k_{\rm Heq}$, with $k_{\rm Heq} \sim 5\Gamma \mathrm{h\,Mpc}^{-1}$ in the $\Omega_b \to 0$ limit. The Ω_b dependencs approximately accounts for the effect of baryons in the transfer function Sugiyama [81], although superposed upon such a T is an oscillation associated with the acoustic oscillations that the baryon-photon fluid participates in, unlike the CDM.

The corresponding *effective* power spectrum slope $n_{\rm eff}(k)$ of the Cold Dark Matter spectrum,

$$n_{\rm eff}(k) \equiv \frac{d\ln P(k)}{d\ln k} \tag{11}$$

drops from the primordial value value $n_{\rm eff} = n$ in the large scale limit $k \downarrow 0$ to $n_{\rm eff} \approx = -3 + (n-1)$ modulo logarithmic corrections at high $k \to \infty$, a direct consequence of the large Hubble drag from radiation, hence slow growth that the high k fluctuations experience. The density power spectrum per e-folding in wavenumber is

$$\mathcal{P}_\rho(k) = \mathrm{d}\sigma_\rho^2/d\ln k \equiv \langle|\delta(k)|^2\rangle k^3/(2\pi^2) \propto k^{n+3}\,.$$

The power progressively drops from small scales to large, defining the hierarchical nature of the power spectrum. (The integrated *rms* density fluctuations up to scale k, $\sigma_\rho(k)$, implicitly defined by (12) is by definition monotonic.)

2.3 Structure Growth

The time evolution of the density perturbation field $\delta(\mathbf{x},t)$ can be inferred from the solution to the three fluid equations. Generally, $|\delta|$ grows with time.

When a cosmic structure reaches virial equilibrium, as in galaxies or clusters, the physical density is constant, but the overdensity relative to the declining $\bar{\rho}_{\rm CDM} \propto a^{-3}$ background still rises. Once the radiation energy density falls off after $z_{\rm eq}$, there is still a long period of growth in the linear regime, defined by density perturbations with $\delta \ll 1$ and velocity perturbations with $(vt_{\rm exp}/d)^2 \ll 1$ (with d the coherence length of the perturbation). For the early phases of growth, it is useful to expand the perturbations in spatial eigenmodes of our three evolution equations. These are simply plane waves, and the Fourier-transformed equations depend only upon k for small δ (mode–mode $k-k'$ couplings occur through the nonlinear $\delta\mathbf{v}$ and $\mathbf{v}\cdot\nabla\mathbf{v}$ terms). The three evolution equations reduce to a single *linearized equation for the growth of density perturbations* $\delta(\mathbf{x},t)$ e.g.,

$$\frac{\partial^2\delta}{\partial t^2} + 2\frac{\dot{a}}{a}\frac{\partial\delta}{\partial t} = \frac{3}{2}\Omega_{m0}H_0^2\,\frac{1}{a^3}\,\delta \tag{12}$$

The general solution to this second order partial differential equation is the sum of a universal growing mode solution $D_1(t)$ and a a decaying mode solution $D_2(t)$,

$$\delta(\mathbf{x},t) = D_1(t)\,\Delta_1(\mathbf{x}) + D_2(t)\,\Delta_2(\mathbf{x}) \tag{13}$$

Because the decaying mode is quickly rendered insignificant in comparison to the growing mode for practical purposes it is usually sufficient to concentrate solely on the growing mode solution.

The density growth factor $D(t)$ is dependent upon the cosmological background: in different FRW Universes the growth of structure will proceed differently. In a matter-dominated FRW Universe $D(t)$ can be solved fully analytically, for more general situations the linear growth factor needs to be evaluated numerically. Ignoring the contribution by radiation, the *linear growth factor* $D(t)$ in a Friedmann-Robertson-Walker Universe containing only matter and a cosmological constant Λ (or equivalent dark energy component), with current density parameters $\Omega_{m,0}$ and $\Omega_{\Lambda,0}$, may be computed from the integral (see Heath [33], Peebles [62], Hamilton [32], Lahav & Suto [49])

$$D(t) = D(t,\Omega_{m,0},\Omega_{\Lambda,0}) = \frac{5\,\Omega_{m,0}H_0^2}{2}\,H(a)\int_0^a \frac{{\rm d}a'}{a'^3 H^3(a')}\,, \tag{14}$$

where the linear density growth factor is normalized to unity at the present epoch, $D(t_0) = 1$. For pure matter-dominated Universes, $\Omega_\Lambda = 0$, one may derive analytical expressions for $D(t)$ (see Peebles [62]). For $\Omega_m = 1$ and no mean curvature, $D = a$. For the general situation including a non-zero cosmological constant, $\Omega_\Lambda \neq 0$, the following fitting formula provides a sufficiently accurate approximation for most purposes [49],

$$D(t) \approx a(t)\,\frac{5\Omega(t)}{2}\,\frac{1}{\Omega(t)^{4/7}\,-\,\Omega_\Lambda(t)\,+\,[1+\Omega_m(t)/2][1+\Omega_\Lambda(t)/70]}\,, \quad (15)$$

in which $\Omega(t) = \Omega_m(t) + \Omega_\Lambda(t)$.

The accompanying growing mode linear velocity perturbations $\mathbf{v}(t)$ are linearly proportional to the generating peculiar gravitational acceleration $\mathbf{g}(t)$,

$$\mathbf{v} \,=\, \frac{2\,f}{3H\Omega}\,\mathbf{g}\,.$$

The deviation from Einstein de Sitter $D = a$ growth is described by the *dimensionless linear velocity growth factor* $f = f(\Omega_m, \Omega_\Lambda)$ encoding how D runs with respect to a:

$$\begin{aligned} f(\Omega_m,\Omega_\Lambda) \,&\equiv\, \frac{a}{D}\frac{\mathrm{d}D}{\mathrm{d}a} \\ &=\, -1\,-\,\frac{\Omega_m}{2}\,+\,\Omega_\Lambda\,+\,\frac{5\Omega_m}{2}\frac{a}{D}\,, \end{aligned} \quad (16)$$

with the implied growth $D_v(t)$ of linear velocity perturbations given by

$$D_v(t) \,=\, a\,D\,H\,f(\Omega_m,\Omega_\Lambda)\,. \quad (17)$$

For a matter-dominated Universe with $\Omega_m \lesssim 1$ Peebles [62] found the famous approximation,

$$f(\Omega_m) \,\approx\, \Omega_m^{0.6}\,. \quad (18)$$

An extension of this approximation for a Universe with both matter and a cosmological constant Λ was given by Lahav et al. [48],

$$f(\Omega_m,\Omega_\Lambda) \approx \Omega_m^{0.6} + \frac{\Omega_\Lambda}{70}\left(1+\frac{\Omega_m}{2}\right) \quad (19)$$

This form clearly shows that the velocity growth being is mainly determined by the matter density Ω_m and is only (very) weakly dependent on the cosmological constant. The latter is to be expected since perturbations in dark energy are expected to damp when k exceeds Ha rather than grow.

Current estimates of the material content of the Universe for tilted ΛCDM models from CMB and large scale structure data are $\Omega_{m,0} \approx 0.27 \pm 0.03$, $\Omega_{\Lambda,0} \approx 0.73 \pm 0.03$ and $\Omega_b \sim 0.045$ [45]. The dark matter to baryon ratio is ~ 5, small enough for acoustic oscillations to be evident in the transfer function, and this effect has now been observed in galaxy redshift surveys. At early times any matter-dominated FRW Universe evolves as the expansion factor $a(t)$, $D(t) = a(t) \propto t^{2/3}$, as in an Einstein-de Sitter Universe (defined by $\Omega_m(a) = \Omega_{m,0} = 1$, $\Omega_\Lambda = 0$).

In the Λ-dominated cosmology favored by current cosmological observations, the universe makes the transition from deceleration to acceleration at

$$a_{m\Lambda} \approx \left(\frac{\Omega_{m,0}}{2\Omega_{\Lambda,0}}\right)^{1/3} . \tag{20}$$

The vacuum energy density associated with the cosmological constant dominates over the mass density of matter at $2^{1/3}a_{m\Lambda}$, hence the Hubble parameter is Λ-dominated and Hubble drag slows the subsequent growth of fluctuations. With $\Omega_{m,0} \approx 0.27$ and $\Omega_{\Lambda,0} \approx 0.73$, this gives $z_{m\Lambda} \approx 0.7$. This freezing out of growth occurs for linear structures on large scales. In nonlinear high-density regions, the local gravity is strong enough for the evolution of structure to continue. As a result, no larger scale weblike patterns will emerge after the Universe gets into exponential expansion, yet within the confines of the existing Cosmic Web structures and objects will continue to evolve as clumps of matter collapse and merge into ever more pronounced and compact halos and features (see Sect. 2.4).

A nice illustration of the evolution is in Fig. 2, showing how the large scale Universe changes in a ΛCDM model from $z = 8$ until the present-day, in a box of size $65\,\mathrm{h}^{-1}$ Mpc. The time proceeds along the length of the two strips, the lateral direction is taken along the length of the box. The developing structure along the two strips shows the emergence of the Megaparsec Cosmic Web out of the nearly uniform and early Universe. Along the lefthand frame time runs from $z \approx 8$ (bottom) until (top) and in the righthand frame from $z \approx 4$ at the bottom to the present-day at $z = 0$ (upper righthand frame).

The cosmic mass distribution is marked by cellular patterns whose characteristic size grows is continuously growing and becomes ever more pronounced up to $z \approx 1.5 - 2$ (centre righthand frame). Clearly recognizable, particular in the lefthand frame, is the hierarchical buildup of the weblike patterns. Both filaments and voids are seen to merge with surrounding peers into ever larger specimen of these features.

Later, as a consequence of the accelerated expansion of the Universe the large scale structure begins to slow at $z \approx 1.5 - 2.0$. As a result the overall spatial distribution of matter remains basically unchanged. Within the existing structures the nonlinear evolution does indeed continue: filaments and clusters remain overdense regions in which gravity continues to mould the clustering and configuration of matter. It results in a continuing sharpening of the weblike features in the Megaparsec universe.

2.4 Nonlinear Clustering

Once the gravitational clustering process has progressed beyond the initial linear growth phase we see the emergence of complex patterns and structures in the density field. Highly illustrative of the intricacies of the structure formation process is that of the state-of-the-art N-body computer simulation,

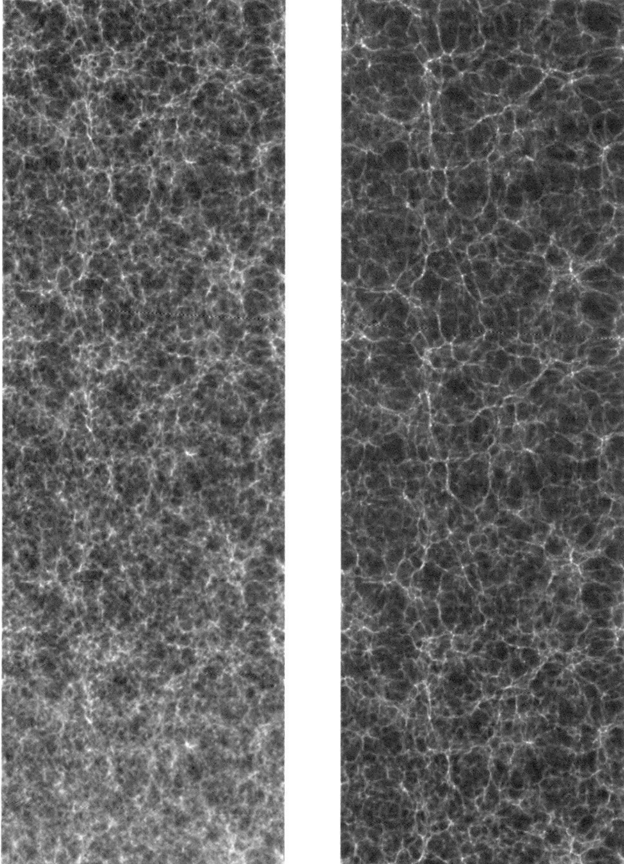

Fig. 2. The development of the large scale Universe from $z = 8$, after the end of the Dark Ages, until the present-day in a timeline proceeding along two strips. The timeline runs from lower lefthand frame (end Dark Ages, $z = 8$) until $z \approx 4$ (*top of lefthand frame*), resuming the latter at the bottom of the righthand frame and running on to the present-day at $z = 0$ at the upper righthand of that frame. The cosmic mass distribution is marked by cellular patterns whose characteristic size grows continuously and becomes ever more pronounced up to $z \approx 1.5 - 2$ (*centre righthand frame*). As a consequence of the accelerated expansion of the Universe the large scale structure freezes at that point: the overall distribution of matter remains basically unchanged, except for the sharpening of the features as a result of the continuing nonlinear evolution within these features. Image courtesy of Aragón-Calvo, also see Aragón-Calvo [1]

the Millennium simulation by Springel et al. [80]. Figure 1 shows two sets of each four time frames out of this massive 10^{10} particle simulation of a ΛCDM matter distribution in a $500\,\mathrm{h}^{-1}$ Mpc box. The time frames correspond to redshifts $z = 8.55$, $z = 5.72$, $z = 1.39$ and $z = 0$ (i.e. at epochs 600 Myr, 1 Gyr,

4.7 Gyr and 13.6 Gyr after the Big Bang). The earliest time frame is close to the epoch when the first dwarf galaxies formed. Current estimates show that the characteristic redshift for reionization of the gaseous IGM by radiation from the first stars, when the so-called *Dark Ages* ended, is at $z_{reh} = 11.4 \pm 2.5$ [45]. (Even with 10 billion particles, the web-like structure that actually exists at $z = 8.55$ is not evident since the waves that have formed it cannot be included in such a simulation.) The first set of frames contains the Dark Matter particle distribution in a $15\,\mathrm{h}^{-1}$ Mpc thick slice of a $125\,\mathrm{h}^{-1}$ Mpc region centered on the central massive cluster of the simulation. The second set zooms and illuminates the details of the emergence of the central cluster in a $31.25\,\mathrm{h}^{-1}$ Mpc sized region.

The first set provides a beautiful picture of the unfolding Cosmic Web, starting from a field of mildly undulating density fluctations towards that of a pronounced and intricate filigree of filamentary features, dented by dense compact clumps at the nodes of the network. The second set of frames depict the evolution of the matter distribution surrounding the central highly dense and compact cluster. In meticulous detail it shows the formation of the filamentary network connecting into the cluster which are the transport channels for matter to flow into the cluster. Clearly visible as well is the hierarchical nature in which not only the cluster builds up but also the filamentary network. At first consisting of a multitude of small scale edges, they quickly merge into a few massive elongated channels. Equally interesting to see is the fact that the dark matter distribution is far from homogeneous: a myriad of tiny dense clumps indicate the presence of the dark halos in which galaxies – or groups of galaxies – will or have formed.

Large N-body simulations like the Millennium simulation and the many others currently available all reveal a few "universal" characteristics of the (mildly) nonlinear cosmic matter distribution. Three key characteristics of the Megaparsec universe stand out:

- Hierarchical clustering
- Anisotropic & Weblike spatial geometry
- Voids

These basic elements exist at all redshifts, but differ in scale, in fact with a growing *nonlinear wavenumber* $k_{NL}(z)$ characterizing the onset of moderate nonlinearity. The linearly-evolving integrated power spectrum defined by (12), $\sigma^2_{\rho L}(k, z) = D^2(z)\sigma^2_{\rho L}(k, 0)$ as a function of redshift. If linear growth were to prevail, formal nonlinearity would occur when $\delta(k, z) \sim 1$, namely at $k = k_{NL}(z)$, where $\sigma_{\rho L}(k_{NL}(z), 0) \equiv D^{-1}(z)$. Monotonicity of $\sigma_{\rho L}$ guarantees $k_{NL}(z)$ increases with decreasing redshift. The cosmic web pattern is developed from waves in a band of wavenumbers just below $k_{NL}(z)$, hence the web-like patterns seen in simulations look somewhat similar at differing redshifts, except the overall scale changes with increasing k_{NL}. (The relevant web-band in $\sigma_{\rho L}$ for the weak to moderate nonlinearity relevant to the web pattern turns out to be about 0.2–0.7 [18, 19], with higher values associated with collapsed density peaks). Because $\Delta\sigma_{\rho L}/\sigma_{\rho L} \propto (n_{\rm eff}+3)\Delta \ln k$ in terms of the effective index of the power spectrum $n_{\rm eff}$, the wavenumber band $\Delta \ln k$ associated with $\sigma_{\rho L}/\sigma_{\rho L} \approx 1/2$ is considerably wider for the flattened spectra associated with higher redshif: that is more waves belong to the web-band around k_{NL}, and the filaments tend to be fatter (more ribbon-like) than at lower redshift [19].

The challenge for any viable analysis tool is to trace, highlight and measure each of the morphological elements of the cosmic web. Ideally it should be able to do so without resorting to user-defined parameters or functions, and without affecting any of the other essential characteristics.

3 Hierarchical Structure Formation

In a simple Einstein-deSitter models of spherical overdense perturbations, when the linear $\delta_L = 1.05$, the flow changes from outward, albeit increasingly lagging the cosmic Hubble flow, to infall, toward complete collapse and virialization by $\delta_L = f_c \approx 1.7$. If so, a typical 2-sigma density peak associated with a scale k will collapsed at $\sigma_{\rho L} \approx f_c/2 \sim 0.8$, the (much) rarer 3-sigma density peaks at $\sigma_{\rho L} \approx 0.6$, hence the collapsed structure band is associated with $\sigma_{\rho L} \gtrsim 0.7$ which defined the upper bound of the web pattern $\sigma_{\rho L}$ described in the last section. A rough relation of characteristic wave number to mass of the collapsed object is $M = (4\pi/3)\bar{\rho}_{\rm m}(2a/k)^3 \approx 10^{12}\Omega_{\rm m}(2k^{-1}/{\rm Mpc})^3\,{\rm M}_\odot$ [17].

Thus, as k_{NL} sweeps down from high redshift, it leaves in its wake first stars which reionize the universe formed in tiny dwarf galaxies with $2k^{-1} \sim 10\,{\rm kpc}$, dwarf galaxies with $2k^{-1} \sim 100\,{\rm kpc}$, large Milky Way like galaxies with $2k^{-1} \sim$ Mpc to rare large clusters with $2k^{-1} \sim 10\,{\rm Mpc}$. The web associated with sligthly lower k's is formed from the front end of the k_{NL}-wake. These features of (zero-dimension) objects embedded in structures of a larger dimension (one-dimensional filaments, two-dimensional sheets) at a lower density is clearly evident in Fig. 3, with the larger encompassing perturbations gradually evolving through the merging and accretion of smaller scale clumps, a process illustrated in Fig. 4.

Aptly described by the concept of *merger tree* (see e.g., Kauffmann & White [42], Lacey & Cole [46]), the precise path that an encompassing perturbation follows towards final collapse and virialization may be highly diverse.

3.1 Hierarchical Structures

Extended features still in the process of collapsing, or collapsed objects which have not yet fully virialized, often contain a large amount of smaller scale substructure at higher density which materialized at an earlier epoch. This

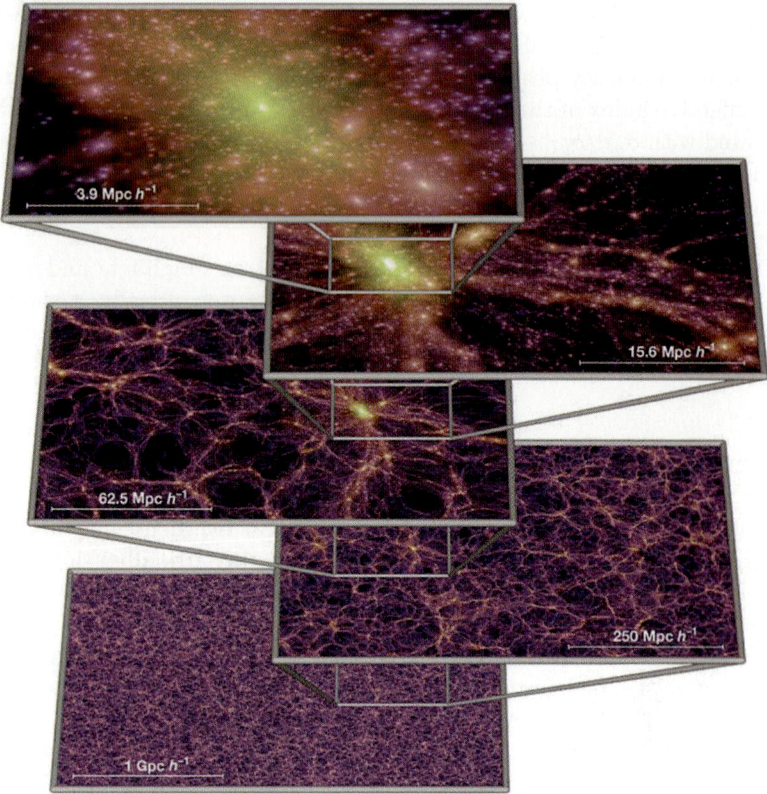

Fig. 3. The hierarchical Cosmic Web: over a wide range of spatial and mass scales structures and features are embedded within structures of a larger effective dimension and a lower density. The image shows how structures in the Millennium simulation are mutually related: at five successive zooms it focusses on a very dense and compact massive cluster at the intersection of a high number of filamentary extensions. Image courtesy of Springel & Virgo consortium, also see Springel et al. [80]. Reproduced with permission of Nature

substructure is a clear manifestation of the hierarchical development of structure in the Universe. This hierarchy of embedded structures is illustrated in Fig. 3, which shows five slices through the Millennium simulation [80], from bottom to top representing successive zoom-ins onto a very dense and compact massive cluster.

Observationally we can recognize traces of the hierarchical formation process in the galaxy distribution on Megaparsec scales. The large unrelaxed filamentary and wall-like superclusters contain various rich clusters of galaxies as well as a plethora of smaller galaxy groups, each of which has a higher density than the average supercluster density. Zooming in on even smaller

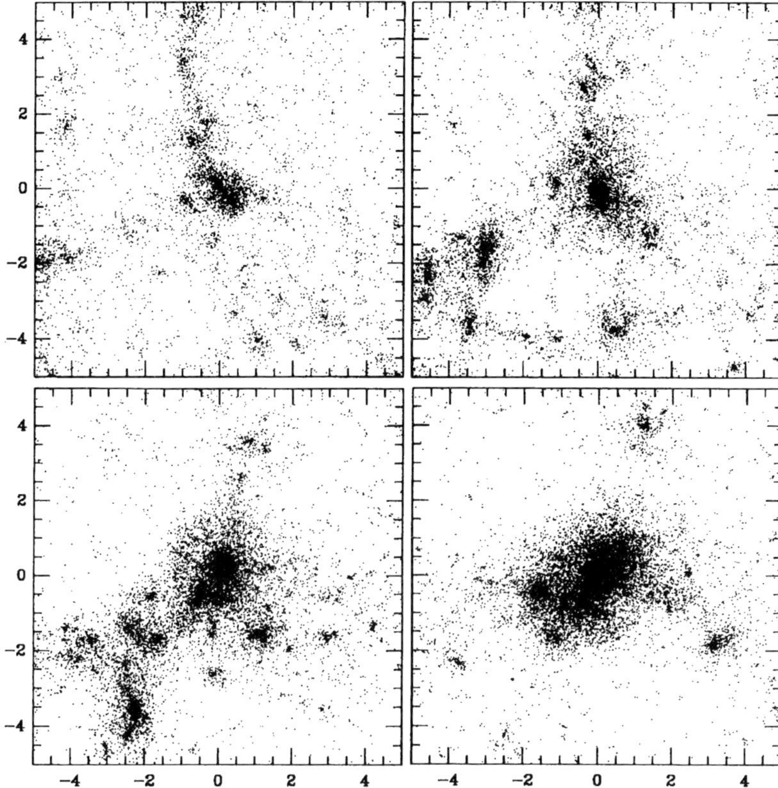

Fig. 4. Illustration of the hierarchical formation of a cluster sized halo. From: van Haarlem & van de Weygaert [83]. Reproduced by permission of the AAS

scales, within groups large galaxies themselves are usually accompanied by a number of smaller satellites and dwarf galaxies. The imprint of hierarchical clustering may also be found in fully collapsed structures, such as clusters and even the halos of galaxies. Even when studying the hot X-ray emitting intracluster gas, more evenly distributed than the galaxies, the majority of clusters appears to display some measure of substructure (e.g. Schücker et al. [71]). Even the Coma cluster appears to be marked by a heavy infalling group [59]. Also galaxies bear the marks of their hierarchical formation. The most visible manifestation concerns the presence of streams in their dark halos, remnants of infalling dwarf galaxies (e.g. Helmi & White [34], Freeman & Bland-Hawthorn [29]).

3.2 Mass Scale Fluctuations

We now generalize the integrated *rms* power $\sigma_{\rho L}(k,z)$ to *rms* fluctuations associated with general filters $W(kR)$ [6]:

$$\sigma_W^2(R) = \int d\ln k\, |\hat{W}(kR)|^2\, d\sigma_{\rho L}^2(k)/d\ln k\,. \tag{21}$$

For example, if the density field is smoothed with a tophat or Gaussian filter, then $\hat{W}(x)$ is

$$\hat{W}(\mathbf{x}) \quad \rightarrow \quad \begin{cases} \hat{W}_{\rm TH} = \dfrac{3}{x^3}(\sin x - x\cos x) & \text{Tophat} \\ \hat{W}_{\rm G} \;= \exp(-x^2/2) & \text{Gaussian} \end{cases}$$

respectively. The Fourier transforms of the filters define smoothing functions W in real space,

$$\begin{cases} W_{\rm TH}(\mathbf{r}) \;=\; \vartheta(R_{\rm TH} - r)/V_{\rm TH}\,, & V_{\rm TH} = 4\pi R^3/3 \\ \\ W_{\rm G}(\mathbf{r}) \;=\; \exp(-r^2/2R_{\rm G}^2)/V_{\rm G}\,, & V_{\rm G} = (2\pi)^{3/2}R^3 \end{cases}$$

Here, $\vartheta(x)$ is the Heaviside function, unity if $x \geq 0$ zero if $x < 0$. The smoothing filter that defines $\sigma_{\rho L}^2(k_R)$ is called the "sharp k-space" filter, simply a top hat in k-space, $\hat{W} = \vartheta(1 - kR_k)$, where $R_k = 1/k_R$. Its Fourier transform is $W(\mathbf{r}) = \hat{W}_{\rm TH}(rk_R)/V_{\rm k}$, with $V_{\rm k} = (4\pi/3)k_R^{-3}/(2\pi)^3$.

The nature of top hat smoothing is clear: around each point $\mathbf{r}$, we volume-average the field over a spherical region of radius R around it. There is a clear mass assignment we can make, $M_{\rm TH} = \bar{\rho}_m a^3 V_{\rm TH}$. For the other filters, the relation between the scale R and an appropriate mass is trickier. The obvious values, $\bar{\rho}_m a^3 V_{\rm G}$ and $\bar{\rho}_m a^3 V_{\rm k}$ turn out not to be applicable to objects.

From the discussion about the nonlinear wavenumber above, it should be clear that $\sigma_{\rho L}(k)$ defines a clock whose ticks march out the development of the hierarchy. Indeed Bond et al. [16] showed that the square,

$$S \equiv \sigma_{\rho L}^2 \tag{22}$$

is the most appropriate. A convenient way to define a filter-independent mass is to determine the "trajectory" $R_{\rm TH}(S)$ by inverting $S = \sigma_{\rm TH}^2(R_{\rm TH})$ and using $M_{\rm TH}$ for every filter. The trajectories $R_{\rm k}(S)$ and $R_{\rm G}(S)$ then define functional relations $R_{\rm k}(M_{\rm TH})$ and $R_{\rm G}(M_{\rm TH})$ among filter scales. There are more sophisticated ways of defining the mass relations among filters using profiles around density field peaks, but this approach gives similar answers. It turns out that the inversion for Gaussian and sharp-k space gives $R_{\rm TH}/R_{\rm G} \approx 2$, with a similar result $R_{\rm TH}/R \approx 2$.

3.3 Collapse and Virialization: Density Barriers

The correspondence between mass and filter scale, $M \propto R^3$, suggests that if one wishes to model (proto)objects of mass M one should study the initial density fluctuation field when it is smoothed on (comoving) spatial scale $R \propto M^{1/3}$, with the exact coefficient depending on filter choice (22):

$$\delta(\mathbf{r}, t|R) = \int \mathrm{d}\mathbf{r}' \, \delta(\mathbf{r}', t) \, W((\mathbf{r} - \mathbf{r}')/R) \,. \tag{23}$$

For the pure power-law spectra $P(k) \propto k^{n_{\rm eff}}$ the fluctuations S on a mass scale M scale as

$$S(M) \propto M^{-(n_{\rm eff}+3)/3} \,. \tag{24}$$

The monotonocity of $S(M)$ with M is generally valid, even with the $n_{\rm eff}(k)$ we have seen arise in ΛCDM and other theories. The Cold Dark Matter spectrum (11) has $n_{\rm eff}(k_{\rm gal}) \approx -2$ on galaxy scales and $n_{\rm eff}(k_{\rm cls}) \approx -1.3$ on clusters scales.

Spherical Haloes: Collapse & Virialization

We now review the extremely instructive nonlinear evolution of a spherically symmetric density peak, which turns around and collapses, with complete collapse to a point predicted to occur when the linearly-extrapolated (primordial) density, $\delta_L(r,t|R) = D(t)/D(t_i)\delta_L(r,t_i|R)$ (12, 13), reaches a critical density excess f_c [30]. No singularity in fact develops, rather shells of mass pass through the origin and oscillate relative to each other finally settling down to a virial equilibrium in which kinetic and gravitational forces are balanced. In more realistic 3D collapses the inevitable non-spherical perturbations enhance the approach to virialization. Thus we can identify smoothed linear overdensities f_{ta}, f_{vir} as well as f_c, as well their nonlinear overdensity counterparts, $\delta_{NL,ta}$, $\delta_{NL,vir}$ as well as $\delta_{NL,c} = \infty$: The collapse and subsequent virialization of a spherical and isolated overdensity is solely dependent on such a critical – and universal – threshold level f_c, and independent of the mass scale M. The same holds true for its decoupling from the Hubble expansion and turnaround. The corresponding characteristic density thresholds for turnaround f_{ta}, collapse f_c and virialization f_{vir} can be derived from the spherical model.

The critical value for an Einstein-de Sitter $\Omega_m = 1$ Universe has the well-known value derived by Gunn & Gott [30],

$$f_c = \frac{3}{20}\,(12\pi)^{2/3} \simeq 1.686 \,, \tag{25}$$

while the corresponding critical nonlinear virialization value is given by

$$\frac{\rho_{vir}}{\rho_u} = 18\pi^2 \simeq 178.0 \,. \tag{26}$$

Similar values can be easily derived for turnaround: the linear turnaround threshold value $f_{\rm ta} = 1.08$, while the nonlinear turnaround density values is $\delta_{NL,ta} = 5.55$.

For a general FRW Universe with $\Omega_{m,0} \neq 1$ and/or $\Omega_{\Lambda,0} \neq 0$ the values depend upon the cosmic epoch at which turnaround, collapse or virialization of the density perturbation takes place, i.e., it is a function of the values of Ω_m and Ω_Λ at the corresponding cosmic epoch. For open cosmologies with $\Lambda = 0$ solutions to the problem were computed by Lacey & Cole [46]. Lahave et al. [48] adressed the issue for FRW universes with a cosmological constant $\Lambda \neq 0$, while Eke et al. [27] computed the explicit solutions for flat $\Omega_m + \Omega_\Lambda = 1$ FRW universes. The general expressions for these situations were summarized by Kitayama & Suto [43]. The case for Dark Energy models with $w \neq -1$ were assessed by Percival (2005). While the linear collapse threshold value f_c does depend somewhat on the cosmological background, the values for plausible cosmologies are only marginally different from those for an Einstein-de Sitter Universe. As may be seen in Fig. 5 the values for f_c in generic-open matter-dominated cosmologies or flat cosmologies with a cosmological constant Λ turn out to have only a weak dependence on $\Omega_{m,0}$: in an open $\Omega_{m,0} = 0.1$ Universe $f_{c,0} \approx 1.615$. We note that the nonlinear virialization threshold $\delta_{NL,vir}$ displays a considerably stronger variation as a function of cosmology.

Useful fitting formulae for the linear spherical model collapse value $\delta_{NL,c}$ were obtained by Bryan & Norman [22] for $\Omega_\Lambda = 0$ FRW universes and for flat Universes:

$$\begin{cases} \delta_{NL,c} = 18\pi^2 + 82(\Omega_m - 1) - 39(\Omega_m - 1)^2 & \Omega_m + \Omega_\Lambda = 1 \\ \delta_{NL,c} = 18\pi^2 + 60(\Omega_m - 1) - 32(\Omega_m - 1)^2 & \Omega_\Lambda = 0 \end{cases} \tag{27}$$

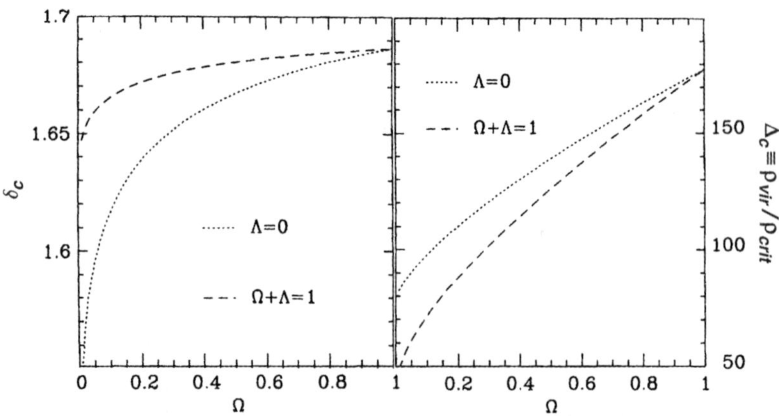

Fig. 5. Left frame: Critical threshold for collapse, f_c, as a function of Ω_m, in the spherical collapse model. Results are plotted for open models with $\Lambda = 0$ (*dotted line*) and flat models with $\Omega_m + \Omega_\Lambda = 1$ (*dashed line*). Righthand panel: the (nonlinear) virial density of collapsed objects in units of critical density. From Eke et al. [27]. Image courtesy of Vincent Eke

Spherical Collapse and Primordial Density Field

Given the primordial density field $\delta_L(\mathbf{x}, t)$, linearly interpolated to the present epoch, at any one cosmic epoch t (redshift z) one may identify the peaks that have condensed into collapsed objects by tracing the regions for whom the filtered primordial density excess

$$\delta_L(\mathbf{x}, t|R) > \frac{f_c(z)}{D(z)} \equiv f_{sc}(z)\,, \tag{28}$$

where the index sc refers to "spherical collapse". For a Gaussian random field, the statisistical distribution of $\delta_L(\mathbf{r}, t|R)$, is

$$\begin{aligned} P(\delta_L)\,\mathrm{d}\delta_L &= \exp[-\delta_L^2(\mathbf{r}, t|R)/2S(R,t)]\,\mathrm{d}\delta_L/\sqrt{2\pi S(R,t)} = \exp[-\nu^2/2]\,\mathrm{d}\nu/\sqrt{2\pi}\,, \\ \nu &= \delta_L(\mathbf{r}, t|R)/\sigma_W(R,t)\,, \quad S(R,t) = \sigma_W^2(R,t)\,. \end{aligned} \tag{29}$$

The number of σ is ν, which is a Gaussian random deviate (i.e., is distributed as the unit-variance normal). The threshold on scale M is therefore achieved when the height ν in σ units is

$$\nu(M) = \frac{f_c}{\sigma_W(M)}\,. \tag{30}$$

High mass objects are very rare because $\sigma_W(M)$ is at low, hence $\nu(M)$ is high.

Collapse and Halo Shape

While the above is based upon spherical collapse, in realistic circumstances primordial density perturbations will never be spherical, nor isolated [6]. In Bond & Myers [17], Sheth & Tormen [76], the dependence upon the shape of the density peak as well as on the tidal influences of the surrounding matter fluctuations were worked out (see Sect. 4.5 for a detailed description of anisotropic ellipsoidal collapse).

In a spherical collapse, the evolution of the outer radius depends only upon the average interior properties of the perturbation, and does not depend upon what the external matter is doing. Non-spherical perturbations such as ellipsoids of course collapse anisotropically. An ellipsoidal overdensity will first collapse along its shortest axis, subsequently along its medium axis and finally along its longest axis. However, the evolution of the outer shell depends upon the details of the interior distribution and on the exterior through the tidal forces acting upon the shell so it is not as clean a case as spherical collapse. There has been a long history of using homogeneous ellipsoids to model anisotropic collapses. Isolated ellipsoids were considered by Icke [38], White & Silk [88]; Peebles [62]. The extension to a cosmological setting where the exterior tidal forces were accurately included formulating it by its relation to the linear deformation tensor of the interior was made by Bond & Myers [17].

This paper showed the collapse along the shortest axis will occur more rapidly than the collapse of comparable spherical peak, that of the medium axis will not differ too much from the spherical value while full collapse along all three axes will be slower than that of its spherical equivalent. This was applied to filtered density peaks by Bond & Myers [17], who determined the critical density threshold $f_{\rm c}$ for complete collapse as a function of the linear tidal field environment or deformation described below, and to random filtered-field points by Sheth et al. [75].

Similar considerations concerning the effect of the non-spherical collapse of density peaks on the mass function of bound objects had been followed in a number of other studies. Monaco [56, 57, 58], Audit et al. [4] and Lee & Shandarin [51] studied models in which the initial (Zel'dovich) deformation tensor was used to find estimates of the collapse time. However, when following the nonlinear evolution of the same configuration by means of a corresponding (homogeneous) ellipsoidal collapse model Bond & Myers [17]; Eisenstein & Loeb [26] found marked differences. Figure 13 in Sect. 4.4 shows a telling comparison between the corresponding collapse time estimates for all three axes of a density peak.

The collapse of a spherical peak depends only upon the density, which is the trace of the deformation tensor, hence $f_c = f_{sc}$ is constant. For a non-spherical peak, the deformation tensor has an anisotopic part as well, with two (normalized) eigenvalues, the ellipticity e and its prolateness p and the collapse threshold depends upon these values, $f_{\rm ec}(e,p)$ [6, 17]. An impression of the sensitivity to e and p of the collapse time $a_{\rm c}(e,p)$ and corresponding collapse threshold $f_{\rm ec}(e,p)$ may be obtained from Fig. 6, which depicts these for an ellipsoidal perturbation in an Einstein-de Sitter Universe. For an ellipsoidal overdensity with the same initial overdensity δ_i the symbols show the expansion factor when the longest axis of the ellipsoid collapses and virializes, as a function of e and p. The axis on the right shows the associated critical overdensity required for collapse. At a given e, the large, medium and small circles show the relation at $p = 0$, $|p| \leq e/2$ and $|p| \geq e/2$, respectively. The solid curve and dashed curves depict the analytical relation specified in Sheth et al. [75] for $p = 0$ and $|p| = e/2$. The time required to collapse increases monotonically as p decreases. The top axis shows the related mass scale $\sigma(m)$ when identified with the value of e as the corresponding most probable value for p=0 (see Sheth et al. [75]).

The main conclusion is that for ellipsoidal collapse the density threshold f_{ec} becomes a "moving barrier", dependent on the ellipticity e and/or the mass scale $\sigma(m)$. On the basis of such ellipsoidal dynamics calculations and normalized by means of N-body simulations, Sheth & Tormen [76] found that the density collapse barrier may be reasonably accurate approximated by the expression

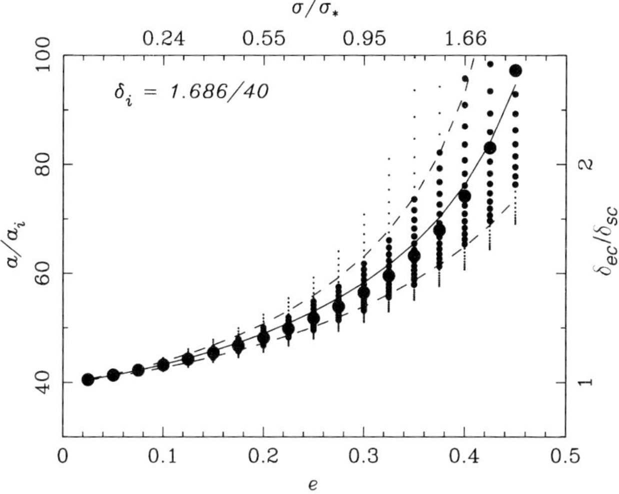

Fig. 6. Evolution of an ellipsoidal perturbation in an Einstein-de Sitter universe. Symbols show the expansion factor when the longest axis collapses and virializes as a function of initial e and p with the same initial overdensity δ_i. The circles correspond to different values of p (see text). The time required to collapse increases monotonically as p decreases. **Right axis:** associated collapse overdensity required for collapse. **Top axis:** estimate of mass resolution $\sigma(m)$ based on the corresponding most probable ellipticity e. From Sheth et al. [75]. Image courtesy of Sheth

$$\begin{aligned} f_{ec}(\sigma, z) \;&\approx\; f_{sc}(z) \left\{ 1 + \beta \left[\frac{f_{sc}^2(z)}{\sigma^2(M)} \right]^{-\alpha} \right\} \\ &= \; f_{sc}(z) \; \left\{ 1 + \beta \, \nu(M,z)^{-\alpha} \right\} \end{aligned} \tag{31}$$

with $\beta \approx 0.485$ and $\alpha \approx 0.615$. Figure 7 shows a few examples of moving barriers for a slightly different context. In this expression, $f_{\rm sc}(z)$ is the critical overdensity required for spherical collapse at a redshift z and $\sigma(M)$ the rms initial density fluctuation smoothed on a mass scale M, both linearly extrapolated to the present epoch. The parameters β and α are determined by ellipsoidal collapse: strictly speaking $\alpha = 0$ and $\beta = 0$ for spherical collapse, yielding the correct asymptotic value $f_{ec} = f_{sc}$. Cosmology enters via the relation f_{sc}, while the influence of the power spectrum enters via $\sigma(M)$. The corresponding modifications have been shown to lead to considerable improvements in the predictive power of the excursion set formalism describing the mass spectrum of condensed objects [76].

Equation 32 shows massive objects with low $\sigma(M)$ have $f_{ec}(z) \approx f_{sc}(z)$, well described by spherical collapse, whereas less massive objects are increasingly affected by external tidal forces as $\sigma(M)$ rises and M decreases. Critical

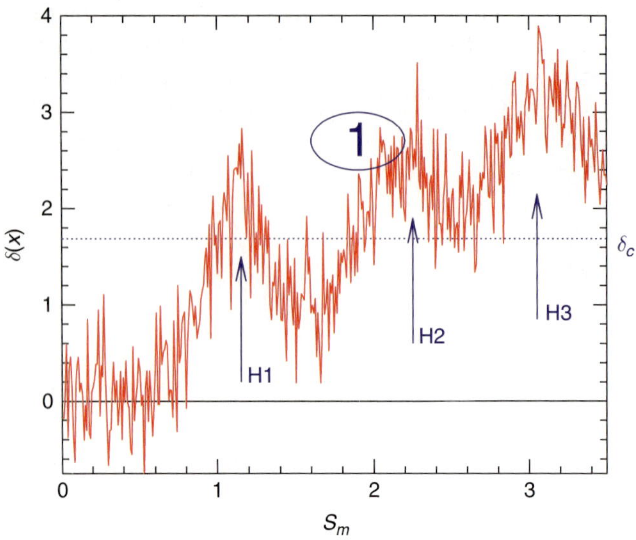

Fig. 7. Excursion Set Formalism, illustrated for the formation of a halo. Random walk exhibited by the average overdensity δ centred on a randomly chosen position in a Gaussian random field, as a function of smoothing scale, parametrized by S_M (large volumes are on the left, small volumes on the right). Dashed horizontal line indicates the collapse barrier f_c. The largest scale (smallest value of S) on which $\delta(S)$ exceeds f_c is an estimate of the mass of the halo which will form around that region. From Sheth & van de Weygaert [77]

thresholds can also be determined for other structural features, such as voids, using 2 thesholds [77] and walls and filaments (Sect. 3.3).

3.4 Halo Excursions

The excursion set formalism, also known as extended Press–Schechter formalism Press & Schechter [67], Peacock & Heavens [61], Bond et al. [16], Sheth [74], evaluates the effects substructure over a range of scales has on the emergence of objects in a cosmic density field. For an early paper on see Epstein [28] and for a recent review see [91].

It elucidates the hierarchical development of structure using just the linear density and tidal fluctuations in combination with the knowledge of their fate once the linear smoothed density exceeds the threshold values $f_{\rm c}$ we have discussed. The idea is that around a point $\mathbf{r}$, $\delta_L \mathbf{r}, t|R)$ defines a trajectory, starting from zero at very large R to larger values at small R. We have seen that $S(M)$ is a convenient clock increasing from high mass to low, hence we can also consider the smoothed field as a function of $S = \sigma_W^2(R)$: $\delta_L(\mathbf{r}, t|S)$. Further, since $S(R,t) = D^2(t) S(R, t_0)$, where t_0 is the current time, $\delta_L(\mathbf{r}, t|S(t))/sqrt S(t)$ is independent of t, a function only of $S(t_0)$ which acts like a pseudo-time. For each $\mathbf{r}$, we have a trajectory in resolution S, $\delta_L(S)$. When δ_L reaches the f_c *barrier*, we identify the scale R with the mass of a collapsed object of mass $M(S)$ at that position. The reader will realize that this prescription is unrealistic in that points very near to each other may have their density fields piercing the barrier at different S, hence be indentified with objects of different mass even though they collapse together. At best the prescription can be statistically valid but not a true real space description. That requires a non-local treatment. Further, since small-scale density peaks are embedded within larger regions which may or may not have pierced the critical collapse threshold. If the larger region has collapsed this will have involved the merging of the small scale peak with its neighbouring halos and surrounding matter while it got absorbed into the more massive entity. Consider the sharp-k filter with its $S_M = S(R_k)$ integrated power. If the linear primordial density field is a homogeneous random Gaussian field, the N-point correlation functions are translation invariant and the Fourier components $\hat{\delta}(\mathbf{k})$ are independent, that is uncorrelated in k. Sliding from a resolution S_M to a higher resolution $S_M + \Delta S_M$, the filtering process in essence involves the increment by a random Gaussian variate $\hat{\delta}(\mathbf{k})$.

Figure 7 shows an example of a typical result: a jagged line representing the linear overdensity centered on a randomly chosen position $\mathbf{r}$ in the initial Gaussian random field as a function of the scale S_M. Because of the independence of each of the Gaussian distributed Fourier components, the process turns into that of a *Brownian random walk*. The density threshold f_c for forming bound virialized objects is given by the dashed line, assumed mass independent here hence the line is horizontal. The largest scales, $S_M = 0$, are those of the homogeneous global FRW Universe so that the random walk will start of at $\delta(S = 0) = 0$. In hierarchical models S_M will increase as we zoom in on to an increasingly resolved mass distribution around the chosen position $\mathbf{x}$. As we move to a higher S_M and smaller R fluctuations of an increasing amplitude will get involved.

The distribution of masses of collapsed and/or virialized objects is equated to the distribution of distances S_M which one-dimensional Brownian motion random walks travel before they first cross a barrier of constant height f_c. In other words, one should find the distribution of the *first upcrossing* of the random trajectory, the lowest value of S for which $\delta(\mathbf{r}|S) = f_c$. The rate of first upcrossings at a threshold was calculated by Chandrasekhar (1943). When the random walk is absorbed by the barrier at the first upcrossing at S, the point $\mathbf{r}$ is identified with a collapsed object of mass $M(S)$. Here rate is per unit psuedo-time, or per unit resolution, dS. In the absence of a barrier, the distribution of trajectories with a density value $\delta_L(S)$ at S is the usual Gaussian distribution:

$$\Pi(\delta_L, S) = \frac{1}{\sqrt{2\pi\, S}} \exp\left\{-\frac{\delta_L^2}{2S}\right\} \tag{32}$$

In the presence of a barrier f_c, the probability distribution $\Pi(\delta_L, S|f_c)$ of trajectories which have a density δ_L at resolution S_M but did not cross the boundary at smaller $S < S_M$) follows from solving the Fokker-Planck equation (see Bond et al. [16], Zentner [91]),

$$\frac{\partial \Pi}{\partial S} = \lim_{\Delta S \to 0} \left\{ \frac{\langle (\Delta\delta)^2 \rangle}{2\Delta S} \frac{\partial^2 \Pi}{\partial \delta^2} - \frac{\langle \delta \Delta\delta \rangle}{\Delta S} \frac{\partial \Pi}{\partial \delta} \right\}, \tag{33}$$

where the next step in the trajectory, $\Delta\delta_L(S) = \delta_L(S+\Delta S) - \delta_L(S)$ as we increment the resolution by ΔS. The critical feature of sharp k-filter is that this step is uncorrelated with the prior value, $\langle \delta_L(S)\Delta\delta_L(S) \rangle = 0$, in which case the drift term vanishes and simple diffusion remains,

$$\frac{\partial \Pi}{\partial S} = \frac{1}{2} \frac{\partial^2 \Pi}{\partial \delta^2}. \tag{34}$$

There is a simple graphical way of determining Π. Consider a trajectory which has reached the threshold for some scale $S < S_M$. Its subsequent path is entirely symmetric and at S_M it is equally likely to be found above as well as below the threshold (see Fig. 7). In other words, for each of these trajectories there is an equally likely trajectory that pierced the barrier at the same scale R but whose subsequent path is a reflection in the barrier, ending up below the threshold. The probability Π that the threshold has never been crossed may be obtained by subtracting the reflected distribution from the overall Gaussian distribution (32),

$$\Pi(\delta_L, S_M|f_c) = \frac{1}{\sqrt{2\pi\, S_M}} \left\{ \exp\left(-\frac{\delta_L^2}{2S_m}\right) - \exp\left(-\frac{(\delta_L - 2f_c)^2}{2S_m}\right) \right\}. \tag{35}$$

Integrating this distribution over all values δ_L yields the probability that the threshold has been crossed at least once, and the corresponding probability that the location is enclosed in an object of mass $\leq M$,

$$P_s(S_M|f_c) = 1 - \int d\delta_L\, \Pi(M|f_c) = 1 - \mathrm{erf}\left\{ \frac{f_c}{\sqrt{2}\sigma(M,t)} \right\} \tag{36}$$

in which $\mathrm{erf}(x)$ is the conventional error function. In an entirely natural fashion this probability takes care of the so-called fudge factor 1/2 which had been missed in the original Press–Schechter result Press & Schechter [67]. They assumed that the fraction of mass in objects of mass $\geq M$ is given by the fraction of mass above the threshold f_c at resolution S_M. This fails to take into account that there are mass fluctuations which did not reach the threshold at mass scale M, yet are part of a collapsed structure on larger mass scale. Indeed, we will see that this is also an essential issue in understanding the development of a *void hierarchy* (see accompanying notes, van de Weygaert & Bond (2005)).

3.5 Halo Mass Distribution

Given the identification of mass, $M(S)$, we may readily infer the number density $n(M)$ of objects of mass M from the mass excursion probability $\Pi(M)$ (35):

$$n(M)\, d\ln M \;=\; \frac{\bar{\rho}_m}{M} \left|\frac{\mathrm{d}P_s}{\mathrm{d}S}\right| \frac{\mathrm{d}S}{d\ln M} d\ln M \tag{37}$$

which translates into

$$n(M)\,\mathrm{d}M \;=\; \sqrt{\frac{2}{\pi}}\, \frac{\rho_u}{M^2}\, \nu(M)\, \exp\left\{-\frac{\nu(M)^2}{2}\right\} \left|\frac{d\ln\sigma(M)}{d\ln M}\right| \tag{38}$$

For a pure power-law power spectrum, $P(k) \propto k^n$, one may readily observe that the mass spectrum of virialized and bound objects in the Universe is a self-similar evolving function

$$n(M)\,\mathrm{d}M \;=\; \sqrt{\frac{1}{2\pi}} \left(1+\frac{n}{3}\right) \frac{\rho_u}{M^2} \left(\frac{M}{M_*}\right)^{(3+n)/6} \exp\left\{-\left(\frac{M}{M_*}\right)^{(3+n)/3}\right\}. \tag{39}$$

The self-similar evolution of the mass distribution is specified via the time development of the characteristic mass $M_*(t)$,

$$M_*(t) \;=\; D(t)^{6/(3+n)}\, M_{*,o}\,. \tag{40}$$

whose present-day value is inversely proportional to f_c,

$$M_{*,0} \;=\; \left(\frac{2A}{f_c^2}\right)^{3/(3+n)} . \tag{41}$$

For a ΛCDM Universe, with $\Omega_m = 0.3$, Fig. 8 depicts the predicted Press–Schechter halo mass functions at several different redshifts [7]: $z = 0$ (solid curve), $z = 5$ (dotted curve), $z = 10$ (short-dashed curve) and $z = 20$ (long-dashed curve).

3.6 Hierarchical Evolution

Smaller mass condensations may have corresponded with genuine physical objects at an earlier phase, while later they may have been absorbed into a larger mass concentration. It is straightforward and insightful to work out the evolving object distribution within the context of the excursion set formalism.

Returning to the graphical representation in Fig. 7 we may easily appreciate what happens as the mass distribution evolves. The linear growth of fluctuations implies a gradual uniform rise of the whole random walk curve as each mass fluctuations increase by the linear growth factor $D(z)$. Going back in

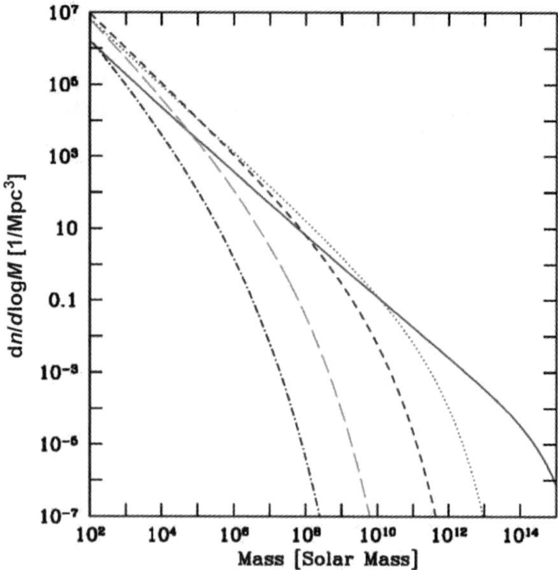

Fig. 8. Press–Schechter halo mass function at several different redshifts: z = 0 (*solid curve*), z = 5 (*dotted curve*), z = 10 (*short-dashed curve*) and z = 20 (*long-dashed curve*). From Barkana & Loeb [7]. Reproduced with permission of Elsevier

time the random walk curve would therefore have had a proportionally smaller amplitude. Linearly translated to the present epoch the density threshold barrier would gradually decrease in amplitude, proportional to $1/D(z)$. Earlier barrier crossings would therefore have occurred at a higher values of $S(R)$, a smaller scale R and a smaller mass M: Location $\mathbf{x}$ would have been incorporated within an object of a correspondingly smaller mass.

As we proceed in time the barrier $f_c(z)$ would descend further. Gradually the random walk path will start to pierce through the barrier at lower S and correspondingly larger values of the mass scale M. The halo into which the point may be embedded will first accrete surrounding matter, thereby gradually growing in mass. Even later the halo may merge with surrounding clumps into a much more massive halo. The corresponding mass scale would reveal itself as the next peak in the random walk. Figure 7 does reveal such behaviour through the presence of three peaks, H1, H2 and H3: H3 corresponds to an early small object that merged with surrounding mass into the more massive peak H2. The latter would merge again with neighbouring peers into the largest clump, object H1.

While the excursion set formalism manages to describe quantitatively the merging and accretion history of halos in a density field, it has opened up the analysis of merging histories of objects in hierarchical scenarios of structure formation [21, 46, 47] and the related construction of the *merger tree* of the population of dark halos [42, 46].

3.7 Extension to the Four Mode Two-Barrier Excursion Set Formalism

We have seen that the hierarchical nature of the cosmic structure formation process plays a prominent role in the nonlinear evolution of and graudal buildup of galaxies, galaxy halos and clusters. In the following sections we will see that it affects all aspects of the nonlinear evolution of large scale structure, including the morphology of filaments and the properties of the void population.

With respect to the void population, we will find that there is a distinct asymmetry between the nonlinear hierarchical evolution of voids and that of haloes (see accompanying review on morphology of the cosmic web). For the evaluation of the hierarchical evolution of voids two processes need to be taken into account: the *void-in-void* process avoids double counting of voids while the *void-in-cloud* process removes voids within encompassing overdensities. What distinguishes voids from their collapsing peers is that clusters will always survive when located within a void, while the reverse is not true: voids within overdense clusters will be rapidly squeezed out of existence.

Sheth & van de Weygaert [77] have shown that the excursion set formalism provides a mathematically properly defined context for describing the asymmetry between void and haloes. The related extension of the formalism to a *two-barrier* formalism culminates in a *four mode* formalism. In this section we summarize these findings, while we refer to [85] for a more proper treatment of the evolution of voids. Figure 9 illustrates the argument. There are four sets of panels. The left-most of each set shows the random walk associated with the initial particle distribution. The two other panels show how the same particles are distributed at two later times.

Cloud-in-Cloud

The first set illustrates the *cloud-in-cloud* process. The mass which makes up the final object (far right) is given by finding that scale within which the linear theory variance has value $S = 0.55$. This mass came from the mergers of the smaller clumps, which themselves had formed at earlier times (centre panel). If we were to center the random walk path on one of these small clumps, it would cross the higher barrier $f_c/D(t) > f_c$ at $S > 0.55$, the value of $D(t)$ representing the linear theory growth factor at the earlier time t.

Cloud-in-Void

The second series of panels shows the *cloud-in-void* process. Here, a low mass clump ($S > 0.85$) virializes at some early time. This clump is embedded in a region which is destined to become a void. The larger void region around it actually becomes a bona-fide void only at the present time, at which time it contains significantly more mass ($S = 0.4$) than is contained in the low mass

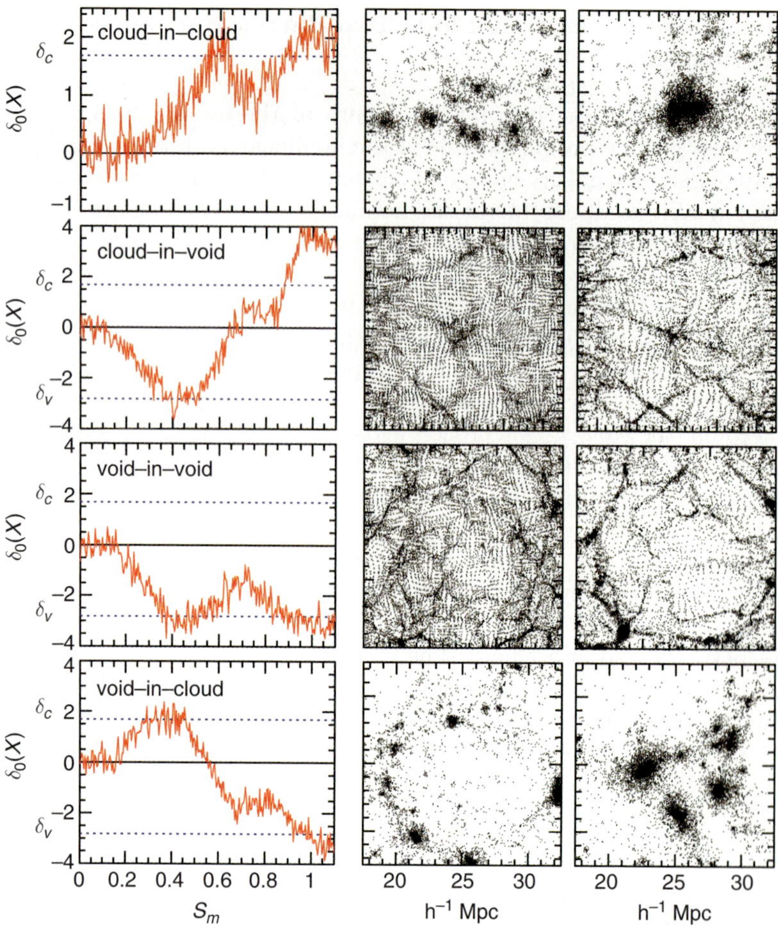

Fig. 9. Four mode two-barrier excursion set formalism. Each row illustrates one of the four basic modes of hierarchical clustering: the *cloud-in-cloud* process, *cloud-in-void* process, *void-in-void* process and *void-in-cloud* process (from **top** to **bottom**). Each mode is illustrated using three frames. Leftmost panels show 'random walks': the local density perturbation $\delta_0(\mathbf{x})$ as a function of (mass) resolution scale S_M (cf. Fig. 7) at an early time in an N-body simulation of cosmic structure formation. In each graph, the dashed horizontal lines indicate the *collapse barrier* f_c and the shell-crossing *void barrier* f_v. The two frames on the right show how the associated particle distribution evolves. Whereas halos within voids may be observable (second row depicts a halo within a larger void), voids within collapsed halos are not (last row depicts a small void which will be squeezed to small size as the surrounding halo collapses). It is this fact which makes the calculation of void sizes qualitatively different from that usually used to estimate the mass function of collapsed halos. From Sheth & van de Weygaert [77]

clump at its centre. Notice that the cloud within the void was not destroyed by the formation of the void; indeed, its mass increased slightly from $S > 0.85$ to $S \sim 0.85$. Such a random walk is a bona-fide representative of $S \sim 0.85$ halos; for estimating halo abundances, the presence of a barrier at f_v is irrelevant. On the other hand, walks such as this one allow us to make some important inferences about the properties of void-galaxies, which we will discussess shortly.

Void-in-Void

The third series of panels shows the formation of a large void by the mergers of smaller voids: the *void-in-void* process. The associated random walk looks very much the inverse of that for the cloud-in-cloud process associated with halo mergers. The associated random walk shows that the void contains more mass at the present time ($S \sim 0.4$) than it did in the past ($S > 0.4$); it is a bona-fide representative of voids of mass $S \sim 0.4$. A random walk path centered on one of these mass elements which make up the filaments within the large void would resemble the cloud-in-void walk shown in the second series of panels. [Note that the height of the barrier associated with voids which are identified at cosmic epoch t scales similarly to the barrier height associated with halo formation: $f_v(t) \equiv f_v/D(t)$.]

Void-in-Cloud

Finally, the fourth series of panels illustrates the *void-in-cloud* process. The particle distribution shows a relatively large void at the early time being squeezed to a much smaller size as the ring of objects around it collapses. A simple inversion of the cloud-in-void argument would have tempted one to count the void as a relatively large object containing mass $S \sim 1$. That this is incorrect can be seen from the fact that, if we were counting halos, we would have counted this as a cloud containing significantly more mass ($S \sim 0.3$), and it does not make sense for a massive virialized halo to host a large void inside.

3.8 Peak Structure

While the extended Press–Schechter excursion set formalism does provide a good description of the mass functions of cosmological objects, it basically involves an intrinsically local description and does not deal with the real internal structure of a genuine collapsed and virialized peak in the mass distribution. Points which would collapse together to form a virialized object of a given mass may be counted as belonging to objects of different mass [16]. Another unappealing aspect is that the derivation of the Press–Schechter formula requires the unphysical sharp k-filter, a rather unphysical form of density smoothing, and a rather arbitrary mass assignment scheme.

It is the nonlocal *peak-patch* description of Bond & Myers [17] that is able to incorporate a more global description of evolving volume elements.

4 Anisotropic and Weblike Patterns

The second key characteristic of the cosmic matter distribution is that of a *weblike geometry* marked by highly elongated filamentary, flattened planar structures and dense compact clusters surrounding large near-empty void regions (see Fig. 1). In this section we focus on the backbone – or skeleton – of the Cosmic Web defined by these anisotropic filamentary and sheetlike patterns.

The recognition of the *Cosmic Web* as a key aspect in the emergence of structure in the Universe came with early analytical studies and approximations concerning the emergence of structure out of a nearly featureless primordial Universe. In this respect the Zel'dovich formalism [90] played a seminal role. It led to view of structure formation in which planar pancakes form first, draining into filaments which in turn drain into clusters, with the entirety forming a cellular network of sheets. As borne out by a large sequence of N-body computer experiments of cosmic structure formation, weblike patterns in the overall cosmic matter distribution do represent a universal but possibly transient phase in the gravitationally driven emergence and evolution of cosmic structure. The N-body calculations have shown that weblike patterns defined by prominent anisotropic filamentary and planar features – and with characteristic large underdense void regions – are a natural manifestation of the gravitational cosmic structure formation process.

Interestingly, for a considerable amount of time the emphasis on anisotropic collapse as agent for forming and shaping structure was mainly confined the Soviet view of structure formation, Zel'dovich's pancake picture, and was seen as the rival view to the hierarchical clustering picture which dominated the western view. Here we intend to emphasize the succesfull synthesis of both elements on the basis of the peak patch description of Bond & Myers [17]. It forms the most elaborate and sophisticated analytical description for the emergence of walls, filaments and fully collapsed triaxial halos in the cosmic matter distribution. Culminating in the Cosmic Web theory [18] it stresses the dominance of filamentary shaped features instead of the dominance of planar pancakes in the pure Zel'dovich theory. Perhaps even more important is its identification of the intimate dynamical relationship between the filamentary patterns and the compact dense clusters that stand out as the nodes within the cosmic matter distribution: filaments as cluster–cluster bridges. To appreciate the intricacies of the *Cosmic Web* theory we need to understand the relation between gravitational tidal forces and the resulting deformation of the matter distribution.

4.1 Anisotropic Collapse

The existence of the Cosmic Web is a result of this tendency of matter concentrations to contract and evolve into anisotropic, elongated or flattened, structures. It is a manifestation of the generic *anisotropic* nature of gravitational collapse, a reflection of the intrinsic anisotropy of the gravitational force in a random density field.

Anisotropic gravitational collapse is the combined effect of *internal* and *external* tidal forces. The *internal* force field of the structure hangs together with the flattening of the feature itself. It induces an anisotropic collapse along the main axes of the structure. The resulting evolution can be most clearly understood in and around a density maximum (or minimum) δ, to first order corresponding to the collapse of a homogeneous ellipsoid [17, 24, 26, 38, 88]. The *external* 'background' force field is the integrated gravitational influence of all external density features in the Universe, as such a manifestation of the inhomogeneous cosmic matter distribution. For most situations the role of the large scale tidal forces in the early phases of the collapse of a feature – the evolutionary phase in which most elements of the cosmic web reside – may be succesfully described by the Lagrangian Zel'dovich formalism [90].

The *peakpatch* formalism embeds the anisotropic tendency of gravitational collapse within the context of a hierarchical mass distribution. It achieves this by combining the nonlinear internal evolution of a particular region in the cosmic mass distribution, and modelling this by means of the homogeneous ellipsoid model, with a reasonably accurate description of the large-scale external tidal influence in terms of the Zel'dovich approximation [17, 75].

4.2 Force Field and Displacement

For the description of the dynamical evolution of a region in the density field – a *patch*- it is beneficial to make a distinction between large scale "background" fluctuations $\delta_{\rm b}$ and small-scale fluctuations $\delta_{\rm f}$,

$$\delta(\mathbf{x}) \;=\; \delta_{\rm b}(\mathbf{x}) \,+\, \delta_{\rm f}(\mathbf{x})\,, \tag{42}$$

in which

$$\begin{aligned} \delta_{\rm f}(\mathbf{x}) \;&=\; \int \frac{{\rm d}\mathbf{k}}{(2\pi)^3}\, \hat\delta(\mathbf{k})\, \hat W_{\rm f}^*(\mathbf{k}; R_b) \\ \delta_{\rm b}(\mathbf{x}) \;&=\; \int \frac{{\rm d}\mathbf{k}}{(2\pi)^3}\, \hat\delta(\mathbf{k})\, \hat W_{\rm b}^*(\mathbf{k}; R_b) \end{aligned} \tag{43}$$

$\hat W_{\rm f}^*(\mathbf{k}; R_b)$ is a high-pass filter which filters out spatial wavenumber components lower than $k < 1/R_b$. $\hat W_{\rm b}^*(\mathbf{k}; R_b)$ is the compensating low-pass filter. The small-scale fluctuating density field $\delta_{\rm f}$ exclusively contributes to the internal evolution of the patch. Predominantly made up of spatial wavenumber components higher than $1/R_b$, it determines the substructure within the patch, sets the corresponding merging times while influencing the overall collapse time of the mass element (see Fig. 22). For our picture to remain valid the scale R_b of the smooth large-scale field should be chosen such that it remains (largely) linear, i.e. the r.m.s. density fluctuation amplitude $\sigma_\rho(R_b,t) \lesssim 1$. Note that the smooth large-scale field $\delta_{\rm b}$ also contributes to the total mass content within the patch.

The small-scale local inhomogeneities induce small-scale fluctuations in the gravitational force field, $\mathbf{g}_{\rm f}(\mathbf{x})$. To a good approximation the smoother background gravitational force $\mathbf{g}_{\rm b}(\mathbf{x})$ (see (2) in and around the mass element includes three components (excluding rotational aspects). The *bulk force* $\mathbf{g}_{\rm b}(\mathbf{x}_{pk})$ is responsible for the acceleration of the mass element as a whole. The divergence $(\nabla \cdot \mathbf{g}_{\rm b})$ encapsulates the collapse of the overdensity while the tidal tensor quantifies its deformation,

$$g_{\rm b,i}(\mathbf{x}) = g_{\rm b,i}(\mathbf{x}_{pk}) + a \sum_{j=1}^{3} \left\{ \frac{1}{3a} (\nabla \cdot \mathbf{g}_{\rm b})(\mathbf{x}_{pk})\, \delta_{\rm ij} - T_{\rm b,ij} \right\} (x_j - x_{pk,{\rm j}})\,. \quad (44)$$

The tidal shear force acting over the mass element is represented by the tidal tensor T_{ij},

$$T_{\rm b,ij} \equiv -\frac{1}{2a} \left\{ \frac{\partial g_{{\rm b},i}}{\partial x_i} + \frac{\partial g_{{\rm b},j}}{\partial x_j} \right\} + \frac{1}{3a} (\nabla \cdot \mathbf{g}_{\rm b})\, \delta_{\rm ij} \quad (45)$$

$$= \frac{1}{a^2} \frac{\partial^2 \Phi_b}{\partial x_i\, \partial x_j} - \frac{3}{2} \Omega H^2\, \delta_{\rm b}(\mathbf{x})\, \delta_{\rm ij}\,, \quad (46)$$

in which the trace of the collapsing mass element, proportional to its overdensity δ_b, dictates its contraction (or expansion).

The force field induces displacements of matter in and around the mass element. The resulting displacement $\mathbf{s}(\mathbf{q},t)$ consists of a superposition of the small-scale and smooth large-scale contributions, $\mathbf{s}_f$ and $\mathbf{s}_b$: matter initially at a (Lagrangian) position $\mathbf{q}$ moves to a location $\mathbf{x}(\mathbf{q},t)$,

$$\mathbf{x}(\mathbf{q},t) = \mathbf{q} + \mathbf{s}(\mathbf{q},t) = \mathbf{q} + \mathbf{s}_{\rm b}(\mathbf{q},t) + \mathbf{s}_{\rm f}(\mathbf{q},t)\,. \quad (47)$$

The smooth large-scale displacement field $\mathbf{s}_{\rm b}$ in and around the patch includes a *bulk displacement* $\mathbf{s}_{pk}$ and a deforming *strain* $\mathcal{E}_{pk,{\rm ij}}$,

$$s_{\rm b,i}(\mathbf{q},t) \approx s_{pk,{\rm i}} + \sum_{j=1}^{3} \mathcal{E}_{pk,{\rm ij}}\, (q_{\rm j} - q_{pk,{\rm j}})\,, \qquad i = 1,\ldots,3\,. \quad (48)$$

The bulk displacement of the (mass) center of the peak

$$\mathbf{s}_{pk} \equiv \mathbf{s}_{\rm b}(\mathbf{q}_{pk})\,, \quad (49)$$

specifies the movement of the mass element as a whole. The large-scale strain field $\mathcal{E}_{b,ij}$ at the location of the patch, $\mathcal{E}_{pk,\mathrm{ij}} \equiv \mathcal{E}_{b,ij}(\mathbf{q}_{pk})$,

$$\mathcal{E}_{\mathrm{b,ij}}(\mathbf{q}) \equiv \frac{1}{2}\left\{\frac{\partial s_{\mathrm{b,i}}}{\partial q_j} + \frac{\partial s_{\mathrm{b,j}}}{\partial q_i}\right\}(\mathbf{q}) \,. \tag{50}$$

embodies the (gravitationally induced) deformation, in volume and shape, of the mass element,

$$\mathcal{E}_{pk,\mathrm{ij}} \equiv \mathcal{E}'_{pk,\mathrm{ij}} + \frac{1}{3a}\,(\nabla\cdot\mathbf{s}_b)(\mathbf{q}_{pk})\,\delta_{\mathrm{ij}} \,. \tag{51}$$

The peak strain's trace $(\nabla\cdot\mathbf{s}_b)(\mathbf{q}_{pk})$ quantifies the shrinking volume of the mass element while the tensor $\mathcal{E}'_{pk,\mathrm{ij}}$ embodies the - mostly externally induced - anisotropic deformation of the region.

The source for the external deformation $\mathcal{E}'_{\mathrm{st,ij}}$ is the external tidal field $T_{\mathrm{b,ij}}$. In the early phases of gravitational collapse the role of the large scale tidal forces is succesfully framed in terms of the by Zel'dovich formalism [90]. The internally induced deformation, a reaction to the nonspherical shape of the mass element, will rapidly enhance along with the nonlinear collapse of the peak.

4.3 Zel'dovich Approximation

In a seminal contribution Zel'dovich [90] found by means of a Lagrangian perturbation analysis that to first order – typifying early evolutionary phases – the reaction of cosmic patches of matter to the corresponding peculiar gravity field would be surprisingly simple. The *Zel'dovich approximation* is based upon the first-order truncation of the Lagrangian perturbation series of the trajectories of mass elements,

$$\mathbf{x}(\mathbf{q},t) \;=\; \mathbf{q} \,+\, \mathbf{x}^{(1)}(\mathbf{q},t) \,+\, \mathbf{x}^{(2)}(\mathbf{q},t) \,+\, \ldots\,, \tag{52}$$

in which the successive terms $\mathbf{x}^m$ correspond to successive terms of the relative displacement $|\partial(\mathbf{x}-\mathbf{q})/\partial\mathbf{q}|$,

$$1 \gg \left|\frac{\partial\mathbf{x}^{(1)}}{\partial\mathbf{q}}\right| \gg \left|\frac{\partial\mathbf{x}^{(2)}}{\partial\mathbf{q}}\right| \gg \left|\frac{\partial\mathbf{x}^{(3)}}{\partial\mathbf{q}}\right| \gg \ldots\,, \tag{53}$$

and embodies the solution of the Lagrangian equations for small density perturbations ($\delta^2 \ll 1$). Assuming *irrotational motion*, in accordance with linear gravitational instability, and restricting the solution to the *growing mode* leads to the plain ballistic linear displacement of the Zel'dovich approximation,

$$\mathbf{x} = \mathbf{q} - D(t)\;\boldsymbol{\nabla}\Psi(\mathbf{q}) = \mathbf{q} - D(t)\;\boldsymbol{\psi}(\mathbf{q})\,. \tag{54}$$

dictated by the Lagrangian *displacement potential* $\Psi(\mathbf{q})$ and its gradient, the Zel'dovich deformation tensor ψ_{mn}. The path's time evolution is specified by the linear density growth factor $D(a)$ [62] (14). An essential aspect of the Zel'dovich approximation is the $1-1$ relation between the *displacement potential* $\Psi(\mathbf{q})$ and the (primordial, linearly extrapolated) gravitational potential $\tilde{\Phi}(\mathbf{q},t)$

$$\Psi(\mathbf{q}) \;=\; \frac{2}{3\,D\,a^2 H^2\,\Omega}\;\tilde{\Phi}(\mathbf{q},t) \;=\; \frac{2}{3\,H_0^2\,\Omega_0}\;\tilde{\Phi}_0(\mathbf{q})\,, \tag{55}$$

where $\tilde{\Phi}_0$ is the linearly extrapolated gravitational potential at the current epoch (a$=$1). The tensor $\psi_{\rm mn}$, directly related to the strain tensor $\mathcal{E}_{\rm mn} = D(t)\psi_{\rm mn}$, describes the deformation of the mass element,

$$\begin{aligned}\psi_{\rm mn} &= \frac{2}{3a^3\Omega H^2}\,\frac{\partial^2\tilde{\Phi}}{\partial q_m\,\partial q_n} = \frac{2}{3\Omega H^2 a}\left(\,\tilde{T}_{\rm mn} + \frac{1}{2}\Omega H^2\,\tilde{\delta}\,\delta_{\rm mn}\,\right)\\ &= \frac{2}{3\Omega_0 H_0^2}\,\tilde{T}_{\rm mn,0} + \frac{1}{3}\tilde{\delta}_0\,\delta_{\rm mn}\end{aligned} \tag{56}$$

The relation establishes the intimate connection between the deformation of an object and the *tidal shear field* $T_{\rm mn}$, expressed in terms of the *linearly extrapolated* primordial values of these quantities, $\tilde{T}_{\rm mn}$ and $\tilde{\delta}$. These evolve according to $\tilde{\delta}(t) \propto D(t)$ and $\tilde{T}_{\rm mn} \propto D/a^3$. On the basis of this relation we immediately see that the (linearly extrapolated) tidal shear field $\tilde{T}_{\rm mn}$ is directly related to the traceless strain tensor $\mathcal{E}'_{\rm mn}$,

$$\tilde{T}_{\rm mn}(t) \;=\; 4\pi G\rho_{\rm u}(t)\,\left\{\mathcal{E}_{\rm mn} - \frac{1}{3}\tilde{\delta}\,\delta_{\rm mn}\right\} \;=\; 4\pi G\rho_{\rm u}(t)\,\mathcal{E}'_{\rm mn}\,. \tag{57}$$

Anisotropic Zel'dovich Collapse

The resulting (mildly nonlinear) local density evolution is entirely determined by the eigenvalues λ_1, λ_2 and λ_3 of the deformation tensor $\psi_{\rm mn}$, ordered by $\lambda_3 \geq \lambda_2 \geq \lambda_1$),

$$\begin{aligned}\frac{\rho(\mathbf{x},t)}{\rho_{\rm u}} &= \left\|\frac{\partial\mathbf{x}}{\partial\mathbf{q}}\right\|^{-1} = \left\|\delta_{\rm mn} - D(t)\psi_{\rm mn}\right\|^{-1}\\ &= \frac{1}{[1-D(t)\lambda_1][1-D(t)\lambda_2][1-D(t)\lambda_3]}\,,\end{aligned} \tag{58}$$

where $\rho(\mathbf{x},t)$ is the local density at time t and $\rho_{\rm u}(t)$ the global (FRW) cosmic density. Dependent on whether one or more of the eigenvalues $\lambda_i > 0$, the feature will collapse along one or more directions. The collapse will proceed along a sequence of three stages. First, collapse along the direction of the strongest deformation λ_3. If also the second eigenvalue is positive, the object will contract along the second direction. Total collapse will occur only if $\lambda_1 > 0$.

The time sequence of four frames in Fig. 10 portraits the success, and shortcomings, of the Zel'dovich scheme. The four frames reveals the gradual morphological procession along *pancake* and *filamentary* stages. A comparison with the results of full-scale N-body simulations shows that in particular at early structure formation epochs the predicted Zel'dovich configurations are accurately rendering the nonlinear matter configurations. The spatial configurations predicted by the Zel'dovich approximation form a reasonably accurate approximation to the linear and mildly nonlinear phases of structure formation. The approximation breaks down when the orbits of migrating matter elements start to cross. Towards this phase the linearly extrapolated gravitational field configuration no longer forms a reasonable reflection of the genuine nonlinear gravitational field. The self-gravity of the emerging structures becomes so strong that the initial "ballistic" motion of the mass elements will get seriously altered, redirected and slowed down.

Fig. 10. Zel'dovich displaced particle distributions inferred from a unconstrained random realization of a primordial matter distribution for a SCDM cosmological scenario in a $50\,h^{-1}$ Mpc. Time sequence from top left to bottom right, frames corresponding to cosmic epochs $a = 0.10, 0.15, 0.20$ and 0.25

4.4 Ellipsoidal Collapse

Full-scale gravitational N-body simulations, and/or more sophisticated approximations, are necessary to deal self-consistently with more advanced nonlinear stages. While the Zel'dovich approximation is relatively accurate in describing the large-scale "background" induced deformation of mass elements, the internal evolution of a mass element quickly assumes a highly nonlinear character and will strongly amplify the externally induced anisotropic shape. Aspects of the subsequent evolution and anisotropic collapse can be reasonably approximated by the *homogeneous ellipsoid model.*

Quintessential is the observation that gravitational instability not only involves the runaway gravitational collapse of any cosmic overdensity, but that it has the additional basic attribute of *inevitably amplifying any slight initial asphericity during the collapse.*

The Ellipsoidal Approximation

The *Homogeneous Ellipsoidal Model* assumes a mass element to be a region with a triaxially symmetric ellipsoidal geometry and a homogeneous interior density, embedded within a uniform background density ρ_{u}.

The early work by Icke [37, 38] elucidated the key aspects of the evolution and morphology of homogeneous ellipsoids within an expanding FRW background Universe, in particular the self-amplifying effect of a collapsing and progressively flattening isolated ellipsoidal overdensity. Translating the formalism of Lynden-Bell [54] and Lin et al. [53] to a cosmological context he came to the conclusion that flattened and elongated geometries of large scale features in the Universe should be the norm. White & Silk [88] managed to provide an elegant analytic approximation for the evolution of the ellipsoid that is remarkably accurate. However, these early studies did not reduce, as they should, to the Zel'dovich approximation in the linear regime. Bond & Myers [17], and Eisenstein & Loeb [26] emphasized that this was because they either ignored any external influences or because they did not include the effects of the external tidal (quadrupolar) influences self-consistently. Once these effects are appropriately included the resulting ellipsoidal collapse model is indeed self-consistent (see also the recent detailed study of Desjacques [24] of the environmental influence on ellipsoidal collapse).

For moderately evolved structures such as a Megaparsec (proto)cluster the ellipsoidal model represents a reasonable approximation at and immediately around the density peak. In the case of highly collapsed objects like galaxies and even clusters of galaxies it will be seriously flawed. One dominant aspect it fails to take into account are the all-important small-scale processes related to the hierarchical substructure and origin of these objects. Nonetheless, the concept of homogeneous ellipsoids has proven to be particularly useful when seeking to develop approximate yet advanced descriptions of the distribution of virialized cosmological objects within hierarchical scenarios of structure formation [17, 72, 75].

In many respects the homogeneous model is a better approximation for underdense regions than it is for overdense ones. Overdense regions contract into more compact and hence steeper density peaks, so that the area in which the ellipsoidal model represents a reasonable approximation will continuously shrink. By contrast, for voids we find that the region where the approximation by a homogeneous ellipsoid is valid grows along with the void's expansion. While voids expand their interior gets drained of matter and develops a flat "bucket-shaped" density profile: the void's natural tendency is to evolve into expanding regions of a nearly uniform density. The approximation is restricted to the interior and fails at the void's outer fringes because of its neglect of the domineering role of surrounding material, such as the sweeping up of matter and the encounter with neighbouring features.

Ellipsoidal Gravitational Potential

The model describes the evolution of a homogeneous ellipsoidal region with a triaxially symmetric geometry, specified by its principal axes $C_1(t)$, $C_2(t)$ and $C_3(t)$. The ellipsoid has a uniform matter density $\rho(t)$, and density excess $\delta(t)$.

In the presence of an external potential contribution the total gravitational potential $\Phi^{(\mathrm{tot})}(\mathbf{r})$ at a location $\mathbf{r} = (r_1, r_2, r_3)$ in the interior of a homogeneous ellipsoid may be decomposed into three separate (quadratic) contributions,

$$\Phi^{(\mathrm{tot})}(\mathbf{r}) \;=\; \Phi_u(\mathbf{r}) \;+\Phi^{(\mathrm{int})}(\mathbf{r}) \;+\Phi^{(\mathrm{ext})}(\mathbf{r})\,. \tag{59}$$

A necessary condition for the ellipsoidal formalism to remain self-consistent is that each of the three separate contributions retains a quadratic form. Higher order contributions, also of the external potential, are ignored. The three separate contributions are:

- *Homogeneous Cosmic Background*

The potential contribution of the homogeneous background with universal density $\rho_{\mathrm{u}}(t)$,

$$\Phi_u(\mathbf{r}) \;=\; \frac{2}{3}\pi G\rho_{\mathrm{u}}\left(r_1^2 + r_2^2 + r_3^2\right)\,. \tag{60}$$

- *Internal Ellipsoidal Potential*

The interior ellipsoidal potential $\Phi^{(\mathrm{int})}(\mathbf{r})$, superimposed onto the homogeneous background,

$$\Phi^{(\mathrm{int})}(\mathbf{r}) \;=\; \frac{2}{3}\pi G\rho_{\mathrm{u}}\,\delta(t)\left(r_1^2 + r_2^2 + r_3^2\right) \;+\; \frac{1}{2}\sum_{m,n} T_{\mathrm{mn}}^{(\mathrm{int})}\, r_m r_n\,,$$

in which $T_{\mathrm{mn}}^{(\mathrm{int})}$ are the elements of the traceless internal tidal shear tensor. The quadratic expression for $\Phi^{(\mathrm{int})}$ assumes a simplified form in the coordinate system defined by the principal axes of the ellipsoid.

$$\Phi^{(\mathrm{int})}(\mathbf{r}) = \pi G \rho_\mathrm{u}\, \delta \sum_m \alpha_\mathrm{m} r_\mathrm{m}^2\,, \tag{61}$$

where the coefficients $\alpha_\mathrm{m}(t)$ are

$$\alpha_\mathrm{m}(t) = \mathcal{R}_1(t)\mathcal{R}_2(t)\mathcal{R}_3(t) \int_0^\infty \frac{\mathrm{d}\lambda}{\left(\mathcal{R}_\mathrm{m}^2(t)+\lambda\right) \prod_{n=1}^3 \left(\mathcal{R}_\mathrm{n}^2(t)+\lambda\right)^{1/2}}\,. \tag{62}$$

The Poisson equation implies the α_m's obey the constraint $\sum_{m=1}^3 \alpha_m = 2$. In the case of a spherical perturbation all three α_m's are equal to $2/3$, reproducing the well-known fact that it does involve a vanishing internal tidal tensor contribution,

$$T_\mathrm{mn}^{(\mathrm{int})} = \frac{\partial^2 \Phi^{(\mathrm{int})}}{\partial r_m\, \partial r_n} - \frac{1}{3}\nabla^2 \Phi^{(\mathrm{int})}\, \delta_\mathrm{mn} = 2\pi G\, \rho_\mathrm{u}\, \delta(t) \left(\alpha_m - \frac{2}{3}\right) \delta_\mathrm{mn}\,.$$

- *External Tidal Influence*

The external gravitational potential $\Phi^{(\mathrm{ext})}$. Assuming that the external tidal field does not vary greatly over the expanse of the ellipsoidal mass element, we may limit the external contribution to its quadrupolar components,

$$\Phi^{(\mathrm{ext})}(\mathbf{r}) = \frac{1}{2} \sum_{m,n} T_\mathrm{mn}^{(\mathrm{ext})}\, r_m r_n\,. \tag{63}$$

$T_{mn}^{(\mathrm{ext})}$ are the components of the external (traceless) tidal shear tensor. By default the latter is limited to its traceless contribtuion, the corresponding (background) density is implicitly included in the (total) internal density, $\rho_u(t)(1+\delta(t))$.

The external field is taken to be the smooth large-scale tidal field $T_{\mathrm{b,mn}}$. The latter is directly related to the traceless large scale (background) strain tensor (57), with eigenvalues τ_m given by (see (71)),

$$\tau_\mathrm{m} = 4\pi G \rho_\mathrm{u}(t)\, \lambda'_\mathrm{vm}(t)\,. \tag{64}$$

where λ'_{vm} are the eigenvalues of the background anisotropic strain tensor $\mathcal{E}'_{pk,\mathrm{ij}}$ at the location of the mass peak.

Ellipsoidal Evolution

The anisotropy of an initially spherically symmetric matter element in the primordial cosmic matter distribution is a direct effect of the external tidal force field. As a result the principal axes of the configuration are the ones defined by the external tidal tensor $T_\mathrm{mn}^{(\mathrm{ext})}$. Both the external large-scale tidal forces inducing the anisotropic collapse and the resulting internal one do strongly enhance the anisotropic shape of the ellipsoid.

The evolution of the ellipsoid is specified by three scale factors $\mathcal{R}_{\rm i}$, one for each of the three principal axes. The boundary of the ellipsoid and the overdensity evolve as

$$C_i(t) = \mathcal{R}_{\rm i}(t)\, R_{pk}\,, \qquad \delta(t) = \frac{a^3}{\mathcal{R}_1\mathcal{R}_2\mathcal{R}_3} - 1\,. \tag{65}$$

in terms of the initial (Lagrangian) radius R_{pk}. The evolution of the scale factors $\mathcal{R}_{\rm i}$ are determined by the gravitational acceleration along each of the principal axes (see (59)). Including the influence of the cosmological constant Λ, this translates into

$$\frac{d^2\mathcal{R}_{\rm m}}{dt^2} = -4\pi G\rho_{\rm u}(t)\left[\frac{1+\delta}{3} + \frac{1}{2}\left(\alpha_m - \frac{2}{3}\right)\delta\right]\mathcal{R}_{\rm m} - \tau_{\rm m}\,\mathcal{R}_{\rm m} + \Lambda R_m\,. \tag{66}$$

with $\alpha_{\rm m}(t)$ the ellipsoidal coefficients specified by the integral (62) and τ_m the eigenvalue of the external (large-scale) tidal shear tensor $T_{\rm mn}^{(\rm ext)}$.

The collapse of the three axes of the ellipsoid will happen at different times. The shortest axis will collapse first, followed by the intermediate axis and finally by the longest axis. The shortest axis will collapse considerably faster than that of the equivalent spherically evolving perturbation while full collapse along all three axis will be slower as the longest axis takes more time to reach collapse. In fact, the longest axis may not collapse at all. An illustration of this behaviour can be found in Fig. 11. It shows the evolution of a slightly overdense *isolated* ellipsoid, with initial axis ratios $a_1 : a_2 : a_3 = 1 : 0.9 : 0.8$, embedded in a background Einstein-de Sitter Universe. Quantitatively the expansion and subsequent contraction of each of the three axes can be followed in Fig. 12. The superimposed blue curve represents the evolution of the equivalent spherical overdensity. The righthand frame shows that this development involves a continuous decrease of both axis ratios.

4.5 Ellipsoidal Collapse and External Influences

In order to properly model the nonlinear collapse of the features in the *Cosmic Web* it is essential to embed the nonlinear anisotropic collapse of mass elements within the large-scale environment. A proper approximation, following Bond & Myers [17], is that of assuming the large-scale tidal influence to be largely linear and assuring that the initial conditions for the ellipsoid asymptotically approach the Zel'dovich equation,

$$\begin{aligned} \mathcal{R}_{\rm m}(t_{\rm i}) &= a(t_{\rm i})\,\{1 - D(t_i)\lambda_{\rm m}\}\,, \\ \frac{d\mathcal{R}_{\rm m}}{dt}(t_{\rm i}) &= H(t_{\rm i})\mathcal{R}_{\rm m}(t_{\rm i}) - a(t_{\rm i})H(t_{\rm i})f(\Omega_{\rm i})\,D(t_{\rm i})\lambda_{\rm m}\,, \end{aligned} \tag{67}$$

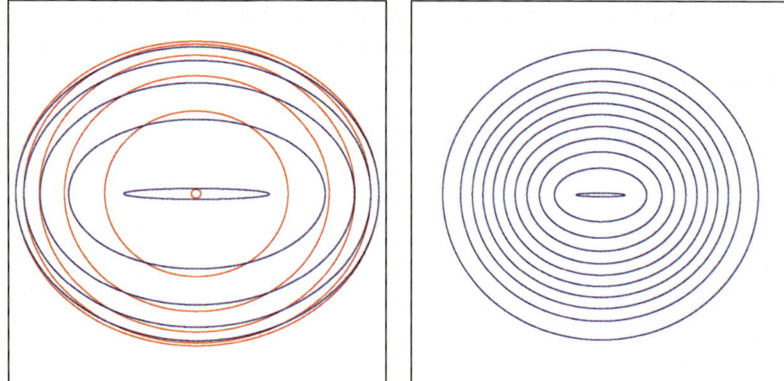

Fig. 11. The evolution of an overdense homogeneous ellipsoid, with initial axis ratio $a_1 : a_2 : a_3 = 1.0 : 0.9 : 0.9$, embedded in an Einstein-de-Sitter background Universe. The two frames show a time sequel of the ellipsoidal configurations attained by the object, starting from a near-spherical shape, initially trailing the global cosmic expansion, and after reaching a maximum expansion turning around and proceeding inexorably towards ultimate collapse as a highly elongated ellipsoid. **Left:** the evolution depicted in physical coordinates. Red contours represent the stages of expansion, blue those of the subsequent collapse after turn-around. **Right:** the evolution of the same object in comoving coordinates, a monologous procession through ever more compact and more elongated configurations

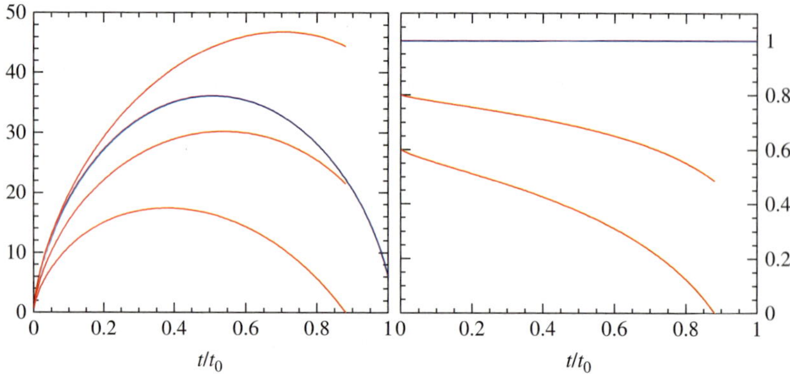

Fig. 12. The evolution of an overdense homogeneous ellipsoid, with initial axis ratio $a_1 : a_2 : a_3 = 1.0 : 0.8 : 0.6$, in an Einstein-de-Sitter background Universe. **Left:** expansion factors for each individual axis; **Right:** axis ratios a_2/a_1 and a_3/a_1. The ellipsoid axes are depicted as red curves. For comparison, in blue, the evolution of an equivalent homogenous spherical overdensity

in which $\lambda_{\rm m}$ are the eigenvalues of the Zel'dovich deformation tensor $\psi_{\rm mn}$, and $D(t)$ is the linear density growth factor and $f(\Omega)$ the corresponding linear velocity factor [62].

By using the implied relation between the eigenvalues of the external tidal tensor $\tau_{\rm m}$ and the large-scale tidal strain tensor $\mathcal{E}_{mn}$ (64) the following equation of motion is obtained,

$$\frac{d^2\mathcal{R}_{\rm m}}{dt^2} = -4\pi G\rho_{\rm u}(t)\left[\frac{1+\delta}{3} + \frac{1}{2}(\alpha_m - \frac{2}{3})\,\delta + \lambda'_{\rm vm}\right]\mathcal{R}_{\rm m} + \Lambda R_m. \quad (68)$$

While the smooth large-scale tidal field induces the anisotropic collapse of the mass element, the subsequent nonlinear evolution differs increasingly from the predictions of the linear Zel'dovich formalism (58). As can be seen in Fig. 13 for nearly all conceivable (external) tidal shear ellipticities the nonlinear ellipsoidal collapse involves a considerably faster collapse along all three axes of an ellipsoid than that following from the Zel'dovich approximation (58). Only for extremely anisotropic tidal configurations the Zel'dovich formalism would find the same collapse time for the longest axis of the mass element.

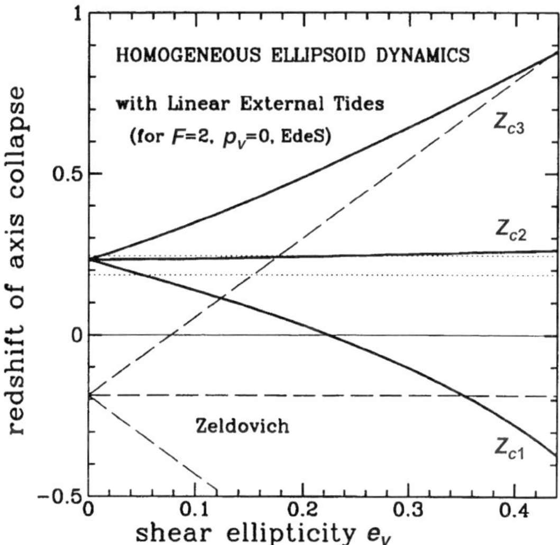

Fig. 13. The collapse redshifts for the three ellipsoidal axes of the initial external tidal shear ellipticity e_v, assuming zero prolaticity p_v, a linear extrapolated density $\delta_0 = 2$ and a linear external tide approximation (68). The dashed curve shows how poorly the Zel'dovich approximation fares: only for the extreme elongations does it get the collapse redshift along the third axis right, while it is far off for the other two directions. Also shown, by dotted lines, are the redshifts at which an equivalent spherical overdensity reaches overdensity 170 (*upper dotted line*) and complete collapse (*lower dotted line*). From Bond & Myers [17]. Reproduced with permission of AAS

4.6 Primordial Structural Morphology

The values of the (Zel'dovich) deformation eigenvalues λ_{v1}, λ_{v2} and λ_{v3} basically determine the (asymptotic) morphology of the resulting features, roughly along the lines specified in Table 1: they function as cosmic shape parameters.

To get insight into the prevailing morphology in the cosmic matter distribution it is necessary to assess the statistical and spatial distribution of the shear eigenvalues. This will determine the overall morphology and geometry of the cosmic density field at the "quasi-linear" stage – i.e. the prominence of mutually interconnected flattened structures, denser elongated filaments and dense compact clumps.

The first assessment of the statistical properties of the deformation tensor in a primordial Gaussian random density fluctuation field is the seminal study by Doroshkevich [25]. He derived the (unconditional) pdf for the eigenvalues λ_1, λ_2 and λ_3,

$$\begin{aligned} P(\lambda_1,\lambda_2,\lambda_3) \;&\sim\; (\lambda_1-\lambda_2)(\lambda_1-\lambda_3)(\lambda_2-\lambda_3) \\ &\times\; \exp\left\{-\frac{15}{2\sigma^2}\left[\lambda_1^2+\lambda_2^2+\lambda_3^2-\frac{1}{2}(\lambda_1\lambda_2+\lambda_1\lambda_3+\lambda_2\lambda_3)\right]\right\}\,. \end{aligned} \tag{69}$$

This yields a probability of 8% that all of the eigenvalues are negative, $\lambda_1 < \lambda_2 < \lambda_3 < 0$, predisposing the formation of a void. The probability that matter elements have one or more positive eigenvalues is filament-dominated weblike morphology is the generic outcome during the moderate quasi-linear evolutionary phase for any scenario with primordial Gaussian perturbations marked by relatively strong perturbations on large scales. The signs of the eigenvalues will determine the (asymptotic) local geometry along the lines specified in Table 1.

For the purpose of understanding the geometry of large scale structure we also should take note of the fact that the values of the deformation tensor eigenvalues are directly constrained by the local density,

Table 1. Asymptotic morphology: deformation eigenvalue conditions for different asymptotic structural morphologies in the Cosmic Web

Structure	Eigenvalue signatures
Peak	$\lambda_1 > 0$; $\lambda_2 > 0$; $\lambda_3 > 0$
Filament	$\lambda_1 > 0$; $\lambda_2 > 0$; $\lambda_3 < 0$
Sheet	$\lambda_1 > 0$; $\lambda_2 < 0$; $\lambda_3 < 0$
Void	$\lambda_1 < 0$; $\lambda_2 < 0$; $\lambda_3 < 0$

$$\tilde{\delta} = (\lambda_{v1} + \lambda_{v2} + \lambda_{v3}), \tag{70}$$

in which $\tilde{\delta}$ is the (linearly extrapolated) initial density contrast. In other words, when we see a supercluster or other interesting feature we should assess the *conditional* probability of the shape parameters for the relevant range of density values. To this end it is helpful to introduce the shear ellipticity e_v and shear prolateness p_v (see Bardeen et al. [6], Bond & Myers [17] 1996a),

$$e_v = \frac{\lambda_{v1} - \lambda_{v3}}{2\sum_i \lambda_{vi}}, \qquad p_v = \frac{\lambda_{v1} - 2\lambda_{v2} + \lambda_{v3}}{2\sum_i \lambda_{vi}}. \tag{71}$$

By implication e_v and p_v are constrained to $e_v \geq 0$ and $-e_v \leq p_v \leq e_v$. The evolution of a patch is spherically symmetric when the shear is isotropic ($\lambda_{v3} = \lambda_{v2} = \lambda_{v1}$), i.e. when $e_v = p_v = 0$. When the collapse is predominantly along one axis ($\lambda_{v3} > 0, \lambda_{v2} \sim \lambda_{v1} < 0$), the initial evolution is towards a classical pancake by $e_v = p_v$. When a second axis is also collapsing ($\lambda_{v3} \sim \lambda_{v2} > 0, \lambda_{v1} < 0$) the result is filamentary, $e_v = -p_v$. In other words, extreme sheet-like structures would have $p_v \approx e_v$, extreme filaments $p_v \approx -e_v$.

Via the quantities e_v and p_v we may get an idea of the prominence of filamentary and sheetlike structures in the cosmic matter distribution by assessing their conditional distribution in the primordial density field for a given $\delta = \nu_f \sigma$. The combined statistical distribution $P(e_v, p_v|\nu_f)$ of e_v and p_v and of the prolaticity, $P(p_v|\nu_f)$, at an arbitrary field location with density are Wadsley & Bond [87] and Bond [14],

$$\begin{aligned} P(\{\lambda_{v1}, \lambda_{v2}, \lambda_{v3}\}|\nu_f) &= P(e_v, p_v|\nu_f) \\ &= \frac{225\sqrt{5}}{\sqrt{2\pi}}\, e_v(e_v^2 - p_v^2)\,\nu_f^3\, e^{-15(\nu_f e_v)^2/2 - 5(\nu_f p_v)^2/2}\, \mathrm{d}e_v\, \mathrm{d}p_v\,. \end{aligned} \tag{72}$$

Figure 14 shows the iso probability contours of $P(e_\mathrm{v}, p_\mathrm{v}|\nu_f)$ for a set of 6 different ν_f values. It manifestly demonstrates the distinct tendency of overdense regions, in particularly those of moderate density, to be *filamentary*: $p_\mathrm{v} < 0$ or, equivalently, eigenvalue signature $(\lambda_1, \lambda_2, \lambda_3) = (-++)$. The figure also underlines the fact that higher peaks tend to be more spherical. This may be quantitatively appreciated from the corresponding expectation values for the the ellipticity and prolaticity of an arbitary field patch with local density $\delta = \nu_f \sigma$ [14],

$$\begin{aligned} \langle e_v|\nu_f, \mathrm{field}\rangle &\approx 0.54\nu_f^{-1}; \qquad \Delta e_v \approx 0.18\nu_f^{-1}, \\ \langle |p_v||\nu_f, \mathrm{field}\rangle &\approx 0.18\nu_f^{-1}; \qquad \Delta p_v \approx 0.22\nu_f^{-1}, \end{aligned} \tag{73}$$

which express the strongly declining nature of ellipticity and prolateness as a function of patch density.

The gross properties of the Cosmic Web may therefore already be found in the primordial density field. In this light it is particularly illuminating to study

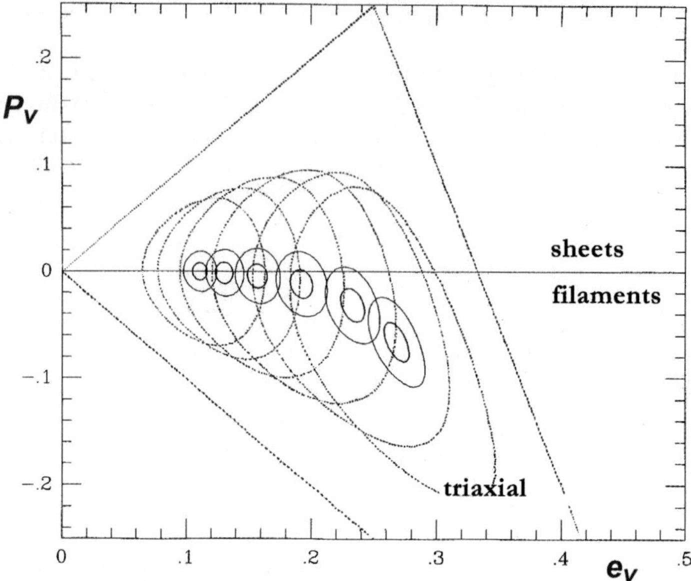

Fig. 14. The 95, 90 and 50% contours of the conditional probability for ellipticity e_v and prolateness p_v subject to the constraint of a given field density value $\nu = \delta/\sigma$. The figure demonstrates that even for high ν the shapes are triaxial and that for lower density values there is a tendency towards filamentary configurations

the distribution of the deformation eigenvalue signatures as a function of density threshold $\nu_f = \delta_f/\sigma$. Figure 15 looks at two aspects of this question [66]. The dependence of structural morphology on the density threshold is given by the probability of the eigenvalue signature on the threshold $\delta = \nu\sigma$, $P(\text{sign}|\delta)$. The left panel of Fig. 15 shows that for Gaussian fields at overdensties above a critical $\delta = 1.56\sigma$ one encounters predominantly spherical-like mass concentrations $(+++)$. By contrast, at lower density contrast $0 < \delta < 1.56\sigma$, most of the initial density enhancements are in elongated filamentary bridges $(-++)$. Planar configurations $(--+)$ are less likely for any positive overdensities $\delta > 0$. The related quantity $P(\delta|\text{sign})$ gives us the density distribution within different types of structure. While the average density of the filaments in the initial configuration is equal to $\delta = 0.6\sigma$, it is the $\delta \sim 1.5 - 2\sigma$ excursions which are precursors of the rare prominent filaments. By contrast, rare planar membrane-like configurations are expected only at lower overdensities of $\delta \sim 0.5 - 1\sigma$. Mean densities for the given shear signatures are $\langle\delta\rangle \approx 1.66\sigma\,, 0.6\sigma\,, -0.6\sigma$, with dispersion $\Delta\delta \approx 0.55\sigma$.

4.7 Evolving Filamentary Morphology

Evidently, the primordial density field analysis only provides a superficial impression of the emerging morphology of the Cosmic Web. What it does

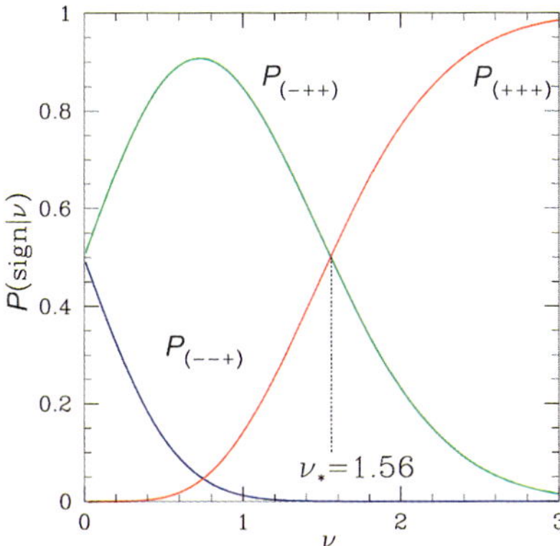

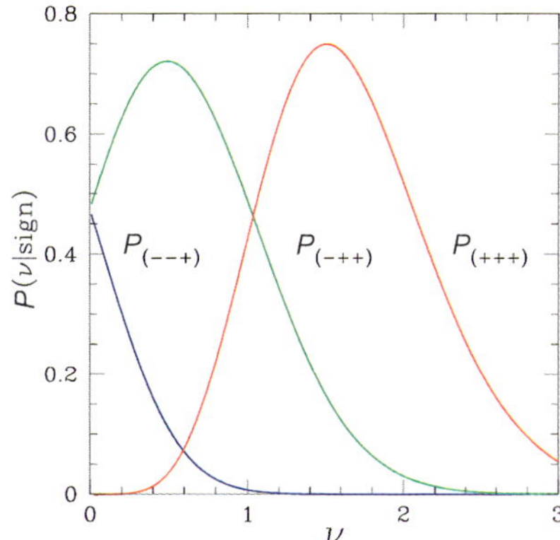

Fig. 15. Left panel: probability of the eigenvalue signature given the overdensity threshold $P(\text{sign}|\nu)$, $\nu = \delta/\sigma$. **Right panel:** density distribution given the signature type of shear tensor, $P(\nu|\text{sign})$. From: Pogosyan et al. [66]

emphasize, and strongly so, is the prevalence of proto-filaments and proto-clusters in the primordial density field.

This impression will only become more pronounced as nonlinear evolution sets in. The salient filamentary nature of the nonlinear mass distribution seen in large N-body simulations (see e.g. Fig. 1) can already be noticed when following the early nonlinear evolution by means of the Zel'dovich mapping (54). A telling illustration of this can be seen in Fig. 16. The left panel shows an initial linear CDM overdensity field δ_L smoothed on a Gaussian scale $R_b = 3.5\,\text{h}^{-1}\,\text{Mpc}$, with $\sigma_\rho = 0.65$. The chosen density threshold is $\delta_L = 1\sigma_\rho$, the level at which $\delta_L(\mathbf{r})$ percolates. The right panel shows $\delta_Z(\mathbf{r}, t)$, the overdensity of the resulting Zel'dovich map at a contour threshold $\delta_Z = 2$, just above where percolation occurs.

The Zel'dovich map in Fig. 16, evolved to $\sigma_8 = 0.7$, clearly shows the dominant filamentary morphology. It disproves the conventional tenet of pancakes representing the dominant overdensity features. Also, it underlines the observation that the prominent filaments already existed in an embryonic – and fattened – form in the initial conditions. As the nonlinear evolution proceeds the cluster regions will collapse even further and occupy even less volume. This will enhance the filamentary character of the cosmic matter distribution even further.

Having argued and illustrated the principal filamentary nature of the Cosmic Web, largely on the basis of a local evaluation of the deformation eigenvalues, we need to assess the apparent coherence of these weblike structures and their mutual relationship. Their overall geometry and topology can be understood by adressing the relationship between the local values of the deformation tensor, responsible for the local morphology, and the global density field.

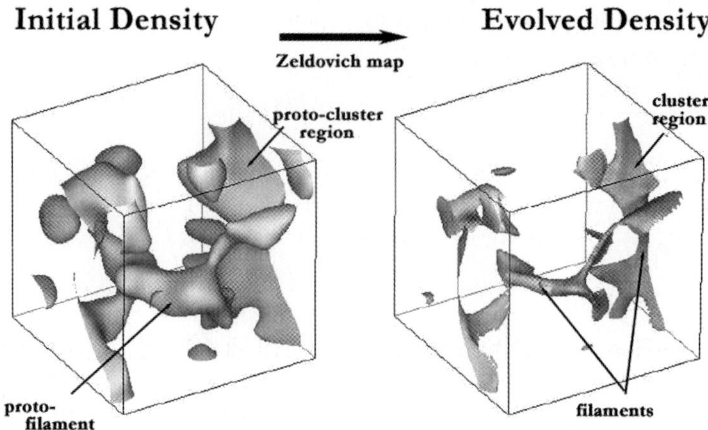

Fig. 16. Cosmic Web and Clusters. (Mean) constrained density field reconstructions $\langle\delta_L|20peaks\rangle$ on the basis of the 20 most massive cluster peaks (patches) in a CDM density field in a $(50\,\mathrm{h}^{-1}\,\mathrm{Mpc})^3$ box with periodic boundary conditions. **Lefthand:** initial linear CDM overdensity field $\delta_L(\mathbf{r})$, smoothed on a Gaussian scale $R_G = 3.5\,\mathrm{h}^{-1}\,\mathrm{Mpc}$ with (iso)density threshold level $\delta_L = 1\sigma_\rho$, with $\sigma_\rho = 0.65$, the level at which δ_L percolates. The location, size and shape of the cluster patches is indicated by means of the black ellipsoids, whose size is proportional to the peak scale R_{pk} and orientation defined by the shear tensor orientation. **Righthand:** the corresponding Zel'dovich map density field δ_Z of the smoothed initial conditions at a contour threshold $\delta_Z = 2$. Based on Bond et al. [18]. Reproduced with permission of Nature

This makes it necessary to turn to the concept of conditional multi-point correlation functions in Lagrangian space (also see Bond [15]), i.e. the statistically averaged density and displacement fields subject to various constraint on the (tidal) shear at multiple points in the cosmic volume. The mathematical language needed for evaluating the implied "protoweb" in the initial density field is that of *constrained random field* theory, first introduced by Bertschinger [10]. In the next Sect. 4.8 we will describe this formalism in some necessary detail.

4.8 Constrained Random Field Formalism

A major virtue of the *constrained random field* construction technique [10, 35, 73, 84] is that it offers the instrument for translating locally specified quantities to the corresponding implied global matter distribution.

Bertschinger [10] described how a set Γ of functional field constraints $C_i[f] = c_i$, $(i = 1, \ldots, M)$ of a Gaussian random field $f(\mathbf{r}, t)$ would translate into field configurations for which these constraints would have the specified values c_i. Any such *constrained field realization* f_c can be written as the sum of a *mean field* $\bar{f}(\mathbf{x}) = \langle f(\mathbf{x})|\Gamma\rangle$, the ensemble average of all field realizations obeying the constraints, and a *residual field* $F(\mathbf{x})$, embodying the field fluctuations characterized and specified by the power spectrum $P(k)$ of the particular cosmological scenario at hand,

$$f_c(\mathbf{x}) = \bar{f}(\mathbf{x}) + F(\mathbf{x}) \tag{74}$$

Bertschinger [10] showed the specific dependence of the mean field on the *nature* $C_i[f]$ of the constraints as well as their *values* c_i. In essence the mean field can be seen as the weighted sum of the field-constraint correlation functions $\xi_i(\mathbf{x})$,

$$\xi_i(\mathbf{x}) \equiv \langle f\, C_i \rangle \tag{75}$$

(where we follow the notation of Hoffman & Ribak [35]). Each field-constraint correlation function encapsulates the repercussion of a specific constraint $C_i[f]$ for a field $f(\mathbf{x})$ throughout the sample volume $\mathcal{V}_s$. For example, the field-constraint correlation function for a constraint on the peculiar velocity or gravity is a dipolar pattern, while a tidal constraint T_{ij} effects a quadrupolar configuration (see van de Weygaert & Bertschinger [84]). The weights for each of the relevant $\xi_i(\mathbf{x})$ are determined by the value of the constraints, c_m, and their mutual cross-correlation $\xi_{mn} \equiv \langle C_m C_n \rangle$,

$$\bar{f}(\mathbf{x}) = \xi_i(\mathbf{x})\, \xi_{ij}^{-1}\, c_j\,. \tag{76}$$

In practice, it is usually beneficial to evaluate the constraint correlation function $\xi_i(\mathbf{r}$, ξ_{ij} and the mean field in Fourier space. For a linear cosmological density field with power spectrum $P(k)$ we have

$$\begin{aligned} \xi_i(\mathbf{r}) &= \int \frac{\mathrm{d}\mathbf{k}}{(2\pi)^3}\, \hat{H}_i(\mathbf{k})\, P(k)\, e^{-i\mathbf{k}\cdot\mathbf{x}} \\ \xi_{ij} &= \int \frac{\mathrm{d}\mathbf{k}}{(2\pi)^3}\, \hat{H}_i^*(\mathbf{k})\, \hat{H}_j(\mathbf{k})\, P(k) \end{aligned} \tag{77}$$

with $\hat{H}_i(\mathbf{k})$ the constraint i's kernel (the Fourier transform of constraint $C_i[f]$) and c_j the value of this constraint.

The additional generation of the residual field F is a nontrivial exercise: the specified constraints translate into locally fixed phase correlations. This renders a straightforward random phase Gaussian field generation procedure unfeasible: the amplitude of the residual field is modified by the local correlation with the specified constraints. Hoffman & Ribak [35] pointed out that for a Gaussian random field the sampling is straightforward and direct, which greatly facilitated the application of CRFs to cosmological circumstances. This greatly facilitated the application of CRFs to complex cosmological issues [44, 55, 69].

Van de Weygaert & Bertschinger [84], following the Hoffman–Ribak formalism, worked out the specific CRF application for the circumstance of sets of local density peak (shape, orientation, profile) and gravity field constraints.

With most calculations set in Fourier space, the constrained field realization for a linear cosmological density field with power spectrum $P(k)$ follows from the computation of the Fourier integral

$$f(\mathbf{x}) = \int \frac{d\mathbf{k}}{(2\pi)^3} \left[\hat{\tilde{f}}(\mathbf{k}) + P(k)\, \hat{H}_i(\mathbf{k})\, \xi_{ij}^{-1}\, (c_j - \tilde{c}_j)\right] e^{-i\mathbf{k}\cdot\mathbf{x}} \tag{78}$$

where the tilde indicates it concerns a regular unconstrained field realization $\tilde{f}$.

One of the major virtues of the *constrained random field* construction technique is that it offers the instrument for translating locally specified quantities into the corresponding implied global matter distributions for a given structure formation scenario. In principle the choice of possible implied matter distribution configurations is infinite. Nonetheless, it gets substantially curtailed by the local matter configuration. The influence of local constraints is set by the coherence scale of matter fluctuations, a function of the power spectrum of fluctuations.

While the CRF formalism is rather straightforward for idealized linear constraints reality is less forthcoming. If the constraints are based on measured data these will in general be noisy, sparse and incomplete. Wiener filtering will be able to deal with such a situation and reconstruct the implied *mean field*, at the cost of losing signal proportional to the loss in data quality (see e.g. Zaroubi et al. [89]). A major practical limitation concerns the condition that the constrained field is Gaussian. For more generic nonlinear clustering situations the formalism is in need of additional modifications. For specific situations this may be feasible [73], but for more generic circumstances this is less obvious (however, see Jones & van de Weygaert 2008).

4.9 Shear Constraints

The Megaparsec scale tidal shear pattern is the main agent for the contraction of matter into the filaments which trace out the cosmic web (see Figs. 18 and 19). For a cosmological matter distribution the close connection between local force field and global matter distribution follows from the expression of the tidal tensor in terms of the generating cosmic matter density fluctuation distribution $\delta(\mathbf{r})$ [84]:

$$\begin{aligned} T_{ij}(\mathbf{r}) =& \frac{3\Omega H^2}{8\pi} \int d\mathbf{r}'\, \delta(\mathbf{r}') \left\{ \frac{3(r_i' - r_i)(r_j' - r_j) - |\mathbf{r}' - \mathbf{r}|^2\, \delta_{ij}}{|\mathbf{r}' - \mathbf{r}|^5} \right\} \\ & - \frac{1}{2}\Omega H^2\, \delta(\mathbf{r}, t)\, \delta_{ij} . \end{aligned}$$

Constrained random field realizations immediately reveal the nature of the density field realizations $\delta(\mathbf{r})$ that would generate a tidal field T_{ij} at particular location $\mathbf{r}_0$. The effect of the local shear constraints on the density profile around a position r_0 may be seen in Fig. 17. The shape of the density contours

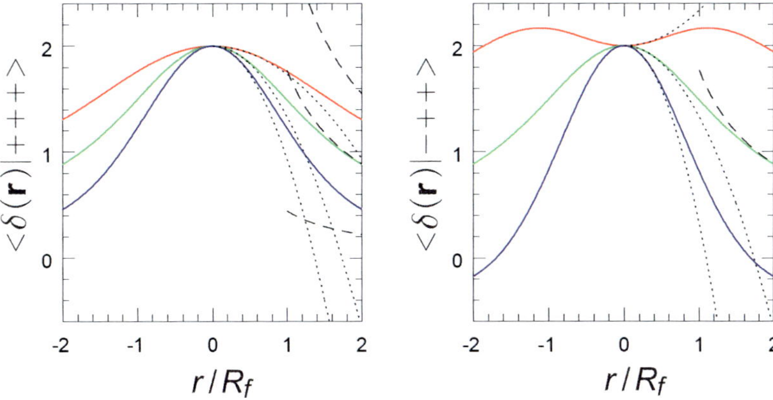

Fig. 17. Constrained primordial density field $\langle\delta(\mathbf{r})|\lambda_1, \lambda_2, \lambda_3\rangle$ as a function of distance $\mathbf{r}$ in units of the filter scale R_f, in the three eigendirections. **Left frame:** shear constraint signature $(+ + +)$. **Right frame:** shear constraint signature $(- + +)$. The "filamentary" behaviour of the density in the neighbourhood of the point manifests itself particularly in the density profile along the x-direction (*top curve*). From: Pogosyan et al. [66]

clearly depends on the signature of the eigenvalues. The righthand frame does reveal an increase in the density along one axis while falling off along the remaining two. This is symptomatic of filamentary bridges that connect the higher density regions where the shape of the density profile is more spherical. In effect, the local shear signature defines the curvature of the density isocontours up to a distance of several filter radii R_f[2].

Pursuing the filamentary configuration implied by the specified $(- + +)$ signature tidal shear, the 3-D density distribution around the location of the specified constraint is shown in Fig. 19. The specified shear tensor is oriented along the box axes. The field is Gaussian filtered on a (rather arbitrary) scale of $2\,h^{-1}$ Mpc. The implied *mean field* $\bar{f}$ is shown in the 3 top panels. Each panel looks along one of the main axes. The constraint clearly works out into perfect global quadrupolar mass distribution. A representative realization of a quadrupolar (CDM) cosmic matter distribution which would induce the specified shear is shown in the second row of panels.

The corresponding maps of the tidal shear in the same region are shown in the bottom row. Included are contour maps of the total tidal field strength. Also we include bars indicating the direction and strength of the tide's compressional component[3]. Along the full length of the filament in Fig. 19

[2] The information contained in the density curvature tensor itself is much more local and less representative of the density behaviour at large distances from the constraint point.

[3] On the basis of the effect of a tidal field, we may distinguish at any one location between "compressional" and "dilational" components. Along the direction of a

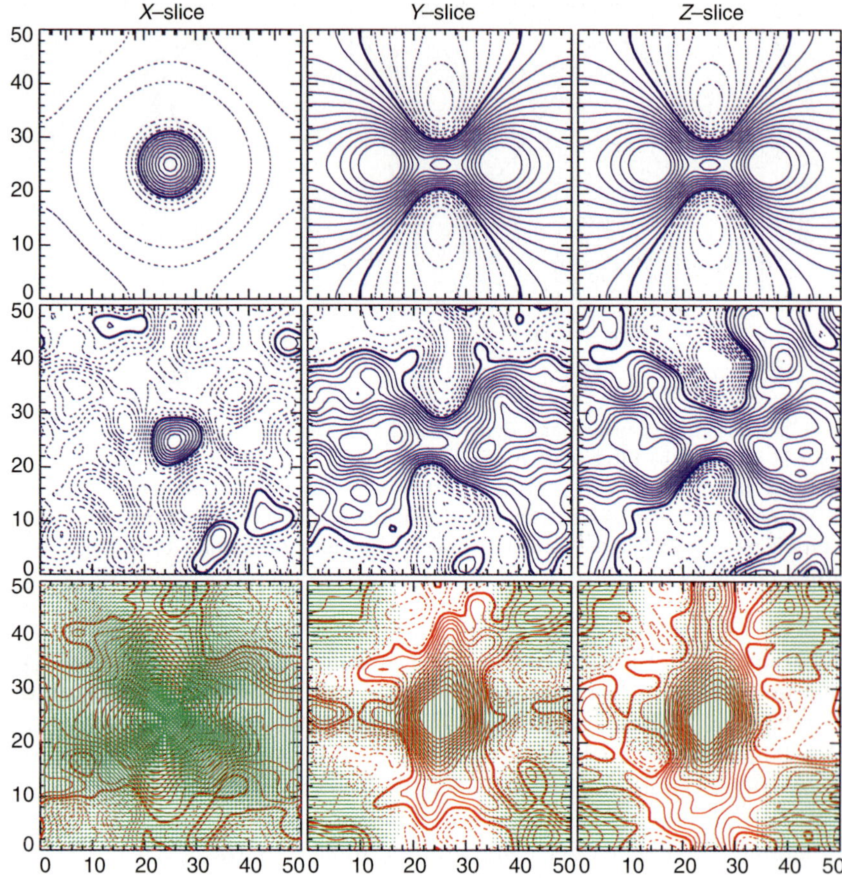

Fig. 18. Constrained field construction of initial quadrupolar density pattern in a SCDM cosmological scenario. The tidal shear constraint is specified at the box centre location, issued on a Gaussian scale of $R_G = 2\,\mathrm{h}^{-1}$ Mpc and includes a stretching tidal component along the x- and y-axis acting on a small density peak at the centre. Its ramifications are illustrated by means of three mutually perpendicular slices through the centre. **Top row:** the "mean" field density pattern, the pure signal implied by the specified constraint. Notice the clear quadrupolar pattern in the y- and z-slice,directed along the x- and y-axis, and the corresponding compact circular density contours in the x-slice: the precursor of a filament. **Central row:** the full constrained field realization, including a realization of appropriately added SCDM density perturbations. **Bottom row:** the corresponding tidal field pattern in the same three slices. The (red) contours depict the run of the tidal field strenght $|T|$, while the (green) tidal bars represent direction and magnitude of the *compressional* tidal component in each slice (scale: $R_G = 2\,\mathrm{h}^{-1}$ Mpc). From van de Weygaert [83]

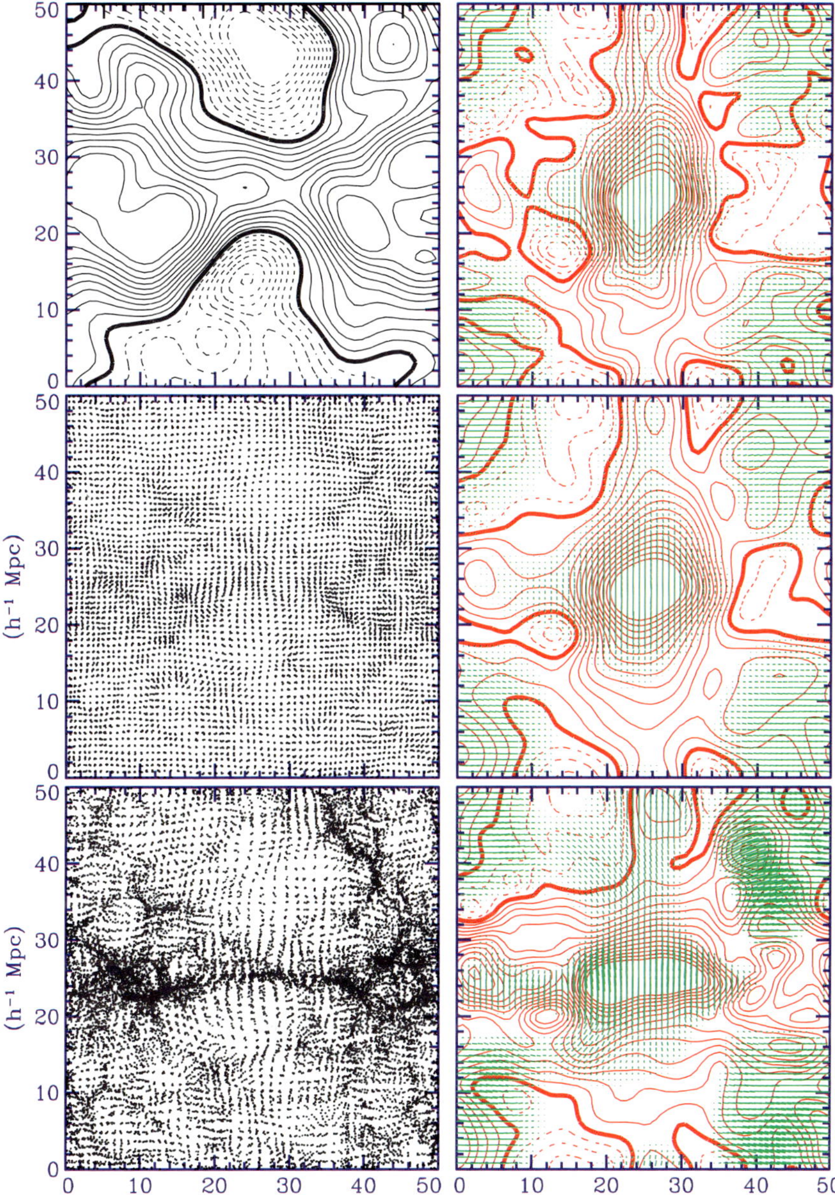

Fig. 19. The emergence of a filament in an SCDM structure formation scenario. **Lefthand column:** density/particle distribution in z-slice through the centre of the simulation box. **Righthand column:** the corresponding tidal field configurations, represented through the full tidal field strength $|T|$ contour maps (red), as well as the corresponding compressional tidal bars (scale: $R_G = 2\,\mathrm{h}^{-1}\,\mathrm{Mpc}$). From top to bottom: primordial field, $a = 0.2$ (visible emergence filament), present epoch. Note the formation of the filament at the site where the tidal forces peaked in strength, with a tidal pattern whose topology remains roughly similar. From van de Weygaert [83]

we observe a coherent pattern of strong compressional forces perpendicular to its axis.

Filaments and Peaks

The dynamical evolution in and around the (proto) filament is depicted in Fig. 19. It shows the emergence of a (CDM) filament with the density/particle distribution along the spine of the emerging filament (lefthand column) and the corresponding tidal configuration (righthand column). The top row corresponds to the primordial cosmic conditions, the central row to $a = 0.2$ and the bottom row to $a = 0.8$. At $a = 0.2$ we recognize the first vestiges of an emerging filament, at $a = 0.8$ it has indeed condensed as the most salient feature in the mass distribution. Also, we see that the filament forms along the ridge seemingly predestined by the primordial tidal configuration (Figs. 19 and 20).

The figure also clarifies the essence of the link between filaments and clusters. At the tip of the evolving filament we observe the emergence of massive cluster patches. They naturally arise in and around the overdense peaks in the primordial quadrupolar mass distribution implied by the tidal shear constraint. These overdense protoclusters were the source of the specified shear. A quadrupolar matter configuration will almost by default evolve into the canonical *cluster-filament-cluster* configuration so prominently recognizable in the observed Cosmic Web.

The two main conclusion from these observations are the embryonic presence of the weblike features in the primordial density field and the intimate link between the cluster distribution and the filigree of filaments as most outstanding structural aspect of the Cosmic Web (see Fig. 20).

4.10 Nodes of the Cosmic Web: Peak Patches

Clusters represent the rare events in the cosmic matter distribution. In the above we have established that they are the ultimate source for the anisotropic contraction of filaments and form the nodes that weave the cosmic web throughout the Universe.

The study of local one-point shear constraints has lead to the conclusion that filaments are indeed the naturally dominant structural feature in the cosmic matter distribution. The remarkable size of the filaments is not, however,

"compressional" tidal component T_c (for which $T_c < 0.0$) the resulting force field will lead to contraction, pulling together the matter currents. The "dilational" (or "stretching") tidal component T_d, on the other hand, represents the direction along which matter currents tend to get stretched as $T_d > 0$. Note that within a plane, cutting through the 3-D tidal "ellipsoid", the tidal field can consist of two compressional components, two dilational ones or – the most frequently encountered situation – of one dilational and one compressional component.

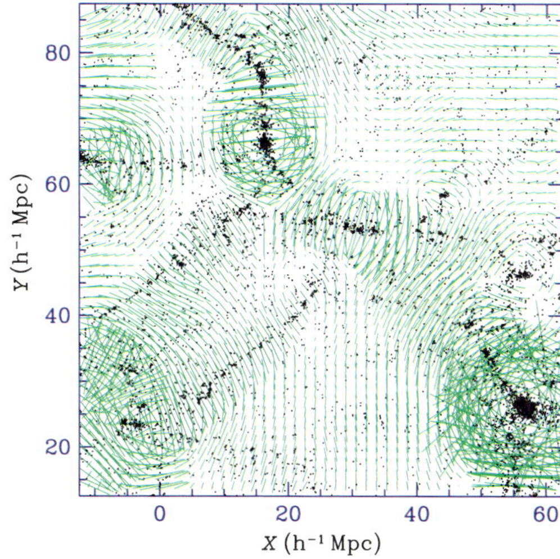

Fig. 20. The relation between the *cosmic web*, the clusters at the nodes in this network and the corresponding compressional tidal field pattern. It shows the matter distribution at the present cosmic epoch, along with the (compressional component) tidal field bars in a slice through a simulation box containing a realization of cosmic structure formed in an open, $\Omega_\circ = 0.3$, Universe for a CDM structure formation scenario (scale: $R_G = 2\,\mathrm{h}^{-1}\,\mathrm{Mpc}$). The frame shows structure in a $5\,\mathrm{h}^{-1}\,\mathrm{Mpc}$ thin central slice, on which the related tidal bar configuration is superimposed. The matter distribution, displaying a pronounced weblike geometry, is clearly intimately linked with a characteristic coherent compressional tidal bar pattern. From: van de Weygaert [83]

derivable from constraints at a given single point. To learn more about the strength, structure and connections of the weblike features we need to investigate their dependence on the location, nature and structure of clusters. For this we need to turn to correlations constrained by at least two rare peak-patches. In order to fully grasp their impact on the overall morphology of the cosmic web we first need to delve into their internal structure.

Clusters at any cosmic epoch are the product of a hierarchical buildup of structure in and around the primordial protocluster, peaks in the primordial mass distribution. In Sect. 3.3 we have discussed in some detail how the anisotropic nature of collapse of (sub)clumps can be included by means of a *moving collapse barrier* in a local extended Press–Schechter description of hierarchical evolution. A more physical image would also try to take into account the matter distribution in and around the primordial peak. This is achieved by the *peak patch* formalism of Bond & Myers [17].

The *peak-patch* formalism exploits the full potential of the peaks formalism [6] by using adaptive spatial information on both small and large scales to construct the hierarchical evolution of collapsing protocluster *peak patches*.

The entire patch moves with a bulk peculiar velocity and is acted upon by external tidal fields, determined by long-wavelength components of the density field.

Peak Patch: Hierarchical & Anisotropic Collapse

The formation of a cluster around an overdensity is approximated as the combination of the linear evolution of a smooth large-scale background field and the coupled nonlinear evolution of the mass element itself, and its substructure. Clusters are identified with the peaks in the primordial Gaussian field on an appropriately large smoothing scale R_G. This scale is determined by filtering the field around a particular peak's location over a range of radii. By means of the ellipsoidal collapse model, including the influence of the external tidal field, the collapse time of the ellipsoidal configuration is determined. At any one cosmic epoch the peak's scale R_{pk} is identified with the largest scale $R_{\rm b}$ on which, according to the homogeneous ellipsoidal model, it has collapsed along all three dimensions.

The mass of the peak is

$$M_{pk} = \frac{4}{3}\pi \rho_{\rm u} a^3 \, R_{pk}^3 \,. \tag{79}$$

Because the formalism works within the spatial mass distribution itself it allows the identification and dissection of overlapping (collapsed) peak patches. Usually this concerns peaks of a different scale. Small-scale peaks may be absorbed/merged with larger peaks with which they largely overlap (*half-exclusions*). If they only partially overlap, with their centers outside each others range, one may seek to define a proper prescription to divide up the corresponding mass (*binary exclusion/reduction*). The resulting mass spectrum of clumps adheres closely to the predictions of the extended Press–Schechter formalism and to the results of N-body simulations.

A major virtue of the peak-patch formalism is that the spatial distribution of the patches may be followed in time. Upon having identified the patches at their original Lagrangian location, they are subsequently displaced towards their Eulerian position (most conveniently by means of the Zel'dovich formalism). A typical result is shown in Fig. 21 (from Platen et al. [64]), a nice illustration of how narrowly collapsed peaks trace the cosmic web.

Anatomy of a Peak

Following the differentiation between nonlinearly evolving short wavelength contributions $\delta_{\rm f}(\mathbf{x})$ and linearly evolving long-wavelength contributions $\delta_{\rm b}(\mathbf{x})$ (see (42)), we can distinguish three contributions to a peak's structure and dynamics,

$$\begin{aligned} \delta(\mathbf{x}) &= \bar{\delta}_{\rm b}(\mathbf{x}) + F_{\rm b}(\mathbf{x}) + F_{\rm f}(\mathbf{x}) \\ &= \bar{\delta}_{\rm b}(\mathbf{x}) + F_{\rm b}(\mathbf{x}) + \delta_{\rm f}(\mathbf{x}) \end{aligned} \tag{80}$$

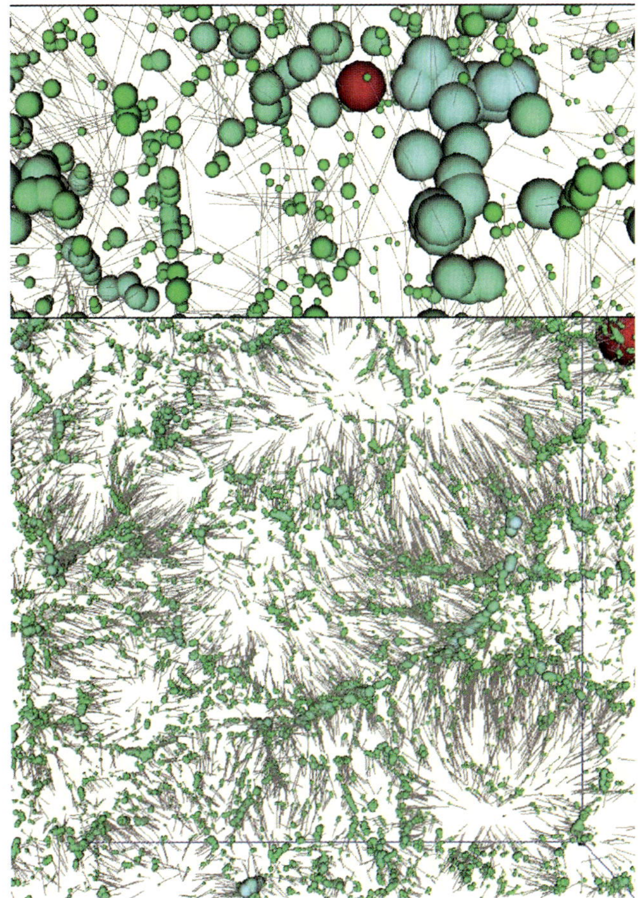

Fig. 21. The distribution of peak patches for a realization of a SCDM density field in a $100\,\mathrm{h}^{-1}$ Mpc box. The lefthand image is a slice through the 3-D matter distribution. The blobs are collapsed peaks, their size related to their spatial extent/mass. Each patch is moved from its Lagrangian position by means of the Zel'dovich formalism. The gray edges are the paths followed by each of the patches. The bottom insert zooms in on one of the regions, offering a more distinct impression of the size of each of the patches. Image courtesy of Erwin Platen

One concerns the *mean field* structure $\bar{\delta}_{\mathrm{b}}(\mathbf{x})$ of the cluster peak specified on a scale R_b and formally corresponds to the ensemble average of all peaks with the specified properties. Because the peak is embedded within a fluctuating (large-scale) field, there is also a *residual* fluctuating large-scale contribution $F_{\mathrm{b}}(\mathbf{x})$. In and around the peak the latter is heavily affected by the peak's presence in that it is hardly existent or at least extremely quiescent in its neighbourhood. The internal substructure of the peak patch mainly consists of the short wavelength contribution $\delta_{\mathrm{f}}(\mathbf{x})$. The latter is hardly affected by the presence of the peak. While formally constrained by the peak's presence, the resulting *residual* contribution $F_{\mathrm{f}}(\mathbf{x})$ is mostly a pure unconstrained Gaussian random field.

The individual components contributing to the total density field around a primordial cluster peak are shown in Fig. 22 (from Bond & Myers [17]). The structure of the peak on a is shown by means of density field contours and peculiar velocity field vectors. The peak's structure was specified on a Gaussian scale of $R_G = 5\,\mathrm{h}^{-1}\,\mathrm{Mpc}$. The solid circle indicates the corresponding peak scale $R_{pk} = 10\,\mathrm{h}^{-1}\,\mathrm{Mpc}$. The overall triaxial structure of the peak is determined by the bakcground mean field shown in the top lefthand panel. The

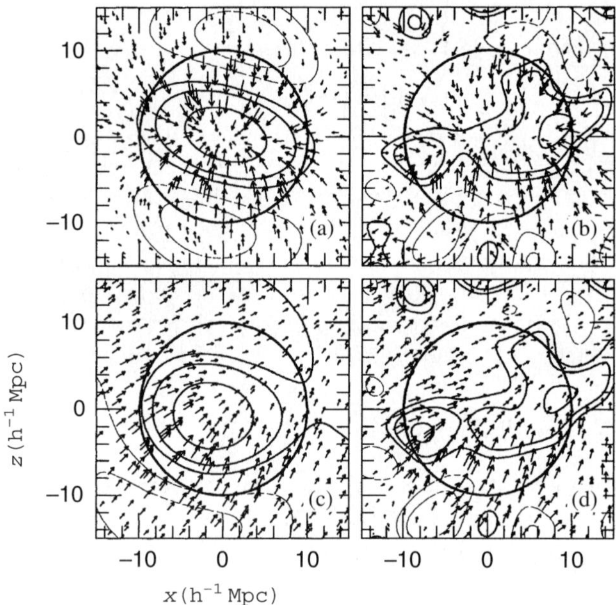

Fig. 22. The individual contributions to the structure (*density field contours*) and peculiar velocity field (*arrows*) in and around a density peak. The first three panels show the (**a**) large-scale mean field δ_{b}, (**b**) the large-scale variance field F_{b}, which is extremely quiescent in the neighbourhood of a peak, and (**c**) the small-scale field δ_{f} responsible for subclumps within the medium. Adding them altogether produces (**d**) the total field around the density peak. In (**a**) and (**b**) the contours increase by factors of 2 from the minimum contour at $f_c/2$, where $f_c = 1.69$ is the critical contour for spherical tophat collapse. The displacement arrows are scaled for appearance, and only one in 12 are sampled. Panels (**c**) and (**d**) start at the f_c contour level for positive densities and at $2f_c$ for negative ones. The peak was constrained to have $\nu_{\mathrm{pk}} = 2.45$, $e_{\mathrm{v,pk}} = 0.14$ and $v_{1,\mathrm{pk}} = 0.46\sigma_v$ on a Gaussian smoothing scale of $R_G = 5\,\mathrm{h}^{-1}\,\mathrm{Mpc}$. The circle at at $10\,\mathrm{h}^{-1}\,\mathrm{Mpc}$ is the average R_{pk} associated with Gaussian peaks at this filter scale. From: Bond & Myers [17]. Reproduced with permission of AAS

velocity vectors delineate the expected shear flow around the peak. Because the specified peak constraints essentially fully specify the structure of the peak on the smoothing scale the background variance field $F_b(\mathbf{x})$ is extremely quiescent (top righthand frame). The small-scale residual field (bottom lefthand frame) includes two subclumps, one of them rather extended. Adding all components together yields the total structure in and around the peak (bottom righthand frame).

The small-scale structure in and around the peak may vary considerably from one realization to another even though the cluster's large scale structure remains the same. The global history and fate of the peak, however, are largely specified by the large-scale anisotropic tidal shear and bulk flow.

4.11 Molecular View of the Cosmic Web

In the observed galaxy distribution "superclusters" are often filamentary cluster-cluster bridges and the most pronounced ones will be found between clusters of galaxies that are close together and which are aligned with each other. Very pronounced galaxy filaments of which the Pisces-Perseus supercluster chain is a telling example are almost inescapably tied in with a high concentration of rich galaxy clusters. The Cosmic Web theory expands the observation of the intimate link between clusters and filaments, described in some detail in Sect. 4.9, to a complete framework for weaving the cosmic web in between the clusters in the cosmic matter distribution.

The Cosmic Web Theory

In the language of the crf formalism discussed in Sect. 4.8 the filamentary bridges in between two peak patches should be regarded as "correlation" bridges. The implied constraint correlation function (or mean field) $\xi_i(\mathbf{r}) = \langle\delta|2\text{pks}\rangle$ defines a protofilament, along the lines seen in Fig. 18. These correlation bridges will be stronger and more coherent as clusters are nearer than the mean cluster separation. Because clusters are strongly clusters and statistically biased [7, 40] there are many cluster pairs evoking strong filamtary bridges.

The filament bridge will break if the separation of the clusters is too large, due to diminishing amplitude of the correlation $\xi_i(\mathbf{r}) = \langle\delta|2\text{pks}\rangle$. Such clusters will be isolated from each other, unless there is a cluster in between to which both have extended their filamentary bridges. As a result, the typical scale of a segment of the filamentary network in a CDM type scenario will be in the order of $\sim 30\,\mathrm{h}^{-1}\,\mathrm{Mpc}$.

This brings us to the aspect of establishing the weblike network characterizing the observed galaxy distribution and matter distribution in computer simulations. Consider laying down the rare cluster peaks in the cosmic matter distribution according to the clustering pattern of peak-patches which become clusters when they evolve dynamically. The correlation bridges arche from cluster to cluster in much of the domain, and tehse dynamically evolve to filaments, creating the network and containing the bulk of the mass.

The order in which the physically significant structures arise is basically the inverse of that in the classical pancake picture: first, high-density peaks, then filaments between them and, possibly, afterwards the walls. The latter should be seen as the rest of the mass between the voids.

Outlining the Web

Figure 23 convincingly demonstrates the viability of the cosmic web theory by illustrating the excellent reconstruction of the primordial density field implied by the presence of a set of selected protocluster peaks. The figure concerns a CDM scenario realization within a comoving region of $50\,h^{-1}$ Mpc (the same box as in Fig. 16). Within this volume the peak patches are identified and rank-ordered in mass.

Of each peak patch the value of the overdensity, the shear tensor $\mathcal{E}_{b,ij}$ and displacement $\mathbf{s}_b$ are measured, at their location $\mathbf{r}_{pk}$ and averaged over the peak-patch size R_{pk}. In addition to the in total $9N$ constraints for N peak patches, the extremum requirement of a vanishing density gradient $\nabla\delta_b = 0$ at $\mathbf{r}_{pk}$ adds a further $3N$ constraints. On the basis of the selection of the N rarest and most massive peak patches the mean (primordial) field realization is determined following the constrained field formalism outlined in Sect. 4.8. The $12N$ peak constraints and the locations of the N peaks result in a mean initial field $\langle\delta_L|N\text{peaks}\rangle$ (76).

We compare the mean field realizations implied by the 5 most massive peak patches, that by the 10 most massive peaks and for the 20 most peaks. In the boxes in the lefthand column of Fig. 23 we have indicated their locations by black ellipsoids of overall size proportional to R_{pk} and shape defined by the shear tensor orientation, with the shortest axis corresponding to the highest shear eigenvalue. The corresponding mean field density field is represented by isodensity contours at a level $\delta_L = 1\sigma_\rho$, where δ_L is smoothed on a scale of $3.5\,h^{-1}$ Mpc. The righthand frames show the Zel'dovich maps of these smoothed initial conditions.

A comparison with Fig. 16 shows the excellent reconstruction obtained by adding in the 20 most massive peaks. Also we see that the reconstruction improves continuously as more and more peaks are added. Some strong bridges seen in the 20 peak reconstruction $\langle\delta_L|20\text{pks}\rangle$ are not as evident in the $\langle\delta_L|10\text{pks}\rangle$ field, although they emerge at lower thresholds.

Web Bridges: Shear, Distance and Orientation

The observations discussed above show that a list of rank-ordered peak-patches is a powerful way to maximally compress the information stored in the initial conditions. They also show what is essential for defining structures on the basis of a modest set of local measurements. That the specification

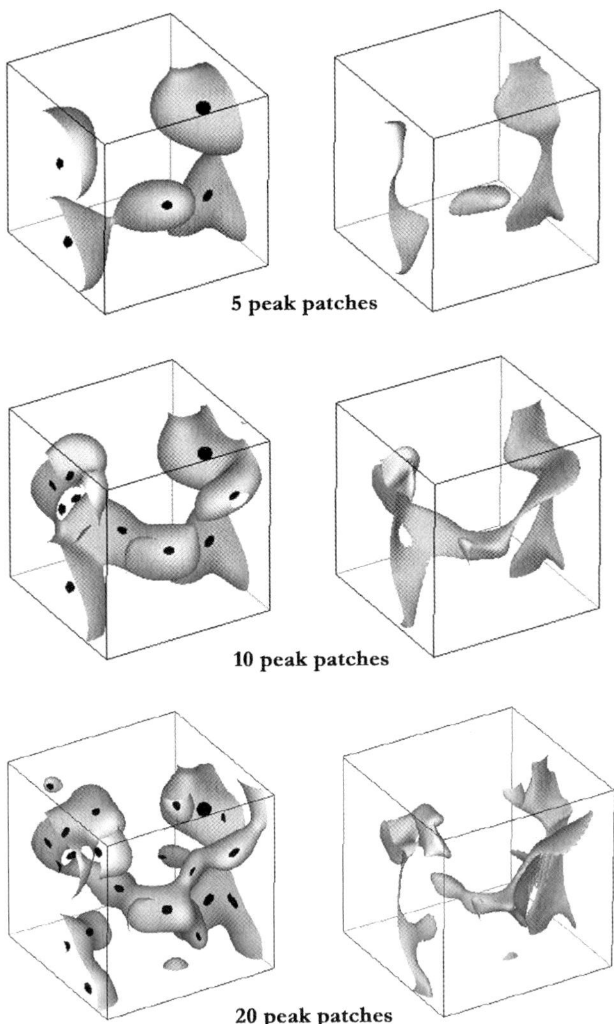

Fig. 23. Building the Cosmic Web with clusters. How adding clusters gradually defines the details of the Cosmic Web. (Mean) Constrained density field reconstructions $\langle \delta_L | Npeaks \rangle$ on the basis of the N most massive cluster peaks (patches) in a CDM model of cosmic structure formation. The volume is a $(50\,\mathrm{h}^{-1}\,\mathrm{Mpc})^3$ box with periodic boundary conditions. The lefthand column frames contain the initial linear CDM overdensity field $\delta_L(\mathbf{r})$, smoothed on a Gaussian scale $R_G = 3.5\,\mathrm{h}^{-1}\,\mathrm{Mpc}$ with (iso)density threshold level $\delta_L = 1\sigma_\rho$, with $\sigma_\rho = 0.65$, the level at which δ_L percolates. The location, size and shape of the cluster patches are indicated by means of the black ellipsoids, whose size is proportional to the peak scale R_{pk} and orientation defined by the shear tensor orientation. The righthand column contain the corresponding Zel'dovich map density field δ_Z of the smoothed initial conditions at a contour threshold $\delta_Z = 2$. **Top row:** the constrained field $\langle \delta_L | 5peaks \rangle$ for 5 peaks, $\langle \delta_L | 10peaks \rangle$ for 10 peaks and $\langle \delta_L | 20peaks \rangle$ for 20 peaks. Based on Bond et al. [18]. Reproduced with permission of Nature

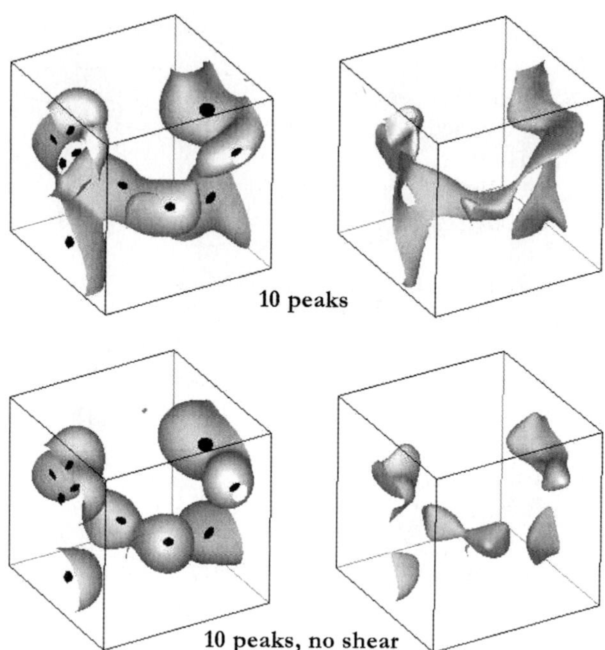

Fig. 24. Cluster Shear and the Cosmic Web. How cluster tidal shear defines the filigree of the Cosmic Web. Comparison between a (mean) cosmic density field generated by the 10 most massive cluster peaks with shear constraints (**top show**) and without shear constraints (**bottom row**), for a CDM simulation in a $(50\,h^{-1}\,\mathrm{Mpc})^3$ box with periodic boundary conditions. Left row: isodensity contours of the linear CDM overdensity field $\delta_L(\mathbf{r})$, smoothed on a Gaussian scale $R_G = 3.5\,h^{-1}\,\mathrm{Mpc}$ with (iso)density threshold level $\delta_L = 1\sigma_\rho$, with $\sigma_\rho = 0.65$. The location, size and shape of the cluster patches are indicated by means of the black ellipsoids, whose size is proportional to the peak scale R_{pk} and orinetation defined by the shear tensor orientation. The righthand column contain the corresponding Zel'dovich map density field δ_Z of the smoothed initial conditions at a contour threshold $\delta_Z = 2$. Both initial density field and Zel'dovich map for the non-shear constraint situation (**bottom row**) do have a more bloblike character, and does hardly contain the matter bridges characterizing the Cosmic Web. Based on Bond et al. [18]. Reproduced with permission of Nature

of the tidal shear at the peak patches is of fundamental importance for the succesfull reconstruction of the Cosmic Web may be appreciated from Figs. 23 and 24. By discarding the tidal shear measurements at the peak patches and only taking into account their overdensity and velocity the implied mean field loses its spatial coherence. Instead of being marked by strong filamentary bridges the mean field will have a more patchy character. It demonstrates our earlier arguments that the tidal shear evoked by the inhomogeneous cosmic mass distribution is of crucial and fundamental importance in defining the Cosmic Web.

The strength and the coherence of the correlation bridges depend strongly on the mutual distance of the clusters and their alignment. The strongest filaments are between close peaks whose tidal tensors are nearly aligned. This may be inferred from the illustration of the 2-point correlation function in Fig. 25: a binary molecule image with oriented peak-patches as the atoms. The initial conditions in this figure have been smoothed and Zel'dovich mapped, producing a telling illustration of the *molecular* picture of large scale structure.

The bridge between two clusters will gradually weaken as the separation between the clusters increases. Strong filaments extend only over a few Lagrangian radii of the peaks they connect. It is in the nonlinear mass distribution that they occur so visually impressive because the peaks have collapsed by about a factor 5 in radius, leaving the long bridge between them, which themselves have also gained more contrast because of the decreases in its transverse dimension.

Another important factor influencing the coherence and strength of the connecting filamentary bridges are the mutual alignments between the shear tensors of the cluster peaks. When we vary the shear orientation from perfect alignment towards misalignment the strong correlation bridge between two clusters will weaken accordingly. The top two panels of Fig. 25 show the difference as two peaks, of equal mass and orientation, are oriented differently. In the lefthand panel they are perfectly aligned, evoking a strong filamentary bridge in between them. When the same clusters are somewhat misaligned, each by $\pm 30°$ with respect to their connecting axis, the bridge severely weakens. The bridge would break at an isodensity level of $\delta_l = 1$, althougn it would remain connected at a lower level, for a misalignment of $\approx \pm 45°$. In the most extreme situation of a misalignment by $\pm 90°$ the bridge would be fully broken, no filament would have emerged between the two clusters. The reason for the strong filaments between aligned peaks is that the high degree of constructive interference of the density waves required to make the rare peak-patches, and to preferentially orient them along their connecting axis leads to a slower decoherence along that axis than along the other axes. This in turn corresponds to a higher density.

Important for the overall weblike structure in the matter distribution is the fact that there is a distinct tendency of clusters to be aligned with each other. The alignment of the orientations of galaxy haloes and clusters with larger scale structures such as clusters, filaments and superclusters have been the subject of numerous studies (see e.g. Binggeli [12], Bond [13], Rhee et al. [68], Plionis & Basilakos [65], Basilakos et al. [8], Trujillo et al. [82], Aragón-Calvo et al. [2], Lee & Evrard [50], Park & Lee [60], Lee et al. [52]). The peak-patch theory [17] offers a natural explanation for these alignments by showing that it is statistically likely that, given a specific orientation of the shear tensor for a cluster peak, neighbouring cluster peaks will be aligned preferentially along its axis and have shear tensors aligned with it.

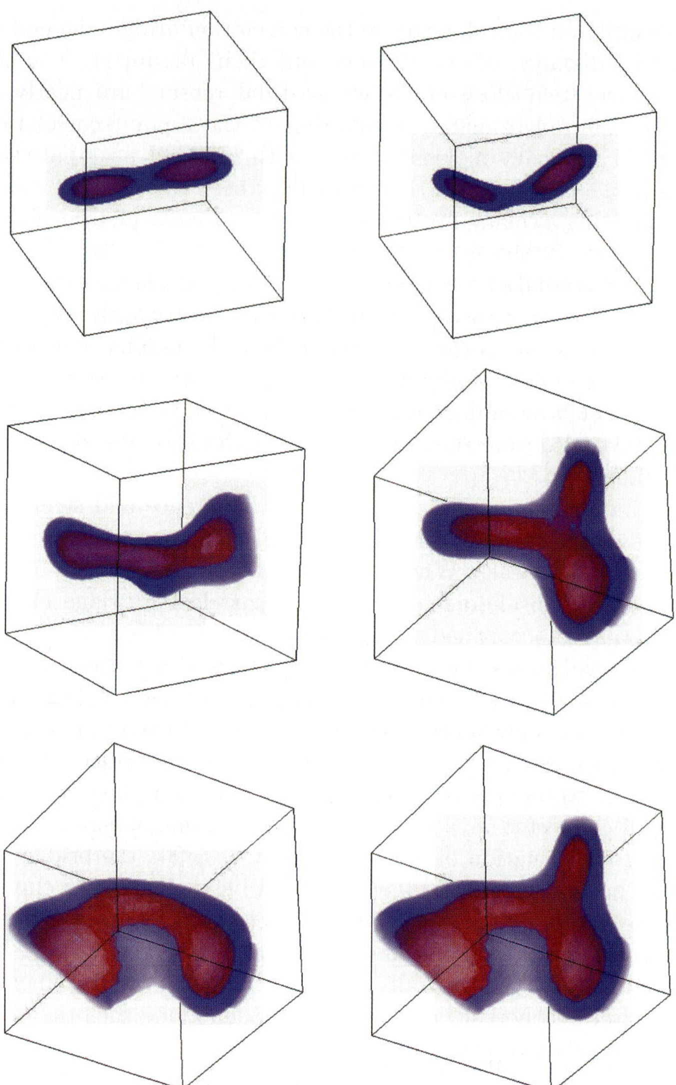

Fig. 25. The molecular picture of large scale structure: "bonds" bridging clusters. Shown are isodensity contours of the Zel'dovich map of the smoothed initial density field. The upper panels show a two-point mean (constrained) field $\langle \delta_L | 2peaks \rangle$ constrained by two oriented clusters separated by $40\,\mathrm{h}^{-1}$ Mpc. Left one is fully aligned, the right pair is partially aligned. The next four panels show three-point (**middle row**) and four-point mean fields for different peak-patch orientations taken from the simulation. Notice the lower density contrast webbing between the filaments. From Bond et al. [19]

Walls and Voids

Upon expanding our inspection in Fig. 25 from 2-peak correlations to three-point and four-point mean fields we see the emergence of low density contrast webbing between the filaments, membrane-like sheets. Stronger membranes will be seen in the regions between the filaments when a number of clusters is close together. Although these are sheetlike structures they are not the classical pancakes. In the molecular view of cosmic structure formation the walls are a mere secondary product.

Voids also do play a significant role in the cosmic web. The formalism is similar, be it reversed, when concentrating on the voids. Void patch constraints create high mean field regions in between them, just where less rare peak patches reside. However, using voids are not as precise a way to get the filamentary structure evoked by the peaks. An upcoming study (Platen et al. [64]) adresses their role and structure in considerably more detail.

Cosmic Scenario

Overall, it is the highly clustered and mutually aligned nature of the cluster distribution which ascertains the salient and coherent weblike nature of the cosmic matter distribution. In turn, this suggests a dependence of the morphology and structure of the cosmic web on the cosmological scenario.

Its pattern and prominence does indeed depend upon the shape of the primordial power spectrum, in particular on the power spectrum index $n(k) = \mathrm{d}\ln P(k)/\mathrm{d}\ln k$. The examples which are shown in the figures concern a CDM spectrum with $n_{\mathrm{eff}} \approx -1.2$ on cluster scales. When the spectrum is steepened clusters become less clustered and the coherence of the web is lost. Although some filaments will remain they will be weaker and shorter. On the other hand, when we flatten the spectrum to $n(k) < -2$, the clusters become more clustered, so that the coherence is more pronounced and the filaments are both strengthened and widened.

4.12 Hierarchical Filament Assembly

In the previous sections we have delved in great depth into the nature and origin of filamentary and sheetlike features in the cosmic web. We have not yet paid a lot of attention to their hierarchical development. In the reality of the nonlinear world the collapse and formation of weblike patterns is considerably more complex. Taking the specific example of an emerging filament, its formation will involve the gradual assembly of small-scale filaments and virialized low mass clumps into a coherent elongated feature.

Figure 26 gives an impression of the intricacies of filament formation Aragón-Calvo [1]. It involves a ΛCDM scenario. The initial configuration consists of a myriad of small-scale filaments, with a large scatter in orientation. As time proceeds these small filaments start to merge into larger filaments,

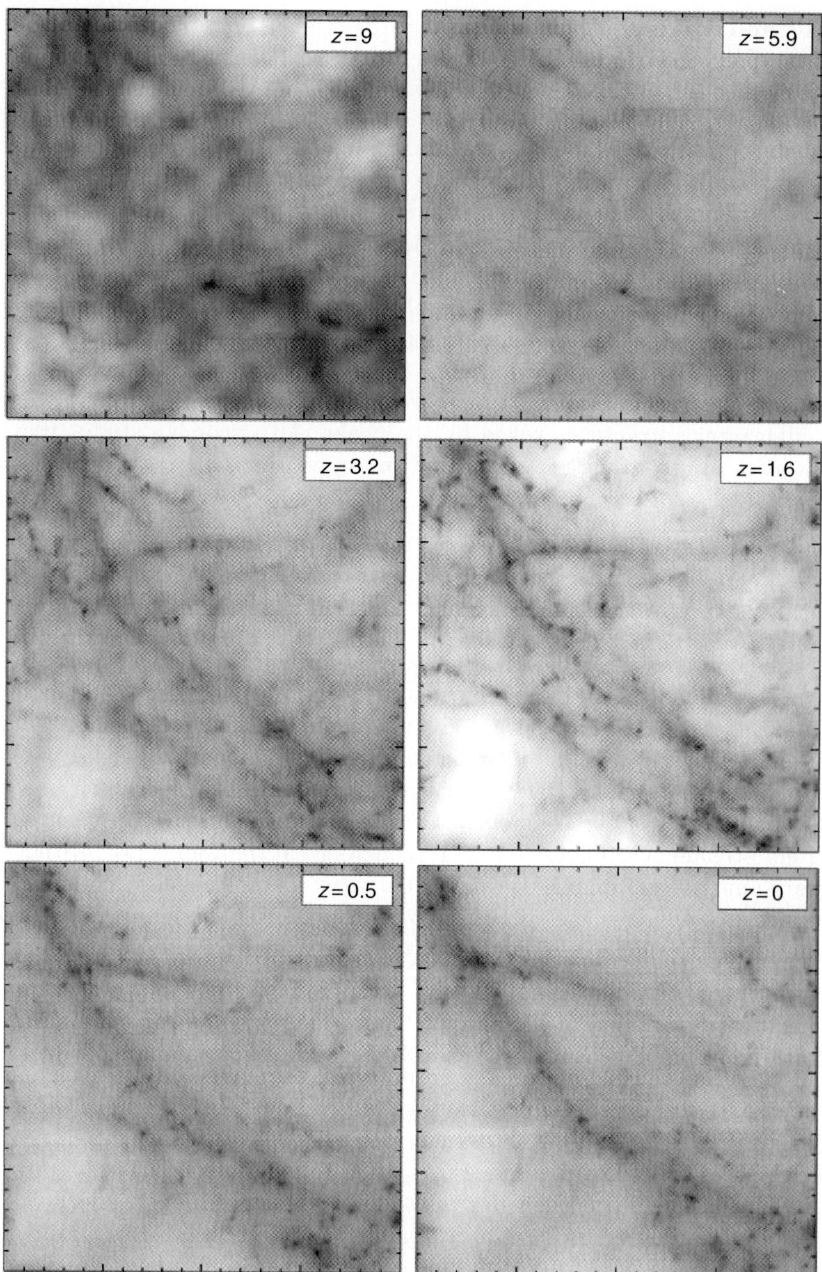

Fig. 26. The hierarchical evolution of weblike features: the formation of a filament in an N-body simulation of structure formation in a LCDM Universe. Following the emergence of small-scale filaments, we observe the gradual merging into ever larger entities, culminating in a large massive and dense filament running along the diagonal of the simulation box. Image courtesy of Miguel Aragón-Calvo

preceded by a gradual change of orientation along that of the gradually unfolding large-scale elongated mass concentration running along the diagonal of the box. Finally, all structure ends up in the massive filamentary feature that emerged out of the initially merely faintly visible large-scale overdensity. The figure not only shows the hierarchical character of the process, but also the dominant tidal influence of the large-scale filament which first appears to orient substructures along its main axis before gradually absorbing them. It illustrates the tendency of matter to contract into a sharp filamentary network already defined in the primordial tidal shear field.

The morphology of the emerging filaments strongly depend on the generalized power spectrum slope $n(k)$ at the corresponding mass scale (also see Sect. 4.11). For high values $n \approx -0.5$ – i.e. for subgalactic scales within the ΛCDM scenario – a rather grainy feature will emerge. Many small scale clumps will have fully collapsed and virialized before they get absorbed into the larger contracting filament. In a scenario with $n(k) = -2$, on the other hand, the contracting filament will be collapsing while the small scale objects within its realm may not yet have fully settled. Often these have not yet even fully virialized and may still reside in a stage with a pronounced anisotropic geometry. Such scenarios will produce a *coherent large-scale filaments* in which the internall small-scale structure will not have a pronounced appearance. Most dramatic will be the situation for $n(k) = -3$, the asymptotic situation in which fluctuations over the full range of scale will undergo contraction and collapse at the same time.

The morphology of filaments, as well as sheets, will also be influenced by an additional effect, that of the *diffusion* of relative dynamical timescales between different mass scales. Anisotropic collapse will involve a speeding up of the one-dimensional collapse of an object, and even often a faster collapse along the medium axis as the object contracts into a filament, but a considerably slower formation time in terms of full three-dimensional collapse and virialization. This will bring the formation time of halos closer to that of the embedding elongated filaments. As a result, the latter will appear to be more coherent than a simple hierarchical analysis on the basis of clump formation would imply.

Finally, the morphology of filaments will also be considerably affected by nonlinear effects. The (extended) Press–Schechter type descriptions involve highly idealized local approximations. They discard the nonlinear interactions between the features forming at different scales.

One particular aspect is that of the consequence of alignments between peaks and the surroundings. The primordial alignments get significantly amplified by the subsequent infall of clumps from the surroundings. A few nonlinear effects may be identified. The filaments act like transport channels of the emerging cosmic web: matter and clumps of matter migrate along the axis of filaments towards highly compact clusters at the nodes of the web. The morphology and nature of filaments – strong, dominating, large and coherent or having the appearance of short, weak, and erratic hairlike extensions connected to nearby peaks – will be of decisive influence over aspects like the angular distribution of clumps which fall into a cluster. van Haarlem & van de Weygaert [86] found that clusters appear to orient themselves towards the direction along which the last substantial clumps fell in. The exclusive and continuous infall of clumps along the spine of dominating filament will therefore induce a strong alignment of cluster orientation, its substructure and the surroundings.

4.13 Anisotropic Excursions

Some aspects of the hierarchical assembly of filaments may be understood within the context of the *excursion set* formalism described above (Sect. 3.4). Shen et al. [72] did seek to extend the excursion formalism to filamentary and planar structures by defining critical density thresholds for the collapse of filaments and walls. In this they invoked the *moving barrier* description for nonspherical collapse of ellipsoidal halos that was introduced by Sheth et al. [75] (see (32)).

Their description invokes the homogeneous ellipsoid model to obtain estimates for the collapse times of walls and filaments. In addition to the full three-dimensional ellipsoidal collapse of halos this involves the specification of collapse times and barriers for the one-dimensional collapse of sheets and two-dimensional filaments. Collapse along the shortest axis of an ellipsoid proceeds more rapidly than the equivalent spherical collapse [38, 88]. The corresponding moving barrier for the formation of a sheet does reflect this in involving the lowest density threshold values (see Fig. 27). The threshold would decrease towards smaller masses, implying the rapid formation of low mass sheetlike objects. By contrast the barrier for filament formation would almost be constant over a sizeable range of mass while the barrier for full three-dimensional collapse does reflect the strong influence of tidal influences for small mass halos: with respect to their higher mass peers they form relatively late (see Fig. 27).

$$\begin{aligned} f_{ec,w}(\sigma,z) &\approx f_{sc}(z)\left\{1 - 0.56\left[\frac{\sigma^2(M)}{f_{sc}^2(z)}\right]^{0.55}\right\} \\ f_{ec,f}(\sigma,z) &\approx f_{sc}(z)\left\{1 - 0.012\left[\frac{\sigma^2(M)}{f_{sc}^2(z)}\right]^{0.28}\right\} \approx f_{sc}(z)\,, \\ f_{ec,f}(\sigma,z) &\approx f_{sc}(z)\left\{1 + 0.45\left[\frac{\sigma^2(M)}{f_{sc}^2(z)}\right]^{0.61}\right\} \end{aligned} \tag{81}$$

Although this description may provide a reasonable impression of the hierarchical buildup of the cosmic web, it almost certainly involves a strong

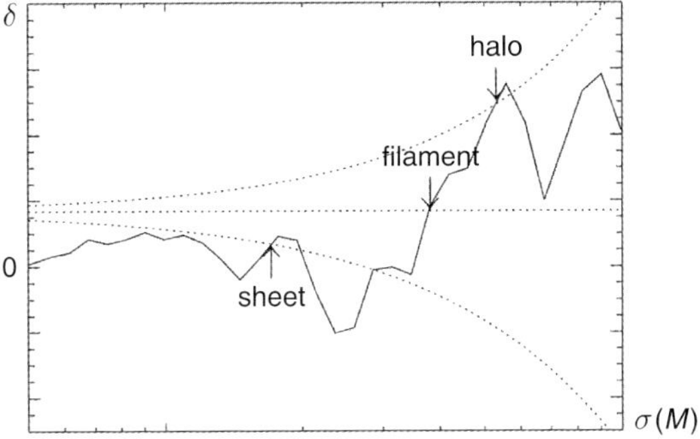

Fig. 27. Example of an excursion random walk (*solid curve*) crossing the barriers (*dotted curves*) associated with sheets, filaments and haloes (bottom to top, see (82)). Plotted is the density perturbation $\delta(M)$ on a mass scale M versus the corresponding $\sigma(M)$ (recall that $\sigma(M)$ is a decreasing function of mass M). The fraction of walks that first cross the lowest (sheet) barrier at $\sigma(M_s)$, then first cross the filament barrier at $\sigma(M_f)$ and finally cross the highest (halo) barrier at $\sigma(M_h)$ represents the mass fraction in halos of mass M_h that are embedded in filaments of mass $M_f > M_h$, which themselves populate sheets of mass $M_s > M_f$. The precise barrier shapes depend on the collapse model. From Shen et al. 2006. Reproduced with permission of AAS

oversimplification. The implicit local description of the excursion set formalism may break down for features whose collapse is thoroughly influenced by the surrounding matter distribution, so strongly emphasized by the *Cosmic Web*. Also the strong nonlinear effects that play a role in the shaping of filamentary features van Haarlem & van de Weygaert [86] may not be sufficiently included in this description. Finally, recent work has shown that a definition of filaments on the basis of density arguments is hazardous: filaments have a considerable range of densities, at least in the present day universe Aragón-Calvo [1], Aragón-Calvo et al. [3], Hahn et al. [31]. An analytical framework that implicitly includes nonlocal effects will offer a better understanding of the hierarchical formation of filaments, bringing us back to the *peakpatch* formalism [17].

4.14 Filaments Versus Walls

In N-body simulations as well as in galaxy redshift distributions it are in particular the filaments which stand out as the most prominent feature of the *Cosmic Web*. It even remains unclear whether *walls* are even present at all. Some argue that once nonlinear clustering sets in the stage in which walls form is of a very short duration or does not occur at all: true collapse would proceed along filamentary structures [11, 36, 70]. Indeed, it may be argued that in the primordial density field overdense regions subject to tidal shear constraints are more filamentary than sheetlike, and become even more so in the quasi-linear regime [18]. There is also a practical problem in identifying them: walls have a considerably lower surface density than filaments. This is exacerbated by the lack of available objective feature detection techniques. Very recent, the analysis of an N-body simulation by means of the new Multiscale Morphology Filter technique did manage to identify walls in abundance [2]. Another indication is that the dissipative gaseous matter within the cosmic web partially aggregates in walls with low overdensities [41]. This argues for the presence of moderate potential wells tied in with dark matter walls.

5 Conclusion: Clusters and the Cosmic Web

In these notes we have reviewed the theoretical framework for the formation of the Cosmic Web in hierarchical scenarios of structure formation. Particular attention was given to the crucial role of clusters within defining the weblike network. They are the main source for the tidal shear field responsible for the spatial outline and dynamical evolution of the prominent filaments and their less pronounced peers, sheetlike membranes.

We wish to conclude our exposé on the connection between the Cosmic Web and the spatial distribution with the quote from Bond & Myers [17] summarizing the essence of what the intrinsic role and identity of clusters is:

> "flowing peak patches at which grand constructive interferences in density and velocity waves mark out the sites of collapse. ... radiating outward from the peak-patch core are filaments and sheets that too are rare. The structure may finally fade into the root-mean-square fluctuations in the medium as coherence in the phases fades into randomness. Or the structure may blend into another peak patch, for rare constructive interferences tend to be clustered."

6 Acknowledgments

We wish to thank Manolis Plionis and Omar López-Cruz for their invitation and the wonderful weeks in Mexico, and for their almost infinite patience regarding our shifting deadlines. RvdW is most grateful to the hospitality of the Canadian Institute for Astrophysics, where we commenced the project leading to these notes. Both authors thank Max-Planck-Institut für Astrophysik in Garching for providing the hospitality and facilities allowing the completion of these lecture notes. In particular we are indebted to Jacqueline van Gorkom, Hans Böhringer and George Rhee. Without their encouragement and the more than helpful assistance and understanding of Sonja Japenga of Springer Verlag we would not have managed to bring these notes to completion. To them we owe a major share of our gratitude ! RvdW is grateful to

Miguel Aragón-Calvo for his permission to use and manipulate various figures from his Ph.D. thesis. He also acknowledges him and Erwin Platen for many inspiring discussions and their contributions towards obtaining insight into the evolution of the Cosmic Web. Most fondly we wish to thank Bernard Jones, for his enthusiastic and crucial support and inspiration, the many original ideas over the years and for his support in completing this manuscript hours past midnight ...

References

1. Aragón-Calvo, M.A.: Morphology and dynamics of the cosmic web. Ph.D. thesis, Groningen University (2007)
2. Aragón-Calvo, M.A., Jones, B.J.T., van de Weygaert, R., van der Hulst, J.M.: Astrophys. J. **655**, L5 (2007)
3. Aragón-Calvo, M.A., Jones, B.J.T., van de Weygaert, R., van der Hulst, J.M.: Mon. Not. R. Astron. Soc., subm. (2007)
4. Audit, E., Teyssier, R., Alimi, J.-M.: Astron. Astrophys. **325**, 439 (1997)
5. Balbus, S.A., Hawley, J.F.: Revs. Mod. Phys. **70**, 1 (1998)
6. Bardeen, J.M., Bond, J.R., Kaiser, N., Szalay, A.S.: Astrophys. J. **304**, 15 (1986)
7. Barkana, R., Loeb, A.: Phys. Rep. **349**, 125 (2001)
8. Basilakos, S., Plionis, M., Yepes, G., Gottlöber, S., Turchaninov, V.: Mon. Not. R. Astron. Soc. **365**, 539 (2006)
9. Bennett, C.L., et al.: Astrophys. J. Suppl. **148**, 1 (2003)
10. Bertschinger, E.: Astrophys. J. **323**, L103 (1987)
11. Bertschinger, E., Jain, B.: Astrophys. J. **431**, 486 (1994)
12. Binggeli, B.: Astron. Astrophys. **107**, 338 (1982)
13. Bond, J.R. Testing Cosmic Fluctuation Spectra. In: Faber, S. (ed.) Nearly Normal Galaxies. Springer, New York, p. 388 (1987)
14. Bond, J.R.: Astrophys. J. preprint (2008)
15. Bond, J.R.: Astrophys. J. preprint (2006)
16. Bond, J.R., Cole, S., Efstathiou, G., Kaiser, N.: Astrophys. J. **379**, 440 (1991)
17. Bond, J.R., Myers, S.T.: Astrophys. J. Suppl. **103**, 1 (1996)
18. Bond, J.R., Kofman, L., Pogosyan, D.Yu.: Nature **380**, 603 (1996)
19. Bond, J.R., Kofman, L., Pogosyan, D.Yu., Wadsley, J.: Theoretical tools for large scale structure. In: Colombi, S., Mellier, Y., (eds.) Wide Field Surveys in Cosmology, 14th IAP meeting. Editions Frontieres, Paris, p. 17 (1998)
20. Bond, J.R., Contaldi, C.R., Kofman, L., Vaudrevange, P.M.: preprint (2008)
21. Bower, R.G.: Mon. Not. R. Astron. Soc. **248**, 332 (1991)
22. Bryan, G.L., Norman, M.L.: Astrophys. J. **495**, 80 (1998)
23. Chandrasekhar, S.: Revs. Mod. Phys. **15**, 1 (1943)
24. Desjacques, V.: Mon. Not. R. Astron. Soc., in press (2008) (`astroph/0707.4670`)
25. Doroshkevich, A.G.: Astrophysics **6**, 320 (1970)
26. Eisenstein, D.J., Loeb, A.: Astrophys. J. **439**, 520 (1995)
27. Eke, V.R., Cole, S., Frenk, C.S.: Mon. Not. R. Astron. Soc. **282**, 263 (1996)
28. Epstein, R.I.: Mon. Not. R. Astron. Soc. **205**, 207 (1983)
29. Freeman, K., Bland-Hawthorn, J.: Ann. Rev. Astron. Astrophys. **40**, 487 (2002)
30. Gunn, J.E., Gott, J.R.: Astrophys. J. **176**, 1 (1972)

31. Hahn, O., Porciani, C., Carollo, M., Dekel, A.: Mon. Not. R. Astron. Soc. **375**, 489 (2007)
32. Hamilton, A.J.S.: Mon. Not. R. Astron. Soc. **322**, 419 (2001)
33. Heath, D.J.: Mon. Not. R. Astron. Soc. **179**, 351 (1977)
34. Helmi, A., White S.D.M.: Mon. Not. R. Astron. Soc. **307**, 495 (1999)
35. Hoffman, Y., Ribak, E.: Astrophys. J. **380**, L5 (1991)
36. Hui, L., Bertschinger, E.: Astrophys. J. **471**, 1 (1996)
37. Icke, V.: Formation of galaxies inside clusters. Ph.D. thesis, Leiden University (1972)
38. Icke, V.: Astron. Astrophys. **27**, 1 (1973)
39. Jones, B.J.T., van de Weygaert, R.: Mon. Not. R. Astron. Soc., to be subm. (2008)
40. Kaiser, N.: Astrophys. J. **284**, 9 (1984)
41. Kang, H., Ryu, D., Cen, R., Song, D.: Astrophys. J. **620**, 21 (2005)
42. Kauffmann, G., White, S.D.M.: Mon. Not. R. Astron. Soc. **261**, 921 (1993)
43. Kitayama, T., Suto, Y.: Astrophys. J. **469**, 480 (1996)
44. Klypin, A., Hoffman, Y., Kravtsov, A., Gottlöber, S.: Astrophys. J. **596**, 19 (2003)
45. Kuo et al.: Astrophys. J. **664**, 687 (2007)
46. Lacey, C., Cole, S.: Mon. Not. R. Astron. Soc. **262**, 627 (1993)
47. Lacey, C., Cole, S.: Mon. Not. R. Astron. Soc. **271**, 676 (1994)
48. Lahav, O., Lilje, P.B., Primack, J.R., Rees, M.: Mon. Not. R. Astron. Soc. **251**, 136 (1991)
49. Lahav, O., Suto, Y.: Living Rev. Relat. **7**, 82 (2004)
50. Lee, J., Evrard, A.E.: Astrophys. J. **657**, 30 (2007)
51. Lee, J., Shandarin, S.F.: Astrophys. J. **500**, 14 (1998)
52. Lee, J., Springel, V., Pen, U-L., Lemson, G.: arXiv0709.1106 (2007)
53. Lin, C.C., Mestel, L., Shu, F.H.: Astrophys. J. **142**, 1431 (1965)
54. Lynden-Bell, D.: Astrophys. J. **139**, 1195 (1964)
55. Mathis, H., White, S.D.M.: Mon. Not. R. Astron. Soc. **337**, 1193 (2002)
56. Monaco, P.: Astrophys. J. **447**, 23 (1995)
57. Monaco, P.: Mon. Not. R. Astron. Soc. **287**, 753 (1997)
58. Monaco, P.: Mon. Not. R. Astron. Soc. **290**, 439 (1997)
59. Neumann, D.M., Lumb, D.H., Pratt, G.W., Briel, U.G.: Astron. Astrophys. **400**, 811 (2003)
60. Park, D., Lee, J.: Astrophys. J. **665**, 96 (2007)
61. Peacock, J.A., Heavens, A.F.: Mon. Not. R. Astron. Soc. **243**, 133 (1990)
62. Peebles, P.J.E.: The Large-Scale Structure of the Universe, Princeton University Press, New York (1980)
63. Percival, W.J.: Astron. Astrophys. **443**, 819 (2005)
64. Platen, E., van de Weygaert, R., Jones, B.J.T.: Mon. Not. R. Astron. Soc., in prep. (2008)
65. Plionis, M., Basilakos, S.: Mon. Not. R. Astron. Soc. **329**, L47 (2002)
66. Pogosyan, D.Yu, Bond, J.R., Kofman, L., Wadsley, J.: Cosmic web: origin and observables. In: Colombi, S., Mellier, Y., (eds.) Wide Field Surveys in Cosmology, 14th IAP meeting. Editions Frontieres, Paris, p. 61 (1998)
67. Press, W.H., Schechter, P.: Astrophys. J. **187**, 425 (1974)
68. Rhee, G., van Haarlem, M., Katgert, P.: Astron. J. **103**, 6 (1991)
69. Romano-Díaz, E., Hoffman, Y., Heller, C., Faltenbacher, A., Jones, D., Shlosman, I.: Astrophys. J. **657**, 56 (2007)

70. Sathyaprakash, B.S., Sahni, V., Shandarin, S.F.: Astrophys. J. **462**, L5 (1996)
71. Schücker, P., Böhringer, H., Reiprich, T.H., Feretti, L.: Astron. Astrophys. **378**, 408 (2001)
72. Shen, J., Abel, T., Mo, H.J., Sheth, R.K.: Astrophys. J. **645**, 783 (2006)
73. Sheth, R.K.: Mon. Not. R. Astron. Soc. **277**, 933 (1995)
74. Sheth, R.K.: Mon. Not. R. Astron. Soc. **300**, 1057 (1998)
75. Sheth, R.K., Mo, H.J., Tormen, G.: Mon. Not. R. Astron. Soc. **323**, 1 (2001)
76. Sheth, R.K., Tormen, G.: Mon. Not. R. Astron. Soc. **329**, 61 (2002)
77. Sheth, R.K., van de Weygaert, R.: Mon. Not. R. Astron. Soc. **350**, 517 (2004)
78. Smoot, G.F., et al.: Astrophys. J. **396**, L1 (1992)
79. Spergel, D.N., Bean, R., Doré, O., Nolta, M.R., Bennett, C.L., Hinshaw, G., Jarosik, N., Komatsu, E., Page, L., Peiris, H.V., Verde, L., Barnes, C., Halpern, M., Hill, R.S., Kogut, A., Limon, M., Meyer, S.S., Odegard, N., Tucker, G.S., Weiland, J.L., Wollack, E., Wright, E.L.: Astrophys. J. Suppl. **170**, 377 (2007)
80. Springel, V., White, S.D.M., Jenkins, A., Frenk, C.S., Yoshida, N., Gao, L., Navarro, J., Thacker, R., Croton, D., Helly, J., Peacock, J.A., Cole, S., Thomas, P., Couchman, H., Evrard, A., Colberg, J.M., Pearce, F.: Nature **435**, 629 (2005)
81. Sugiyama, N.: Astrophys. J. Suppl. **100**, 281 (1995)
82. Trujillo, I.,Carretero, C., Patiri, S.G.: Astrophys. J. **640**, L111 (2006)
83. van de Weygaert, R.: Froth across the universe, dynamics and stochastic geometry of the cosmic foam. In: Plionis, M., Cotsakis, S., (eds.) Modern Theoretical and Observational Cosmology, Proceedings of the 2nd Hellenic Cosmology Meeting. ASSL **276** Kluwer, Dordrecht, pp. 119–272 (2002)
84. van de Weygaert, R., Bertschinger, E.: Mon. Not. R. Astron. Soc. **281**, 84 (1996)
85. van de Weygaert, R., Bond, J.R.: Observations and morphology of the cosmic web. In: Plionis, M., López-Cruz, O., Hughes, D., (eds.) A Pan-Chromatic View of Clusters of Galaxies and the LSS, Springer New York 2008
86. van Haarlem, M., van de Weygaert, R.: Astrophys. J. **418**, 544 (1993)
87. Wadsley, J.W., Bond, J.R.: preprint (1996)
88. White, S.D.M., Silk, J.: Astrophys. J. **231**, 1 (1979)
89. Zaroubi, S., Hoffman, Y., Fisher, K.B., Lahav, O.: Astrophys. J. **449**, 446 (1995)
90. Zel'dovich, Ya.B.: Astron. Astrophys. **5**, 84 (1970)
91. Zentner, A.R.: IJMPD, **16**, 763 (2007)

Observations and Morphology of the Cosmic Web

Rien van de Weygaert[1] and J. R. Bond[2]

[1] Kapteyn Astronomical Institute, University of Groningen, P.O. Box 800, 9700 AV Groningen, The Netherlands
weygaert@astro.rug.nl
[2] Canadian Institute for Theoretical Astrophysics, University of Toronto, TorontoWe describe here the essential properties and elements of the cosmic web as revealed in observations and simulations. ON M5S 1A7, Canada
bond@cita.utoronto.ca

1 Introduction

One of the most striking examples of a physical system displaying a salient geometrical morphology, and the largest in terms of sheer size, is the Universe as a whole. The past few decades have revealed that on scales of a few up to more than a hundred Megaparsec, the galaxies conglomerate into intriguing cellular or weblike patterns that pervade the observable cosmos.

The key structural components of the galaxy and cosmic mass distribution (see Fig. 1),

- Clusters
- Filaments
- Sheets/Walls
- Voids

are not merely randomly and independently scattered features. On the contrary, they have arranged themselves in a seemingly highly organized and structured fashion, the *Cosmic Foam* or *Cosmic Web*. They are woven into an intriguing *foamlike* tapestry that permeates the whole of the explored Universe. The vast under-populated *void* regions in the galaxy distribution represent both contrasting as well as complementary spatial components to the surrounding *planar* and *filamentary* density enhancements. At the intersections of the latter we often find the most prominent density enhancements in our universe, the rich *clusters* of galaxies (see Fig. 1).

In these notes we will delve into the observational and morphological aspects of the Cosmic Web. In the accompanying manuscript (van de Weygaert and Bond, [189]) we have presented the theory behind the emergence of the

R. Van de Weygaert and J. R. Bond: *Observations and Morphology of the Cosmic Web*, Lect. Notes Phys. **740**, 409–467 (2008)
DOI 10.1007/978-1-4020-6941-3_11

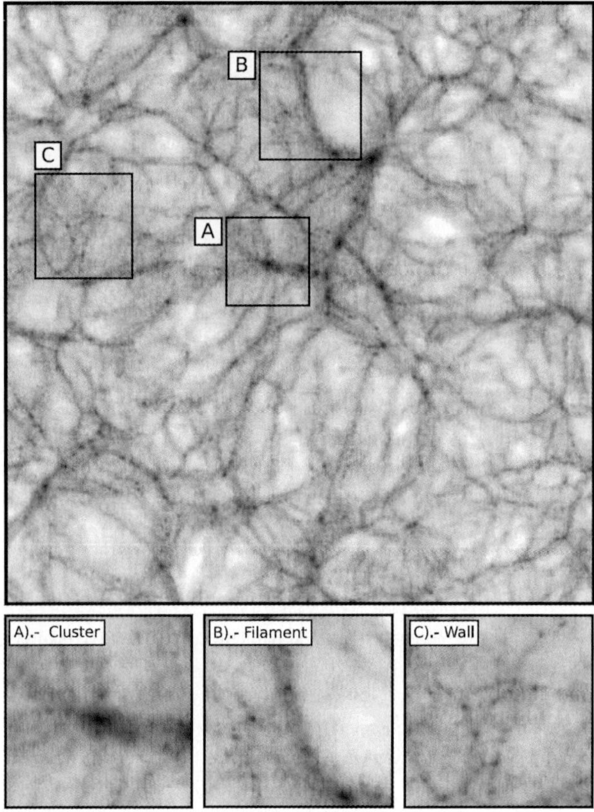

Fig. 1. The Cosmic Web. The image shows the weblike patterns traced by the Dark Matter distribution, at the present epoch, in a Universe based on a ΛCDM scenario. It concerns an N-body simulation in a box of $200\,h^{-1}$ Mpc size. The three boxes indicate examples of the main structure components of the Cosmic Web. Amongst others, the image clarifies the mutual spatial relationship between these elements. Low-density and low contrast walls are less prominent than the outstanding filamentary channels which define the texture of the Cosmic Web. Near the intersection points of filaments and sheets we find high-density cluster nodes. The figures demonstrates the significance of the concept "Cosmic Web". Image courtesy of Miguel Aragón-Calvo, see Aragón-Calvo [5]

Cosmic Web from the pristine near-uniform Universe. The theoretical framework of *Cosmic Web* has to be confronted with the information obtained from a variety of sources. On the observational side the cosmic web has first been seen in redshift maps of the spatial galaxy distribution. The recent success in mapping the spatial weblike dark matter distribution by means of weak lensing observations forms a breakthrough for our understanding of the large-scale dynamics. Equally important sources of information concern the Lyα forest and the WHIM, the imprint of the gaseous material that fell into the weblike structures defined by the dark matter distribution. Two aspects of the large scale universe do play a special role in our study. Clusters of galaxies are the prime objects in defining the pattern and structure of the Cosmic Web. On the other hand we have the large voids as important structural and dynamical components. They are of prime significance for the morphology of the large scale Universe.

2 The Emergence of the Cosmic Web

Towards the end of the seventies a set of new observations did start to unveil the existence of coherent structures larger than that of clusters of galaxies. With the review of [124] the supercluster paradigm established itself as the new view of the large scale distribution of matter and galaxies in the Universe. It had gradually emerged as a result of various early galaxy redshift surveys of nearby regions in the Universe (e.g. Chincarini and Rood [33], Gregory and Thompson [73], Einssto, Joeveer and Saar [55]) and put on a firm fotting with the completion of the first systematic and large redshift survye, the CfA1 survey [43]. Along with these efforts came the unexpected finding of the first example of large cosmic voids, the Bootes void [96].

2.1 Galaxies and the Cosmic Web

It was the celebrated map of the first CfA redshift slice [45] that showed the connection between the basic elements of the Cosmic Web that was going to emerge in the more complete picture. While it provided an initial hint of the existence of the Cosmic Web it was so thin that it was not immediately clear what its true nature was, whether it were bubbles, pancakes, or something else. In recent years this view has been expanded dramatically to the present grand vistas offered by the 100,000s of galaxies in the 2dF – two-degree field – Galaxy Redshift Survey, the 2dFGRS (e.g. Colless et al. [39]), and SDSS [180] galaxy redshift surveys.[1] These and many other redshift surveys have unequivocally established that galaxies are located in dense, compact clusters, in less dense filaments, and in sheetlike walls surrounding vast, almost empty regions called voids, the structural components of the Cosmic Web.

The first impressions of a weblike galaxy distribution seen in the shallow CfA2 redshift slices got firmly established as a universal cosmic phenomenon through the publication of the results of the Las Campanas redshift survey (LCRS [166]). Its chart of 26,000 galaxy locations in six thin strips on the sky, extending out to a redshift of $z \sim 0.1$, did provide the first impression of structure in a truely cosmologically representative volume of space. The Las Campanas redshift survey confirmed the ubiquity and reality of weblike patterns over vast reaches of our Universe. Also important was that it did not show any strong evidence of inhomogeneities surpassing sizes of $100 - 200\,\mathrm{h}^{-1}$Mpc.

[1] See http://www.mso.anu.edu.au/2dFGRS/ and http://www.sdss.org/

This is most dramatically illustrated by the map 2dFGRS and SDSS maps. The published maps of the distribution of nearly 250,000 galaxies in two narrow "slice" regions on the sky yielded by the 2dFGRS surveys reveal a far from homogeneous distribution. Instead, we recognize a sponge-like arrangement, with galaxies aggregating in striking geometric patterns such as prominent filaments, vaguely detectable walls and dense compact clusters on the periphery of giant voids.[2] The three-dimensional view emerging from the SDSS redshift survey provides an even more convincing image of the intricate patterns defined by the cosmic web (Fig. 2). A careful assessment of the galaxy distribution in our immediate vicintiy reveals us how we ourselves are embedded and surrounded by beautifully delineated and surprisingly sharply defined weblike structures. In particular the all-sky nearby infrared 2MASS survey (see Fig. 3) provides us with a meticulously clear view of the web surrounding us.

The cosmic web is outlined by galaxies populating huge *filamentary* and *wall-like* structures, the sizes of the most conspicuous one frequently exceeding $100\,h^{-1}$ Mpc. The closest and best studied of these massive anisotropic matter concentrations can be identified with known supercluster complexes, enormous structures comprising one or more rich clusters of galaxies and a plethora of more modestly sized clumps of galaxies. A prominent and representative nearby specimen is the Perseus-Pisces supercluster, a $5\,h^{-1}$ wide ridge of at least $50\,h^{-1}$ Mpc length, possibly extending out to a total length of $140\,h^{-1}$ Mpc. While such giant elongated structures are amongst the most conspicuous features of the Megaparsec matter distribution, filamentary features are encountered over a range of scales and seem to represent a ubiquitous and universal state of concentration of matter. In addition to the presence of such filaments the galaxy distribution also contains vast planar assemblies. A striking local example is the *Great Wall*, a huge planar concentration of galaxies with dimensions that are estimated to be of the order of $60\,h^{-1} \times 170\,h^{-1} \times 5\,h^{-1}$ Mpc [66]. In both the SDSS and 2dF surveys even more impressive planar complexes were recognized, with dimensions substantially in excess of those of the local Great Wall. At the moment, the socalled *SDSS Great Wall* appears to be the largest known structure in the Universe (see Fig. 4).

2.2 Cosmic Nodes: Clusters

Within and around these anisotropic features we find a variety of density condensations, ranging from modest groups of a few galaxies up to massive compact *galaxy clusters* (see eg. Fig. 5). The latter stand out as the most massive, and most recently, fully collapsed and virialized objects in the Universe.

[2] It is important to realize that the interpretation of the Megaparsec galaxy distribution is based upon the tacit yet common assumption that it forms a a fair reflection of the underlying matter distribution. While there are various indications that this is indeed a reasonable approximation, as long as the intricate and complex process of the formation of galaxies has not been properly understood this should be considered as a plausible yet heuristic working hypothesis.

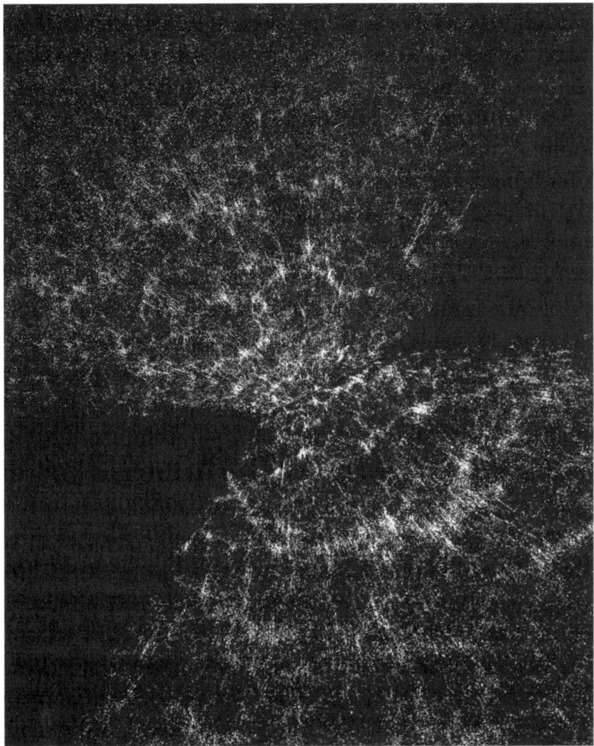

Fig. 2. SDSS is the largest and most systematic sky survey in the history of astronomy. It is a combination of a sky survey in 5 optical bands of 25% of the celestial (northern) sphere. Each image is recorded on CCDs in these 5 bands. On the basis of the images/colours and their brightness a million galaxies are subsequently selected for spectroscopic follow-up. The total sky area covered by SDSS is 8452 square degrees. Objects will be recorded to $m_{\text{lim}} = 23.1$. In total the resulting atlas will contain 10^8 stars, 10^8 galaxies and 10^5 quasars. Spectra are taken of around 10^6 galaxies, 10^5 quasars and 10^5 unusual stars (in our Galaxy). Of the 5 public data releases 4 have been accomplished, ie. 6670 square degrees of images is publicly available, along with 806,400 spectra. In total, the sky survey is now completely done (107%), the spectroscopic survey for 68%. This image is taken from a movie made by Subbarao, Surendran and Landsberg (see website: http://astro.uchicago.edu/cosmus/projects/sloangalaxies/). It depicts the resulting redshift distribution after the 3rd public data release. It concerns 5282 square degrees and contained 528,640 spectra, of which 374,767 galaxies

Approximately 4% of the mass in the Universe is assembled in rich clusters. They may be regarded as a particular population of cosmic structure beacons as they typically concentrate near the interstices of the cosmic web, *nodes* forming a recognizable tracer of the cosmic matter distribution [23]. Clusters not only function as wonderful tracers of structure over scales of dozens up to

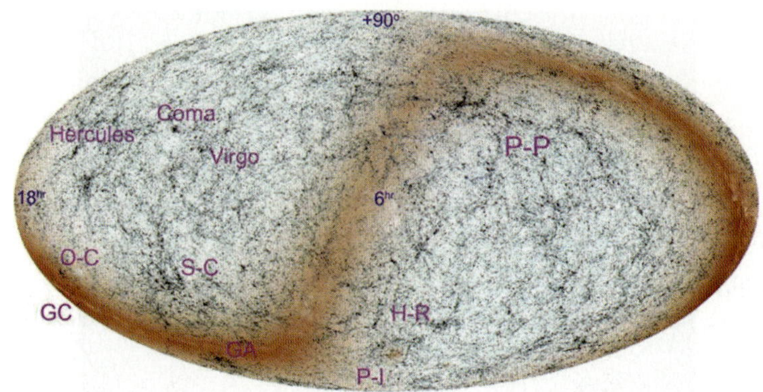

Fig. 3. Equatorial view of the 2MASS galaxy catalog (6h RA at centre). The greyscale represents the total integrated flux along the line of sight – the nearest (and therefore brightest) galaxies produce a vivid contrast between the Local Supercluster (**centre-left**) and the more distant cosmic web. The dark band of the Milky Way clearly demonstrates where the galaxy catalog becomes incomplete due to source confusion. Some well known large-scale structures are indicated: P-P=Perseus-Pisces supercluster; H-R=Horologium-Reticulum supercluster; P-I=Pavo-Indus supercluster; GA='Great Attractor'; GC=Galactic Centre; S-C=Shapley Concentration; O-C=Ophiuchus Cluster; Virgo, Coma, and Hercules=Virgo, Coma and Hercules superclusters. The Galactic 'anti-centre' is front and centre, with the Orion and Taurus Giant Molecular Clouds forming the dark circular band near the centre. Image courtesy of J.H. Jarrett. Reproduced with permission from the Publications of the Astronomical Society of Australia 21(4): 396–403 (T.H. Jarrett). Copyright Astronomical Society of Australia 2004. Published by CSIRO PUBLISHING, Melbourne Australia

hundred of Megaparsec but also as useful probes for precision cosmology on the basis of their unique physical properties.

The richest clusters contain many thousands of galaxies within a relatively small volume of only a few Megaparsec size. For instance, in the nearby Virgo and Coma clusters more than a thousand galaxies have been identified within a radius of a mere $1.5\,h^{-1}$ Mpc around their core (see Fig. 6). Clusters are first and foremost dense concentrations of dark matter, representing overdensities $\Delta \sim 1000$. In a sense galaxies and stars only form a minor constituent of clusters. The cluster galaxies are trapped and embedded in the deep gravitational wells of the dark matter. These are identified as a major source of X-ray emission, emerging from the diffuse extremely hot gas trapped in them. While it fell into the potential well, the gas got shock-heated to temperatures in excess of $T > 10^7$ K, which results in intense X-ray emission due to the bremsstrahlung radiated by the electrons in the highly ionized intracluster gas. In a sense clusters may be seen as hot balls of X-ray radiating gas. The amount of intracluster gas in the cluster is comparable to that locked into

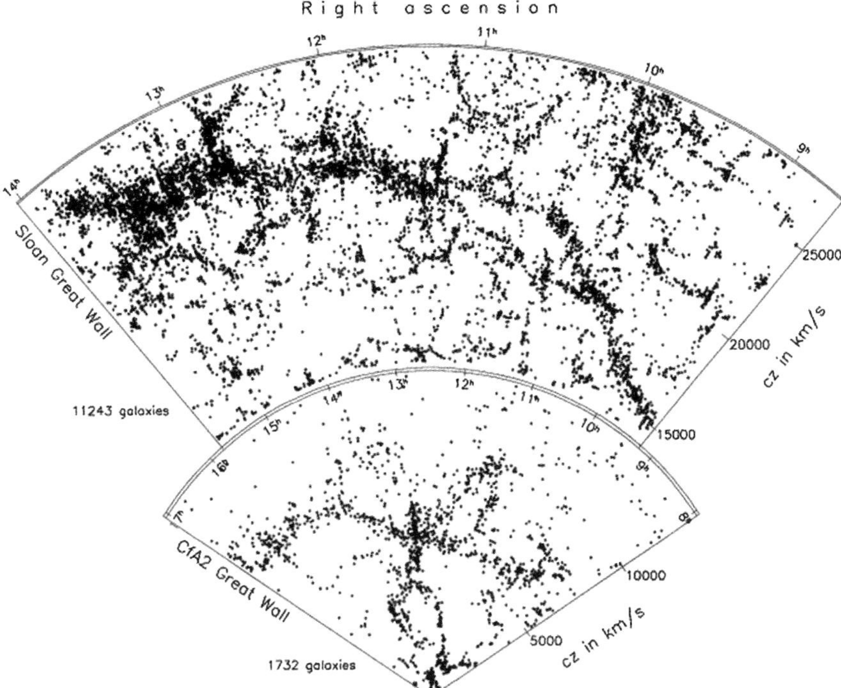

Fig. 4. The CfA Great Wall (bottom slice, Geller and Huchra [66]) compared with the Sloan Great Wall (**top slice**). Both structures represent the largest coherent structural in the galaxy redshift surveys in which they were detected, the CfA redshift survey and the SDSS redshift survey. The (CfA) Great Wall is a huge planar concentration of galaxies with dimensions that are estimated to be of the order of $60\,h^{-1} \times 170\,h^{-1} \times 5\,h^{-1}$ Mpc. Truely mindboggling is the Sloan Great Wall, a huge conglomerate of clusters and galaxies. With a size in the order of 400 h^{-1} Mpc it is at least three times larger than the CfA Great Wall. It remains to be seen whether it is a genuine physical structure or mainly a stochastic arrangement and enhancement, at a distance coinciding with the survey's maximum in the radial selection function. Image courtesy of M. Jurić, see also Gott et al. [70]. Reproduced by permission of the AAS

stars, and stands for $\Omega_{\rm ICM} \sim 0.0018$ [63]. The X-ray emission represents a particularly useful signature, an objective and clean measure of the potential well depth, directly related to the total mass of the cluster (see e.g. Reiprich and Böhringer, [146]). Through their X-ray brightness they can be seen out to large cosmic depths. The deep gravitational dark matter wells also strongly affects the path of passing photons. While the resulting strong lensing arcs form a spectacular manifestation, it has been the more moderate distortion of background galaxy images in the weak lensing regime [88, 89] which has opened up a new window onto the Universe. The latter has provided a direct

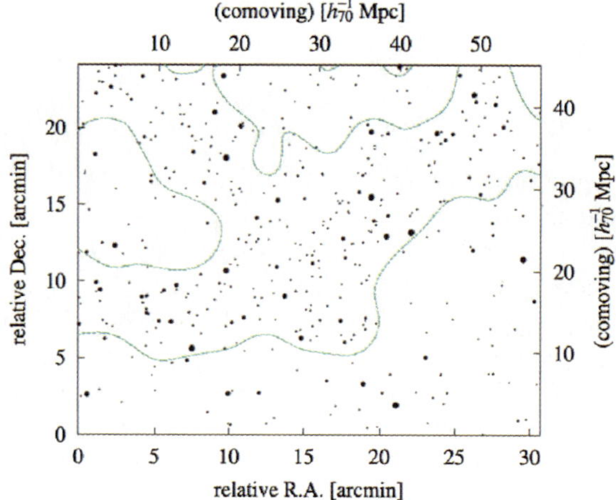

Fig. 5. The cosmic web at high redshifts: a prominent weblike features at a redshift $z \sim 3.1$ found in a deep view obtained by the Subaru telescope. Large scale sky distribution of 283 strong Lyα emitters (*black filled circles*), the Lyα absorbers (*red filled circles*) and the extended Lyα emitters (*blue open squares*). The dashed lines indicate the high-density region of the strong Lyα emitters. From Hayashino et al. 2004. Reproduced by permission of the AAS

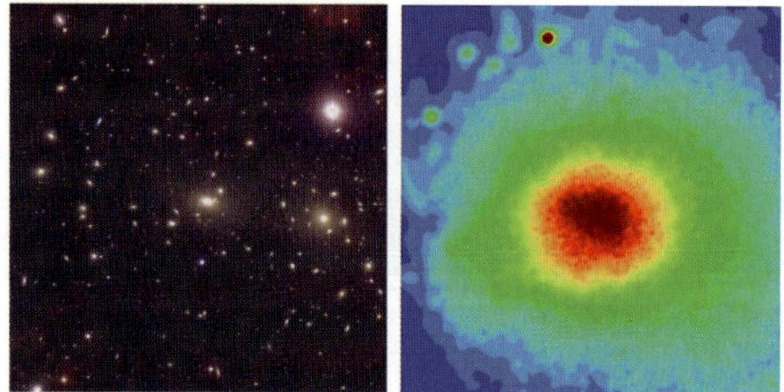

Fig. 6. Comparison of optical and X-ray images of the Coma cluster, A1656. The cluster is at a distance of $\approx 70\,\mathrm{h}^{-1}$ Mpc. **Left:** optical image of the galaxies in the centre of the Coma cluster. The Coma cluster contains more than 1000 galaxies within a central region of $\approx 1.5\,\mathrm{h}^{-1}$ Mpc, mostly elliptical and SO galaxies. Clearly visible are the two dominant giant elliptical galaxies, NGC4878 and NGC4889. The colour image was created from 3 separate exposures taken in blue, red and near-infrared, with the KPNO 0.9 m telescope (courtesy of Omar López-Cruz). **Right:** ROSAT X-ray image at 0.5–2.0 keV.of the central region of the Coma cluster (courtesy: S.L. Snowden, NASA/GSFC). The image is $\approx 1^\circ \times 1^\circ$, corresponding to a size of $1.2\,\mathrm{h}^{-1}$ Mpc at the cluster's redshift $z = 0.0232$

probe of the dark matter content of clusters and the large scale universe (for a review see e.g. Mellier [118], Refregier [144]) (also see Sect. 2.4).

Cluster Catalogs

The Abell catalogue of optically identified galaxy clusters [2, 3] has fulfilled a central role for the study of clusters and their large scale matter distribution on scales of several tens of Megaparsec (see Bahcall [9]). With the arrival of large new galaxy redshift surveys deep and objectively identified cluster samples have opened a plethora of elaborate, detailed and systematic studies of the cluster population. Cluster samples extracted from the SDSS survey [15, 119, 150] will continue to play a large role. New and objective cluster detection techniques have improved the range and completeness of the cluster samples while minimizing projection effects [67, 95, 119]. Projection effects may evoke false detections and contaminate studies of the cluster large scale distribution. Amongst the most promising methods for optical or NIR cluster detection is that of *red-sequence detection* [67], in which clusters are simultaneously detected as overdensities in projected angular position, colour and magnitude. It uses the observational fact that the bulk of the early-type galaxies in rich clusters lie along a linear and narrow colour-magnitude relation [109, 196]. The Red-Sequence Cluster Survey (RCS) seeks to exploit this observation to compose a large catalog of clusters. Extrapolating cluster detection towards the NIR, Kochanek et al. [100] assembled a cluster catalog from the 2MASS galaxy sample. Other cluster samples are selected through their X-ray emission, believed to represent a more robust manner for selecting mass-limited samples. Particularly noteworthy is the ROSAT-ESO Flux Limited X-ray catalog (REFLEX [21]), which contains all clusters brighter than an X-ray flux of 3×10^{-12} ergs^{-1} cm^{-2} over a large part of the southern sky. In addition there is the RASS X-ray selected SDSS cluster sample [141], combining both optical and X-ray selection criteria. Recently, within the context of the Deep Lens Survey, Wittman et al. [195] presented the first cluster sample on the basis of their weak gravitational lensing signature. Perhaps potentially most promising is the use of the Sunyaev-Zel'dovich effect, the small CMB spectral distortion caused by the scattering of the CMB photons off the high-energy intracluster electrons [175, 176]. Carlstrom et al. [28] and Vale and White [185] proposed the construction of cluster catalogs using the SZ effect. While the Planck satellite mission will certainly be a major step forward in the detection of SZ clusters, optimism has been slightly tempered by the recent result of Lieu et al. [104]. Within the WMAP observations centered on 31 clusters they found a CDM decrement which was at least a factor 4 smaller than expected.

Cluster Clustering

Through their high visibility clusters can be traced out to vast distances in the Universe. Following the basic assumption that they are a fair and direct, be it sparse, tracer of the underlying matter distribution clusters are ideally suited for probing the spatial matter distribution over large regions of space. Maps of their distribution contain information on spatial clustering on scales of up to hundreds of Megaparsec. A large range of observational studies, mostly based on optically or X-ray selected samples, display a substantial level of clumping of clusters on scales where clustering in the galaxy distribution has diminished below detectability levels. A wide range of observational studies on the basis of such optically selected samples have shown that the clustering of clusters is significantly more pronounced than that of galaxies. Their two-point correlation function has a shape similar to that of galaxies, but with a substantially higher amplitude and detectable out to distances of at least $\sim 50\ h^{-1}$ Mpc.

A good impression of the spatial distribution of rich clusters may be obtained from Fig. 7 (from Borgani and Guzzo, [23]). It shows the spatial distribution of the clusters in the REFLEX galaxy cluster catalogue [21]. Maps such as these confirm that clusters are highly clustered [9, 23]. They aggregate to form huge supercluster complexes, coinciding with the filaments, walls and related features in the galaxy distribution. These superclusters are

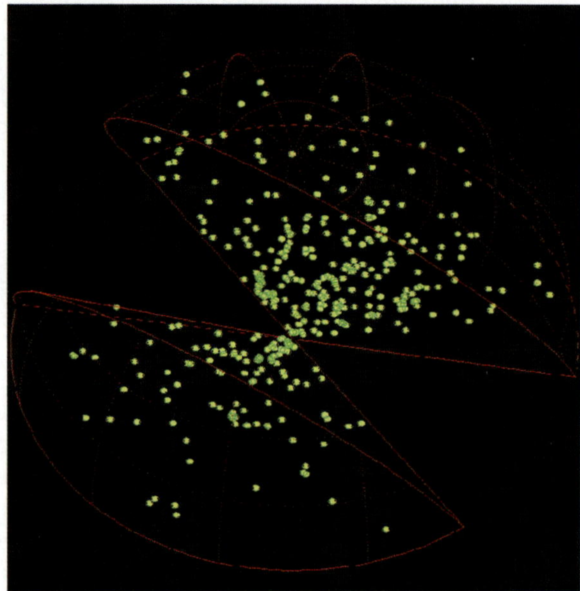

Fig. 7. The spatial cluster distribution. The full volume of the X-ray REFLEX cluster survey within a distance of $600\,h^{-1}$Mpc. The REFLEX galaxy cluster catalogue [21], contains all clusters brighter than an X-ray flux of $3 \times 10^{-12}\,\mathrm{ergs}^{-1}\mathrm{cm}^{-2}$ over a large part of the southern sky. The missing part of the hemisphere delineates the region highly obscured by the Galaxy. Courtesy: Borgani and Guzzo [23]. Reproduced by permission of Nature

moderate density enhancements on scale of tens of Megaparsec, typically in the order of a few times the average density. Either they are still co-expanding with the Hubble flow, be it at a slightly decelerated rate, or they just started contracting. Within these structures clusters reside at the dense intersections of filaments, along which mass drains into the massive clusters [190].

Cluster Dipole

Clusters may also provide a better and more extensive view of the contributions to the local gravitational force field by comparing the inferred Local Group motion to the CMB dipole. Scaramella et al. [159] and Plionis and Valdarnini [137] sought to establish by means of the cluster distribution within a distance of $r \approx 300\ \mathrm{h}^{-1}$Mpc whether the origin of our cosmic motion should be located within this volume, or whether there are indications for even larger cosmic structures. Interestingly, the results of Plionis & Kolokotronis [138] and Kocevski and Ebeling [99] appears to suggest that X-ray selected clusters in the nearby Universe indicate a significantly larger dynamical influence of structures over scales of 150 h^{-1} Mpc than previously indicated by similar dipole studies on the basis of the IRAS Point Source Catalog Redshift survey (PSCz, see e.g. Branchini et al. [26]) and the dipole anisotropy of the 2MASS Redshift survey [59]. The latter find that mass structures beyond a distance of 140 h^{-1} Mpc only induce a negligible acceleration on the Local Group. Using the combined X-ray REFLEX, eBCS [54] and CIZA samples, [99] came to the conclusion that only 44% of the local motion is due to infall into the Great Attractor region while 56% is induced by more distant mass concentrations between 130 h^{-1} Mpc and 180 h^{-1} Mpc away. The Shapley supercluster, one of the largest concentrations of clusters out to $z = 0.12$, is responsible for at least 30% of the acceleration induced by structures beyond 130 h^{-1} Mpc. Also the Horologium-Reticulum supercluster is found to have a substantial impact. The schematic dipole profile (Fig. 8) indeed provides an enticing insight into the implied local cosmic dynamics. Also interesting is the presence of a significant underdensity in the cluster distribution on the nother hemisphere, at a distance $\sim 150\ \mathrm{h}^{-1}$ Mpc.

Cluster Bias

The results on the strong clustering of clusters motivated theoretical arguments for the idea of them forming a biased tracer of the matter distribution. The first simple linear biasing prescriptions were justified by the idea that clusters form from high-density peaks in the primordial density field, filtered over an appropriately large scale [10, 87]. Biasing prescriptions may incorporate or quantify an array of complex and usually ununderstood *"gastrophysical"* processes [49]. However, to understand the influence on clustering it may suffice to derive a heuristic bias factor of function. The value of a simple (linear) *bias*

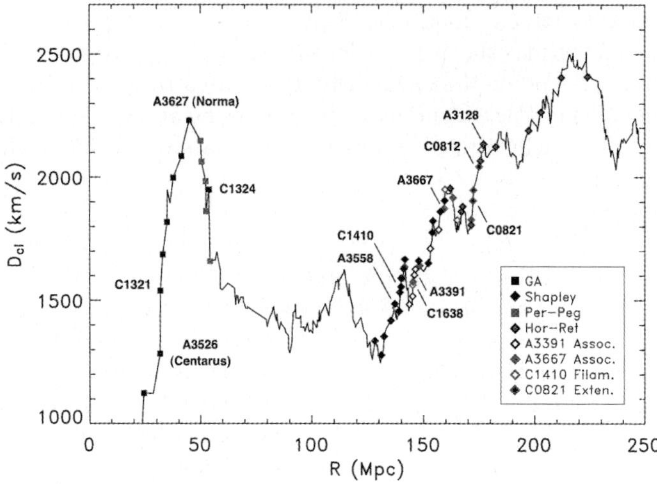

Fig. 8. Schematic X-ray cluster dipole profile. Clusters associations are grouped by symbol shading to highlight their impact on the overall dipole amplitude. Abell and CIZA clusters begin with letters "A" and "C". Acronyms are: GA/Great Attractor, Hor-Ret/Horologium-Reticulum, Per-Peg/Perseus-Pegasus. Image courtesy: Kocevski and Ebeling [99]. Reproduced by permission of AAS

factor would be a function of cluster mass, structure formation scenario and cosmic epoch. Following up on the original *peak bias* idea [10, 87], an array of more sophisticated theoretical bias model have been proposed. Seeking to describe and analyze the bias of different species of galaxies as well as of clusters, these modifications elaborated upon this idea and increased the realism of the approximation [13, 48, 115, 120, 179].

2.3 Cosmic Depressions: the Voids

Complementing this cosmic inventory leads to the existence of large *voids*, enormous regions with sizes in the range of 20–50 h^{-1} Mpc that are practically devoid of any galaxy, usually roundish in shape and occupying the major share of space in the Universe. Forming an essential ingredient of the *Cosmic Web*, they are surrounded by elongated filaments, sheetlike walls and dense compact clusters.

Voids have been known as a feature of galaxy surveys since the first surveys were compiled [33, 55, 73]. Following the discovery by Kirshner et al. [96, 97] of the most dramatic specimen, the Boötes void, a hint of their central position within a weblike arrangement came with the first CfA redshift slice [45]. This view has been dramatically endorsed and expanded by the redshift maps of the 2dFGRS and SDSS surveys [1, 39]. They have established voids as an integral component of the Cosmic Web. The 2dFGRS maps and SDSS maps (see e.g. Figs. 2 and 10), and the void map of the 6dF survey in Fig. 9, are

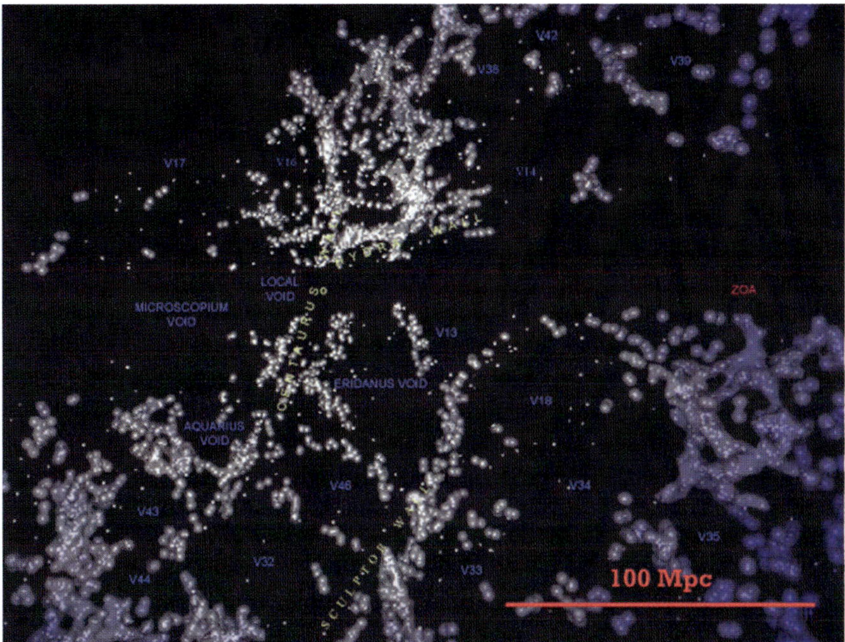

Fig. 9. A region of the 6dF redshift survey marked by the presence of various major voids. The image concerns a 3D rendering of the galaxy distribution in a 1000 km/s thick slice along the supergalactic SGX direction, at SGX = −2500 km/s. Image courtesy of A. Fairall

telling illustrations of the ubiquity and prominence of voids in the cosmic galaxy distribution.

For the most systematic and complete impression of the cosmic void population the Local Universe provides the most accessible region. Recently, the deep view of the 2dFGRS and SDSS probes (see e.g. Fig. 12) has been supplemented with high-resolution studies of voids in the nearby Universe. Based upon the 6dF survey [77], Fairall (private communication) identified nearly all voids within the surveyed region out to 35,000 km s^{-1}. It is the 2MASS redshift survey [83] – the densest all-sky redshift survey available – which has provided a uniquely detailed census of large scale structures in our Local Universe [59]. Partially including 6dF redshifts, the 2MASS redshift survey entails a complete and systematic survey of structure in the nearby Universe up to 14,000–16,000 km s^{-1}. This includes a complete sample of voids, directly identifiable from the density and velocity field reconstruction by Erdoğdu et al. [59] does contain a nice complete sample of voids in our Local Universe, although though some measure of bias and upper-limit to the size of identifiable voids is introduced via the substantial level of spatial smoothing going along with the Wiener filter processing. A nice impression of the typical structure, geometry and size of voids is given by shell section through the local Cosmic Web seen in the Aitoff sky projection in Fig. 11.

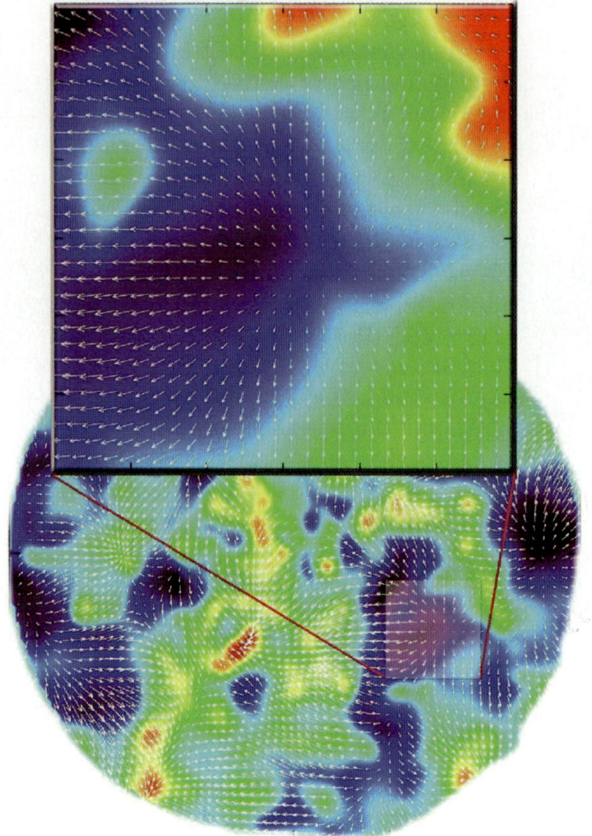

Fig. 10. Gravitational impact of the Sculptor Void. The righthand frame shows the inferred velocity field in and around the Sculptor void near the Local Supercluster. The colour map represents the density values, with dark blue at $\delta \sim -0.75$ and cyan near $\delta \sim 0.0$. The vectors show the implied velocity flow around the void, with a distinct nearly spherically symmetric outflow. It is a zoom-in onto the indicated region in the density and velocity map in the Local Universe (lefthand) determined on the basis of the PSCz galaxy redshift survey. The peculiar velocities of the galaxies in the PSCz galaxy redshift catalogue were determined by means of a linearization procedure [25], the resulting galaxy positions and velocities have been translated by DTFE into the depicted density and velocity flow maps. The Local Group is at the centre of the map of our Local Universe (lefthand). To the left we see the Great Attractor region extending out towards the Shapley supercluster. To the righthand side we can find the Pisces-Perseus supercluster. The density values range from ~ 4.9 (red) down to ~ -0.75 (darkblue), with cyan coloured regions having a density near the global cosmic average ($\delta \sim 0$). The velocity vectors are scaled such that a vector with a length of $\approx 1/33$rd of the region's diameter corresponds to 650 km/s. The density and velocity field have an effective Gaussian smoothing radius of $R_G \sim \sqrt{5}$ h^{-1} Mpc. The top righthand insert zooms in on the Local Supercluster and Great Attractor complex. From: Romano-Díaz and van de Weygaert [149]

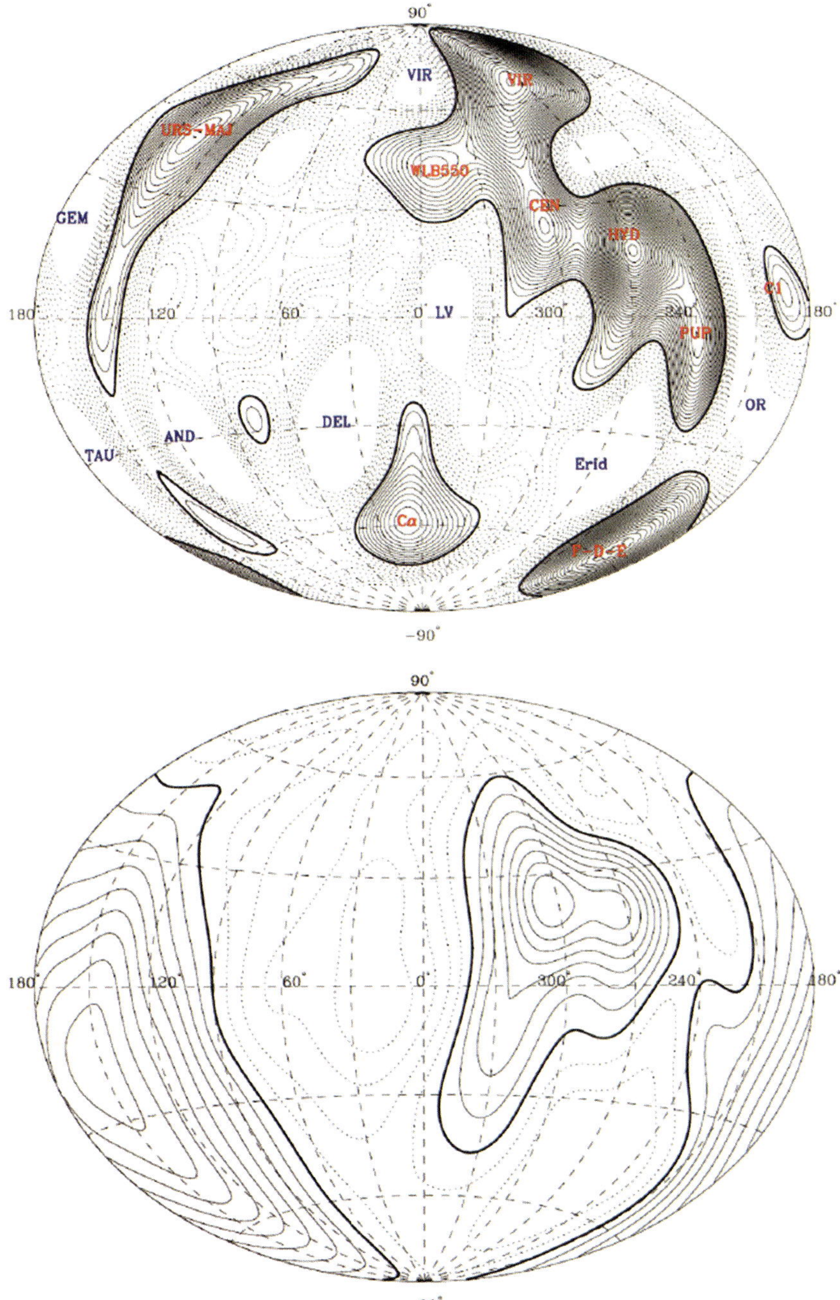

Fig. 11. (continued)

Void Sizes

Voids in the galaxy distribution account for about 95% of the total volume (see Kauffmann and Fairall [93], El-Ad, Piran and da Casta [56], El-Ad and Piran [57], Hoyle and Vogeley [81], Plionis and Basilakos [139], Rojas et al. [148], Platen, van de Weygaert and Jones [137]).

The typical sizes of voids in the galaxy distribution depend on the galaxy population used to define the voids. Voids defined by galaxies brighter than a typical L_* galaxy tend to have diameters of order 10–20 h^{-1}Mpc, but voids associated with rare luminous galaxies can be considerably larger; diameters in the range of 20–50 h^{-1} Mpc are not uncommon (e.g Hoyle and Vogeley [81], Plionis and Basilakos [139]). These large sizes mean that only now we are beginning to probe a sufficiently large cosmological volume to allow meaningful statistics with voids to be done. Firm upper limits on the maximum void size have not yet been set. Recently there have been claims of the existence of a supersized void, in the counts of the NVVS catalogue of radio sourcs, and of its possible imprint on the CMB via the ISW effect in the form of a 'cold spot'. If this will be confirmed it will pose an interesting challenge to any cosmological scenario (see Rudnick et al. [151]).

At the low end side of the void size distribution a very detailed survey of the Local Volume, the very nearby Universe in and immediately around our Local Supercluster, does provide some tentative information. At this close range a few studies claim to have found what may be the smallest genuine voids in existence. In his Catalog and Atlas of Nearby Galaxies Tully [183] noted the presence of the *Local Void* in the Local Supercluster. The Local Void begins directly from the boundaries of the Local Group and extends in the direction of the north pole of the LSC by $\sim$ 14 h^{-1} Mpc. Similar and even smaller minivoids have recently been found by the analysis of Tikhonov and Karachentsev [181] of the galaxy distribution in the Catalog of

Fig. 11. (continued) 2MASS view of the Local Void outflow. The reconstructed density (**top frame**) and velocity field (**bottom frame**) of the 2MASS redshift survey, evaluated on a thin shell of 2000 km s^{-1}, shown in Aitoff projection. From Erdoğdu et al. [59]. Top: the reconstructed density field in the thin shell, providing a telling section through the Local Cosmic Web. Dashed lines show $\delta < 0$, solid lines $\delta \geq 0$, with contour spacing of $\Delta\delta = 0.1$. Easily identifiable overdensities are Ura Major, the Virgo cluster, the Centaurus cluster, Hydra cluster and the Fornax-Doradus-Eridanus (F-D-E) supercluster complex. Most interestingly are the locations of local voids: Gemini (Gem), Taurus (Tau), Andromeda (And), Delphinus (Del), Virgo (Vir), Eridanus (Erid), Orion (Ori), and the Local Void (LV). Bottom: Dashed lines show infall velocities, solid lines outflow. First solid line is for $v_{\rm rad} =$ 0 km s^{-1}, and contour spacing is $|\Delta v_{\rm rad}| = 50$ km s^{-1}. Clearly visible is the strong outflow from the Local Void, reflected in the strong central patch. From: Erdoğdu et al. [59]

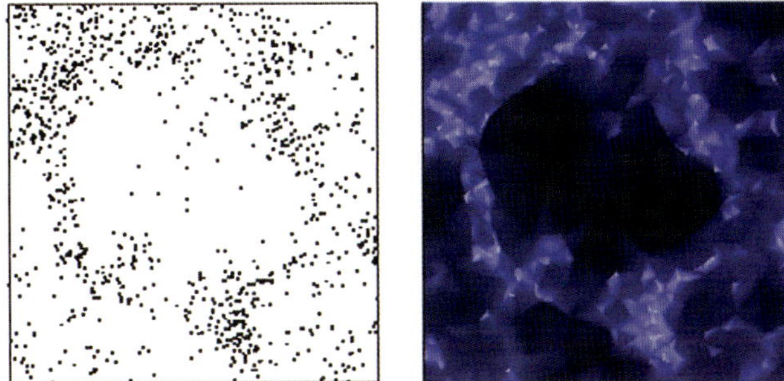

Fig. 12. Void region in de 2dFGRS survey. From: Schaap [160]

Neighbouring Galaxies [92]. Because the latter entails a meticulously detailed view of the true spatial distribution of galaxies out to 5 h^{-1} Mpc, it allowed the identification and mapping of minivoids in the Local Volume. Tikhonov and Karachentsev [181] and Tikhonov and Klypin [182] claim to have found a total of some 30 minivoids, completely free of galaxies, with sizes of 0.7–3.5 h^{-1} Mpc.

The Meaning of Voids

There are a variety of reasons why the study of voids is interesting for our understanding of the cosmos.

- Firstly, because they are a prominent aspect of the Megaparsec Universe it is necessary to understand the structure of evolution of voids in order to get a proper and full understanding of the formation and dynamics of the Cosmic Web.
- Secondly, voids may contain a considerable amount of information on the underlying cosmological scenario and on global cosmological parameters.
- Thirdly, their pristine low-density environment implies them to be interesting regions for studying the influence of cosmic environment on the formation of galaxies.

We will address the last two aspects in more detail below, along with a discussion of the available observational information on the dynamics of voids. A more focussed discussion of void evolution and dynamics within the context of the Cosmic Web is the subject of Sect. 4.

Void Dynamics

The essential role of voids in the organization of the cosmic matter distribution was recognized soon after their discovery [84]. This also includes their dynamical influence. As a result of their underdensity voids represent a region of weaker gravity, resulting in an effective repulsive peculiar gravitational influence. Various studies have indeed found strong indicatations for their imprint in the peculiar velocity flows of galaxies in the Local Universe.

Bothun et al. [24] made the first claim of seeing pushing influence of voids when assessing the stronger velocity flows of galaxies along a filament in the first CfA slice. Stronger evidence came from the extensive and systematic POTENT analysis of Mark III peculiar galaxy velocities [194] in the Local Universe [19, 46]. POTENT found that for a fully selfconsistent reconstruction of the dynamics in the Local Universe, it was inescapable to include the dynamical influence of voids (see e.g. Dekel [47]). The DTFE maps by Romano-Díaz and van de Weygaert [149] of the density and velocity field in the Local Universe obtained from the PSCz redshift sample [25] do provide a very clear visual image of the influence of such voids in the Local Universe, with the pushing influence of the Sculptor void at the Local Supercluster as most outstanding example (see Fig. 10).

With the arrival of new and considerably improved data samples the dynamical influence of voids in the Local Universe has been investigated and understood in greater detail. The reconstruction of the density and velocity field in our local cosmos on the basis of the 2MASS redshift survey has indeed resulted in a very interesting and complete view of the dynamics on Megaparsec scales. As one may infer from Fig. 11 the repulsive influence of the Local Void is impressively strong and outstanding. This conclusion goes along with the conclusions reached on the basis of an extensive and careful analysis of the peculiar velocity of the Local Group by Tully et al. [184]. They are lead to the conclusion that the Local Void is responsible for a considerable repulsive influence, accounting for ~ 259 km s^{-1} of the ~ 631 km s^{-1} Local Group motion with respect to the CMB. While partly dependent on the details of the analysis, it seems hard to avoid the conclusion that we do not feel the presence of voids in our universe.

Voids and the Cosmos

Voids may function as probes of global cosmological parameters and on the underlying cosmology. Their intrinsic structure and shape, the outflow velocities and the corresponding redshift distortions are related to various aspects of the underlying cosmology. The outflow from the voids depends on the matter density parameter Ω_m, the Hubble parameter $H(t)$ and possibly on the cosmological constant Λ (see e.g. Van de Weygaert and van Kampen [187], Martal and Wassermann [111], Dekel and Rees [50], Bernardeau et al. [17], Fliche and Triay [61]). These parameters also dictate their redshift space distortions [154, 163].

Another interesting link between void structure and cosmology has recently been emphasized by Park and Lee [127] and Lee and Park [103]. They found that the intrinsic structure and shape of voids are sensitive to various aspects of the power spectrum of density fluctuations, including the imprint of dark energy.

The cosmological ramifications of the reality of a supersized void akin to the identified by Rudnick et al. [151] in the NVVS radio source counts would obviously be far-reaching.

Void Galaxies

A major point of interest concerns the galaxies within the voids, the void galaxies. Voids provide a unique and still largely pristine environment for studying the evolution of galaxies [80, 105, 131] and may represent a major challenge for current scenarios of structure formation. Peebles [131] pointed out that the observed salient and total absence of dwarf galaxies in nearby voids – for example the absence of dwarfs in the Local Void noticed by Karachentseva et al. [91] - could possibly involve strong ramifications for the viability of the ΛCDM cosmology on small scales.

A clear picture of the relation between void galaxies and their surroundings is just becoming available, be it there is still a lot of uncertainty concerning the physics which drives the observed correlations. The simplest models of biased galaxy formation (e.g Little and Weinberg [105]) predict that voids would be filled with galaxies of low luminosity, or galaxies of some other uncommon nature [80]. More sophisticated models have recently been developed [14, 64, 78, 116]; in these models the properties of galaxies are determined by the halos they inhabit. The recent interest in environmental influences on galaxy formation has prodded substantial activity in this direction [29, 74, 75, 81, 91, 101, 128, 140, 148, 178, 181].

2.4 Cosmic Shear and the Cosmic Web

The cosmic web is first and foremost defined and outlined by the dark matter distribution, the gravitationally dominant component which sets the corresponding gravitational potential. Galaxies are assumed to trace the underlying dark matter distribution. Even though the galaxies do indeed seem to provide a reasonable impression of the matter distribution, a direct map of the dark matter itself would obviously allow a real and unbiased view of the dynamics of the cosmic web.

A recent study has indeed managed to reveal the spatial dark matter distribution through its effect on the paths of the photons as they move through the Universe, meanwhile confirming that galaxies and starlight are in fact good tracers. Massey et al. [114] succeeded in producing the first truely three-dimensional map of the dark matter distribution. Their study is based on (weak) gravitational lensing data from the Cosmic Evolution Survey (COSMOS), and concerns a total region of 1637 square degrees meticulously observed by the ACS camera onboard the HST. An accurate and detailed two-dimensional map of the projected mass distribution clearly reveals the filamentary features connecting the high-density clusters (Fig. 13). Until recently,

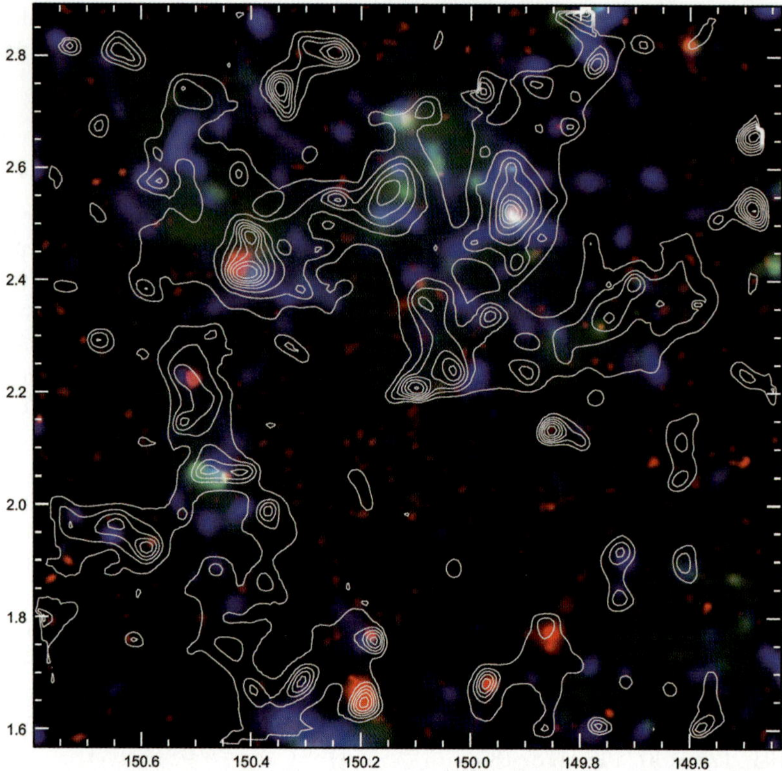

Fig. 13. The total projected matter density inferred from the large weak lensing study of the COSMOS data, shown in contours. The projected mass is dominated by dark matter. For comparison the matter surface density contours are superimposed on tracers of the baryonic matter distribution: (1) blue, the stellar mass (within $\Delta z \approx 0.1$) (2) yellow: the galaxy number density (within $\Delta z \approx 0.1$) and (3) red: hot dense gas, seen by deep X-ray observations with the XMM satellite. The X-ray emission by point sources has been removed. The dark matter reveals filamentary overdense regions that are topologically connected but insufficiently dense to generate X-ray emission: a loose network of filaments tracing the Cosmic Web. Within the filamentary network we recognize the dense compact cluster nodes. The most prominent peak in all four tracers is a single cluster of galaxies at $z = 0.73$ (α, δ=149 h,55 min, 2°31'). Courtesy of Richard Massey, also see Massey et al. [114]. Reproduced by permission of Nature

such weak lensing mass reconstructions were confined to the high-density regions in and around clusters because of the outstanding strength of their lensing signal. With the COSMOS map probing the more moderately dense regions of the cosmic web it turns out that stellar mass and galaxy number density do indeed accurately follow the dark matter distribution while the correlation with the X-ray emission – confined to the inner regions of clusters – is significantly less pronounced.

By complementing the lensing data with redshifts of the galaxy sources a tomographic analysis of the data, involving the assessment of the differential growth of the lensing signal between many thin slices separated by $\Delta z = 0.05$, made it possible to reconstruct the full three-dimensional matter distribution (Fig. 14). It did reveal that the massive cluster at $z = 0.73$ ($\alpha, \delta = 149$ h, 55 min, 2°31') is embedded within a giant three-dimensional structure which includes at least one filament.

The 3-D dark matter map is truely historical in that it uncovered for the first time the reality of a weblike pattern in the dark matter underlying the one that we see in the galaxy distribution. The potential for this new light

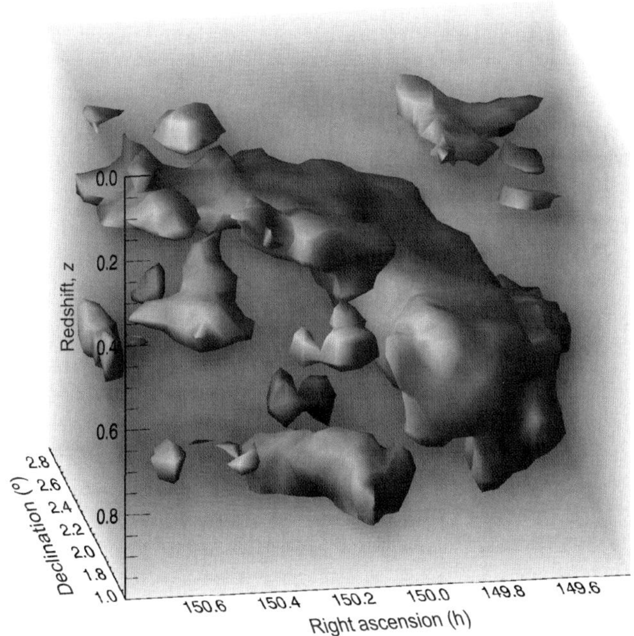

Fig. 14. Three-dimensional reconstruction of the dark matter distribution. The three axes correspond to right ascension, declination and redshift: with distance increasing towards the bottom. The redshift scale is highly compressed and the survey volume is really an elongated cone. The isodensity contour corresponds to a level of $1.4 \times 10^{13} M_\odot$ within a circle of radius 700 kpc and $\Delta z = 0.05$, arbitrarily chosen to highlight the filamentary structure. The 3-D map has been inferred from the tomographic analysis of the COSMOS weak lensing data, involving the assessment of the differential growth of the lensing signal between many thin slices separated by $\Delta z = 0.05$. The 3-D map reveals that the massive $z = 0.73$ cluster is indeed part of much larger 3-D structure, including a filament partially aligned along the line of sight. Courtesy of Richard Massey, also see Massey et al. [114]. Reproduced by permission of Nature

on the dark side of the Universe is tremendous. The detected filamentary DM network provides a direct and transparent link to theories of structure formation, directly tying in with collisionless dark matter and gravity without the necessity to involve complex and as yet not fully understood hydrodynamic, radiative and starformation processes.

2.5 The Gaseous Cosmic Web

Galaxies are assumed to trace the underlying dark matter distribution, and their spatial distribution (still) represents the most detailed and clearest outline of its intricate weblike features which we have available. Nonetheless, stars and galaxies do in fact represent only a minor fraction of all the baryons in the Universe. As far as baryons are concerned the cosmic web is first and foremost an intricate network of diffuse gaseous lanes pervading the Universe (see Fig. 16).

In other words, while in practice galaxies are used as tracers, it is the diffuse intergalactic medium (IGM) which forms the main baryonic constituent of the cosmic web. At high redshift ($z \gtrsim 2$) the overwhelming majority of baryons are in a diffuse, photoionized intergalactic medium, partly enriched by the products of stellar nucleosynthesis. This gas is observable as HI absorption lines in the spectra of distant background quasars (see Rauch [143], Cen et al. [30]). The resulting redshifted Lyman α (Lyα) absorption along their line of sight produces the Lyα forest, which represents a highly sensitive *one-dimensional* probe of the (gaseous) cosmic web (see Fig. 15). By the current epoch, hierarchical structure formation has produced deep potential wells into which the baryons accrete, thereby moving a significant portion of the baryons from the IGM into stars, galaxies, groups and clusters. Hydrodynamical simulations of cosmic structure formation have indicated that a significant fraction of the baryons at $z \sim 0$ are found in a gaseous form. The gas around emerging clusters falls into their potential wells and turns into hot highly ionized X-ray emitting intracluster gas. Most of the gas, with a temperature between 10^5 and 10^7 K, is found in regions of moderate overdensities $\delta \sim 10$–100. Part of this gas is associated with the virial regions around galaxies, accounting for around $\Omega_b \sim 0.024$ of the total $\Omega_b = 0.045$ contributed by baryons to the density of the Universe [63]. The remaining component of this diffuse Warm-Hot Intergalactic Medium (WHIM) mostly traces out the filamentary features in the cosmic web. It may account for a significant fraction of the missing baryons at low redshifts (Fukugita et al. [62] Fukugita & Peebles [63]). Probably there is also a significant amount of low temperature WHIM with $T < 10^5$ K, distributed mostly as sheet-like structures (Kang et al. [90]). The WHIM may even account for up to 30%–40% of the baryonic mass in the Universe (Davé et al. [42]). Its evolution is driven primarily by shock heating as the gas falls into the gravitationally generated potential wells, mainly those defined by the nonequilibrium large-scale structures such as filaments. For the heating of the

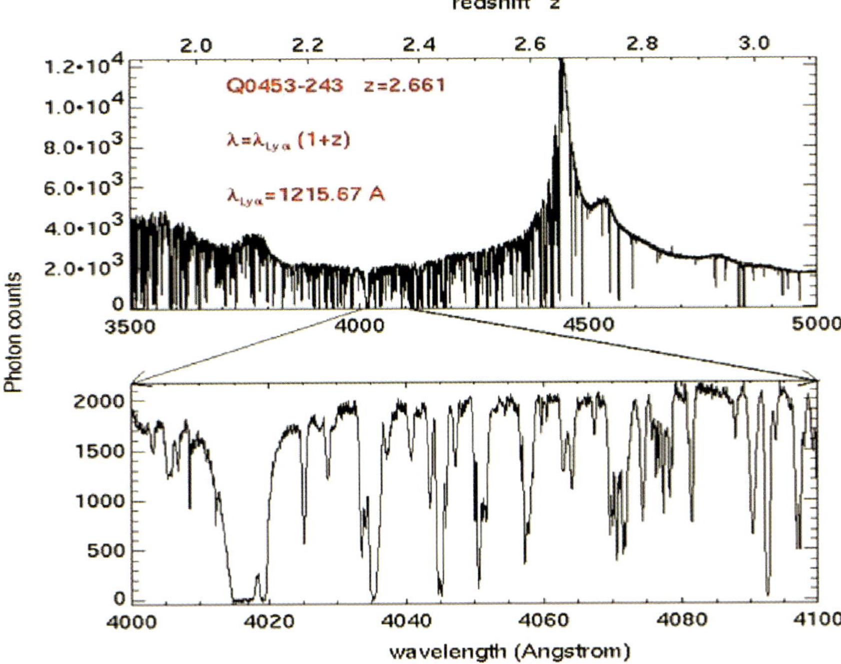

Fig. 15. The spectrum of the quasar Q0453-243 obtained with the HIRES spectrograph on the Keck I telescope. The quasar has an emission redshift of $z = 2.661$. To the left of the Lymanα emission line you see the "forest" of HI absorption lines produced by intervening, tenuous intergalactic clouds. The lower panel zooms in on the region between 4000 and 4100 A. The particularly strong line at 4020 A is a "damped Lymanα absorption feature produced in a cloud which is optically thick in HI. Image courtesy of Matteo Viel

gas, processes like supernova feedback, radiative cooling and photoionization are only of secondary importance.

The shock-heated WHIM gas in filaments and sheets is manifested best through emissions and absorptions in soft X-ray and far UV. It will make significant contributions to the soft X-ray background, and can be detected through absorptions of highly ionized species suchs as OVII and OVIII in AGN spectra and line emissions from OVII and OVIII ions. Detection of WHIM absorption in X-ray observations were reported by various groups (Kaastra et al. [86], Nicastro et al. [121]), while there was also a report of a possible detection of WHIM emission from a filament around Coma (Finoguenov [60]).

The study of the IGM represents an impressively rich source for our understanding of the cosmic web. Potentially the intricate structure can be traced in much more detail than by means of the discrete galaxy distribution. Different chemical species and ionization stages probe different density and temperature regimes within the cosmic web, which in turn may be related

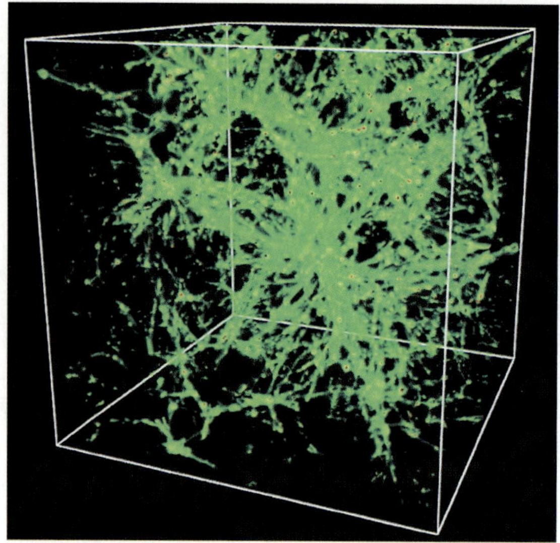

Fig. 16. Spatial distribution of the warm/hot intergalactic gas (WHIM) with temperature in the range $10^5 - 10^7$ K, at $Z = 0$, in a box of 85 h^{-1}Mpc. The image reveals the striking pattern of the Cosmic Web into which the WHIM gas has settled itself as it flowed into the potential wells set by the dark matter distribution. The green regions have densities about 10–20 times the mean baryon density of the universe at $Z = 0$; the yellow regions have densities about 100 times the mean baryon density, while the small isolated regions with red and saturated dark colours have even higher densities reaching about 1000 times the mean baryon density and are sites for current galaxy formation. Image courtesy of Renyue Cen, also see Cen and Ostriker [31]. Reproduced by permission of the AAS

to different regimes and stages of galaxy formation. However, in particular the most directly accessible study of the IGM, that of absorption line studies, is confined to one-dimensional probes. This renders it difficult to translate these to a three-dimensional image (yet, a constrained inversion is not entirely unfeasible, see Pichon et al. [132]). Emission line studies of the WHIM would offer the exciting potential of mapping the cosmic web through its gaseous contents. A meticulous detailed mapping comparable to that traced by the galaxy distribution remains as yet only a remote possibility.

3 Spatial Structure, Pattern Analysis and Object Identification

Many attempts to describe, let alone identify, the features and components of the Cosmic Web have been of a mainly heuristic nature. There are various relevant issues. The primary issue is that of defining a technique that sensitively probes the properties of the Cosmic Web. Another major point of concern involves the sampling of the web patterns, by default limited in scope. Cosmological theories generally describe the development of structure in terms of continuous (dark matter) density and velocity fields. To a large extent our knowledge stems from a discrete sampling of these fields.

In the real world it is impossible to get exhaustive values of data at every desired point of space. The product of astronomical observations, physical experiments and computer simulations often concern data sets in two, three or more dimensions. This may involve the value of some physical quantity: the galaxy density field, the dark matter density field or the peculiar velocity field are amongst the best known examples. Often these are measured or determined from an irregularly distributed set of reference points.

The principal task for any formalism that seeks to process the sampled data on the cosmological matter distribution is to optimally retain or extract the required information on the Cosmic Web. Dependent on the purpose of a study, various different strategies may be followed:

- *Statistical Analysis*

  One strategy is to distill various statistical measures, or other sufficiently descriptive cosmological measures, which characterize specific aspects of the large scale matter distribution. In essence this involves the compression of the available information into a restricted set of parameters or functions, with the intention to compare or relate these to theoretical predictions.

- *Feature Identification*

  The identification and isolation of features and objects in the cosmic matter distribution – clusters, filaments and voids – is essential for understanding the nature of structures which form in the Universe and provides an important link between observation and theoretical models. On the one hand this may involve a cosmographic study of individual structures in our Cosmic neighbourhood. Their detail usually forms a welcome complement to surveys of large samples of similar objects, while sometimes they highlight the extremes in the cosmological zoo. Perhaps most important is the necessity of well-defined feature identifiers for proper statistical studies of cosmic structure formation.

- *Structure Reconstruction*

  For the determination of various statistical characterizations of cosmic structure it is imperative to define an optimal reconstruction of cosmic density and velocity fields. Demanding in itself, such a reconstruction is often complicated by the usually *discrete* nature of the sample point distribution and the highly inhomogeneous nature of the sample point distribution. The translation into a continuous field which optimally reflects reality is a far from trivial procedure and forms the subject of an extensive literature in computer science, visualization and applied sciences.

3.1 Statistics of the Cosmic Web

There is a variety of statistical measures characterizing specific aspects of the large scale matter distribution (for an extensive and complete review see Martínez & Saar, [112]). Below we list a selection of methods for structure characterisation and finding. It is perhaps interesting to note two things about this list:

(a) each of the methods tends to be specific to one particular structural entity
(b) there are no explicit wall-finders.

Both issues emphasize an important property that generic techniques for tracing structural features should possess (see 3.2). The skeleton formalism [122, 171, 172] accomplished this by tracing the mathematically well-defined skeleton of the Cosmic Web and arguing its close relationship to its filamentary constituents. More generic is the Scale Space approach adopted by Aragón-Calvo [5] (also see Aragón-Calvo et al. [6]): it provides a uniform approach to finding Blobs, Filaments and Walls as individual objects that can be catalogued and studied.

Structure from Higher Moments

The clustering of galaxies and matter is most commonly described in terms of a hierarchy of correlation functions. The two-point correlation function (and its Fourier transform, the power spectrum) remains the mainstay of cosmological clustering analysis and has a solid physical basis. However, the nontrivial and nonlinear patterns of the cosmic web are mostly a result of the phase correlations in the cosmic matter distribution [32, 38, 153]. While this information is contained in the moments of cell counts [44, 65, 130] and, more formally so, in the full hierarchy of M-point correlation functions ξ_M, their measurement has proven to be impractical for all but the lowest orders [85, 130, 177].

The Void probability Function [102, 193] provided a characterisation the "voidness" of the Universe in terms of a function that combined information from many higher moments of the point distribution. But, again, this has not provided any identification of individual voids.

Topological Methods

The shape of the local matter distribution may be traced on the basis of an analysis of the statistical properties of its inertial moments [8, 12, 110]. These concepts are closely related to the full characterization of the topology of the matter distribution in terms of four Minkowski functionals [117, 162]. They are solidly based on the theory of spatial statistics and also have the great advantage of being known analytically in the case of Gaussian random fields. In particular, the *genus* of the density field has received substantial attention as a strongly discriminating factor between intrinsically different spatial patterns [69, 82].

The Minkowski functionals provide global characterisations of structure. An attempt to extend its scope towards providing locally defined topological measures of the density field has been developed in the SURFGEN project defined by Sahni and Shandarin and their coworkers [158, 164]. The main problem remains the user-defined, and thus potentially biased, nature of the continuous density field inferred from the sample of discrete objects. The usual filtering techniques suppress substructure on a scale smaller than the filter radius, introduce artificial topological features in sparsely sampled regions and diminish the flattened or elongated morphology of the spatial patterns. Quite possibly the introduction of more advanced geometry based methods to trace the density field may prove a major advance towards solving this problem.

Importantly, Martínez et al. [113] and Saar et al. [155] have generalized the use of Minkowski Functionals by calculating their values in a hierarchy of scales generated from wavelet-smoothed volume limited subsamples of the 2dF catalogue. This approach is particularly effective in dealing with non-Gaussian point distributions since the smoothing is not predicated on the use of Gaussian smoothing kernels.

3.2 Structure Finding

In addition to the statistical characterization of the cosmic matter density field, a major effort goes into identifying and isolating features and individual structures in the cosmic matter distribution. The vast majority of these studies have focussed on the detection of clusters of galaxies. Tracing filamentary has gained relatively little attention, and with the exception of a few rare outstanding concentrations – the Great Wall [66] and the SDSS Great Wall [70] – the detection of sheets is a virtually nonexistent activity.

Cluster Finding

In the context of analyzing distributions of galaxies we can think of cluster finding algorithms. There we might define a cluster as an aggregate of neighbouring galaxies sharing some localised part of velocity space. Algorithms like HOP attempt to do this. However, there are always issues arising such as how to deal with substructure: that perhaps comes down to the definition of what a cluster is. Here we focus on defining coherent structures based on particle positions alone. The velocity space data is not used since there is no prior prejudice as to what the velocity space should look like.

Filament Finding

The connectedness of elongated supercluster structures in the cosmic matter distribution was first probed by means of percolation analysis, introduced and emphasized by Zel'dovich and coworkers [197], while a related graph-theoretical construct, the minimum spanning tree of the galaxy distribution, was extensively probed and analysed by Bhavsar and collaborators [11, 34, 72] in an attempt to develop an objective measure of filamentarity.

Finding filaments joining neighbouring clusters has been tackled, using quite different techniques, by Colberg, Krughoff and Connolly [35] and by Pimbblet [133]. More general filament finders have been put forward by a number of authors. Stoica et al. [173] use a generalization of the classical Candy model to locate and catalogue filaments in galaxy surveys. This approach has the advantage that it works directly with the original point process and does not require the creation of a continuous density field. However, it is very computationally intensive.

The mathematically most rigorous program for filament description and analysis is that of the skeleton analysis of density fields by Novikov, Colombi and Doré [122] (2-D) and Sousbie et al. [17] (3-D). Based on Morse theory (see Colombi, Pogosyan and Souradeep [40]) the skeleton formalism analyzes continuous density fields and detects morphological features – maxima and saddle points in the density field – by relating density field gradients to the Hessian of the density field (also see Doré et al. [52]). It results in an elegant and effective tool with a particular focus towards tracing the filamentary structures in the cosmic web. However, it is computationally intensive and may be sensitive to the specific method of reconstruction of the continuous density field. The Hessian of the density field also forms the basis of the MMF analysis developed by Aragon-Calvo [5] (see Fig. 17) although MMF embeds this within a formalism that explicitly adresses the multiscale character of the cosmic density field and includes the shape conserving abilities of the tessellation based density field reconstruction Schaap and van de Weygaert [161].

Void Finding

Voids are distinctive and striking features of the cosmic web, yet identifying and tracing their outline within the complex spatial geometry of the Cosmic Web has proven to be far from trivial. There have been extensive searches for voids in galaxy catalogues [81, 139] and in numerical simulations [4, 7].

Several factors contribute to making systematic void-finding difficult. One major obstacle is that there is not an unequivocal definition of what a void is and as a result there is considerable disagreement on the precise outline of such a region (see e.g. Shandarin et al. [165]). The fact that voids are almost empty of galaxies means that the sampling density plays a key role in determining what is or is not a void [163]. Moreover, void finders are often predicated on building void structures out of cubic cells [93] or out of spheres (e.g: Patiri et al. [129]). Because of the vague and diverse definitions, and the diverse interests in voids, there is a plethora of void identification procedures [4, 7, 36, 56, 76, 81, 93, 129, 139, 165]. For example, there are methods that attempt to synthesize voids from the intersection of cubic or spherical elements and

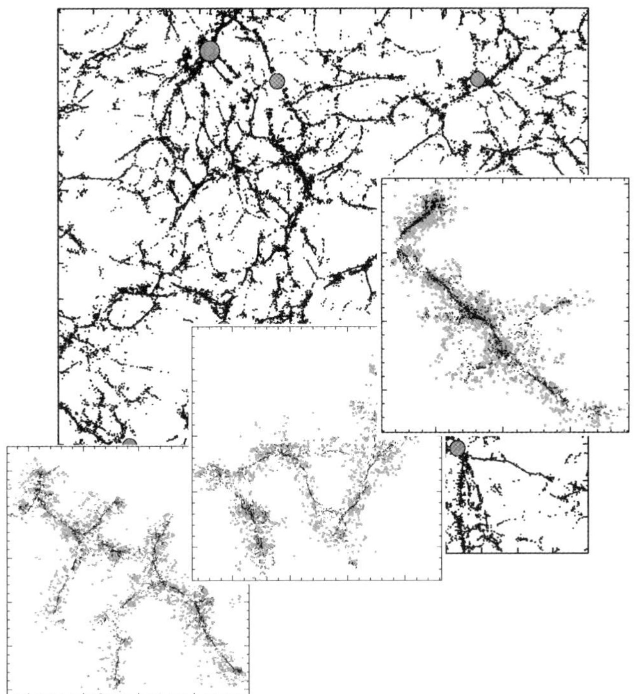

Fig. 17. The filamentary network in a GIF ΛCDM simulation. The filaments were identified by means of the MMF technique of Aragón-Calvo et al. [6]. The filled (grey) circles correspond to clusters with a mass above 10^{14} $M_{\odot}$. The inserts contain three specific examples of filaments. The gray dots represent the original (simulation) dark matter particles. The spine of the filaments (black particles) is the result of the filament compression algorithm of Aragón-Calvo [5]. Image courtesy M. Aragón-Calvo, also see Aragón-Calvo [5]

do so with varying degrees of success. The *Aspen-Amsterdam Void Finder Comparison Project* of Colberg et al. [37] will clarify many of these issues. The Watershed-based algorithm of Platen, van de Weygaert and Jones [135] aims to avoid issues of both sampling density and shape.

3.3 Reconstruction of the Cosmic Web

For a meaningful analysis and interpretation of spatial data it is often necessary to obtain estimates of the related field values throughout the sample volume. The *reconstructed* continuous field may subsequently be processed in order to yield a variety of interesting parameters. Ideally, reconstruction procedures should be based upon solid statistical foundations. The complex reality of the cosmic web – marked by asymmetric and anisotropic features and a large range of densities– renders it very difficult to develop and infer statistical methods from first principle. The work by Erdoğdu et al. [58] and Kitaura and Enßlin [98] represent examples of possibly rewarding strategies.

In the observational reality galaxies are the main tracers of the cosmic web and it is mainly through the measurement of the redshift distribution of galaxies that we have been able to map its structure. Another example is that of the related study of cosmic flows in the nearby Universe, based upon the measured peculiar velocities of a sample of galaxies located within this cosmic volume. Likewise, simulations of the evolving cosmic matter distribution are almost exclusively based upon N-body particle computer calculation, involving a discrete representation of the features we seek to study. Both the galaxy distribution as well as the particles in an N-body simulation are examples of *spatial point processes* in that they are

- *discretely sampled*
- have an *irregular spatial distribution.*

A major part of any reconstruction procedure is the filtering and interpolation of the measured data.

3.4 Spatial Data: Filtering and Interpolation

Issues of *smoothing* and *spatial interpolation* of the measured data over the sample volume are of considerable importance and interest in many different branches of science. Interpolation is fundamental to graphing, analysing and understanding of spatial data. Key references on the involved problems and solutions include those by Ripley [147], Sibson [170], Watson [192], Cressie [41]. While of considerable importance for astronomical purposes, many available methods escaped attention. A systematic treatment and discussion within the astronomical context is the study by Rybicki and Press [152], who focussed on linear systems as they developed various statistical procedures related to linear prediction and optimal filtering, commonly known as Wiener filtering. An extensive, systematic and more general survey of available mathematical methods can be found in a set of publications by Lombardi and Schneider [106–108].

DTFE: Delaunay Tessellation Field Estimator

A particular class of spatial point distributions is the one in which the point process forms a representative reflection of an underlying smooth and continuous density/intensity field. The spatial distribution of the points itself may then be used to infer the density field. This forms the basis for the interpretation and analysis of the large scale distribution of galaxies in galaxy redshift surveys. The number density of galaxies in redshift survey maps and N-body particles in computer simulations is supposed to be proportional to the underlying matter density.

One noteworthy example of a technique which uses this fact is the *DTFE* method, a linear version of *natural neighbour* interpolation. The DTFE technique [161, 188] recovers fully volume-covering and volume-weighted continuous fields from a discrete set of sampled field values. The method has been developed by Schaap and van de Weygaert [161] and forms an elaboration of the velocity interpolation scheme introduced by Berhardeau and van de Weygaert [16]. It is based upon the use of the Voronoi and Delaunay tessellations of a given spatial point distribution to form the basis of a natural, fully self-adaptive filter in which the Delaunay tessellations are used as multidimensional interpolation intervals. An example is the void density and velocity field in Fig. 24.

The primary ingredient of the DTFE method is the Delaunay tessellation of the particle distribution. The Delaunay tessellation of a point set is the uniquely defined and volume-covering tessellation of mutually disjunct Delaunay tetrahedra (triangles in 2D). Each is defined by the set of four points whose circumscribing sphere does not contain any of the other points in the generating set [51]. The Delaunay tessellation and the Voronoi tessellation of the point set are each others *dual.* The Voronoi tessellation is the division of space into mutually disjunct polyhedra, each polyhedron consisting of the part of space closer to the defining point than any of the other points [123, 191].

DTFE exploits three properties of Voronoi and Delaunay tessellations [160, 188]. The tessellations are very sensitive to the local point density. DTFE uses this to define a local estimate of the density on the basis of the inverse of the volume of the tessellation cells. Equally important is their sensitivity to the local geometry of the point distribution. This allows them to trace anisotropic features such as encountered in the cosmic web. Finally, DTFE exploits the adaptive and minimum triangulation properties of Delaunay tessellations in using them as adaptive spatial interpolation intervals for irregular point distributions. In this way it is the first order version of the *Natural Neighbour method* [27, 169, 174, 184].

Within the cosmological context a major – and crucial – characteristic of a processed DTFE density field is that it is capable of delineating three fundamental characteristics of the spatial structure of the megaparsec cosmic matter distribution. It outlines the full hierarchy of substructures present in the sampling point distribution, relating to the standard view of structure in the Universe having arisen through the gradual hierarchical buildup of matter concentrations. DTFE also reproduces any anisotropic patterns in the density distribution without diluting their intrinsic geometrical properties. This is particularly important when analyzing the the prominent filamentary and planar features marking the Cosmic Web. A third important aspect of DTFE is that it outlines the presence and shape of voidlike regions. Because of the interpolation definition of the DTFE field reconstruction voids are rendered as regions of slowly varying and moderately low density values.

Multiscale Morphology Filter

Recently a variety of methods have been developed towards a complete morphological analysis of the cosmic web in the cosmic matter distribution. Perhaps the most rigorous program, with a particular emphasis on the description and analysis of filaments, is that of the *skeleton* analysis of density fields by Novikov, Colombi and Doré [122] (2-D) and Sousbie et al. [172] (3-D) (see Sect. 3.2). Another strategy has been followed by Hahn et al. [76]. They identify clusters, filaments, walls and voids in the matter distribution on the basis of the tidal field tensor $\partial^2\phi/\partial x_i \partial x_j$, determined from the density distribution filtered on a scale of $\approx 5\,\mathrm{h}^{-1}$ Mpc. Here we shortly focus on the Multiscale Moprhology Filter (MMF), introduced by Aragón-Calvo et al. [5]. The MMF dissects the cosmic web on the basis of the multiscale analysis of the Hessian of the density field.

Figure 18 contains a schematic overview of the Multiscale Morphology Filter (MMF) to isolate and extract elongated filaments (dark grey), sheet-like walls (light grey) and clusters (black dots) in the weblike pattern of a cosmological N-body simulation Aragón-Calvo et al. [6]. The first stage is the translation of a discrete particle distribution (top lefthand frame) into a DTFE density field (top centre). This guarantees a morphologically unbiased and optimized density field retaining all features visible in a discrete galaxy

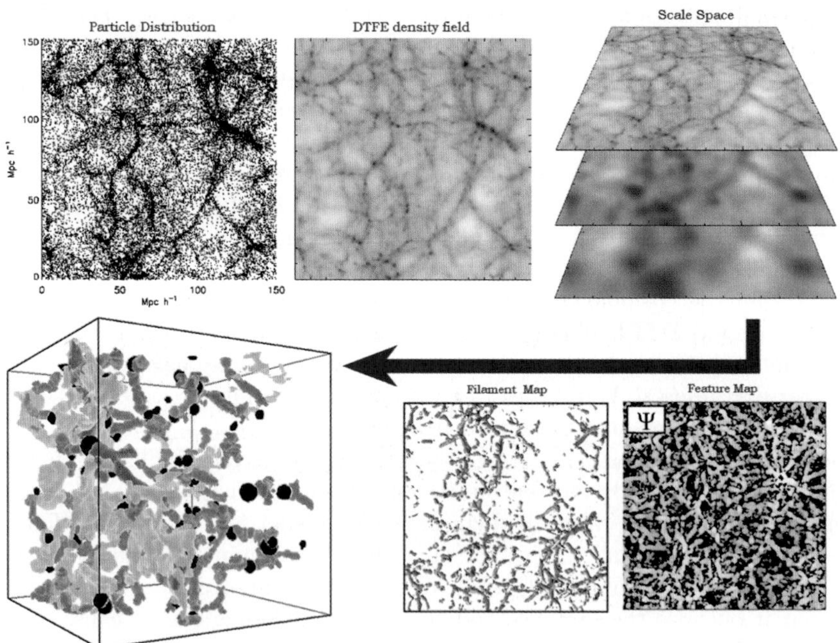

Fig. 18. Scheme of the Multiscale Morphology Filter for extracting weblike morphologies. See text for explanation. From van de Weygaert and Schaap [188]

or particle distribution. The DTFE field is filtered over a range of scales (top righthand stack of filtered fields). By means of morphology filter operations defined on the basis of the Hessian of the filtered density fields the MMF successively selects the regions which have a bloblike (cluster) morphology, a filamentary morphology and a planar morphology, at the scale at which the morphological signal is optimal. This produces a feature map (bottom lefthand). By means of a percolation criterion the physically significant filaments are selected (bottom centre). Following a sequence of blob, filament and wall filtering finally produces a map of the different morphological features in the particle distribution (bottom lefthand). The 3-D isodensity contours in the bottom lefthand frame depict the most pronounced features (also see Fig. 17).

MMF and the Cosmic Web

Two noteworthy recent results obtained by MMF concerns the inventory of mass and volume content of the Cosmic Web [5], shown in Fig. 19. The results relate to the present-day epoch in a ΛCDM N-body simulation. Clusters occupy the smallest volume fraction in the cosmic web, accounting for only 0.4%. They do, however, represent a major share of the mass (28%), making them by far the densest components of the Cosmic Web. Most mass (39%) in the Universe resides in filaments, tracing out almost 10% of the total volume. Sheet contain only a small fraction of the mass, $\approx 5.5\%$ and occupy a relatively small volume (4.9%), making them the most tenuous structures in the Cosmic Web.

Also highly relevant is the issue of the connection between filaments and clusters. The number of filaments emanating from a cluster turns out to be a strong function of the cluster mass (see Fig. 20). More massive clusters are connected to considerably more filaments: MMF analysis indicates that

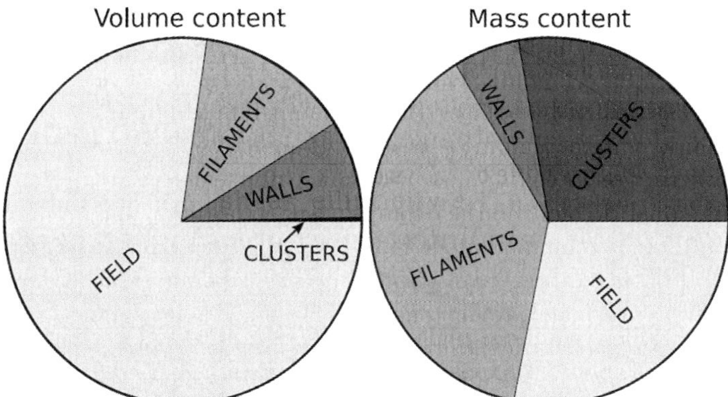

Fig. 19. Pie diagram showing an inventory of the Cosmic Web in terms of volume (**left**) and mass (**right**). Image courtesy M. Aragón-Calvo, also see Aragón-Calvo [5]

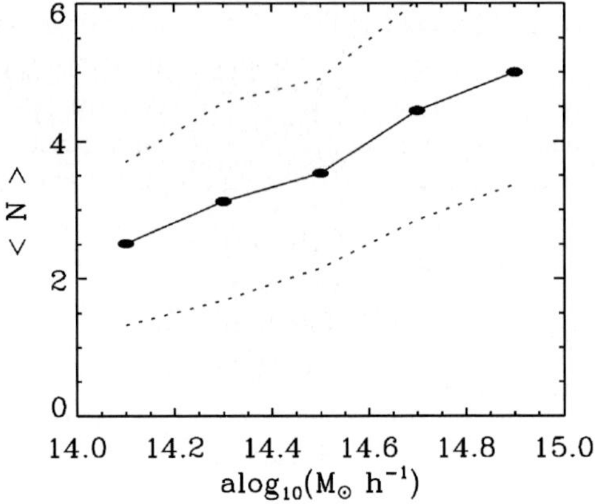

Fig. 20. Mean number of filaments as a function of the mass of the clusters to which they are connected (*solid line*). *Dotted line*: 1σ dispersion. Image courtesy M. Aragón-Calvo, also see Aragón-Calvo [5]

clusters with a mass $M \sim 10^{14}$ $M_\odot$ have on average 2 filaments connected to them, clusters with a mass $M \sim 10^{15}$ $M_\odot$ more than five filaments. Other studies have found a similar relation based on intracluster filaments found in N-body simulations [35] and visually identified filament-cluster connections from the 2dF galaxy redshift survey [133].

4 Voids

A manifest and prominent morphological aspect of the Megaparsec matter and galaxy distribution is the marked and dominant presence of large under-dense regions, the *Voids.* A proper and full understanding of the formation and dynamics of the Cosmic Web is not possible without understanding the structure and evolution of voids. With respect to their role in the structure and buildup of the Cosmic Web we need to address three crucial aspects of void evolution:

- *Formation and Evolution of Voids*
  Voids form in and around density troughs in the primordial density field. As a result of the corresponding weaker internal gravity matter matter streams out of the interior of voids while the void as a whole will expand with respect to the background Universe.
- *Void Dynamics and Void Outflow*
  As a result of their underdensity voids represent a region of weaker gravity. This results in an effective repulsive gravitational influence. Various galaxy redshift surveys and studies of galaxy peculiar velocities have indeed uncovered this imprint in the cosmic velocity flow in the Local Universe (see Sect. 2.3).

- *Void Hierarchy and Substructure*
  Not only galaxies, galaxy halos and clusters of galaxies get assembled in a hierarchical fashion. Also the buildup of voids proceeds via a complex and intricate process of hierarchical evolution. Insight into this evolution is essential for understanding the overall geometry and structure of the Cosmic Web. The remnants of the hierarchical void evolution can still be seen when studying the observed spatial galaxy distribution or when analyzing N-body simulations of structure formation. It should also form the basis for the study of properties of the void galaxy population and the dependence on environment.

In the subsequent sections we will address each of these issues in some detail.

4.1 Formation and Evolution of Voids

Voids emerge out of the density troughs in the primordial Gaussian field of density fluctuations. Early theoretical models of void formation concentrated on the evolution of isolated voids [18, 20, 79, 84]. Initially underdense regions expand faster than the Hubble flow, and thus expand with respect to the background Universe. If they are not embedded within overdense regions, such regions eventually form voids which are surrounded by dense void walls. At any cosmic epoch the voids that dominate the spatial matter distribution are a manifestation of the cosmic structure formation process reaching a non-linear stage of evolution.

In a void-based description of the evolution of the cosmic matter distribution, voids mark the transition scale at which density perturbations have decoupled from the Hubble flow and contracted into recognizable structural features. On the basis of theoretical models of void formation one might infer that voids may act as the key organizing element for arranging matter concentrations into an all-pervasive cosmic network [84, 145, 168, 186]. As voids expand, matter is squeezed in between them, and sheets and filaments form the void boundaries. This view is supported by numerical studies and computer simulations of the gravitational evolution of voids in more complex and realistic configurations [36, 53, 68, 111, 125, 145, 187]. A marked example of the evolution of a typical large and deep void in a ΛCDM scenarios is given by the time sequence of six frames in Fig. 21.

Void Characteristics: an Inventory

The formation and evolution of voids involves a range of interesting and intricate processes and aspects. A listing of a dozen characteristic properties may elucidate this.

- *Voids expand.*
  The underdensity of a void corresponds to a weaker interior gravitational field. With respect to the global universe this leads to an effective (peculiar) gravity inducing a general flow out of the void region.

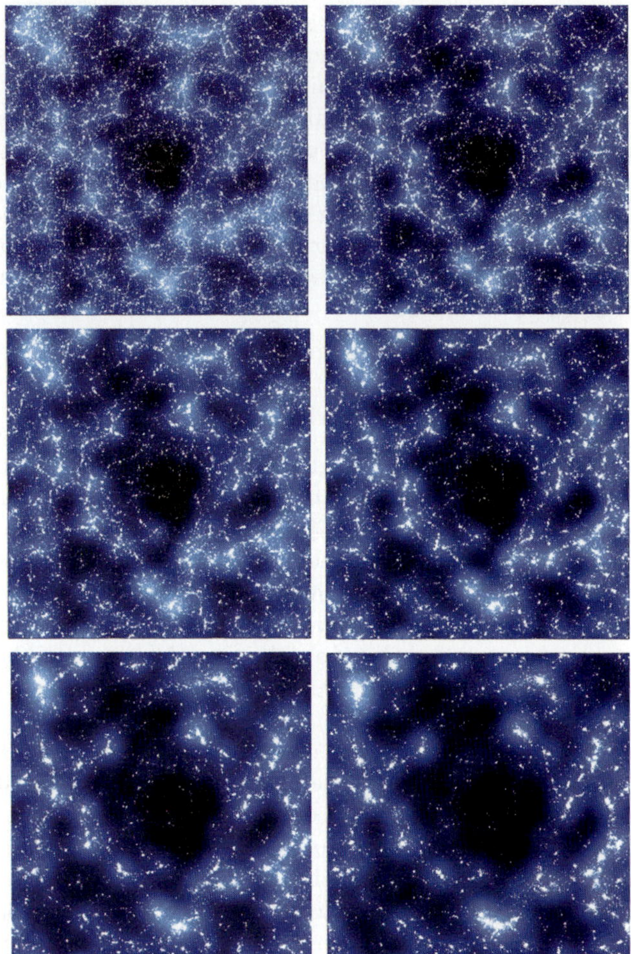

Fig. 21. Simulation of evolving void (LCDM scenario). Image courtesy of Erwin Platen

- *Voids empty.*
  As matter streams out of the void, the density within the void decreases. Isolated voids will asymptotically evolve towards an underdensity $\delta = -1$, pure emptiness.

- *Voids form ridges.*
  As the density within voids gradually increases outward, the corresponding peculiar (outward) gravitational acceleration decreases outward: void matter in the centre moves outward faster than void matter towards the boundary. As a result matter accumulates in ridges surrounding the void

(see Fig. 22). The steepness of the resulting density profile depends on the protovoid depression [126].

- *"Bucket" density profile*
  Voids assume a "bucket" shape – marked by a uniform interior density depression and a steep outer boundary – as a result of the fast outflow from the "flat" centre in a primordial underdensity. While their matter content accumulates near and around steep density ridges, the interior involves into a region resembling a low-density homogeneous FRW Universe (see Fig. 22).

- *Superhubble void expansion*
  Related to the uniform density interior of mature voids the corresponding peculiar velocity field is that of a "Superhubble" flow [84]: the interior flowfield of voids is marked by a uniform velocity divergence [160]. For a spherically symmetric void model it is rather straightforward to analytically infer that this is the expected natural tendency for voids (see Fig. 22). It is a manifestation of Birkhoff's theorem, according to which a void region can be described as an isolated lower Ω FRW universe [68, 187]. An analysis of N-body simulations by means of the DTFE technique has shown this also to be the case for the more complex situation of hierarchical structure formation (see Sect. 4.3).

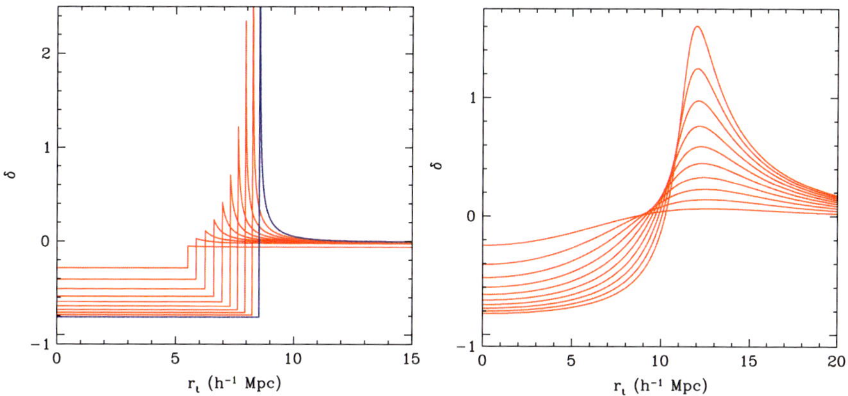

Fig. 22. Spherical model for the evolution of voids. **Left**: a pure (uncompensated) tophat void evolving up to the epoch of shell-crossing. Initial (linearly extrapolated) density deficit was $\Delta_{\text{lin},0} = -10.0$, initial (comoving) radius $\widetilde{R}_{i,0} = 5.0$ h^{-1} Mpc. **Right**: a void with an angular averaged SCDM profile. Initial density deficit and characteristic radius are same as for the tophat void (**left**). The tendency of this void to evolve into a tophat configuration by the time of shell crossing is clear. Shell-crossing, and the formation of a ridge, happens only if the initial profile is sufficiently steep

- *Characteristic void and shellcrossing*
  Overdense spherical peaks have a characteristic and time of collapse, coincident with a linearly extrapolated density $\delta_c = 1.69$. Voids have a similar globally valid characteristic epoch of evolution, that of *shellcrossing.* This happens when interior shells of matter take over initially exterior shells. It happens when a primordial density depression attains a linearly extrapolated underdensity $\delta_v = -2.81$ (for EdS universe). A perfectly spherical "bucket" void will have expanded by a factor of 1.72 at shellcrossing, and therefore have evolved into an underdensity of $\sim 20\%$ of the global cosmological density, ie. $\delta = -0.8$.

- *Identity observed voids*
  Bertschinger's thesis work demonstrated that once voids have passed the stage of shellcrossing they enter a phase of self-similar expansion [18]. Subsequently, their expansion will slow down with respect to the earlier linear expansion. This impelled Blumenthal et al. [20] to identify voids in the present-day galaxy distribution with voids that have just reached the stage of shell-crossing.

- *Void shapes: spherical tendencies*
  Icke [84] pointed out that any (isolated) aspherical underdensity will become more spherical as it expands. The effective gravitational acceleration is stronger along the short axis than along the longer axes. For overdensities this results in a stronger inward acceleration and infall, producing increasingly flattened and elongated features. By contrast, for voids this translates into a larger *outward* acceleration along the shortest axis so that asphericities will tend to diminish. For the interior of voids this tendency has been confirmed by N-body simulations [187]. In reality, voids will never reach sphericity as a result of large scale dynamical and environmental factors [136].

- *Nonlinearity of voids*
  While by definition voids correspond to density perturbations of at most unity, $|\delta_v| \leq 1$, mature voids in the nonlinear matter distribution do represent *highly nonlinear* features. This may be best understood within the context of Lagrangian perturbation theory [157]. Overdense fluctuations may be described as a converging series of higher order perturbations, the equivalent perturbation series is less well behaved for voids. The successive higher order terms of both density deficit and corresponding velocity divergence alternate between negative and positive (see Fig. 23).

- *Dilution of void substructure*
  In hierarchical scenarios of structure formation void regions contain substantial amounts of infrastructure (see Sect. 4.4). The low-density environment of voids slows the growth of structure (for a thorough analytical

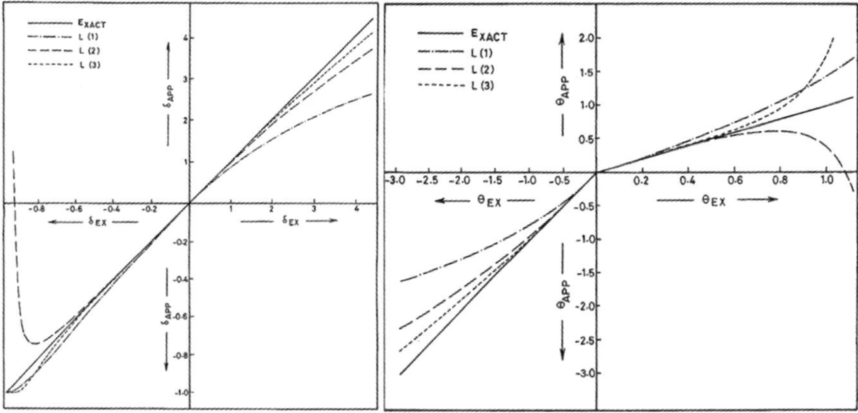

Fig. 23. Void nonlinearity in Lagrangian perturbation theory, from Sahni and Shandarin [157]. **Lefthand Frame**: density contrast δ_{APP} in Lagrangian perturbation series $L(n)$ plotted against the exact tophat solution δ_{EX} for underdense regions (**lower left**) and overdense regions (**upper right**). Whereas the accuracy of $L(n)$ increases with n when describing the behaviour of overdense regions, $L(n)$ wiht $N > 1$ do not fare as well when applied to underdense regions. Although for voids $L(n)$ with $n = 2, 3$ are initially more accurate than $L(1)$ (Zel'dovich approximation), their accuracy becomes poorer with time. Moreover, $L(2)$ shows pathological behaviour at late times when $\delta_{\mathrm{EX}} < -0.7$. Righthand frame: The dimensionless velocity divergence field θ_{APP} in Lagrangian perturbation series $L(n)$ is shown plotted against the exact solution θ_{EX}, for overdense regions (**lower left**) and underdense regions (**upper right**). $l(n)$ with $n = 2, 3$ give better results than $l(a)$ for overdense but not for underdense regions. From Sahni and Shandarin [157]. Image courtesy of Sergei Shandarin

treatment see Goldberg and Vogeley [68]). The net result is a diluted and diminished infrastructure which remains visible, at ever decreasing density contrast, as cinders of the earlier phases of the *void hierarchy* in which the substructure stood out more prominent (see Sect. 4.8).

- *Collapse of voids*
  Instead of expanding, voids embedded in a larger scale environment of sufficient overdensity, or surrounded by structures effecting a strong enough tidal force field, may tend to collapse. This process of void collapse is especially relevant for small (sub)voids near the boundaries of large dominating voids. The process is of crucial importance in the hierarchical evolution of voids (see Sect. 4.8).

4.2 Void Identity and Maturity

One question of relevance is that of the identity of the observed voids. In other words, what defines a mature void? A reasonable answer may be found on the basis of the spherical model. This teaches us that voids may be assigned a characteristic dynamical time, corresponding to a threshold of the linearly extrapolated primordial density field. A reasonable suggestion is that of a void reaching maturity at the moment of *shell-crossing*, ie. the stage at which the

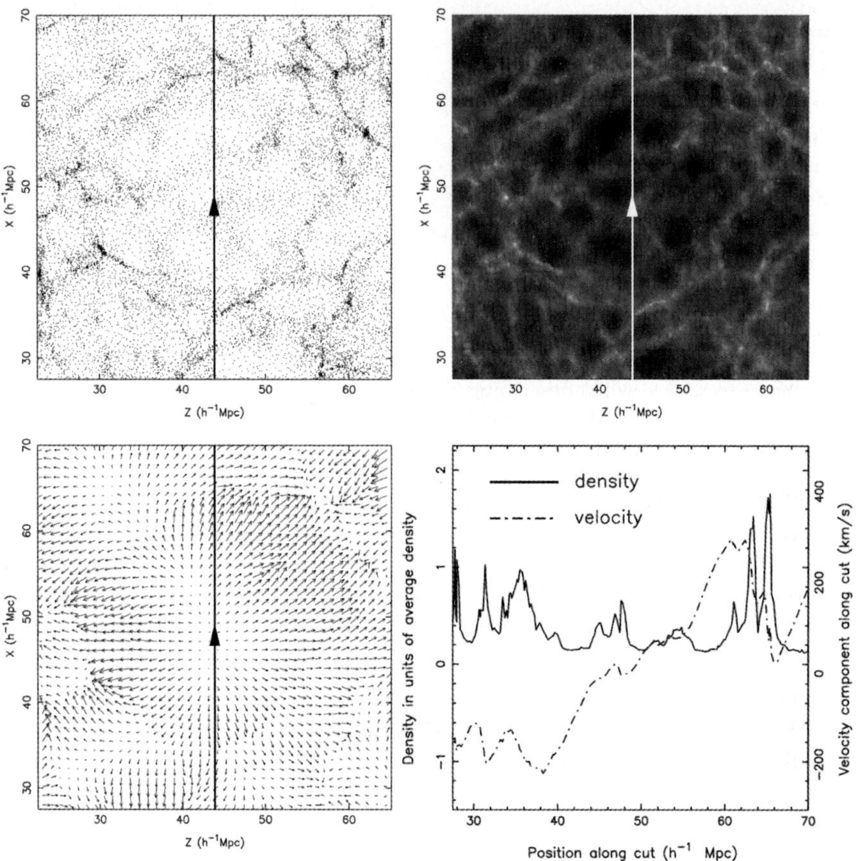

Fig. 24. The density and velocity field around a void in the GIF LCDM simulation. The top righthand panel shows the N-body simulation particle distribution within a slice through the simulation box, centered on the void. The top righthand panel shows the grayscale map of the DTFE density field reconstruction in and around the void, the corresponding velocity vector plot is shown in the bottom lefthand panel. Notice the detailed view of the velocity field: within the almost spherical global outflow of the void features can be recognized that can be identified with the diluted substructure within the void. Along the solid line in these panels we determined the linear DTFE density and velocity profile (**bottom righthand frame**). We can recognize the global "bucket" shaped density profile of the void, be it marked by substantial density enhancements. The velocity field reflects the density profile in detail, dominated by a global super-Hubble outflow. From Schaap [160]

inner shells of a void do overtake the outer shells as a result of their larger peculiar acceleration. Bertschinger [18] pointed out that a void would assume a self-similar expansion and propagate at a slower rate through the surrounding medium Bertschinger [18]. On the basis of this observation, Blumenthal et al. [20] suggested that the voids observed in galaxy redshift surveys, or in N-body simulations, should be identified with such shell-crossing voids.

The void threshold that corresponds to shell-crossing of a spherical tophat void, $\delta_v = -2.81$ (for a $\Omega_m = 1$ Einstein-de Sitter Universe, and for a growing-mode perturbation). Once the (fictitious) linear growth of a density trough in the primordial density field has reached the *void barrier* the depression will have evolved into a genuine void. Given the primordial density field $\delta(\mathbf{x})$, linearly interpolated to the present epoch, at any one cosmic redshift z one can identify the voids that have evolved beyond the shell-crossing phase and emerged as mature voids,

$$\Delta(\mathbf{x}) < \delta_{\rm ssc}(z, \Omega_m, \Omega_\Lambda) \approx \frac{\delta_{\rm v}}{D(z)}, \tag{1}$$

where the index *ssc* refers to “spherical shell crossing”.

4.3 A Void in LCDM

Soon after their discovery, various studies pointed out their essential role in the organization of the cosmic matter distribution (e.g. icke [84], Regős and Geller [145]). Their effective repulsive influence over their surroundings has been recognized in various galaxy surveys in the Local Universe (see Sect. 2.3).

Here we address the void's dynamical influence by means of a case study of the structure and outflow from a void selected from a ΛCDM GIF N-body simulation Kauffmann et al. [94]. Figure 25 shows a typical void-like region in a ΛCDM Universe. It concerns a 256^3 particles GIF N-body simulation, encompassing a ΛCDM ($\Omega_m = 0.3, \Omega_\Lambda = 0.7, H_0 = 70\,\mathrm{km/s/Mpc}$) density field within a (periodic) cubic box with length 141 h^{-1} Mpc and produced by means of an adaptive $\mathrm{P}^3\mathrm{M}$ N-body code.

The top left frame shows the particle distribution in and around the void within this 42.5 h^{-1} Mpc wide and 1 h^{-1} Mpc thick slice through the simulation box. In the same figure we include panels of the density and velocity field in the void, determined by means of a DTFE reconstruction (see Schaap [160] van de Weygaert and Schaap [188]). Both form a nice illustration of the capacity of the tessellation-based DTFE interpolation and reconstruction technique to translate the inhomogeneous particle distribution into highly resolved continuous and volume-filling fields and even follow the density field as well as velocity flow throughout diluted void regions.

Void Infrastructure

The void region appears as a slowly varying region of low density (top right-hand frame). Notice the clear distinction between the empty(dark) interior

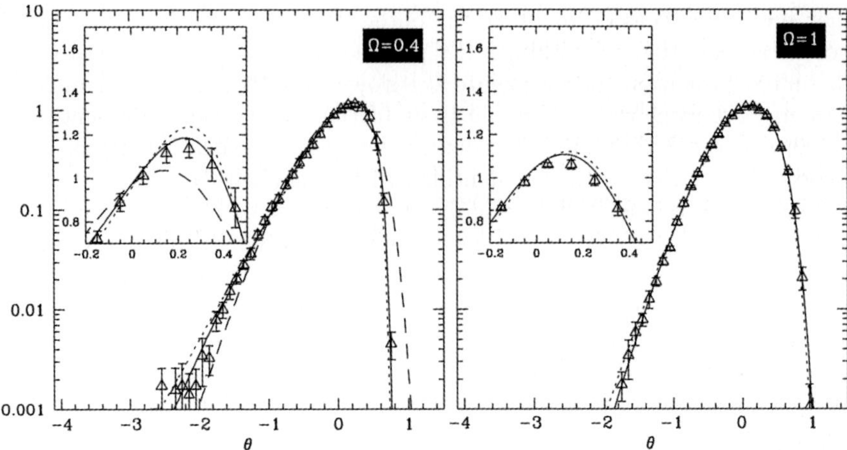

Fig. 25. The imprint of voids on the cosmic velocity field. The velocity divergence pdf for a matter-dominated $\Omega = 0.4$ and a $\Omega = 1.0$ universe, determined from a CDM N-body simulation by means of the DTFE technique. The pdf has a sharp high-value edge, defined by the outflow from voids. **Lefthand frame**: $\Omega = 0.4$, with superimposed (*dashed*) the pdf for a $\Omega = 1.0$ Universe. **Righthand frame**: $\Omega = 1.0$. From Bernardeau et al. [17]

regions of the void and its edges. In the interior of the void several smaller *subvoids* can be distinguished, with boundaries consisting of low density filamentary or planar structures.

The general characteristics of the expanding void are most evident when following the density and velocity profile along a one-dimensional section through the void. The bottom-left frame of Fig. 24 shows these profiles for the linear section along the solid line indicated in the other three frames. The first impression is that of the *bucket-like* shape of the void, be it interspersed by a rather pronounced density enhancement near its centre. This profile shape does confirm to the general trend of low-density regions to develop a near uniform interior density surrounded by sharply defined boundaries. Because initially emptier inner regions expand faster than the denser outer layers the matter distribution gets evened out while the inner layers catch up with the outer ones.

Void Velocity Field

The flow in and around the void is dominated by the outflow of matter from the void, culminating into the void's own expansion near the outer edge. The comparison with the top two frames demonstrates the strong relation with features in the particle distribution and the density field. Not only is it slightly elongated along the direction of the void's shape, but it is also sensitive to some prominent internal features of the void. Towards the "SE" direction the flow appears to slow down near a ridge, near the centre the DTFE reconstruction identifies two expansion centres.

The void velocity field profile is intimately coupled to that of its density field. The linear velocity increase is a manifestation of its general expansion. The near constant velocity divergence within the void conforms to the *super-Hubble flow* expected for the near uniform interior density distribution. Because voids are emptier than the rest of the universe they will expand faster than the rest of the universe with a net velocity divergence equal to

$$\theta = \frac{\nabla \cdot \mathbf{v}}{H} = 3(\alpha - 1)\,, \tag{2}$$

$$\alpha = H_{\rm void}/H\,, \tag{3}$$

where α is defined to be the ratio of the super-Hubble expansion rate of the void and the Hubble expansion of the universe.

Expanding Voids and the Cosmos

Evidently, the highest expansion ratio is that for voids which are completely empty, ie. $\Delta_{\rm void} = -1$. The expansion ratio α for such voids may be inferred from Birkhoff's theorem, treating these voids as empty FRW universes whose expansion time is equal to the cosmic time. For a matter-dominated Universe with zero cosmological constant, the maximum expansion rate that a void may achieve is given by

$$\theta_{\rm max} = 1.5\,\Omega_m^{0.6}\,, \tag{4}$$

with Ω_m the cosmological mass density parameter. For empty voids in a Universe with a cosmological constant a similar expression holds, be it that the value of α will have to be numerically calculated from the corresponding equation. In general the dependence on Λ is only weak. Generic voids will not be entirely empty, their density deficit $|\Delta_{\rm void}| \approx 0.8-0.9$ (cf. eg. the linear density profile in Fig. 25). The expansion rate $\theta_{\rm void}$ for such a void follows from numerical evaluation of the expression

$$\theta_{\rm void} = \frac{3}{2}\,\frac{\Omega_m^{0.6} - \Omega_{m,\rm void}^{0.6}}{1 + \frac{1}{2}\Omega_{m,\rm void}^{0.6}}\,; \qquad \Omega_{m,\rm void} = \frac{\Omega_m(\Delta_{\rm void}+1)}{(1+\frac{1}{3}\theta)^2} \tag{5}$$

in which $\Omega_{m,\rm void}$ is the effective cosmic density parameter inside the void.

When assessing the statistics of the velocity field divergence, using appropriate tools, one may indeed find a sharp positive divergence cutoff marking the maximum expansion rate of void regions. On the basis of their tessellation based technique, an early velocity field oriented version of DTFE, Bernardeau and van de Weygaert [16] and Bernardeau et al. [17] demonstrated that potentially one may indeed infer information on $\Omega_{m,0}$ from the expansion of voids.

4.4 Void Sociology

Computer simulations of the gravitational evolution of voids in realistic cosmological environments do show a considerably more complex situation than that described by idealized spherical or ellipsoidal models (see Martel and Wassermann [111], Regős and Geller [145], Dubinski et al. [53], van de Weygaert and van Kampen [187], Goldberg and Vogeley [68], Colberg et al. [36], Padilla et al. [125]). In recent years the huge increase in computational resources has enabled N-body simulations to resolve in detail the intricate substructure of voids within the context of hierarchical cosmological structure formation scenarios [7, 36, 68, 71, 78, 116, 125]. They confirm the theoretical expectation of voids having a rich substructure as a result of their hierarchical buildup (see e.g. Fig. 21).

Sheth and van de Weygaert [168] treated the emergence and evolution of voids within the context of *hierarchical* gravitational scenarios. It leads to a considerably modified view of the evolution of voids. The role of substructure within their interior and the interaction with their surroundings turn out to be essential aspects of the *hierarchical* evolution of the void population in the Universe. An important guideline are the heuristic void model simulations by Dubinski et al. [53], and the theoretical void study by Sahni et al. [156] within the context of a Lagrangian adhesion model approach by Sahni et al. [156]. Sheth and van de Weygaert [168] showed that the hierarchical development of voids, akin to the evolution of overdense halos, may be described by an *excursion set* formulation [22, 142, 167]. In some sense voids have a considerably more complex evolutionary path than overdense halos. This prodded the development of a two-barrier excursion set formalism (see Sect. 3.7 in accompanying lecture notes on the theory of the Cosmic Web). The two barriers refer to two processes that dictate the evolution of voids: their *merging* into ever larger voids as well as the *collapse* and disappearance of small ones embedded in overdense regions.

Void Merging

First, consider a small region which was less dense than the critical $\delta_{\rm v}$. It may be that this region is embedded in a significantly larger underdense region which is also less dense than the critical density. Many small primordial density troughs may exist within the larger void region. Once small voids located within the larger embedding underdensity have emerged as true voids at some earlier epoch, their expansion tends to slow down. Subsequently, they merge and get absorbed into the larger void emerging from the embedding underdensity as it reaches its shell-crossing phase. Therefore, we should identify the larger region as a large void today, while the smaller subvoids should not anymore be counted as such (see Fig. 26 bottom row).

Void Collapse

A *second* void process is responsible for the radical dissimilarity between void and halo populations. If a small scale minimum is embedded in a sufficiently high large scale density maximum, then the collapse of the larger surrounding region will eventually squeeze the underdense region it surrounds: the small-scale void will vanish when the region around it has collapsed completely. Alternatively, though usually coupled, they may collapse as a result of the tidal force field in which they find themselves. If the void within the contracting overdensity has been squeezed to vanishingly small size it should no longer be counted as a void (see Fig. 26 bottom row).

The collapse of small voids is an important aspect of the symmetry breaking between underdensities and overdensities. In the primorial Universe, Gaussian primordial conditions involve a perfect symmetry between

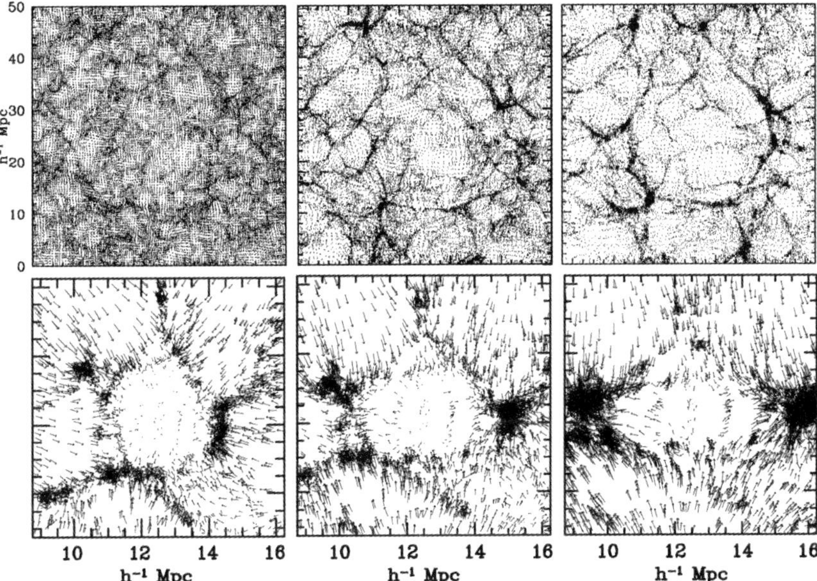

Fig. 26. The two modes of void evolution: void merging (**top row**) and void collapse (**bottom row**). **Top**: three timesteps of evolving void structure in a 128^3 particle N-body simulation of structure formation in an SCDM model ($a_{\rm exp} = 0.1, 0.3, 0.5$). The sequence shows the gradual development of a large void of diameter ≈ 25 h^{-1} Mpc as the complex pattern of smaller voids and structures which had emerged within it at an earlier time, merge with one another. It illustrates the *void-in-void* process of the evolving void hierarchy. **Bottom**: a choice of three collapsing voids in a constrained N-body simulation, each embedded within an environment of different tidal shear strength. The arrows indicate the velocity vectors, showing the infall of outer regions onto the void region. As a result the voids will be crushed as the surrounding matter rains down on them

under- and overdense. Any inspection of a galaxy redshift map or an N-body simulation shows that there is a marked difference between matter clumps and voids. While the number density of halos is dominated by small objects, void collapse is responsible for the lack of small voids.

4.5 Void Excursions

The excursion set formalism allows an elegant formulation and evaluation of the complex evolution of voids outlined above in terms of a *two-barrier* excursion set formalism. The *merging* and *collapse* barriers have been indicated by horizontal bars in the Brownian random walk diagram of Fig. 27. In the formalism developed by Sheth and van de Weygaert [168] the maturing/merging threshold is set to a fixed threshold value, independent of scale: the *shell-crossing* value $\delta_v = -2.81$ of spherical voids. The *void collapse* of an underdensity embedded within a contracting overdensity is set by the collapse barrier δ_c (for halos).

Void-in-Void and Void-in-Cloud

Since many small voids may coexist within one larger void, we must not count all of the smaller voids as distinct objects, lest we overestimate the

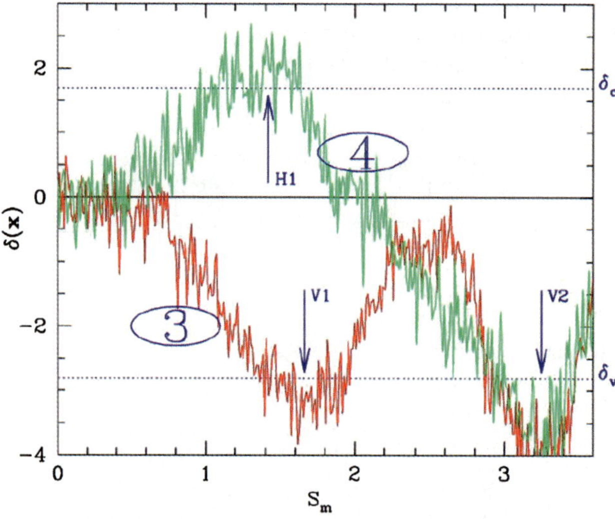

Fig. 27. Two-barrier excursion set formalism for the two void processes: void merging (red) and void collapse (green). Random walk exhibited by the average overdensity δ centred on a randomly chosen position in a Gaussian random field, as a function of smoothing scale, parametrized by S_M (large volume are on the left, small volumes on the right). Dashed horizontal lines indicate the collapse barrier δ_c and the void shell-crossing barrier δ_v

number of small voids and the total volume fraction in voids. This is called the *void-in-void* problem. In this case small voids from an early epoch merge with one another to form a larger void at a later epoch. It is analogous to the well-known *cloud-in-cloud* problem associated with the number density of initially overdense peaks. To account for the impact of voids disappearing when embedded in collapsing regions, we must also deal with the *void-in-cloud* problem. Also see Fig. 9 in van de Weygaert and Bond (2005).

By contrast, the evolution of overdensities is governed only by the *cloud-in-cloud* process; the *cloud-in-void* process is much less important, because clouds which condense in a large scale void are not torn apart as their parent void expands around them. This asymmetry between how the surrounding environment affects halo and void formation is incorporated into the *excursion set approach* by using one barrier to model halo formation and a second barrier to model void formation (Fig. 9 in van de Weygaert and Bond (2005)). Only the first barrier matters for halo formation, but both barriers play a role in determining the expected abundance of voids.

Brownian Void Walks

Figure 27 depicts two different random walks, each illustrative examples of the void evolution processes. The red Brownian random walk relates to the *void-in-void* trajectory of void formation through the merging of voids. The green Brownian random walk depicts the fateful events unfolding for a collapsing void, a *void-in-halo* trajectory. The *void-in-void* random walk looks very much the inverse of that for the cloud-in-cloud process associated with halo mergers. The associated random walk shows that the present-day void V1 contains more mass ($S \sim 1.6$) than the smaller void V2 ($S \sim 3.2$) which merged into $V1$. The red random walk concerns a location which at early times was found within a small void $V2$. This void, however, is embedded on a mass larger mass scale within an overdense halo $H1$. Once this entity collapses into a massive virialized halo, V2 will have disappeared.

If a walk first crosses δ_c and then crosses δ_v on a smaller scale, then the smaller void is contained within a larger collapsed region. Since the larger region has collapsed, the smaller void within it no longer exists, so it should not be counted. The only bona-fide voids are those associated with walks which cross δ_v without first crossing δ_c. The problem of estimating the fraction of mass in voids reduces to estimating the fraction of random walks which first crossed δ_v at S, and which did not cross δ_c at any $S' < S$: the description of the void hierarchy requires solution of a *two-barrier* problem.

4.6 Void Spectrum

The analytical evaluation of the two-barrier random walk problem leads directly to a prediction of the distribution function $n_v(M)$ for voids on a mass scale M. With respect to the linear extrapolated density field the matured void on a mass scale m corresponds to a fractional relative underdensity $\sqrt{\nu_v(M)}$,

$$\nu_v(M) \equiv \frac{|\delta_v|}{\sigma(M)}\,, \tag{6}$$

with the dependence on the mass scale M entering via the rms density fluctuation on that scale, $\sigma(M)$. According to the Sheth and van de Weygaert [168] the resulting void mass spectrum may be approximated by

$$n_v(M)\,\mathrm{d}M \approx \tag{7}$$

$$\sqrt{\frac{2}{\pi}}\frac{\rho_u}{M^2}\,\nu_v(M)\,\exp\left(-\frac{\nu_v(M)^2}{2}\right)\left|\frac{\mathrm{d}\ln\sigma(M)}{\mathrm{d}\ln M}\right|\,\exp\left\{-\frac{|\delta_\mathrm{v}|}{\delta_\mathrm{c}}\frac{\mathcal{D}^2}{4\nu_v^2}-2\frac{\mathcal{D}^4}{\nu_v^4}\right\}.$$

which for a pure power-law power spectrum yields

$$\begin{aligned} n_v(M)\,\mathrm{d}M \approx & \sqrt{\frac{1}{2\pi}}\left(1+\frac{n}{3}\right)\frac{\rho_u}{M^2}\left(\frac{M}{M_{v,*}}\right)^{(3+n)/6}\exp\left\{-\left(\frac{M}{M_{v,*}}\right)^{(3+n)/3}\right\} \\ & \times\exp\left\{-\frac{\mathcal{D}^2}{2}\left(\frac{|\delta_\mathrm{v}|}{4\delta_\mathrm{c}}+\mathcal{D}^2\left(\frac{M}{M_{v,*}}\right)^{-(3+n)/3}\right)\left(\frac{M}{M_{v,*}}\right)^{-(3+n)/3}\right\}. \end{aligned} \tag{8}$$

The quantity $\mathcal{D}$ is the "*void-and-cloud parameter*",

$$\mathcal{D} \equiv \frac{|\delta_\mathrm{v}|}{(\delta_\mathrm{c}+|\delta_\mathrm{v}|)}\,. \tag{9}$$

It parameterizes the impact of halo evolution on the evolving population of voids: the likelihood of smaller voids being crushed through the *void-in-cloud* process decreases as the relative value of the collapse barrier δ_c with respect to the void barrier δ_v becomes larger.

Along with the derived void distribution a variety of related interesting observations may be made. One aspect concerns the fraction of mass contained in voids on mass scale M,

$$f(M) = \frac{M\,n_v(M)}{\rho_u}\,. \tag{10}$$

The resulting distribution is also peaked. The top lefthand frame of Fig. 28 shows that most of the void mass is indeed to be found in voids of characteristic mass $M_{v,*}$. At any given time the mass fraction in voids is approximately thirty percent of the mass in the Universe.

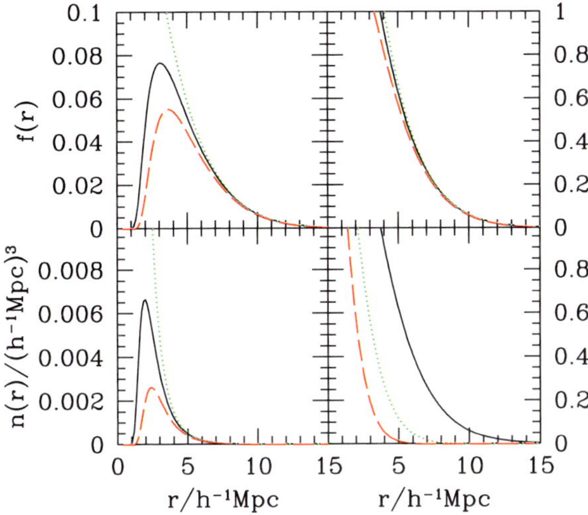

Fig. 28. Distribution of void radii predicted on the basis of (9), in an Einstein de-Sitter model with $P(k) \propto k^{-1.5}$, normalized to $\sigma_8 = 0.9$ at $z = 0$. Top left panel shows the mass fraction in voids of radius r. Bottom left panel shows the number density of voids of radius r. Note that the void-size distribution is well peaked about a characteristic size provided one accounts for the void-in-cloud process. Top right panel shows the cumulative distribution of the void volume fraction. Dashed and solid curves in the top panels and bottom left panel show the two natural choices for the importance of the void-in-cloud process discussed in the text: $\delta_c = 1.06$ and 1.686, with $\delta_v = -2.81$. Dotted curve shows the result of ignoring the *void-in-cloud* process entirely. Clearly, the number of small voids decreases as the ratio of $\delta_c/|\delta_v|$ decreases. Bottom right panel shows the evolution of the cumulative void volume fraction distribution. The three curves in this panel are for $\delta_c = 1.686(1+z)$, where $z = 0$ (*solid*), 0.5 (*dotted*) and 1 (*dashed*)

Characteristic Void Size

Expression (9) shows clearly that $n(M)$ cuts-off sharply at both small and large values of ν_v. This becomes clear when inspecting The number density $n_v(R)$ of voids of radius R^3 in Fig. 28 (bottom lefthand frame). It shows that the distribution of void masses is reasonably well peaked about $\nu \approx 1$, corresponding to a characteristic mass scale of order $\sigma_0(M) \approx |\delta_v|$.

The above implies that at any one cosmic epoch there is a *characteristic void size* which increases with time: the larger voids present at late time formed from mergers of smaller voids which formed at earlier times. For pure

[3] The conversion of the void mass scale to equivalent void radius R is done by assuming the simplest approximation, that of the spherical tophat model. According to this model a void has expanded by a factor of 1.7 by the time it has mature, so that $V_v = (M/\rho_u) * 1.7^3$.

power-law power spectra this means that this *self-similar* evolution of the void population centers around the evolving void mass $M_{v,*}$:

$$M_{v,*}(t) \propto D(t)^{6/(3+n)} M_{v,*,o} , \tag{11}$$

in which the present-day characteristic void mass, inversely proportional to $|\delta_v|$, is

$$M_{v,*,o} = \left(\frac{2A}{\delta_v^2}\right)^{3/(3+n)} . \tag{12}$$

Self-similar Void Evolution

In an Einstein de-Sitter universe, $\delta_{\rm c}$, $\delta_{\rm v}$ and $\sigma(m)$ all have the same time dependence, so (9) evolves self-similarly, parameterized by the characteristic *"void mass"* $M_{v,*}$. Also for more general world-models the approximation of self-similar void evolution should be quite accurate as the time dependences are only slightly different.

4.7 Void Evolution

The population of large voids is insensitive to the *void-in-cloud* process. The large mass cutoff of the void spectrum is similar to the ones for clusters and reflects the Gaussian nature of the fluctuation field from which the objects have condensed. The gradual merging of voids into ever larger ones is embodied in the self-similar shift of the peak of the void spectrum, ie. of $M_{v,*}$. The abundance of voids which larger than the typical initial comoving sizes of clusters is therefore reasonably described by peaks theory [10, 168].

While the two-barrier excursion set formalism offers an attractive theoretical explanation for the distinct asymmetry between clumps and voids and for the peaked void size distribution, we need to identify where the disappearing small-scale voids are to be found in a genuine evolving cosmic matter distribution. Using the GIF N-body simulations of various CDM scenarios, Platen [134] has managed to trace various specimen of this unfortunate void population. Using the new *Watershed Void Finder* technique [135] identified small-scale voids at high redshift ($z = 3$) and subsequently followed their evolution. Figure 29 shows the void distribution in and around a large central underdensity at four cosmic epochs, $z = 3.0, 2.0, 1.5$ and 0.5. The fate of the subvoids within the large present-day void is clearly visible: the interior ones tend to merge with surrounding peers while the ones near the boundary get squeezed out of existence. Close inspection shows that the small voids are not collapsing isotropically. Instead they tend to get sheared by their surroundings.

This image of void formation in the dark matter distribution has been elaborated by Furlanetto and Piran [64] to describe the implications for voids

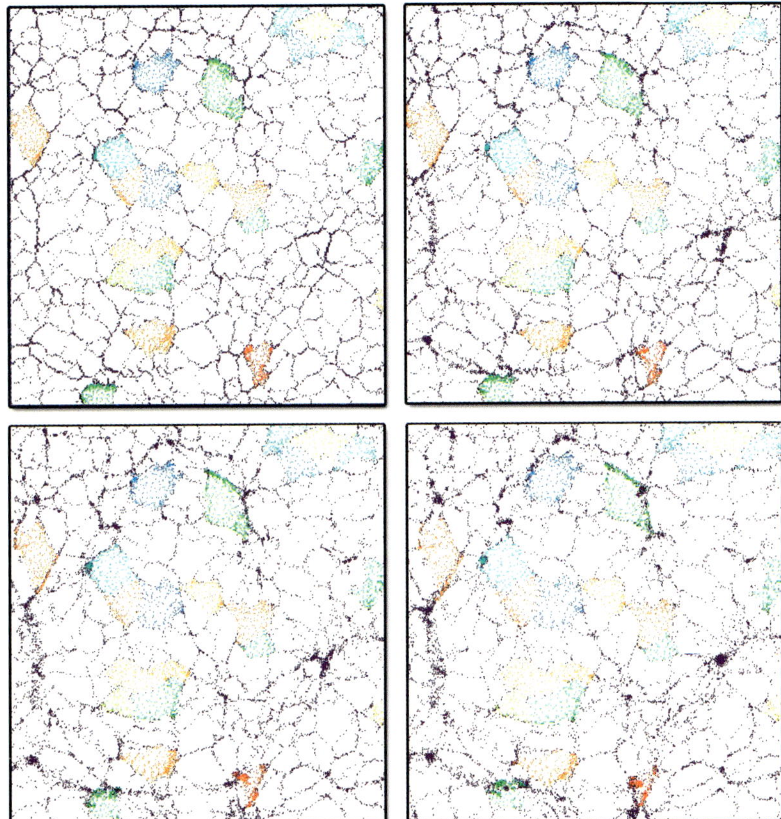

Fig. 29. Evolving Void Hierarchy: the structure in and around a large central void in a GIF ΛCDM simulation. At $z = 3$ the watershed WVF voidfinder [135] has been applied to trace the outline of voids in the matter distribution. Particles at the surrounding ridges (boundaries) are subsequently followed. The four frames depict the resulting particle distribution in a 5 h^{-1} Mpc thick and ≈ 60 h^{-1} Mpc wide slice, at 4 successive time intervals: $z = 3.0, 2.0, 1.5$ and 0.5. Clearly visible is the fate of subvoids within the large present-day void: either they merge into the background or they get squeezed out of existence near the boundary. From Platen et al. [136]. Courtesy GIF simulation: J. Colberg and Virgo consortium

in the galaxy distribution while it forms the starting point for various ongoing investigations.

The demise of small voids near the boundaries of large voids, touching the surrounding filaments and sheets, is a clear indication for the importance of tidal influences on the developing subvoid. Tidal stresses induced by the large scale vicinity will be of major importance for their final fate. One may argue that tidal influences are more important for voids than they are for halos. Because their underdensity is naturally limited ($\delta \geq -1$) and because their size is expanding the environment retains a dominant dynamical influence, in particular over the outer region of the voids. The accompanying force field will in general be anisotropic and if strong enough enforce a shearing collapse. It is entirely in line with the recent observation by Park and Lee [127] and Platen, van de Weygaert and Jones [136], that the shape of voids is significantly affected by the tidal influence of the surrounding matter distribution.

4.8 Soapsud of Voids

An important aspect of the implied void population is that it is approximately *space-filling*. It underlines the adagio that the large scale distribution of matter may be compared to a *soapsud of expanding bubbles*. This follows from evaluation of the cumulative integral

$$f_V(M) \equiv \int_M^\infty (1.7)^3 \, \frac{M' \, n_v(M')}{\rho_u} \, \mathrm{d}M' \, . \tag{13}$$

where the factor 1.7 is an estimate of the excess expansion of the void based upon the spherical model for void evolution (see footnote). The resulting (current) cumulative void volume distribution is shown in the top righthand panel of Fig. 28. For a finite value of void radius R the whole of space indeed appears to be occupied by voids. Even more impressive is the corresponding self-similar evolution of the culumative void volume distribution $f_V(M,t)$. The bottom righthand frame of Fig. 28 shows the gradual shift of the cumulative volume distribution towards larger voids. The correct image appears to be that of a gradually unfolding bubbly universe in which the average size of the voids grows as small voids merge into ever larger ones.

5 Conclusion: Morphology of the Cosmic Web

The Megaparsec scale galaxy distribution defines one of the most intriguing spatial patterns in nature, the Cosmic Web. In these notes we have looked into the many diverse aspects of the available observational information. For a considerable period the spatial analysis of weblike structures has been based on rather ill-defined heuristic concepts, difficult to interpret within the context of existing theories. We have provided a review of the recent activity towards this direction. A set of techniques has opened the path towards a meaningful quantitative analysis. Morphologically, the most distinct elements of the Cosmic Web are filaments and voids. Filaments have figured prominently in the accompanying theoretical treatise van de Weygaert and Bond, 2005 on the formation of the web. In these lecture notes we have put special emphasis on the voids in cosmic matter and galaxy distribution.

6 Acknowledgments

We wish to thank Manolis Plionis and Omar López-Cruz for their invitation and the wonderful weeks in Mexico, and for their almost infinite patience regarding our shifting deadlines. RvdW is most grateful to the hospitality of the Canadian Institute for Astrophysics, where we commenced the project leading to these notes. Both authors thank Max-Planck-Institut für Astrophysik in Garching for providing the hospitality and facilities allowing the completion of these lecture notes. In particular we are indebted to Jacqueline van Gorkom, Hans Böhringer and George Rhee. Without their encouragement and the more than helpful assistance and understanding of Sonja Japenga of Springer Verlag we would not have managed to bring these notes to completion. To them we owe a major share of our gratitude ! RvdW is grateful to Miguel Aragón-Calvo for his permission to use and manipulate various figures from his Ph.D. thesis. He also acknowledges him and Erwin Platen for many inspiring discussions and their contributions towards obtaining insight into the evolution of the Cosmic Web. Most fondly we wish to thank Bernard Jones, for his enthusiastic and crucial support and inspiration, the many original ideas over the years and for his support in completing this manuscript hours past midnight ...

References

1. Abazajian, K., et al. (The SDSS collaboration): Astron. J. **126**, 2081 (2003)
2. Abell, G.: Astrophys. J. Suppl. **3**, 211 (1958)
3. Abell, G., Corwin, G.O., Olowin, R.P.: Astrophys. J. Suppl. **70**, 1 (1989)
4. Aikio, J., Mähönen, P.: Astrophys. J. **497**, 534 (1998)
5. Aragón-Calvo, M.A.: Morphology and dynamics of the Cosmic Web, Ph.D. thesis, Groningen University (2007)
6. Aragón-Calvo, M.A., Jones, B.J.T., van de Weygaert, R., van der Hulst, J.M.: Astron. Astrophys. **474**, 315 (2007)
7. Arbabi-Bidgoli, S., Müller, V.: Mon. Not. R. Astron. Soc. **332**, 205 (2002)
8. Babul, A., Starkman, G.D.: Astrophys. J. **401**, 28 (1992)
9. Bahcall, N.A.: Ann. Rev. Astron. Astrophys. **26**, 631 (1988)
10. Bardeen, J.M., Bond, J.R., Kaiser, N., Szalay, A.S.: Astrophys. J. **304**, 15 (1986)
11. Barrow, J.D., Bhavsar, S.P., Sonoda, D.H.: Mon. Not. R. Astron. Soc. **216**, 17 (1985)
12. Basilakos, S., Plionis, M., Rowan-Robinson, M.: Mon. Not. R. Astron. Soc. **323**, 47 (2001)
13. Basilakos, S., Plionis, M.: Astrophys. J. **550**, 522 (2001)
14. Benson, A.J., Hoyle, F., Fernando, T, Vogeley, M.S.: Mon. Not. Roy. Astron. Soc. **340**, 160 (2003)
15. Berlind, A.A., Frieman, J., Weinberg, D.H., Blanton, M.R., Warren, M.S., Abazajian, K., Scranton, R., Hogg, D.W., Scoccimarro, R., Bahcall, N.A., Brinkmann, J., Gott, III J.R., Kleinman, S.J., Krzesinski, J., Lee, B.C.,

Miller, C.J., Nitta, A., Schneider, D.P., Tucker, D.L., Zehavi, I.: Astrophys. J. Suppl. **167**, 1 (2006)
16. Bernardeau, F., van de Weygaert, R.: Mon. Not. R. Astron. Soc. **279**, 693 (1996)
17. Bernardeau, F., van de Weygaert, R., Hivon, E., Bouchet, F.: Mon. Not. R. Astron. Soc. **290**, 566 (1997)
18. Bertschinger, E.: Astrophys. J. **295**, 1 (1985)
19. Bertschinger, E., Dekel A, Faber, S.M., Dressler, A., Burstein, D.: Astrophys. J. **364**, 370 (1990)
20. Blumenthal, G.R., Da Costa, L., Goldwirth, D.S., Lecar, M., Piran, T.: Astrophys. J. **388**, 324 (1992)
21. Böhringer, H., Schuecker, P., Guzzo, L., Collins, C., Voges, W., Schindler, S., Neumann, D.M., Cruddace, R.G., DeGrandi, S., Chincarini, G., Edge, A.C., MacGillivray, H.T., Shaver, P.: Astron. Astrophys. **369**, 826 (2001)
22. Bond, J.R., Cole, S., Efstathiou, G., Kaiser, N.: Astrophys. J. **379**, 440 (1991)
23. Borgani, S., Guzzo, L.: Nature **409**, 39 (2001)
24. Bothun, G.D., Geller, M.J., Kurtz, M.J., Huchra, J.P., Schild, R.E.: Astrophys. J. **395**, 347 (1992)
25. Branchini, E., Teodoro, L., Frenk, C.S., Schmoldt, I., Efstathiou, G., White, S.D.M., Saunders, W., Sutherland, W., Rowan-Robinson, M., Keeble, O., Tadros, H., Maddox, S., Oliver, S.: Mon. Not. R. Astron. Soc. **308**, 1 (1999)
26. Branchini, E., Freudling, W., Da Costa, L.N., Frenk, C.S., Giovanelli, R., Haynes, M.P., Salzer, J.J., Wegner, G., Zehavi, I.: Mon. Not. R. Astron. Soc. **326**, 1191 (2001)
27. Braun, J., Sambridge, M.: Nature **376**, 655 (1995)
28. Carlstrom, J.E., Holder, G.P., Reese, E.D.: Ann. Rev. Astron. Astrophys. **40**, 643 (2002)
29. Ceccarelli, L., Padilla, N.D., Valotto, C., Lambas, D.G: Mon. Not. Roy. Astron. Soc. **373**, 1440 (2006)
30. Cen R., Miralda-Escudé J., Ostriker J.P., Rauch M.: Astrophys. J. **437**, 9 (1994)
31. Cen R., Ostriker J.: Astrophys. J. **650**, 560 (2006)
32. Chiang, L.-Y., Coles, P.: Mon. Not. R. Astron. Soc. **311**, 809 (2000)
33. Chincarini, G., Rood, H.J.: Nature **257**, 294 (1975)
34. Colberg, J.M.: Mon. Not. R. Astron. Soc. **375**, 337 (2007)
35. Colberg, J.M., Krughoff, K.S., Connolly, A.J.: Mon. Not. R. Astron. Soc. **359**, 272 (2005)
36. Colberg, J.M., Sheth, R.K., Diaferio, A., Gao, L., Yoshida, N.: Mon. Not. R. Astron. Soc. **360**, 216 (2005)
37. Colberg, J.M., Pearce, F., Foster, C., Platen, E., Brunino, R., Basilakos, S., Fairall, A., Feldman, H., Gottlöber, S., Hahn, O., Hoyle, F., Müller, V., Nelson, L., Neyrinck, M., Plionis, M., Porciani, C., Shandarin, S., Vogeley, M., van de Weygaert, R.: Mon. Not. R. Astron. Soc., subm. (2008)
38. Coles, P., Chiang, L.-Y.: Nature **406**, 376 (2000)
39. Colless, M., et al.: Astro-ph/0306581 (2003)
40. Colombi, S., Pogosyan, D., Souradeep, T.: Phys. Rev. Let. **85**, 5515 (2000)
41. Cressie, N.: Statistics for Spatial Data, rev. edn. John Wiley & Sons, Chichester (1993)
42. Davé R., et al.: Astrophys. J. **552**, 473 (2001)

43. Davis, M., Huchra, J., Latham, D.W., Tonry, J.: Astrophys. J. **253**, 423 (1982)
44. de Lapparent, V., Geller, M.J., Huchra, J.P.: Astrophys. J. **369**, 273 (1991)
45. de Lapparent, V., Geller, M.J., Huchra, J.P.: Astrophys. J. **302**, L1 (1986)
46. Dekel, A., Bertschinger, E., Faber, S.M.: Astrophys. J. **364**, 349 (1990)
47. Dekel, A.: Ann. Rev. Astron. Astrophys. **32**, 371 (1994)
48. Dekel, A., Lahav, O.: Astrophys. J. **520**, 24 (1999)
49. Dekel, A., Rees, M.J.: Nature **326**, 455 (1987)
50. Dekel, A., Rees, M.J.: Astrophys. J. **433**, L1 (1994)
51. Delaunay, B. N.: Bull. Acad. Sci. USSR Clase. Sci. Mat. **7**, 793 (1934)
52. Doré, O., Colombi, S., Bouchet, F.R.: Mon. Not. R. Astron. Soc. **344**, 905 (2003)
53. Dubinski, J., da Costa, L.N., Goldwirth, D.S., Lecar, M., Piran, T.: Astrophys. J. **410**, 458 (1993)
54. Ebeling, H., Edge, A.C., Allen, S.W., Crawford, C.S., Fabian, A.C.: Mon. Not. Roy. Astron. Soc. **318**, 333 (2000)
55. Einasto, J., Joeveer, M., Saar E.: Nature, **283**, 47 (1980)
56. El-Ad, H., Piran, T., da Costa, L.N.: Astrophys. J. **462**, L13 (1996)
57. El-Ad, H., Piran, T.: Astrophys. J. **491**, 421 (1997)
58. Erdoğdu, P., Lahav, O., Zaroubi, S., Efsathiou, G., Moody, S., Peacock, J.A., Colless, M., Baldry, I.K., Baugh, C.M., Bland-Hawthorn, J., 2dFGRS Team: Mon. Not. Roy. Astron. Soc. **352**, 939 (2004)
59. Erdoğdu, P., Huchra, J.P., Lahav, O., Colless, M., Cutri, R.M., Falco, E., George, T., Jarrett, T., Jones, D.H., Kochanek, C.S., Macri, L., Mader, J., Martimbeau, N., Pahre, M., Parker, Q., Rassat, A., Saunders, W.: Mon. Not. Roy. Astron. Soc. **368**, 1515 (2006)
60. Finoguenov A., Briel U.G., Henry J.P.: Astron. Astrophys. **410**, 777 (2003)
61. Fliche, H.H., Triay, R.: gr-qc/0607090 (2006)
62. Fukugita M., Hogan C., Peebles P.J.E.: Astrophys. J. **503**, 518 (1998)
63. Fukugita, M., Peebles, P.J.E.: Astrophys. J. **616**, 643 (2004)
64. Furlanetto, S.R., Piran, T.: Mon. Not. Roy. Astron. Soc. **366**, 467 (2006)
65. Gaztañaga, E.: Astrophys. J. **398**, L17 (1992)
66. Geller, M., Huchra, J. P.: Nature **246**, 897 (1989)
67. Gladders, M.D., Yee, H.K.C.: Astron. J. **120**, 2148 (2000)
68. Goldberg, D.M., Vogeley, M.S.: Astrophys. J. **605**, 1 (2004)
69. Gott, J.R. III, Dickinson, M., Melott, A.L.: Astrophys. J. **306**, 341 (1986)
70. Gott, J.R. III, Jurić, M., Schlegel, D., Hoyle, F., Vogeley, M., Tegmark, M., Bahcall, N., Brinkmann, J.: Astrophys. J. **624**, 463 (2005)
71. Gottlöber, S., Łokas, E.L., Klypin, A., Hoffman, Y.: Mon. Not. Roy. Astron. Soc. **344**, 715 (2003)
72. Graham, M.J., Clowes, R.G.: Mon. Not. R. Astron. Soc. **275**, 790 (1995)
73. Gregory, S.A., Thompson, L.A.: Astrophys. J. **222**, 784 (1978)
74. Grogin, N.A., Geller, M.J.: Astron. J. **118**, 256 (1999)
75. Grogin, N.A., Geller, M.J.: Astron. J. **119**, 32 (2000)
76. Hahn, O., Porciani, C., Carollo, M., Dekel, A.: Mon. Not. R. Astron. Soc. **375**, 489 (2007)
77. Heath Jones, D. et al.: Mon. Not. R. Astron. Soc. **355**, 747 (2004)
78. Hoeft, M., Yepes, G., Gottlöber, S., Springel, W.: Mon. Not. Roy. Astron. Soc. **371**, 401 (2006)
79. Hoffman, Y., Shaham, J.: Astrophys. J. **262**, L23 (1982)

80. Hoffman, Y., Silk, J., Wyse, R.F.G.: Astrophys. J. **388**, L13 (1992)
81. Hoyle, F., Vogeley, M.: Astrophys. J. **566**, 641 (2002)
82. Hoyle, F., Vogeley, M.: Astrophys. J. **580**, 663 (2002)
83. Huchra, J., et al. *Nearby large-scale structures and the zone of avoidance*, ASP Conf. Ser. Vol. 239, Fairall, K.P., Woudt, P.A., (Astron. Soc. Pac., San Francisco), p. 135 (2005)
84. Icke, V.: Mon. Not. R. Astron. Soc. **206**, 1P (1984)
85. Jones, B.J.T., Martínez, V.J., Saar, E., Trimble, V.: Rev. Mod. Phys. **76**, 1211 (2005)
86. Kaastra J.S., Lieu R., Tamura T., Paerels F.B.S., den Herder J.W.: Astron. Astrophys. **397**, 445 (2003)
87. Kaiser, N.: Astrophys. J. **284**, 9 (1984)
88. Kaiser, N.: Statistics of gravitational lensing 2: weak lenses. In: Martinez, V.J., Portilla, M., Saez, D. (eds.) New Insights into the Universe, Lecture Notes in Physics 408, p. 279. Springer-Verlag Berlin, Heidelberg, New York (1992)
89. Kaiser, N., Squires, G.: Astrophys. J. **404**, 441 (1993)
90. Kang H., Ryu D., Cen R., Song D.: Astrophys. J. **620**, 21 (2005)
91. Karachentseva, V.E., Karachentsev, I.D., Richter, G.M.: Astron. Astrophys. **135**, 221 (1999)
92. Karachentsev, I.D., Karachentseva, V.E., Huchtmeier, W.K., Makarov, D.I.: Astron. J. **127**, 2031 (2004)
93. Kauffmann, G., Fairall, A.P.: Mon. Not. R. Astron. Soc. **248**, 313 (1991)
94. Kauffmann, G., Colberg, J.M., Diaferio, A., White, S.D.M.: Mon. Not. R. Astron. Soc. **303**, 188 (1999)
95. Kim, R. S. J., Kepner, J.V., Postman, M., Strauss, M.A., Bahcall, N.A., Gunn, J.E., Lupton, R.H., Annis, J., Nichol, R.C., Castander, F.J., Brinkmann, J., Brunner, R.J., Connolly, A., Csabai, I., Hindsley, R.B., Izević, Ž., Vogeley, M.S., York, D.G.: Astron. J. **123**, 20 (2002)
96. Kirshner, R.P., Oemler, A., Schechter, P.L., Shectman, S.A.: Astrophys. J. **248**, L57 (1981)
97. Kirshner, R.P., Oemler, A., Schechter, P.L., Shectman, S.A.: Astrophys. J. **314**, 493 (1987)
98. Kitaura, F.S., Enßlin, T.A.: arXiv:0705.0429 (2007)
99. Kocevski, D.D., Ebeling, H.: Astrophys. J. **645**, 1043 (2006)
100. Kochanek, C.S., White, M., Huchra, J., Macri, L., Jarrett, T.H., Schneider, S.E., Mader, J.: Astrophys. J. **585**, 161 (2003)
101. Kuhn, B., Hopp, U., Elsässer, H.: Astron. Astrophys. **318**, 405 (1997)
102. Lachieze-Rey, M., da Costa, L.N., Maurogordata, S.: Astrophys. J. **399**, 10 (1992)
103. Lee, J., Park, D.: Phys. Rev. Lett (subm), arXiv0704.0881L (2007)
104. Lieu, R., Mittaz, J.P.D., Zhang, S.-N.: Astrophys. J. **648**, 176 (2006)
105. Little, B., Weinberg, D.H.: Mon. Not. Roy. Astron. Soc. **267**, 605 (1994)
106. Lombardi, M., Schneider P.: Astron. Astrophys. **373**, 359 (2001)
107. Lombardi, M., Schneider P.: Astron. Astrophys. **392**, 1153 (2002)
108. Lombardi, M., Schneider P.: Astron. Astrophys. **407**, 385 (2003)
109. López-Cruz, O.: Photometric properties of low-redshift galaxy clusters. Ph.D. thesis, Univ. Toronto (1997)
110. Luo, R., Vishniac, E.: Astrophys. J. Suppl. **96**, 429 (1995)
111. Martel, H., Wasserman, I.: Astrophys. J. **348**, 1 (1990)

112. Martínez, V., Saar E.: Statistics of the Galaxy Distribution, Chapman & Hall/CRC Press, USA (2002)
113. Martínez, V., Starck, J.-L., Saar, E., Donoho, D.L., Reynolds, S.C., de la Cruz, P., Paredes, S.: Astrophys. J. **634**, 744 (2005)
114. Massey, R., Rhodes, J., Ellis, R., Scoville, N., Leauthaud, A., Finoguenov, A., Capak, P., Bacon, D., Aussel, H., Kneib, J.-P., Koekemoer, A., McCracken, H., Mobasher, B., Pires, S., Refregier, A., Sasaki, S., Starck, J.-L., Taniguchi, Y., Taylor, A., Taylor, J.: Nature **445**, 286 (2007)
115. Matarrese, S., Coles, P., Lucchin, F., Moscardini, L.: Mon. Not. Roy. Astron. Soc. **286**, 115 (1997)
116. Mathis, H., White, S.D.M.: Mon. Not. Roy. Astron. Soc. **337**, 1193 (2002)
117. Mecke, K.R., Buchert, T., Wagner, H.: Astron. Astrophys. **288**, 697 (1994)
118. Mellier, Y.: Ann. Rev. Astron. Astrophys. **37**, 127 (1999)
119. Miller, C.J., Nichol, R.C., Reichart, D., Wechsler, R.H., Evrard, A.E., Annis, J., McKay, T.A., Bahcall, N.A., Bernardi, M., Böhringer, H., Connolly, A.J., Goto, T., Kniazev, A., Lamb, D., Postman, M., Schneider, D.P., Sheth, R.K., Voges, W.: Astron. J. **130**, 968 (2005)
120. Mo, H.J., White, S.D.M.: Mon. Not. Roy. Astron. Soc. **282**, 347 (1996)
121. Nicastro K., et al.: Nature **421**, 719 (2003)
122. Novikov, D., Colombi, S., Doré, O.: Mon. Not. R. Astron. Soc. **366**, 1201 (2006)
123. Okabe, A., Boots, B., Sugihara, K., Chiu, S. N.: Spatial Tessellations : Concepts and Applications of Voronoi Diagrams, 2nd ed. John Wiley & Sons, Chichester, Toronto (2000)
124. Oort, J.H.: Ann. Rev. Astron. Astrophys. **21**, 373 (1983)
125. Padilla, N.D., Ceccarelli, L., Lambas, D.G.: Mon. Not. Roy. Astron. Soc. **363**, 977 (2005)
126. Palmer, P.L, Voglis, N.: Mon. Not. R. Astron. Soc. **205**, 543 (1983)
127. Park, D., Lee, J.: Phys. Rev. Lett. **98**, 081301 (2007)
128. Patiri, S.G., Betancort-Rijo, J.E., Prada, F., Klypin, A., Gottlöber, S.: Mon. Not. R. Astron. Soc. **369**, 335 (2006)
129. Patiri, S.G., Prada, F., Holtzman, J., Klypin, A., Betancort-Rijo, J.E.: Mon. Not. R. Astron. Soc. **372**, 1710 (2006)
130. Peebles, P.J.E.: The Large-Scale Structure of the Universe, Princeton University Press (1980)
131. Peebles, P.J.E.: Astrophys. J. **557**, 495 (2001)
132. Pichon C., Vergely J.L., Rollinde E., Colombi S., Petitjean P.: Mon. Not. R. Astron. Soc. **326**, 597 (2001)
133. Pimbblet, K.A.: Mon. Not. R. Astron. Soc. **358**, 256 (2005)
134. Platen, E.: Segmenting the Universe. M.Sc. thesis, Groningen University (2005)
135. Platen, E., van de Weygaert, R., Jones, B.J.T.: Mon. Not. R. Astron. Soc. **380**, 551 (2007)
136. Platen, E., van de Weygaert, R., Jones, B.J.T.: Mon. Not. R. Astron. Soc., subm, arXiv:0711.2480 (2008)
137. Plionis, M., Valdarnini, R.: Mon. Not. R. Astron. Soc. **249**, 46 (1991)
138. Plionis, M., Kolokotronis, V.: Astrophys. J. **500**, 1 (1998)
139. Plionis, M., Basilakos, S.: Mon. Not. R. Astron. Soc. **330**, 399 (2002)
140. Popescu, C.C., Hopp, U., Elsässer, H.: Astron. Astrophys. **325**, 881 (1997)
141. Popessu, P., Böhringer, H., Brinkmann, J., Voges, W., York, D.G.: Astron. Astrophys. **423**, 449 (2004)

142. Press, W.H., Schechter, P.: Astrophys. J. **187**, 425
143. Rauch M.: Ann. Rev. Astron. Astrophys. **36**, 267 (1998)
144. Refregier, Y.: Ann. Rev. Astron. Astrophys. **41**, 645 (2003)
145. Regős, E., Geller, M.J.: Astrophys. J. **373**, 14 (1991)
146. Reiprich, T.H., Böhringer, H.: Astron. Nachr. **320**, 296 (1999)
147. Ripley, B.D.: Spatial Statistics, Wiley & Sons, Chichester 1981 (1981)
148. Rojas, R.R., Vogeley, M.S., Hoyle, F., Brinkmann, J.: Astrophys. J. **624**, 571 (2005)
149. Romano-Díaz, E., van de Weygaert, R: Mon. Not. Roy. Astron. Soc. **382**, 2 (2007)
150. Rozo E, Wechsler, R.H.. Koester, B.P., McKay, T.A., Evrard, A.E., Johnston, D., Sheldon, E.S., Annis, J., Frieman, J.A.: arXiv:astro-ph/0703571 (2007)
151. Rudnick, L., Brown, S., Williams, L.R.: Astrophys. J. **671**, 40 (2007)
152. Rybicki, G.B., Press, W.H.: Astrophys. J. **398**, 169 (1992)
153. Ryden, B.S., Gramann, M.: Astrophys. J. **383**, 33 (1991)
154. Ryden, B.S., Melott, A.L.: Astrophys. J. **470**, 160 (1996)
155. Saar, E., Martiínez, V.J., Starck, J.-L., Donoho, D.L.: Mon. Not. Roy. Astron. Soc. **374**, 1030 (2007)
156. Sahni, V., Sathyaprakash, B.S., Shandarin, S.F.: Astrophys. J. **431**, 20 (1994)
157. Sahni, V., Shandarin, S.F.: Mon. Not. R. Astron. Soc. **282**, 641 (1996)
158. Sahni, V., Sathyaprakash, B.S., Shandarin, S.F.: Astrophys. J. **495**, 5 (1998)
159. Scaramella, R., Vettolani, G., Zamorani, G.: Astrophys. J. **376**, L1 (1991)
160. Schaap, W. E.: The delaunay tessellation field estimator. Ph.D. thesis, Groningen University (2007)
161. Schaap, W. E., van de Weygaert, R.: Astron. Astrophys. **363**, L29 (2000)
162. Schmalzing, J., Buchert, T., Melott, A.L., Sahni, V., Sathyaprakash, B.S., Shandarin, S.F.: Astrophys. J. **526**, 568 (1999)
163. Schmidt, J.D., Ryden, B.S., Melott, A.L.: Astrophys. J. **546**, 609 (2001)
164. Shandarin, S.F., Sheth, J.V., Sahni, V.: Mon. Not. R. Astron. Soc. **353**, 162 (2004)
165. Shandarin, S., Feldman, H.A., Heitmann, K., Habib, S.: Mon. Not. R. Astron. Soc. **376**, 1629 (2006)
166. Shectman, S.A., Landy, S.D., Oemler, A., Tucker, D.L., Lin, H., Kirshner, R.P., Schechter, P.L.: Astrophys. J. **470**, 172 (1996)
167. Sheth, R.K.: Mon. Not. R. Astron. Soc. **300**, 1057 (1998)
168. Sheth, R. K., van de Weygaert, R.: Mon. Not. R. Astron. Soc. **350**, 517 (2004)
169. Sibson, R.: Math. Proc. Cambridge Phil. Soc. **87**, 151 (1980)
170. Sibson, R.: A brief description of natural neighbor interpolation. In: *Interpreting Multi-variate Data*, ed. by V. Barnett (Wiley, Chichester 1981) pp. 21–36
171. Sousbie, T.: Le squelette de l'Univers: un nouvel outil d'analyse topologique des grandes structures. Ph.D. thesis l'Ecole Normale Supérieure de Lyon (2006)
172. Sousbie, T., Pichon, C., Colombi, S., Novikov, D., Pogosyan, D.: astroph/07073123 (2007)
173. Stoica, R.S., Martínez, V.J., Mateu, J., Saar, E.: Astron. Astrophys. **434**, 423 (2005)
174. Sukumar, N.: The natural element method in solid mechanics. Ph.D. thesis, Northwestern University (1998)
175. Sunyaev, R.A., Zel'dovich, Y.B.: Comments Astrophys. Space Phys. **2**, 66 (1970)

176. Sunyaev, R.A., Zel'dovich, Y.B.: Comments Astrophys. Space Phys. **4**, 173 (1972)
177. Szapudi, I.: Astrophys. J. **497**, 16 (1998)
178. Szomoru, A., van Gorkom, J.H., Gregg, M.D., Strauss, M.A.: Astron. J. **111**, 2150 (1996)
179. Tegmark, M., Peebles, P.J.E.: Astrophys. J. **500**, L79 (1998)
180. Tegmark, M., SDSS Collaboration: Astrophys. J. **606**, 702 (2004)
181. Tikhonov, A.V., Karachentsev, I.D.: Astrophys. J. **653**, 969 (2006)
182. Tikhonov, A.V., Klypin, A.: arXiv:0708.2348v1 (2007)
183. Tully, R.B.: Nearby Galaxy Catalog, Cambridge University Press, Cambridge, (1988)
184. Tully, R.B., Shaya, E.J., Karachentsev, I.D., Courtois, H., Kocevski, D.D., Rizzi, L., Peel, A.: arXiv:0705.4139v1 (2007)
185. Vale, C., White, M.: New Astronomy. **11**. 207 (2006)
186. van de Weygaert, R.: Voids and the geometry of large scale structure. Ph.D. thesis, Leiden University (1991)
187. van de Weygaert, R., van Kampen, E.: Mon. Not. R. Astron. Soc. **263**, 481 (1993)
188. van de Weygaert, R., Schaap, W.E: The Cosmic Web: geometric analysis. In: Mart nez, V., Saar, E., Mart nez-Gonzalez, E., Pons-Borderia, M. (eds.) Data Analysis in Cosmology, lectures summerschool Valencia 2004, p. 129. Springer-Verlag (2007)
189. van de Weygaert, R., Bond, J.R.: Clusters and the theory of the Cosmic Web. In: Plionis, M., López-Cruz, O., Hughes, D., (eds.) A Pan-Chromatic View of Clusters of Galaxies and the LSS, Springer (2008)
190. van Haarlem, M., van de Weygaert, R.: Astrophys. J. **418**, 544 (1993)
191. Voronoi, G.: J. Reine Angew. Math. **134**, 167 (1908)
192. Watson, D.F.: *Contouring: A guide to the analysis and display of spatial data*, Pergamom Press, Oxford (1992)
193. White, S.D.M.: Mon. Not. R. Astron. Soc. **186**, 145 (1979)
194. Willick, J.A., Courteau, S., Faber, S.M., Burstein, D., Dekel, A., Strauss, M.A.: Astrophys. J. Suppl. **109**, 333 (1997)
195. Wittman, D., Dell'Antonio, I.P, Hughes, J.P., Margoniner, V.E., Tyson, J.A., Cohen, J.G., Norman, D.: Astrophys. J. **643**, 128 (2006)
196. Yee, H.K.C., López-Cruz, O.: Astron. J. **117**, 1985 (1999)
197. Zel'dovich, Ya.B., Einasto, J., Shandarin, S.F.: Nature **300**, 407 (1982)

Index

Abundance, 5, 6, 12, 16, 17, 33, 34, 177–184, 186, 187, 189, 196, 197, 199, 203, 204, 321, 365, 404, 455, 458
AGN, 10, 16–18, 32, 35–38, 40–43, 49–51, 59, 157, 172, 198–201, 248, 287, 431
Anisotropic collapse, 355, 366, 367, 372, 374, 375, 377, 390, 401

Baryon fraction, 13, 15, 63, 64, 201, 206, 244, 275, 280, 288, 321
Black hole, 16, 32, 38, 40, 46, 49, 51, 90, 114, 172, 342
Bremsstrahlung, 4, 5, 6, 15, 17, 25, 32, 178, 302, 309, 311, 414

Cavities, 37–41, 43–45, 48–50, 143, 171
Chemical evolution, 177, 187, 198, 204–206
Cluster environment, 36, 56, 61, 143, 171, 172, 183
Cluster mass, 10, 12, 14, 31, 33, 62–64, 153, 213, 229, 232, 236–237, 246, 288, 289, 300, 302, 303, 306–308, 310–318, 327, 329, 420, 441
Cluster merger, 2, 18, 21, 22, 27, 34, 38, 52, 54, 152, 155–159, 167, 169, 170, 172
CMB, 62, 64, 146, 218, 255, 259–264, 267, 268, 271, 272, 276, 277, 280–283, 291, 306, 312, 321, 336, 342, 345, 417, 419, 424, 426
Cold fronts, 2, 7, 8, 19, 21, 22, 24, 26–29, 52, 54, 55, 160
Cooling, 2, 9, 14, 15, 17, 23, 32, 35, 37–40, 42, 49–51, 61, 230, 310, 327, 431
Cooling cores, 2, 24–26, 38, 50
Cooling flows, 23, 24, 26, 38, 51, 276, 312
Cooling time, 17, 23–25, 37, 38, 49, 61
Cosmological constraints, 246, 247, 289, 315, 318, 319
Cosmological parameters, 36, 62–65, 213, 246, 248, 280, 287, 293, 297, 298, 301, 315, 323, 324, 326, 327, 336, 425, 426

Dipole, 156, 261, 419, 420
Dynamical friction, 72, 95–99, 103, 108, 112, 114, 115, 202

Emissivity, 4–6, 12, 26, 144, 274, 275, 277, 303, 320, 321
Equipartition, 3, 21, 22, 143, 146–151, 166

Faraday rotation, 2, 160, 161, 163, 166
Feedback, 17, 32, 35, 37, 38, 51, 198, 201, 287, 431
Filaments, 34, 36, 41, 52, 165, 167, 201, 213, 248, 282, 336, 337, 346, 349, 358, 363, 365, 366, 378–381, 388, 393, 394, 397–404, 409–412, 418–420, 428, 430, 431, 433, 434, 436, 437, 440–443, 459, 460

Galactic dynamics, 71, 99, 104, 116
Galaxy distribution, 10, 11, 126, 127, 132, 134, 303, 393, 409–412, 418, 421, 424, 429, 431–432, 436, 438, 442–443, 446, 459–460
Gas dynamics, 1, 230
Gas fraction, 13, 63, 186
Gaussian random fields, 340, 341, 355, 358, 359, 382, 383, 392, 341, 355
Gravitational lensing, 23, 213–217, 227, 229, 243, 248, 275, 280, 281, 288, 417, 427

Heating, 2, 9, 14, 15–19, 21, 22, 24, 25, 32, 38, 42, 48, 54, 105, 107, 198, 200–202, 207, 280, 430
Hydrostatic equilibrium, 1, 9, 10, 12, 14, 18, 24, 25, 33, 37, 62, 307–309, 311, 313, 314, 316, 327, 328

ICM, 1, 2, 3, 4, 7, 8, 14, 15, 16, 17, 18, 32, 33, 34, 36, 37, 39, 50, 53–61, 143, 149, 155, 158–161, 164–169, 172, 177–189, 196–207, 287, 302, 305, 306, 308, 309, 311, 312, 316, 318, 327
Inverse Compton, 18, 44, 52, 146, 151, 157, 161, 165, 255, 256, 259, 262
ISM, 46, 48, 58–61, 198, 205

Kelvin-Helmholtz instabilities, 29, 56, 59

Lensing mass, 220, 230, 235, 236, 237, 239–244, 248, 428
Line emission, 4, 6, 17, 32, 431

Mach number, 18–20, 22, 25–28, 46, 158
Magnetic Field, 1, 2, 7, 8, 21, 28, 29, 52, 56, 143, 144, 146–149, 151, 157, 160–167
Mass function, 62, 64, 65, 140, 287, 288, 289, 296–301, 315–322, 324, 329, 356, 361–362, 364, 365
Mass profile, 10, 11, 16, 25, 36, 54, 109, 222, 223, 230, 232, 233, 236–239, 244
Metallicity, 90, 177, 180–181, 187, 188, 195, 196, 199–205, 207, 274
Metal production, 177, 187, 188, 202
Mini-halos, 155, 156, 160
Morphology, 32, 36, 122, 168, 169, 225, 245, 246, 248, 335–337, 363, 372, 378, 380, 381, 389, 399, 401, 404, 409, 411, 435, 440, 441, 460, 461

Outburst, 32, 37–40, 42, 43, 45, 46, 48–51

Preheating, 15, 16, 198, 200–202, 207, 340, 342

Radio bubbles, 42
Radio emission, 1, 19, 37, 38, 41, 43, 45, 46, 48, 52, 59, 143, 147–152, 154–158, 160, 164, 165, 167, 171, 172
Radio galaxies, 1, 16, 143, 157, 164, 167–172
Radio halos, 52, 149–153, 157–159
Radio relics, 153, 154, 159, 160
Radio sources, 31, 46, 48, 59, 62, 143, 149, 151, 153, 155, 160, 161, 163, 168, 171, 172, 261, 264–266, 270
Reacceleration, 144, 157–159
Red sequence, 125, 132, 136–140, 302, 417

Shear, 29, 167, 216, 217, 220–221, 227–229, 232, 233, 235–237, 239, 241, 242, 275, 368, 370, 373–375, 377–382, 384–386, 388, 393–397, 401, 403, 404, 427, 453, 458, 460
Shocks, 2, 3, 4, 7, 9, 14, 17–22, 24–26, 32, 37–43, 48, 50–53, 156, 158, 160, 167, 168
SN Ia, 64, 183, 188–192, 194, 198–200
Sound speed, 1, 9, 18, 26, 34, 39
Strong lensing, 213, 215–217, 219, 229–230, 236–239, 246, 415
Structure formation, 159, 287, 288, 289, 336, 337–366, 371, 373, 384, 387, 389, 395, 399–400, 404, 420, 427, 430, 433, 443, 445, 446, 452–453
Supernova, 14, 16–18, 21, 22, 32, 62, 157, 158, 177, 182, 184, 188–191, 195–200, 248, 287, 328, 431
Synchrotron, 18, 52, 143–146, 148, 149, 151, 157, 160, 161, 165, 166, 263

SZ effect, 240, 244, 255–256, 257–275, 277–281, 288, 306, 312, 325, 417

Temperature gradient, 8, 28, 156, 181, 318
Temperature profile, 11, 34–36, 42, 63, 244, 261, 308
Tessellation, 124, 132, 133, 436, 438, 439, 449, 451
Thermal conduction, 2, 8, 10, 28, 38, 55, 61, 165
Tidal (interactions, force, radius), 24, 37, 56, 72, 88, 99, 100–103, 105, 107–115, 355, 357, 366, 367–370, 374, 387, 390, 401, 404, 447, 453
Tidal stripping, 107, 237
Turbulence, 21, 52, 158–160, 167, 168

Velocity dispersion, 16, 49, 75, 80–82, 85–87, 97, 98, 120, 198, 231, 237, 240, 247, 289, 295, 308–309, 315–316
Voids, 34, 336–338, 340, 342, 346, 348, 358, 360, 363–366, 373, 378, 394, 399, 409, 411, 412, 420–427, 434, 436, 437, 439, 440, 442–460

Walls, 335, 341, 358, 366, 394, 399, 402–404, 409–412, 418, 420, 434, 440, 443
Weak lensing, 23, 54, 62, 119, 213, 216, 218, 219, 227, 229, 232, 233, 236, 237, 239, 243, 244, 410, 415, 428–429

X-ray luminosity, 6, 15, 32, 43, 49, 53, 151–154, 156, 184, 198, 201, 240, 263, 279, 302–303, 311–313, 318, 320, 321, 325